The Quiet Revolution

CALIFORNIA STUDIES IN THE HISTORY OF SCIENCE

J. L. HEILBRON, Editor

The Galileo Affair: A Documentary History, edited and translated by Maurice A. Finocchiaro

The New World, 1939–1946 (A History of the United States Atomic Energy Commission, volume 1), by Richard G. Hewlett and Oscar E. Anderson, Jr.

Atomic Shield, 1947–1952 (A History of the United States Atomic Energy Commission, volume 2), by Richard G. Hewlett and Francis Duncan

Atoms for Peace and War, 1953–1961: Eisenhower and the Atomic Energy Commission (A History of the United States Atomic Energy Commission, volume 3), by Richard G. Hewlett and Jack M. Holl

Lawrence and His Laboratory: A History of the Lawrence Berkeley Laboratory, volume 1, by J. L. Heilbron and Robert W. Seidel

Scientific Growth: Essays on the Social Organization and Ethos of Science, by Joseph Ben-David, edited by Gad Freudenthal

Physics and Politics in Revolutionary Russia, by Paul R. Josephson

From c-Numbers to q-Numbers: The Classical Analogy in the History of Quantum Theory, by Olivier Darrigol

The Quiet Revolution: Hermann Kolbe and the Science of Organic Chemistry, by Alan J. Rocke

The Quiet Revolution

Hermann Kolbe and the Science of Organic Chemistry

Alan J. Rocke

UNIVERSITY OF CALIFORNIA PRESS
Berkeley / Los Angeles / London

University of California Press
Berkeley and Los Angeles, California

University of California Press
London, England

Library of Congress Cataloging-in-Publication Data

Rocke, Alan J., 1948–
The quiet revolution : Hermann Kolbe and the science of organic chemistry / Alan J. Rocke.
p. cm.—(California studies in the history of science)
Includes bibliographical references and index.
ISBN 0–520–08110–2 (cloth : alk. paper)
1. Kolbe, Hermann, 1818–1884.
2. Chemists—Germany—Biography.
3. Chemistry, Organic—Germany—History—19th century. I. Title.
II. Series.
QD22.K578R63 1993
540′.92—dc20
[B] 92–28190
CIP

Printed in the United States of America

1 2 3 4 5 6 7 8 9

The paper used in this publication meets the minimum requirements of American National Standard for Information Sciences—Permanence of Paper for Printed Library Materials, ANSI Z39.48–1984 ♾

For Aaron Ihde and Cristine Rom
My mentor and my muse

Contents

Acknowledgments

Every book is in a sense a collaborative venture. The debts that I have incurred during this project are more numerous than I can list and deeper than I can properly recognize. On the institutional level, grants from the National Endowment for the Humanities and the Ohio Board of Regents greatly eased my path, and I have been continually assisted by librarians, administrators, and colleagues at Case Western Reserve University. I thank staff at Freiberger and Sears Libraries, the Library of Congress, and Memorial Library of the University of Wisconsin. I have also been aided far beyond reasonable duty by William B. Jensen and the Oesper Collection in the History of Chemistry at the University of Cincinnati.

Frederic L. Holmes, Elizabeth Knoll, and Kathryn Olesko each read the entire manuscript with care and gave me the benefit of their considerable expertise; the book is much better for their interest and advice. For helpful comments on chapter drafts, I wish to thank Evan Bukey, Theodore Hamerow, Susannah Heschel, Kenneth Ledford, Colin Russell, and Fritz Stern. Some of the themes of this work were presented in 1990 at a conference on Chemical Sciences in the Modern World sponsored by the Beckman Center for the History of Chemistry; I have much benefited from conversations there (and elsewhere) with O. Theodor Benfey, Robert Fox, Roald Hoffmann, Seymour Mauskopf, Mary Jo Nye, Yakov Rabkin, Hans-Werner Schütt, and Arnold Thackray. Others who have given valuable counsel include Mathias Hartmann, Martin Helzle, Christoph Meinel, Peter Salm, Oliver Schwarz, and Stephen Weininger. I also thank Michelle Nordon and

Kathy Walker for professional assistance at the University of California Press.

I owe an immense debt to German scholars and archivists. The largest single component of the Kolbe Nachlass—524 important letters—is held by Vieweg Verlag in Wiesbaden, and staff members there have been extraordinarily generous in their help, both on and off site; I particularly thank Albrecht Weis, Michael Langfeld, and Ilse Dobslaw. Not far behind in importance for this project were the rich Sondersammlungen of the Deutsches Museum in Munich, and my work there over the course of several visits was much aided by the friendly ministrations of Otto Krätz and Margret Nida-Rümelin. I owe a particular debt to Elisabeth Vaupel, director of the Abteilung Chemie, who answered an immodestly long series of queries with unfailing patience and made a number of important suggestions. Lis Renner at the Bayerische Staatsbibliothek, Inge Auerbach at the Hessisches Staatsarchiv, Günther Beer at the University of Göttingen, Gertrud Schwendler and Klaus Sühnel at the University of Leipzig, Helmut Rohlfing at the Niedersächsische Staats- und Universitätsbibliothek, and Klaus Hafner and Susanna Priebe at the August-Kekulé-Sammlung in Darmstadt all assisted me in important ways. For details about Kolbe's family and early life, I am grateful to Pastors Karl Heinz Thiel, Peter Dortmund, and Hermann Charbonnier, of Elliehausen/Göttingen, Stöckheim, and Lutterhausen/Hardegsen, respectively; I also thank Karl Heinz Bielefeld of the Kirchenkreisarchiv Göttingen and Herr Leenders of the Landeskirchlichesarchiv Hannover. Eric, Otto, and Alexander von Baeyer graciously made their grandfather's (resp. great-grandfather's) letters available to me, extending warm hospitality as well as professional courtesy.

I am particularly grateful to Professor Colin Russell and Mr. and Mrs. Raven Frankland, who granted me generous use of the microfilms of Edward Frankland's large correspondence. I also thank Professor Albert Menne and the Hugo-Dingler-Stiftung for permission to use the Erlenmeyer letters held in the Hofbibliothek Aschaffenburg. Other institutions who permitted access and use of valuable materials include the Justus-Liebig-Museum in Giessen, the Niedersächsische Staats- und Universitätsbibliothek in Göttingen, the Zentrales Archiv der Akademie der Wissenschaften der DDR (recently renamed the Akademie der Wissenschaften zu Berlin Brandenburg), the library of the Freie Universität and the Staatsbibliothek Preussischer Kulturbesitz also in Berlin, the Bayerische Staatsbibliothek in Munich, the Universitätsbibliothek Marburg, the Royal Society, the Royal Society of Chemistry, the British Library, the Royal Institution and University

College Archives in London, and the Bibliotheque Nationale and the Archives de l'Académie des Sciences in Paris.

A version of parts of chapters 5, 9, and 13 is appearing with permission concurrently as a paper in *Osiris*, [2] **8** (1993), 51–79. Some of the themes and the language in this book will also appear as an essay in S. Mauskopf, ed., *The Chemical Sciences in the Modern World* (Philadelphia: University of Pennsylvania Press, 1993). A version of the material on chauvinism in chapter 14 will appear in the *Bulletin for the History of Chemistry*. Finally, the material on antisemitism, also in chapter 14, was first presented in a Festschrift session in 1989 in honor of O. T. Benfey and is to be published in a volume on the Intersection of Jewish and Scientific Cultures edited by Yakov Rabkin and Ira Bernstein.

Every author owes his greatest debt of gratitude to those who helped in the least tangible but most important ways: his family and close friends. I could not have gotten to the end of this road without them.

Note: Unless otherwise indicated, units of measurement and currencies, atomic weights, and chemical formulas are always reproduced as in the original sources. Readers are referred to the Glossary for assistance with German and chemical terminology.

Introduction

The title of this book contains three entities: a quiet revolution, a man named Hermann Kolbe, and the science of organic chemistry. A few words of orientation are necessary about each of these subjects.

I use the phrase "quiet revolution" to denote a series of changes in chemical science during the 1850s, which centered on reforms of atomic weights and molecular formulas and on the subdiscipline of organic chemistry. Preparation for these changes began in the 1830s and 1840s with the work of Justus Liebig, Jean-Baptiste Dumas, Auguste Laurent, and Charles Gerhardt. The essential elements of this extended event included the following: the decline of Jacob Berzelius' electrically based theory of atomic–molecular combination, the rise of so-called type theories based on reactions wherein chlorine substitutes for the hydrogen of organic substances, the establishment of consistent ("two-volume") molecular magnitudes spanning organic as well as inorganic chemistry, the return from conventional chemical equivalents to a modified version of Berzelian atomic weights, and finally, the introduction of the theory of "atomicity of the elements," which led to what became known as valence and structure theory. The fundamental ideas of structure theory were delineated by August Kekulé in 1857–1858 and independently by A. S. Couper in 1858; they were given a particularly clear form and development by A. M. Butlerov from 1861. A largely equivalent and partly independent formulation of these ideas was due to Edward Frankland and Kolbe in 1857–1860.

There are a number of justifications for emphasizing the magnitude of these developments. Many quantitative indicators of the size and importance of the profession—number of chemists, number of chemistry students, number of journals, number of papers published, increase in technological applications, and growth of chemical industries—suggest an inflection point shortly after the middle of the century.[1] In the years before 1850, the discipline was well-developed and fully professionalized, but by any reasonable measure it was quite small in size and low in public profile. The situation was dramatically different a generation later. One example illustrating the change is the Deutsche Chemische Gesellschaft, which experienced phenomenal growth immediately following its founding in 1867. Within a few years, its *Berichte* ballooned in size from a thin volume in 1868 to giant and unwieldy tomes that required repeated subdivision. Another example is even more telling, namely, the total number of chemical compounds known. In 1860 there were about 3,000 well-characterized substances in the chemical literature; this number had grown steadily during the preceding several decades, with a consistent doubling period of about twenty years. But just about 1860 the trend dramatically accelerated, so that the doubling time thereafter was about nine years. It has remained so ever since.[2]

Also relevant is the centrality of organic chemistry in this revolution—nearly all of the new compounds just mentioned were organic—and the centrality of structure theory within organic chemistry. Events in the story center on the increasing ability of chemists to discern the internal arrangements of the atoms within molecules (what in the 1850s and 1860s were usually called molecular *constitutions* and what are today called *structures*[3]). This theoretical knowledge went hand in hand with the new and ever more powerful ability to synthesize both familiar and novel organic substances. The new synthetic chemistry resulted in important applications in the production of dyes, pharmaceuticals, and other products of the burgeoning chemical industry of the late nineteenth century. Excluding such standard heavy industrial indicators as coal, steel, and railroads, the fine chemicals industry took the leading role in Germany, and it became increasingly dependent on the scientific investigation of chemical substances, especially organic compounds studied via structure theory. In summary, it can be argued that organic chemistry was the earliest pure science to have a massive impact on technology and on a national economy.

Thus, in both the theoretical and applied realms structure theory provided a virtually unlimited and highly fertile field for academic chemists. To use the Deutsche Chemische Gesellschaft as an example once again, the early volumes of the *Berichte* are packed with structure-theoretical investigations, far outweighing all other kinds of papers put together. The number of chemists rapidly expanded in the 1850s and 1860s, and the German states suddenly began to compete with one another in building palatial laboratories, especially for *organikers* (events that are discussed below). By 1880 nearly all German universities had such institutes, and nearly all were led by organic chemists.[4]

If structure theory was vital for the growth of chemistry in the second half of the century, the Gerhardt–Laurent reforms were clearly the harbinger of structure theory. The reforms were proposed during a period of theoretical anarchy; both contemporary and subsequent commentators have emphasized the confusion that prevailed during this period. Water could be written HO, ~~H~~O, H_2O, or H_2O_2, and in urging reform, Laurent and Kekulé each gleefully filled an entire page of text with currently defended formulas for acetic acid.[5] Four major concurrent systems of atomic weights and formulas, and many variations thereof, were widely accepted in the 1840s and 1850s.

Neither Laurent nor Gerhardt was able to produce strong new evidence for his views, and the situation was only clarified when Alexander Williamson (1824–1904) developed the first compelling experimental chemical evidence for the central element of the reform. Williamson's work (1850–1854) was followed by similar investigations by A. W. Hofmann, Gerhardt, Adolphe Wurtz, William Odling, and Kekulé; by the time of Gerhardt's death in 1856, the reform was clearly winning the day, at least in Germany and England. This is not to deny the traditional key role accorded to the Karlsruhe conference of 1860 and to its canonical hero, Stanislao Cannizzaro. But to a degree that has not previously been appreciated,[6] the revolution had already been largely consummated by the time of the conference, at least in Germany. The relatively invisible nature of these changes has led me to choose the name "quiet revolution." But however quiet it may have appeared to later observers, the participants themselves saw this period as revolutionary and looked to the 1850s as the critical decade. Evidence for this statement is provided in chapter 6.

Considerations such as these have convinced me that this quiet revolution was the last major transition leading to modern chemistry. One purpose for writing this book, then, is to highlight and chronicle this important period in more depth than has hitherto been attempted. A convenient vehicle for this purpose is the life and career of Hermann Kolbe (1818–1884).

Kolbe was one of the most outstanding and successful German chemists during the remarkable period in which German chemistry, like the wider manifestations of German industrial and political power, rose to a position of world leadership. A student of Friedrich Wöhler and Robert Bunsen, Kolbe succeeded Bunsen at Marburg in 1851. In the mid-1860s two major German universities, Leipzig and Bonn, had vacancies, and Kolbe received a call (or offer) to each of them in succession; it was only after he declined the latter offer that Kekulé received and accepted the invitation to Bonn. Kolbe was Liebig's favorite chemist among the younger generation and was initially favored to succeed Liebig in Munich, although that offer never materialized. At the time of his transfer to Leipzig, no one in his generation except perhaps Hofmann could touch his reputation and standing in the field of chemistry.

Kolbe was a leading figure in two major interrelated inflection points in the history of science. One was the series of scientific events described above; the other was institutional and pedagogical in nature, namely, the phenomenal growth of academic laboratories starting in Germany after mid-century. Around mid-century, a laboratory accommodating 20 to 40 student workers at any one time was considered large and was qualitatively distinct from the eighteenth- and early nineteenth-century model of a small handful of elite students working with the professor on the basis of personal patronage. By contrast, Kolbe's lab in Leipzig (designed by Kolbe and dedicated in 1868) was built to accommodate 130 student workers, and within a short time even this huge capacity was heavily oversubscribed. Among the first of what Jeffrey Johnson properly calls "second-generation institutes,"[7] it was for a decade by far the largest and most modern academic laboratory in the world and was much imitated. Its capacity exceeded that of Hofmann's well-publicized new institutes at Berlin and Bonn *together*.

Despite his preeminence, Kolbe has until now escaped any substantial biographical treatment in any language, and this fact in itself indicates a certain scholarly opportunity. But even more importantly, much can be learned about the rise of nineteenth-century German chemical science and industry from a careful study of Kolbe's career, in part because there are very few modern full-length biographies of German chemists of this period, and none at all in English. Despite a recent intensification of historical interest, the rise of German laboratory science and its societal implications are still very imperfectly understood, due largely to this dearth of detailed studies focusing on particular individuals and localities.

A prominent German economic historian has recently drawn attention to the need for studying the development of German industrialization from a *regional* perspective.[8] More generally, German historiography in this century has suffered from an overemphasis on the larger states, especially Prussia, which on occasion has created distortions and omissions. In a similar sense, it can be argued that the history of nineteenth-century German science has focused too much on events in Prussia, neglecting the smaller states before 1870.[9] This is particularly ironic in the case of chemistry, as Prussia was among the most backward of the German states in this respect until Hofmann's call to Berlin in 1865; the leading personalities of the previous generation had been Liebig in Giessen (in the Grand Duchy of Hesse), Wöhler in Göttingen (in the Kingdom of Hanover), and Bunsen in Marburg (in Electoral Hesse) and Heidelberg (in the Grand Duchy of Baden). Following Kolbe's career as we do in this book provides a kind of grand tour of many of these smaller states: he was raised in Hanover, educated in Hanover and Electoral Hesse, employed for four years in the Duchy of Braunschweig, succeeded Bunsen in Marburg, then finally became Ordinarius in Leipzig in the Kingdom of Saxony. In each case, I am careful to establish some sense of the regional context of politics, culture, and *Wissenschaftspolitik* and to relate fiscal, administrative, and pedagogical details of the management of various German scientific institutes—including those of Kolbe's colleagues and rivals.

In conclusion, Kolbe's career provides an unexcelled lens through which to view transformations in pedagogy, theory construction, research practice, and administration of academic science in the context of German society. This is the case even though—or rather, precisely *because*—Kolbe did not think of himself as a "modern" (i.e., structural) chemist. Not far beneath the surface of a self-professed conservative and empiricist lay the activities of an imaginative and fruitful theorist, but a theorist with a curious twist. Kolbe had his own agenda and pursued it with single-minded intensity. He viewed himself as bearing the mantle of the great classical chemists, especially that of Berzelius. Although most of the collegial community regarded Kolbe as having helped establish the basis of structure theory, he attacked the structuralists so continuously and so viciously after 1870 that much of his posthumous notoriety—and, I believe, the dearth of serious biography—was due to his quirky obstreperousness. Within a decade, he succeeded in transforming himself from the leader of his field into an embarrassing and obnoxious crank. Kolbe's tragic fate was to have ended his life in opposition to most of what was then (and is today) regarded as the essential principles of organic chemistry.

As a consequence, by looking at the field through Kolbe's eyes we see the transformations of this period from the perspective of a man who was at once both an insider and an outsider. I have devoted much effort to understanding and motivating Kolbe's idiosyncratic viewpoint, but I have also striven to depict the wider context fully and fairly, not only in Germany, but also in France and Great Britain. It is only by such a combination of individual attention and contextual sweep that any biographical account becomes coherent and meaningful. Moreover, this kind of approach and the intriguing quirks of this particular case study highlight a number of broader philosophical and historiographic issues, including generational change in scientific methodology, the nature of scientific controversies, the role of empirical evidence in theory assessment, and psychological and social factors in the contexts of discovery and justification. These issues are explored in the chapters that follow and in the concluding observations.

The natural audience for the story I have to tell consists of historians of chemistry and of nineteenth-century German science, but the novel aspects and historiographic richness of my case study give me the hope that the book will have more general interest for the history of science community. Accordingly, I have attempted to deal gently with the reader, assuming no particular knowledge of chemistry beyond simple definitions of atoms and molecules and a nodding acquaintance with the outlines of the science. The neophyte in chemistry will need to have recourse to the glossary, which covers some basic information and technical vocabulary regarding the modern (post-1860) theory of the structure of organic molecules.

Ever since the period described in this book, organic chemistry has had an intimidating reputation with laymen and students, and today it is often regarded as the principal distasteful hurdle to be overcome in a premedical curriculum. There is no question that organic chemistry in its full dimensions is a large and complex science, having important affiliations with other branches of chemistry as well as with physics and biology. However, the essential theory of the science for over a century—founded as it is on the principles of valence and structure and comprising compounds of a mere handful of chemical elements—is simple and clear. That structural basis, a kind of Tinkertoy-like set of schematic rules that can be taught literally in a few minutes, can carry a student (or a reader) remarkably far. Consequently, it is my hope and expectation that the following account will be widely accessible. I have tried to be equally kind to readers who may know little about the arcana of nineteenth-century German history. Thus, I hope that chemists with a taste for history will enjoy reliving these old battles and learning about the lives and social context of some of the figures whose eponymous reactions they teach daily to undergraduates.

I would like to conclude with a few personal comments. My first academic specialization was in organic chemistry, and I have never lost my fascination with this discipline and with its structuralist basis as characterized in the previous paragraph. I cannot help but feel that the great spurt in the growth, success, and standing of the field during the middle decades of the nineteenth century was to some degree a measure of the intellectual coherence and beauty, not to mention heuristic power, of structure theory as it was formulated and developed. There is thus a certain irony in the circumstance that I have chosen as my protagonist the principal *opponent* of this lovely theory that I so much admire.

Moreover, I confess to having no admiration for the personality and character of Hermann Kolbe. Kolbe was capable of acting with generosity, selflessness, and love, and for most of his career he was a superb scientist, but many of his actions late in life displayed an array of prejudices and an intensity of malevolence that cannot easily be excused. Such feelings of a biographer carry a certain advantage, in that one is thereby inoculated against the danger of hagiography that is so endemic to the biographical genre. The corresponding issue on the other side is that one must be scrupulously fair, and I hope my readers will judge that I have been.

Without wishing to draw an exact parallel, I find myself in a position similar to that of the Stalin biographer Robert Tucker, who detests his protagonist, and depicts him as a neurotic personality without any consciousness of his own wickedness. To my mind, Kolbe also appears to have developed a mental illness. Not being a psychologist, I have come to no conclusions on the exact nature and etiology of Kolbe's neurosis or personality disorder, if such it was. I have, however, attempted to provide empirical grist for whomever might wish to turn the psychohistorical mill, by investigating Kolbe's upbringing, physical health, family life, relationships with students and colleagues, use of language and metaphor, and stresses and strains of daily life. As with Tucker's Stalin, it is clear to me that Kolbe's internal motivations and self-justifications, although twisted, were absolutely pure. Finally, like Tucker, I conclude by noting that my motivation for choosing this particular subject was simply that here was a fine historical opportunity: "when you seek to penetrate the heart of darkness, it is very important that the project be interesting from the scholarly point of view."[10] And whatever else one might say on the topic, Kolbe is a fascinating subject.

1

Academic Chemistry in Early Nineteenth-Century Germany

University Reform

In 1840, Justus Liebig lamented that at the time of the liberation of the German states from French domination "there were no longer any scientists in Germany."[1] A certain discount for hyperbole must be applied to this, as to many of Liebig's programmatic assertions. Among the active workers in German academic chemistry at the time of Napoleon's final defeat were Friedrich Stromeyer, Johann Wolfgang Döbereiner, Heinrich Klaproth, J. B. Trommsdorff, and Karl Kastner—all justly famous chemists in their day. Chemistry was being practiced by an identifiable professional community of capable scholars and researchers, which, although less prominent than the corresponding communities in France or England, was far from negligible in importance. Allied with the academic chemical community in various ways were the guild-based but state-licensed and increasingly university-educated apothecaries who compounded drugs for the medical profession. Finally, largely outside this professional community were the practical chemical entrepreneurs and workers who made the soda, potash, acids, gunpowder, soap, paints, and dyes needed in their society by largely empirical methods in small shops. Recognition of the practical utility of chemistry in medicine, pharmacy, technology, mining, metallurgy, agriculture, and most other arts of material civilization resulted in a certain minimum level of social support for chemists and their community, even in the agrarian German states of the early *Vormärz* period (1815–1848).[2]

But despite the importance of pharmacists and technical chemists, the real institutional home for the chemical community in Germany (more so than in any other country) was the university.[3] The universities had arisen in the middle ages essentially as guilds of teachers and students; indeed, the Latin word *universitas* from which the modern term derives was used by the medieval guilds to refer to the totality of workers in a given craft. The guild model survived into modern German history. After satisfying entrance requirements (after 1834 this meant only graduating from a Gymnasium—a neohumanist secondary school—with the leaving certificate called the *Abitur*), the student (apprentice) worked with a professor (master) until he had satisfied the minimum demands of his intellectual craft. Students typically studied a fairly narrow curriculum—usually in one of the three professional faculties of law, medicine, or theology—having already spent nine years in Gymnasium learning the classical languages, religion, history, and German, as well as a modicum of modern foreign languages, mathematics, and natural science. But the matriculated student had full freedom to take whatever university courses he chose. The student could also travel from one university to another, and due to the fragmentation of the German states, there were many from which to choose. The baccalaureate and master's degrees having fallen into disuse, the doctorate was the degree that mattered, and it could be earned after two or three years of hard work. In addition to examinations, a dissertation was often (but not always) required; in the sciences during the nineteenth century, this was on the order of a substantial (and often published) article.

Most university graduates entered one of the traditional professions or became civil servants. If the student desired an academic career, it was customary to spend one or more years after the doctorate at other universities, sometimes foreign ones, as a kind of *Wanderjahr*, or journeyman period. One then normally needed to obtain a second degree, the *venia legendi*, or teaching certification. The Germans still refer to this process as *Habilitation*, or enablement. A second dissertation was required, analogous to the journeyman's certifying masterpiece; this was called the *Habilitationsschrift*. Some universities in the scholastic mode also required a public disputation on a theme chosen by the aspirant. The successful applicant now had the right to teach courses at the university that had certified him; but his income derived solely from fees collected directly from students electing his courses, nothing from the university itself. Those who had passed this certification were known as *Privatdozenten*, or private lecturers; uncommon in the eighteenth century, Privatdozenten proliferated in the nineteenth.

After a few years one could hope to be named an *ausserordentlicher* (extraordinary) professor, also called *Extraordinarius*. These were poorly paid positions, however. The ultimate goal was to receive a "call" to an *ordentlicher* (ordinary) professorship, which was a "chair" or *Ordinarius* position for which one was paid a real salary, as well as being entitled to student fees for all except the required weekly public lecture. As in the guild model, the universities were largely self-regulated by the "masters," that is, the ordentlicher professors. From their number they elected deans of constituent faculties, the rector (president) of the university, and a senate. They also had the power, at least in theory, to select new members of their body, i.e., to fill vacant chairs.

In eighteenth-century Germany, this corporative structure was particularly strong, and many small universities lived almost entirely from their own distressingly meager financial means. Even ordentlicher professors often earned ridiculously small salaries, and student fees usually could not make up the difference since enrollments declined throughout the century. As a consequence, an academic profession barely existed at all, and what did exist tended toward mediocrity. Much of the instruction in the fourth or "philosophical" faculty (designed to prepare students to enter one of the professional faculties) was painfully elementary and hidebound. Publishing novel research was not considered necessary or even desirable and hence was not normally done. The universities, having become essentially narrow professional schools, left scientific research largely to the academies of sciences and to wealthy or patronized amateurs.

The University of Göttingen was a notable exception to this pattern. Founded on a consciously "modernist" and liberal model, well funded and endowed with what soon became the best library in the German states, it attracted an international and often wealthy student body and served as an intellectual pipeline between the kingdoms of Hanover and Great Britain (whose sovereign was the same throughout most of the eighteenth century). In the second half of the century, it became the "first university" of Germany and the central locus of the movement to recast the universities from their stodgy medieval outlook toward the model of progressive teaching institutions, incorporating a strong research mandate. A. G. Kästner wrote toward the end of the century that those professors who failed this mandate were as "mouse turds among peppercorns."[4] But most of the Göttingen professoriate were of the best odor. During its first century, the university was graced by Albrecht von Haller and J. F. Blumenbach in medicine and natural history; Kästner, Tobias Meyer, Wilhelm Weber, G. C. Lichtenberg, and Carl Friedrich Gauss in mathematics and physics; and Stromeyer and Friedrich Gmelin in chemistry. The humanities were equally well represented.

The other German universities changed dramatically after the turn of the century, when their sad state reached a point of crisis and the chaos and humiliation of the French wars provided an effective excuse for a new beginning. The German reforms were influenced to a degree by Napoleon's own progressive ideas about education, but even more by a neohumanist revival that was more typical of German Romanticism and constituted a reaction *against* Napoleon. Using the University of Göttingen as a model, such Prussian reformers as Wilhelm von Humboldt, J. G. Fichte, and Friedrich Schleiermacher worked to recast the universities of their state in a more "modern" mold. One side of the eighteenth-century resistance to the research ethic had been the Enlightenment ideals of utility, universalism, gradual progress, and encyclopedic breadth. It turned out that the new *Wissenschaftsideologie* was consistent with more particularist and individualist instincts, as well as with a certain iconoclastic dynamism and a strong disavowal of crass materialist utilitarianism—in other words, an ideology consonant with Romantic notions in general. Moreover, placing a neohumanist rhetorical overlay on university reform had the salutary consequence of rescuing the universities from their predominant eighteenth-century role of narrow professional and civil service training and providing the basis for their new mission of broadly based education of the mind and spirit. Finally, Göttingen's practice of hiring scholarly "stars" active in research and publication had been very successful in attracting relatively large numbers of wealthy, noble, and foreign students. Hanover reaped the benefits of the resulting prestige and indirect income, and envious university administrators in other German states wanted a piece of that action. In this way, the research mandate became associated with neohumanist reforms.

Taking stronger control over the corporative universities and hiring a larger and more eminent faculty meant both larger financial investments and higher expectations on the part of the state ministries of culture. The faculties were still expected to prepare a prioritized list of (usually three) candidates to fill any vacant chair, but the state could, and sometimes did, overrule the faculty to follow a program of its own. Once the research mandate became an expected standard, the German states often competed with one another in attempting to land the biggest academic stars for their universities and attract the highest number of wealthy and foreign students. Salaries rose, at least for the Ordinarien, and the professoriate became a real career option. The leading role was taken after 1817 by Prussian Kultusminister Karl Freiherr vom Stein zum Altenstein, who put the reforms into effect, especially at the newly founded universities of Berlin and Bonn, and in the process revived the moribund philosophical faculties.

In the meantime, the Ordinarien were competing for student fees with the other faculty ranks, the Extraordinarien and especially the increasingly numerous Privatdozenten. One consequence of this phenomenon was the gradual raising of the standards for admission to the teaching staff, that is, for the *venia legendi*. Thus, vigorous competition among states, among academic ranks, and among the members of each rank for preferment, as well as the exercise of greater control by the states for their own reasons, resulted in a general strengthening of the research mandate, ever higher standards for research excellence, and increasing professionalization of academic careers. Institutionalized first in the German states during the *Vormärz*, the dualist professorial standard of teaching *and* research was widely copied by countries around the world during the second half of the nineteenth century.

Pedagogical Reform

Associated with the reform of the German universities in the early nineteenth century was a gradual change of opinion that came about in how best to *teach* in the universities. The change was most dramatic in the natural sciences and medicine, where an empiricist–sensationalist epistemology derived from leading Enlightenment ideas led to less reliance on lectures and lecture demonstrations and eventually dictated laboratory-based instruction for most students in the sciences. In this new movement, some historians have emphasized the role of eighteenth-century pedagogical reformers such as Johann Heinrich Pestalozzi,[5] while others have tried to trace the change to early philological and historical seminars at late eighteenth-century modernist universities such as Göttingen. The humanist seminar has been an appealing historical model because it was there that the new research-oriented scholars tried to create a monism from their dualist activities, by interesting their students in research problems and making research a routine aspect of pedagogy, at least for selected advanced students.

Recent work has, however, cast some doubt on a simple evolutionary model from the humanist seminar to the scientific laboratory-based university institute.[6] One problem in making this connection has been reconciling it with the atmosphere of the aggressively idealist, nonutilitarian and nonmaterialist neohumanist rhetoric of the day and considering the popularity in Germany of a speculative and nonempirical Naturphilosophie movement. Another problem has been the recognition that most of the early institutes and seminars (i.e., those before about 1820) were not in fact research oriented but were instead largely propaedeutic in nature.[7] Moreover, a distinction needs to be made between the introduction of laboratory research for the professionalizing elite student, which occurred early, and laboratory work for all, including average students studying ancillary disciplines.[8] This latter pattern of the scientific teaching laboratory did not emerge until relatively late; the first models were only created in the 1830s. The pioneering field in this development was chemistry.

A useful way to begin to describe the way these events came about is to discuss briefly four of the leading protagonists in this sea change. In addition to some curious commonalities in their backgrounds, all had a decisive influence on the young Hermann Kolbe—who will be more formally introduced in the next chapter.

Jacob Berzelius

Berzelius (1779–1848)[9] was the son and stepson of Lutheran country pastors, from whom he imbibed a dose of theism strongly moderated by late eighteenth-century rationalism and materialism. His education was fully in the spirit of the Gustavian Swedish Enlightenment and was much influenced by French ideas. During his years as a medical student at the University of Uppsala, he found himself "irrevocably gripped" by the love of chemical experimentation. Although to a certain extent an autodidact in chemistry, he learned the French antiphlogistic chemistry of Lavoisier and Fourcroy from his professors, in particular A. G. Ekeberg and Pehr Afzelius. He also consorted with liberal and progressive circles at the university. He graduated from Uppsala in 1804, at which time he became an adjunct (similar to Privatdozent) at the Stockholm School of Surgery—later the Karolinska Mediko-Kirurgiska Institutet—and a physician for the poor. In his meager spare time, he pursued an increasingly successful experimental research program. In January 1807 Berzelius was appointed professor at the school, and he remained in this position for the rest of his life. It was significant that Berzelius spent his career in a professional school, not a university; it required him to remain close to the practical and empirical level, consistent with his innate inclinations. In the 1820s Berzelius fought unsuccessfully for legal equivalence of the Karolinska Institutet with the Swedish universities and urged a more utilitarian and modernist university curriculum. He also fought the phosphorist school, a Swedish version of Naturphilosophie that was influential in the first quarter of the nineteenth century.[10]

Berzelius' first love was physiological chemistry, but he soon discovered stoichiometry and atomic theory. His utter brilliance as a bench chemist, his theoretical talent, and his extraordinary capacity for work led to a variety of fundamental contributions in both of these areas by 1812, although most were published only in Swedish and so were little known outside of the country. In the summer of 1812 he spent four and a half months in England; in 1818–1819 he spent almost a year in Paris, also visiting Great Britain and Germany. Other trips followed. These foreign travels, and the consequent translations of his books and papers into the major European languages, effectively spread Berzelius' ideas, and by 1820 he was recognized internationally as one of the greatest of living chemists. It is largely to Berzelius (and mainly during these years) that we owe the successful elaboration of atomic theory, as well as the initial stages of development of experimental and theoretical organic chemistry from the base established by Scheele, Lavoisier, Fourcroy, and Gay-Lussac.[11]

Berzelius' reputation was even further enhanced during the 1820s. It was only after 1820 that reliable and full translations of his monumental textbook began to appear, which was first begun in Swedish in 1808. A new edition, edited from 1825 on by Friedrich Wöhler, appeared first in German, the Swedish version following along behind. This edition was the first by Berzelius to treat organic chemistry in detail; the two organic volumes can be viewed as the first full-length organic chemistry textbook in history. Moreover, Berzelius' new position as Secretary of the Swedish Academy of Sciences brought with it an obligation to write an annual report on the progress of chemistry in all subfields throughout the world. The first of these reports, for the calendar year 1820, was quickly translated into German, and a pattern was established that persisted until Berzelius' death. The size of these reports, usually known by their German name *Jahresberichte*, gradually increased until they were the size of substantial books. In the 1820s, Berzelius was at the top of his form and he knew it. His magisterial judgments of his colleagues' work in the reports were closely followed and highly respected. Berzelius had become the supreme judge and legislator in his science, the one man whose word mattered to all.

An additional factor promoting Berzelius' high standing in Germany was his practice of accepting selected applicants for advanced work in his laboratory. His second guest worker, and the first non-Scandinavian, was C. G. Gmelin in 1814–1815, who later became a professor at Tübingen. Closely following Berzelius' first visit to Germany came the young Eilhard Mitscherlich and the brothers Heinrich and Gustav Rose, all from Berlin and all in 1820–1821. Mitscherlich's visit resulted from the circumstance that Altenstein had offered Berzelius the chair of chemistry at the university, vacated by Klaproth's death in 1816. Berzelius declined; at least six others were then offered or were considered for the chair, including Stromeyer and Leopold Gmelin. Asked his advice during his visit to Germany in 1819, Berzelius recommended Mitscherlich, whom he had only just befriended. Altenstein agreed, but with the proviso that Mitscherlich should first study with Berzelius in Stockholm. The Rose brothers' trips were also related to this connection, as well as to their friendship with Mitscherlich.

At most, Berzelius accommodated two or three guests in his laboratory, and he often had none at all. Systematic instruction was not given; rather, Berzelius allowed visitors to follow their own research ideas, simply giving advice whenever it was desired. In addition to the four Germans just mentioned, Wöhler came to Stockholm in 1823–1824 and Gustav Magnus, another Berliner, in 1827–1828. By 1830 Justus Liebig could be considered a disciple of Berzelius, even though he had not studied directly with the master; a few years later, Robert Bunsen joined the Berzelians, again in spirit if not in the flesh in Stockholm. German chemistry was thus strongly infused with Berzelian ideas in the 1820s and 1830s, through both direct and indirect channels.

By the late 1830s, however, Berzelius was in a theoretical retreat, most noticeably in the field of organic chemistry, largely because of the experiments and ideas of upstart French chemists such as J. B. Dumas, Auguste Laurent, and Charles Gerhardt. A sense that Berzelian chemistry was passé gradually took root in Germany as well, certainly well before Berzelius' death in 1848. Berzelius himself, who gave up most laboratory work by about 1835, grew increasingly inflexible and cantankerous. His opinions were always freely and openly expressed in his *Jahresberichte*, and they created much ill will among those whom he attacked—above all, the French chemists, but also his hitherto devoted admirer Liebig. His most loyal disciple and friend, however, was Wöhler.

Friedrich Wöhler

Wöhler (1800–1882)[12] was the son of a Hessian agronomist and veterinarian, and he grew up near Frankfurt. After attending the Gymnasium there, he entered Marburg University in 1821 with the intent of studying medicine. His passion from early childhood, however, had been chemical experimentation and mineral collecting, and Ferdinand Wurzer's lectures did not attract him. Accordingly, he transferred to Heidelberg to study with Leopold Gmelin. But Gmelin judged that Wöhler already knew too much chemistry to profit from his courses; he advised Wöhler to study with Berzelius after receiving his M.D. degree. In the fall of 1823, upon receiving a favorable response to his inquiry from Stockholm, Wöhler took this step. It determined the course of his life.

Wöhler not only learned Berzelian techniques in his year in Stockholm, he also learned fluent Swedish and formed an extremely close friendship with the older Swede that lasted until Berzelius' death. Berzelius eventually urged the familiar form of address upon his student—for the time, an unusually strong mark of regard of an older for a younger man. Back in Germany, Wöhler sought habilitation at Heidelberg, but was instead hired for the new Berlin Gewerbeschule (trade school) at a salary of 400 thalers. Three years later, he had become a highly respected chemist and was earning 1200 thalers.[13] By this time he had also become Berzelius' viceroy in Germany by translating and editing the German editions of Berzelius' *Jahresberichte* and *Lehrbuch*—an average of about one and a half large volumes of text per year for over twenty years.

Late in 1831 Wöhler accepted for personal reasons a call to the newly founded Technische Hochschule (Institute of Technology) in Kassel at a diminished salary of 800 thalers plus free rent. There he continued the experimental work that he had begun so well in Berlin. Wöhler's work on cyano compounds, beryllium, yttrium, and aluminum had already brought him fame; his synthesis of urea in 1828 was particularly dramatic, in its implications both for organic synthesis and organic isomerism. Wöhler's models were the sober empiricist Gmelin and the incomparable Berzelius; he was a superb and enormously prolific experimental chemist. Disinclined toward philosophy or even chemical theory, Wöhler fit in well with the rationalist and practical traditions of the Berlin Gewerbeschule and Kasseler Technische Hochschule and, later, the University of Göttingen. He impressed everyone with his kind and unassuming character. In their correspondence, he and Berzelius often ridiculed the Naturphilosophen and the Hegelian philosophers.[14]

Wöhler first met Liebig in 1825, and the two young chemists always seemed to be stepping on each others' toes in their research during the middle to late 1820s. To avoid future problems, in 1829 they began to collaborate occasionally on topics of interest to both of them. By this time they were already close friends, and they maintained this friendship until Liebig's death in 1873. Wöhler learned from Liebig the newly improved method for elemental organic analysis one year after Liebig developed it in the fall of 1830.[15] When Wöhler suffered the death of his young wife in 1832, Liebig invited him to Giessen for companionship and for the distraction from grief that hard work could offer.[16] This was the period in which the two chemists completed their work on the benzoyl series, an article that galvanized the chemical world and is usually regarded as the best single contribution of either Liebig or Wöhler.

In the spring of 1836, Wöhler transferred to Göttingen as Stromeyer's successor. Gmelin had declined the offer, and Wöhler, supported by Gmelin and Berzelius, was preferred over his close rival and friend Liebig. Liebig wrote Wöhler that he, like Gmelin, would have declined, but regretted that he did not get the offer to use to good effect with his administration.[17] Stromeyer had inherited from J. F. Gmelin a small but well-equipped teaching and research laboratory on the ground floor of an old but spacious dwelling; a director's residence was provided upstairs. Shortly after his arrival, Wöhler wrote Berzelius describing many details, quite pleased with his new environs.[18] He quickly became an ornament of the faculty and taught a phenomenal number of students during his forty-six year tenure there.

Since soon after his arrival in Göttingen Wöhler provided Kolbe with his first detailed introduction to chemistry, and since Wöhler's early teaching career has never been closely studied, a discussion of the latter is warranted. Every semester Wöhler taught general theoretical (inorganic) chemistry at 9:00 A.M. six days a week and a laboratory practicum every Monday, Tuesday, Thursday, and Friday at 11:00 to 1:00. In the summer semester he also taught pharmacy Monday through Friday at 6:00 A.M. Advanced students were allowed to work all day every day in the lab. He used the laboratory left him by Stromeyer, although he reappointed it in "Berzelian" style during his first semester and refloored and repainted it two years later. It appears that Wöhler had reasonable demand for his lectures and practicum during his first two years, but precise numbers and names are not available.[19]

Whatever the initial numbers were, Wöhler's correspondence provides evidence for a noticeable increase in his enrollments beginning in summer semester 1838—coincidentally, the semester that Hermann Kolbe entered the university. Although exact information is sketchy here, too, we do know that he had twenty-eight students taking the practicum by the spring of 1840 and forty by the end of 1841, a remarkable level that continued to be maintained thereafter. After the influx started, it appears that Wöhler began to assign special projects to his most advanced students, investigations that might yield publishable results. There is no evidence that he ever did this in Berlin or Kassel, although it appears that he did have a practicum in Kassel.[20] The first Wöhler pupils whose names appear as authors of published papers were pharmacy students: August Stürenburg and Friedrich Weppen, who enrolled at Göttingen in May 1838, and Georg

Schnedermann, who came in the fall of 1839. Apparently all were from bourgeois Hanoverian families. Schnedermann worked with Wöhler for no less than six years, published a number of short papers, and became Wöhler's assistant in his last semester. The first Wöhler chemistry Ph.D. was earned by Friedrich Carl Voelckel, the son of a Bavarian merchant, who enrolled for summer semester 1839 and received his degree three years later. He later became professor of chemistry in Solothurn, Switzerland.[21]

In these early years, Wöhler's approach to publication of his student's work varied according to its significance and the student's precise role. He did not hesitate to use student results in his own papers, often without even naming the student, if the work was simply straightforward or mechanical assistance. If more skill or persistence had been needed, Wöhler was careful to acknowledge the assistance by name. Finally, there are a few examples in these years of Wöhler supervising what was essentially independent original research, and in such cases, the student published in his name alone.[22]

From 1838 until 1841, Wöhler appears to have had only a very small number of these select advanced students working on such projects at any given time—one, two, or three per semester. By summer semester 1841—again, ironically, the very semester Kolbe became an all-day Praktikant—a real research cohort of eight advanced workers emerged for the first time. In addition to Kolbe, Voelckel, Schnedermann, and Weppen, the group included the medical students Otto Griepenkerl and August Vogel and the philosophy students August Beringer and Wilhelm Knop.[23]

The pattern for the future was now set. Wöhler's increasing popularity and fame, and the rising profile of the chemistry profession itself, ensured that Wöhler would have substantial and rising enrollments ever after. Having inherited Stromeyer's assistant, H. A. Wiggers (later professor of pharmacology at Göttingen), Wöhler successfully petitioned for a second (Schnedermann), hired for the winter semester 1841/42, to help him with the now heavy numbers. That semester he had another increase, now past forty Praktikanten, more than the space could really accommodate. No fewer than fourteen of these men were doing advanced projects and working not just the scheduled four or eight hours per week but morning to evening in the lab. The number of these advanced Praktikanten, however, seems to have been rather variable, for in winter semester 1843/44 he had only three, while for each of the following two semesters the number jumped back up to twelve.[24]

Some additional information on Wöhler's students can be obtained from careful study of the Göttingen matriculation registry. During the period before Kolbe left Göttingen (fall 1842), a total of twenty-one students can be identified who were known (or can safely be presumed) to have studied with Wöhler. Eight listed chemistry as their field of study; the third of these was Kolbe. Kolbe's preparation apparently was not as thorough—or his progress not as swift—as that of Voelckel or Schnedermann, who arrived slightly later but were earlier in publishing articles from the lab. Nonetheless, it is interesting to note that Wöhler's most famous student was also very nearly his first. The rest of these twenty-one Wöhler students from Kolbe's days in Göttingen are divided by discipline roughly equally among pharmacy, medicine, and philosophy.[25]

This cohort represents but a small fraction of the total number passing through Wöhler's lab during these years. The rest cannot be identified by name, but it is probable that they were mostly students of pharmacy or medicine. All told, the number of students who passed through his laboratory from his arrival in Göttingen until Kolbe left for Marburg six years later was probably between 100 and 200.

Early in the spring of 1842, ground was broken for a new laboratory extension in Hospitalstrasse, immediately adjacent to the old lab. Directed closely by Wöhler, construction of the "magnificent building" was completed by that fall; the two sections could now accommodate up to fifty workers.[26] (This building, since destroyed, sufficed until 1859–1860, when a completely new and much larger laboratory was constructed for the Chemical Institute.) Unfortunately, Kolbe could not personally enjoy the expanded facilities since he left Göttingen in the fall of 1842. He must have watched with interest, however, as the facility slowly rose and became fitted for chemical research, while working daily in the old lab that summer.

Justus Liebig

Liebig (1803–1873)[27] was the son of a wholesale materials supplier in Darmstadt. Like Wöhler and Berzelius, from an early age Liebig desperately wanted to be a chemist, and he was largely self-taught. In his autobiography, Liebig described his youthful passion for reading every chemistry book he could find and his utter devotion to reproducing every experiment possible in his father's makeshift laboratory—and moreover, to performing the same experiment many times until he had absolutely mastered it. He also got to know all the local artisans in tanning, dyeing, soapmaking, and metallurgy, and he thoroughly learned their empirical chemical arts. However, he was a poor pupil. When asked by his teacher what would become of him and he replied, "a chemist," both his classmates and the teacher exploded in laughter because, Liebig said, no one then considered chemistry as a possible career.[28]

An unsuccessful pharmaceutical apprenticeship in Heppenheim (1817–1818) was followed by enrollment at the Universities of Bonn and Erlangen (1820–1822). Liebig was attracted by the lectures of Schelling and Kastner, the former of whom was the leader of Naturphilosophie and the latter reputed to be one of the best chemists in Germany. In 1821 he purchased and read an early (partial) German edition of Berzelius' textbook. At this point, he followed Kastner's advice to study for a time in Paris, and through Kastner's connections he was granted a stipend from the Grand Duke of Hesse-Darmstadt that enabled him to do so.[29]

During his years in Paris (1822–1824) Liebig heard lectures by Arago, Dulong, Thenard, and Gay-Lussac, which had for him an "indescribable charm." Alexander von Humboldt, then residing in Paris, made the acquaintance of the young man and recommended him to Gay-Lussac, who took Liebig into his laboratory for collaborative research. The two men produced several important contributions—Liebig's first successful research. These events, as Liebig wrote in letters home and recollected in his memoirs of old age, came as revelations of a hitherto unknown world of true science. He was overwhelmed by the sophistication of the methods, the experimentalist commitment, and above all, the conscientious avoidance of unnecessary hypotheses. This was in marked contrast to the speculative approach of Schelling, L. Oken, G. H. Schubert, and other Naturphilosophen he had previously so admired in Germany. As Wöhler was forging a personal bond with Berzelius, Liebig was simultaneously becoming warm friends with Berzelius' principal French rival.[30]

While still at Erlangen, Liebig had hoped eventually to gain a chair at a German university and, as Kastner had suggested to him, to open a laboratory-based pharmaceutical–chemical institute similar to existing models (especially that of J. B. Trommsdorff at Erfurt).[31] This possibility was realized in the spring of 1824, when the Grand Duke (without consulting the faculty) offered Liebig an ausserordentlicher professorship at Hesse-Darmstadt's single (and tiny) university at Giessen. That fall, Liebig began his teaching career with twelve eager listeners and two Praktikanten in the university's improvised chemical laboratory, newly established in the guardroom of an abandoned army barracks at the edge of town. It appears that he shared this space, none too amicably, with the ordentlicher professor of chemistry, Ludwig Wilhelm Zimmermann (1782–1825). The following summer semester both men advertised chemistry courses, but only Liebig got customers. Despondent, Zimmermann committed suicide in the Lahn

River, and Liebig ascended to the rank of Ordinarius at the amazing age of twenty-two. He also was given a raise to the extremely poor annual salary of only 800 florins (equivalent to around $300 U.S. at the time).[32]

In the summer of 1826, Liebig, in conjunction with two colleagues, achieved his goal of opening a pharmaceutical–chemical institute. Since the university refused financial support for such a narrow professionalist endeavor, it was initially run as a private venture. They were allowed to use the university's chemical laboratory, however. Although the proprietors rated the institute's capacity as twenty to thirty students, it appears that in its first decade, usually only about ten worked there at any one time. Recent work suggests that the institute was initially successful in its *advertised* function of pharmaceutical training but not in its *ultimate* function of advanced chemical education and research.[33] However, from winter semester 1826/27 on, Liebig insisted that all students in his chemical institute spend an entire semester working all day every day in the laboratory. Hence, his later claim that his pedagogical philosophy that later became so famous originated at the beginning of his Giessen years may well be accurate, in spirit if not in detail.[34]

During the fall of 1830, Liebig invented an apparatus for elemental analysis that was to revolutionize organic chemistry—his *Kaliapparat*, or potash-bulb apparatus. Berzelius' third trip to Germany took place just at this time, and he was able to spend a few days with Liebig. Until this time, Berzelius had regarded Liebig as having been infected with the "geschwind aber schlecht" (fast but sloppy) methods of the French. Berzelius now changed his mind, and for the next decade, he and Liebig formed an extremely close bond.[35] Liebig became an ardent Berzelian, but without yielding in his regard for Gay-Lussac. His friendships with Wöhler as well as Berzelius marked the time during which Liebig's loyalties were consolidated in the emerging German experimentalist school, of which he, Wöhler, and (slightly later) Bunsen were the most prominent members and of which Berzelius was the honorary dean. All of these men (with the possible exception of Bunsen) began to regard French chemistry of the new generation led by Dumas as sloppy, superficial, and self-aggrandizing.

By the early 1830s, Liebig had lost patience with his penurious administration, and he complained bitterly and tenaciously about the lack of financial and material resources in Giessen. Serious real and imagined illnesses added to the strain. Liebig's peremptory complaints, and even more so, his growing fame, made the administrators listen. In 1835 Liebig's laboratory institute was finally brought under the official aegis of the university, and funds were approved to renovate and expand it. Soon thereafter, Liebig himself received a substantial raise in salary, the laboratory was given a proper annual budget, and a new lecture hall was built. In 1838 the number of laboratory workers rose significantly—it reached twenty for the first time—and the composition of the practicum students shifted suddenly from nearly exclusively pharmacy to a mixture of pharmacy and chemistry majors. Within a few years they were nearly all chemists. Simultaneously, the lab began to attract foreign students. Liebig later suggested that the idea of practical chemical instruction "was at that time in the air," which constituted his explanation for the sudden popularity of his lab in the late 1830s.[36]

Liebig's first students of note were Friedrich Knapp, who came to Giessen in 1835, and Heinrich Will and Hermann Fehling, who arrived in 1837. All three had successful academic careers in chemistry, Will becoming Liebig's successor. Up until then, only a handful of Ph.D. chemists had emerged from Liebig's laboratory; from this time on, large numbers of promising students were to come. This divide is also marked by a change in the sort of research projects that Liebig assigned. For years, Liebig had been using students to carry out analyses or isolated fragments of research in a similar way to what Wöhler did from about 1838. But for the first time in the late 1830s, we see Liebig organizing projects for multiple workers that were well articulated and coordinated around a single problem area of interest to Liebig. F. L. Holmes has plausibly suggested that this shift may have been related to a change in Liebig's own research orientation. From about 1836 Liebig became intensely involved in a variety of writing and editing projects, and in 1840 he gave up theoretical organic research entirely. As he began to find it difficult to maintain his personal research agenda, he began to trust that agenda more to his students.[37]

He had more and more students to whom to turn. In 1839 Liebig's lab was again expanded and further renovated—the architect was J. P. Hofmann, whose son August Wilhelm entered Liebig's lab that same year—and his salary and budget were again increased. The renovation must have had something to do with a remarkable further increase in the quantity and quality of his students, traceable to that year. In addition to Hofmann, Hermann Kopp (already a Ph.D.) as well as Franz Varrentrapp, Lyon Playfair, and John Stenhouse all entered Liebig's lab that year, and Adolf Strecker came the next year. All became prominent academic chemists of the new generation. By 1841 Liebig had fifty workers, and in 1843 there were sixty-eight.[38] In the latter year, a new branch laboratory for beginners was constructed to handle the crowds, which was physically separate from the old one, and it was placed under Will's directorship.[39]

The great watershed in student demand for Liebig's lab was approximately simultaneous with that for the lab of his close friend Wöhler. The widespread assumption that Liebig must have enjoyed high popularity from the start appears not to survive careful scrutiny, for the transition to the pattern that became so famous, as we have seen, does not much predate 1840.[40]

Robert Bunsen

Bunsen (1811–1899)[41] was the son of a professor of modern languages at the University of Göttingen. He studied the sciences there, completing his Ph.D. under Stromeyer in 1831. From the beginning of 1832 to the fall of 1833, he traveled on a government grant in Germany, France, and Austria, seeking especially to investigate practical, technical, and geological subjects. A nine-month residence in Paris enabled him to learn from Gay-Lussac, Dumas, Chevreul, Pelouze, Regnault, and others, and he had the good fortune to spend a month in Giessen just when Wöhler and Liebig were collaborating on their benzoyl work. He also visited Mitscherlich and the Rose brothers in Berlin. On his return home, he habilitated at Göttingen, teaching technical chemistry and stoichiometry. Upon Stromeyer's death (18 August 1835) he took over the general lectures and the laboratory practicum. In the spring of 1836, he was appointed Wöhler's successor at the Kassel Technische Hochschule at the reduced salary of 650 thalers.[42]

Even before his transfer to Kassel, Bunsen had begun to establish an enviable research reputation. His papers merited favorable mention in Berzelius' *Jahresberichte* from 1835 on, and that fall he had the pleasure of accompanying Berzelius on a journey from Kassel to Göttingen.[43] From the late 1830s Berzelius began to extol Bunsen's work in the highest terms, both for its technical virtuosity and for its relevance to Berzelius' theoretical positions.[44] Bunsen later described Berzelius as "my truest friend and counselor."[45] To the extent that he had any theoretical commitment it was thoroughly Berzelian, maintaining the Swede's notational and formula styles to at least the 1880s, long after all others had abandoned it.[46]

In Kassel, Bunsen continued to teach students in the small and primitive laboratory inherited from Wöhler and began what were to become world-famous researches into eudiometry and the cacodyl radical. His growing eminence led to his call in 1839 to the University of Marburg, initially as Extraordinarius (at 650 thalers) but raised to the Ordinarius rank (800 thalers) two years later. His predecessor, Ferdinand Wurzer, had since 1811 been teaching selected students in his laboratory, but had done so according to the older pedagogical pattern, namely, on a small scale and on the basis of personal patronage. Wöhler, who spent a semester studying with Wurzer before transferring to Heidelberg, had found him intolerably old-fashioned.[47]

By contrast, in the spring of 1840 Bunsen initiated a chemical practicum in the consciously "modern" (Liebig–Wöhler) style, namely, in the words of Christoph Meinel, a "planned, didactically self-contained, and consistently structured unit of instruction." Moreover, the practicum was publicly advertised in the university course list, and the professor took a standardized fee for it.[48] The course was advertised as eight hours a week during precisely the same hours as Wöhler's practicum at Göttingen, but well-motivated students had no trouble persuading Bunsen to allow them to work all day every day. Bunsen was also careful to remit all fees for students who could not afford them—or even for those students who were exceptionally well motivated, regardless of their means.[49] In 1852 he was called to Heidelberg as Leopold Gmelin's successor, and there he formed the nucleus of what was the liveliest academic chemical community—though not a "school" in the usual sense—in all of Germany until the 1860s.

Bunsen was a man of uncommon benevolence, kindness, and humor, universally admired by his peers and revered by his students. He had a straightforward practical and empirical orientation, was sensitive to technological implications of his work (without, on a point of principle, ever taking out a patent), and was uninterested in theory to a degree bordering on outright hostility. There is probably no great chemist in history who was more averse to hypotheses and theoretical structures of all kinds, nor one more skilled in laboratory operations. After his great cacodyl investigation, which ended in 1841, he never returned to organic chemistry; later in Heidelberg, he totally excluded this field from his activities. His students and biographers have plausibly suggested that his aversion to organic chemistry was directly related to his aversion to theory, for it was just at the time of his one and only sizable organic-chemical project that organic chemistry became and remained intensely theoretical—and correspondingly, intensely disputatious, a quality that was also anathema to the gentle Bunsen.[50]

Another significant gap in Bunsen's career appears to have been directly related to his theoretical aversion: he never founded a school, despite his eminence and despite teaching a phenomenal number of students over no less than sixty-five years. Indeed, testimony from many sources agrees that he lavished his greatest attention on beginners and that once the tyro began to show initiative and independence, Bunsen lost interest in further guiding him. In Heidelberg, Privatdozenten were not directed or counseled by Bunsen, nor were they even allowed to work in his university facility. Instead, they were forced to cobble together improvised labs in their lodgings or in specially rented spaces. This was due less to lack of consideration by Bunsen than to his adamant refusal to create a coherent chemical school. It was enough for him to continue to induct new members into the fraternity of academic chemists and to pursue his own always fruitful research agenda in physical, inorganic, analytical, and geological chemistry.[51]

The New Culture of Science in Germany

Liebig, Wöhler, and Bunsen, and behind them the guiding spirit of Berzelius, constituted the chief representatives of the founding generation of post-Napoleonic German academic chemistry, a field that grew explosively and attained world preeminence by the second half of the century. Among them, they created during the Vormärz new models of chemical research, as well as pedagogical reforms that were much emulated in the second half of the century, not only in Germany but elsewhere as well. But these models were not created in a vacuum. Some context and biographical detail has already been established here. Several questions now need to be addressed. What specific predecessor institutions and patterns did our founders have in mind in pursuing their innovations? What clienteles were they attempting to serve, in what fashion and for what purposes? And what wider cultural and psychological models were influential for them during this period?

With regard to the first of these questions, it should be clear by now that our founders were not the first to develop laboratory-based instruction in German universities.[52] Breslau established a chemical institute in 1811, which was led after 1815 by N. W. Fischer. It seems that Döbereiner had a practicum at Jena after 1810; enrollment figures are not known, but twenty are reported in 1828. Even earlier, a practicum had been established at Göttingen by J. F. Gmelin, which was continued by his successor Stromeyer. Stromeyer is a particularly interesting figure since it appears that his practicum was the first to be advertised in a university course list. His subject was inorganic quantitative analysis, and his clientele mostly medical students; among his students were Wöhler's teacher Leopold Gmelin, as well as Mitscherlich and Bunsen.[53] A few other examples of early chemical practica exist. All of these were, however, small affairs, rarely sanctioned or supported by university authorities and usually available only on the basis of personal patronage of the laboratory's director. Apparatus and chemical stocks were as a rule the personal property of the director. The principal institutionalized purposes of academic laboratories during this period were for the personal research of the professor and for the preparation of lecture demonstrations.

Early nineteenth-century academic laboratories in other countries followed the same pattern. The situation in France has been examined for the cases of Gay-Lussac and Dumas. There the situation was slowest to change among the principal European countries because of the degree of centralization of academic culture in Paris and a certain insularity that took hold.[54] Insularity was even more evident in Britain early in the century, though British academic institutions were gradually transformed by the influence of Liebig and other German chemists after about 1840.[55]

The University of Berlin exemplifies the difficulties that German academic chemists experienced in mounting successful experimental institutes. One recent study of the ideological battles over institution building in the early Vormärz depicts the situation in Prussia as characterized by a conflict between disciples of Schelling and Kant.[56] Humboldt, the architect of the university in 1809–1810, was inclined toward speculative idealist philosophy and the intellectual intuition of Schelling; as Liebig so luridly portrayed a generation later, Naturphilosophie had not yet spent its popular force. But already in 1810, Humboldt was succeeded by Friedrich Schuckmann, and after 1817, the ministry of culture was directed for more than two decades by Altenstein.

Schuckmann and Altenstein were much more favorably influenced by the empirical attitude of Kant and the latter's emphasis on sense intuition. Many of the German experimental scientists of the nineteenth century were enrolled under a Kantian banner. Kantian physical scientists hired early for Berlin include Klaproth, E. G. Fischer, Christian Weiss, and Paul Erman, and H. F. Link arrived in 1815. Altenstein sought to acquire the sober experimentalist Berzelius as the successor to Klaproth; when that failed, he hired Link's student Mitscherlich and the brothers Rose. In the following few years, he brought in J. C. Poggendorff, Heinrich Dove, and Gustav Magnus. The same pattern transpired at other universities: C. G. Bischof at Bonn, N. W. Fischer at Breslau, L. W. Gilbert at Halle, and Kastner at Halle, Bonn, and Erlangen were all self-professed Kantians or clearly influenced by his ideas. German physicists during this period and later were even more oriented toward Kant.[57]

Despite the experimentalist stamp of nearly all of Altenstein's acquisitions, however, laboratory instruction was virtually unavailable to Berlin students. Mitscherlich had a laboratory not at the university but at the Prussian Academy of Sciences, where he was a member. Following the example of Stromeyer, Heinrich Rose began a practicum in his first semester of teaching in the fall of 1822. This was at first in Mitscherlich's lab, then in his own residence, for the university refused to pay expenses or provide a space. Magnus was also forced to set up a "university" lab in his lodgings, but fortunately, he came from a wealthy family and could afford to provide proper equipment and supplies. None of these men had more than a small handful of students at any one time, even as late as the 1840s and 1850s. Mitscherlich, Magnus, and Rose all complained to the university administration without effect about the lack of support. However, as late as 1841 all three men also asserted a preference for the use of academic laboratories only for personal research, a few advanced students, and lecture demonstrations—not for general education. The first large-scale university-supported laboratory institute at Berlin came only upon A. W. Hofmann's arrival in 1865.[58]

Representatives of other sciences were also having trouble in initiating and maintaining healthy laboratory-based practica in German academia. In physics, for instance, Wilhelm Weber introduced practical exercises in his lab at Göttingen as early as winter semester 1833/34, and Franz Neumann established his mathematical–physical seminar at Königsberg the following year. In addition, Magnus is often cited as having established an early physics practicum at Berlin in 1843. However, these cannot be compared to the Liebig–Wöhler style of chemical laboratories. Lacking the argument of importance for technology, medicine, and pharmacy and lacking the student clienteles that chemistry could lay claim to, these physics practica remained small and undernourished. Participants were mostly prospective Gymnasium teachers of mathematics and physics, a clientele whose numbers were very limited due to the relatively small amount of science taught in the German Gymnasien of the Vormärz.[59] The field of physiology has also been examined for signs of early laboratory-based education. Johannes Müller at Berlin and Jan Purkyně at Breslau in the 1830s, and Jakob Henle at Heidelberg in the 1840s are candidates for early innovators in this respect.[60] It is curious that these developments happened simultaneously with those in the field of chemistry.

The crucial change in both chemistry and physiology was from the older private-research cum student patronage-research cum lecture-demonstration model of academic labs, which extends well back into the eighteenth century, into the modern mass-education cum mass-research model. What made all the difference was student demand, with the symptom of the crucial change being the induction of large numbers of students into the practica. Once the numbers became available, entrepreneurial professors could then extort conditions from their administrations for support of their institutes and make use of the students in implementing their research programs. From here on, the intense competitiveness of German academia could provide the engine of change throughout the rest of the century.

Partial answers to the question of where the clientele for chemistry was located are beginning to appear. A revealing study of 105 European universities and technical schools (65 of them from German-speaking countries) has established the overall shifts that took place in chemical professorships during the eighteenth and the first half of the nineteenth centuries.[61] University chemistry posts before about 1790 were almost entirely confined to medical faculties. Significant growth in the number of academic chemists had occurred during the eighteenth century, at first ensconced in positions that embraced several fields and were filled by physicians, but culminating in the second half of the century in a trend toward *dedicated* chemistry professorships in the medical schools. Between 1790 and 1845, the total number of university chemistry posts remained approximately constant (at around sixty in these 105 universities), but a trend began toward moving these professorships from the medical to the philosophical faculties. By 1845 the majority of academic chemists were teaching the subject as an independent science, not as a purely ancillary medical art.

The circumstance that two of our founders, Berzelius and Wöhler, were educated as physicians and taught in medical school faculties is emblematic of the medical orientation of academic chemistry. Döbereiner (at Jena), Klaproth (at Berlin), Kastner (at Erlangen), and Zimmermann and Liebig (at Giessen) were early examples of chemists in German philosophical faculties; Bunsen was moved from the Marburg medical to the philosophical faculty upon his promotion to Ordinarius in 1841. This was the position to which Kolbe succeeded. The image of chemistry as a practical and empirical art rather than a true science such as physics provided the resistance against movement into the philosophical faculty. This resistance was gradually overcome as evidence of the maturation of chemical science percolated into the consciousness of university administrators, faculty colleagues, and students. Clearly, though, even at mid-century and even in modernist chemistry curricula in philosophical faculties, a substantial student contingent group came from medicine.

Pharmacy, however, was a rival to medicine as a source of chemistry students, and it formed the locus of the critical transition period during the early nineteenth century.[62] Since the late eighteenth century, pharmacists in Germany had been trying to raise professional standards, above all by transforming pharmaceutical training from craft apprenticeship to academic education. Because prospective pharmacists were largely excluded from the universities by the fact that few attended Gymnasium and hence did not possess the necessary Abitur certificate, there arose several chemical boarding schools that emphasized laboratory practica. These were led by such chemists as J. C. Wiegleb, S. F. Hermbstaedt, J. F. A. Göttling, C. T. Göbel, and especially J. B. Trommsdorff. Many of these proprietors were well-known university professors. The best of these schools offered well-rounded scientific instruction and attracted clientele that included budding chemical technologists and civil servants as well as pharmacists.

But an alternative route around the difficulty was simultaneously being created. The philosophical faculties in German universities accepted students without the Abitur since these faculties had traditionally served as preparatory education for the "professional" faculties. This was true even after their status was raised and their propaedeutic function dropped in the wake of the neohumanist reforms. As chemistry moved increasingly into the philosophical faculties, a new and fruitful source of student clientele became available in the form of pharmacy students without Abiturs. As German academic chemists discovered this rich vein, the Trommsdorff-style schools lost business and were forced to close. An obvious conflict existed at the neohumanist universities between what looked like narrow and practical professional training in the new institutes and the widely accepted Romantic convictions regarding the pure and elevated character of university education, but the chasm could be bridged by appropriate argumentation.

There is evidence that at least some chemists deliberately pursued this strategy.[63] Liebig founded his laboratory school self-consciously on Trommsdorff's model, but aggressively defended his pedagogy with Humboldtian neohumanist rhetoric. Although for many years most of his (and Wöhler's) students were in pharmacy or medicine, the purpose of his all-day practica, he argued, was not to show students how to boil soap or to compound drugs but rather to educate the mind and teach the student how to think. Chemistry was a true science, independent of other sciences but complementary to them—including such humanist sciences as philology and history. At least in his rhetoric (which as we will see is an important qualification), Liebig was contemptuous of the eighteenth-century emphasis on utility and application. The way to learn any discipline, he argued, was to concentrate first on the study of pure knowledge and theory, but always in conjunction with laboratory manipulations. Indeed, applications would emerge fastest and first under the hands of those who could think clearly and could logically develop their pure understanding, leaving those who learned their craft merely by rote far behind.

It may seem that Liebig was trying to have it both ways—appealing to pure theory and neohumanist *Bildung* while at the same time pushing intense laboratory practice for large numbers of students as well as the *ultimate* utilitarian value of his pedagogical philosophy. This ambivalence can be understood by looking at some more distant models that may have helped guide the predilections and psychology of Liebig and our other founders. The first point to be noted is that these men were autodidacts to a striking degree. A rather sharp generational change occurred in Germany that resulted especially from the French wars, and it would not be surprising to see new and foreign elements of thought in the first Vormärz generation, nor would perfect self-consistency be expected.

Indeed, one factor that has been too little emphasized is the curious prevalence of French and specifically French Enlightenment ideas in this group of German "Romantic" scientists: namely, an ardent empiricist commitment coupled with a marked orientation toward practical or technological utility. These ideas flourished even in the face of the strong currents of nativist Francophobia, neohumanism, and speculative and idealist philosophies prevalent in Germany at this time. Berzelius was educated in the 1780s and 1790s under the spell of French Enlightenment ideas suffusing Gustavian-age Sweden. He maintained a lifelong commitment to physiological chemistry and to practical and empirical medicine due to his convictions of utility, he taught his entire career at a professional medical school rather than a university, and he tenaciously fought the neohumanists, Naturphilosophen, and other idealist philosophers of the Schelling–Hegel school.[64]

The situation was similar with Liebig and Wöhler, and not just because of Berzelius' influence. Both were raised in technically oriented families, and as students, neither man found in Germany the experimentalist style of science he was seeking: Wöhler traveled to Sweden, and Liebig to Paris, where each discovered a new world of precision and empiricism. As we have seen, Liebig's proximate goal was to become a professor neither in the older *nor* in the modern (neohumanist) mold, but rather to found a professional laboratory school for pharmaceutical chemists. After a decade and a half of what felt to him like mucking around in theoretical organic chemical matters, in 1840 Liebig threw it all aside and for the rest of his life devoted himself to practical and technological applications in agriculture, physiology, and medicine. To the end of his life, what especially troubled him about the modern "structural" chemistry was its distance from practical (physiological) application.[65]

Wöhler's first position was at the Berlin Trade School, his second at the Kassel Technische Hochschule. Only at the age of thirty-six did he go to a university, and even then it was to Göttingen, which was the most prominent representative of the *Aufklärung* in Germany, little touched by idealist or romantic modes of thought. He maintained his personal and institutional ties to medicine and pharmacy all his life.

Bunsen's early environment was also Göttingen, and as a professor's son, he must have gotten a strong dose of the university's rationalist flavor. His teacher Stromeyer was educated partly at the École Polytechnique under Vauquelin. Similarly, Bunsen's *Wanderjahr* was spent largely in Paris; it also included a stay with Liebig and Wöhler. The strongest influence on his scientific development, however, may well have been Berzelius, even if indirectly. After serving as Privatdozent at Göttingen, he went, as Wöhler had, to the Kassel Technische Hochschule; only in 1839 did he become professor at Marburg. He had the most pronounced technological orientation of any of our founders. In 1841 he invented a battery that proved revolutionary both for practical applications and for pure science, and in 1855 he developed his eponymous laboratory gas burner. The major advance in eudiometry connected with his name also had great practical significance.

We have summarized the origins of systematic chemical laboratory-based instruction at the German universities at the hands of Liebig in particular, but also Wöhler and Bunsen slightly later. It was Liebig's model to which most people looked and many emulated, especially after 1850. To be sure, Coleman has suggested that the physiologist Purkyně was a principal innovator in Germany, and he has argued for the relevance of the empiricist Enlightenment pedagogue Pestalozzi, which would suggest parallel development in physiology from similar factors. And yet, Coleman himself noted that one essential element is absent from Purkyně's early activities, and that is the *number* and *type* of students in the laboratory. It seems that except for Stromeyer's relatively uninfluential and smaller-scale efforts, Liebig really was the first to offer systematic lab instruction to large numbers of students, many of whom were of only average abilities or were studying ancillary disciplines.[66] The model also carried over into physics, but as we have seen, those developments came significantly later.

It is Coleman who has most clearly indicated, if not investigated in detail, what he calls "a veritable culture of science" in early nineteenth-century Germany—an outgrowth of Enlightenment ideas that was in some essential respects at odds with the dominant romantic culture.[67] If neohumanism represented a break with the past in Germany, this scientific culture represents a more cosmopolitan element of continuity, an influence that looks toward rather than away from France. For Ernst Bischoff, a professor of medicine at Bonn and a representative of the older patterns, Liebig and his style of research and mass teaching of students in the laboratory represented a modern academic cancer, which he (correctly) saw as "an imported French evil."[68]

The pedagogical orientation does have a definite French feel to it. We have stressed the predominance of professional and trade schools rather than universities in the backgrounds of our founding generation. Many of these institutions were patterned directly or indirectly on French models, especially the École Polytechnique (where Liebig himself studied), although eighteenth-century German predecessor institutions existed as well.[69] By contrast, the universities and their administrators, under the thrall of anti-utilitarian neohumanism, tended to view laboratories as connected with the materialistic world of commerce and technology, alien to their sacred refuge of Wissenschaft. This seems to have worked especially effectively at Berlin, the fount of neohumanist reform, to delay the introduction of routine laboratory instruction. Gradually this negative perception faded, and after 1848 it was replaced by positive promotion of laboratory sciences in the universities of most of the German states. This new movement was based partly on the conviction in the state ministries that experimental science could provide a means for promoting industrialization, economic modernization, and social stability.[70]

Overall, the commonalities in the interests and careers of our founders explored in the past few pages provide exemplification and further support for Coleman's point and for the claims made by other recent scholars regarding a certain disjunction between neohumanism and the development of science at the German universities. Yet it would be foolish to deny that these men were also an intimate part of their culture and shared many nativist, romantic, and neohumanist values. One area in which this can be seen is the sometimes diffuse tradition of nineteenth-century Kantianism, to which I have already referred.

Elsewhere I have argued for a distinct, though perhaps indirect, influence of Kantian philosophy on Berzelius.[71] I know of no evidence of the direct influence of Kant on Bunsen. However, Debus, noting inter alia Bunsen's orientation toward empiricism and precision, speculated that although Bunsen may never have studied Kant's works, he always acted and thought in a Kantian spirit.[72] There are some suggestive connections between Kant and Liebig, who was, after all, a student of the Kantian Kastner. I have noted his passionate experimentation even as a boy. In his autobiography, he argued that this activity produced in him a characteristic that is found particularly in successful chemists: the ability to "think in phenomena." This talent, he said, can only be developed by constant exercise of the senses, and it increased in him to the point of becoming a photographic *visual* memory of compounds and reactions.[73] Debus remarked that this "concrete" manner of thinking as described by Liebig and exemplified by Bunsen is sharply opposed to a more abstract mathematical style and precisely illustrates Kantian sense intuition.[74]

That Liebig thought this talent was vital to his great success is indicated by the fact that he spent the first several pages of his autobiography describing it. Those scientists that lack such an ability, he thought, do not reach or maintain their potential. Soon before Berzelius' death, he expressed to Wöhler the opinion that as great as Berzelius was, he was too wedded to mere data, that he never mastered the vital art of "creation through thoughts,"[75] which is another way of saying the same thing. But Liebig was underestimating Berzelius. In a letter to Wöhler, Berzelius explained that he could "sense" incorrect theories even without being able to propose a better alternative, the same way a musician can sense delicate differences of intonations.[76]

Wöhler stated something similar, in his characteristic self-deprecating way:

> My imagination is fairly active, but I am somewhat slow in my thinking. No one is less oriented to be a critic than I. The organ for philosophical thought is entirely missing in me, as you know so well, just as that for mathematics. Only for observation do I imagine that I have a passable facility in my brain, which may be connected with a sort of instinct to be able to predict empirical relationships.[77]

Such a predilection for the *anschaulich* and the intuitive found in Liebig, Bunsen, Berzelius, and Wöhler is strongly reminiscent of Kantian sense intuition. This tendency is also (at least rhetorically) discernible in Kolbe, who constantly stressed his mission to teach students to "think like a chemist," by which he meant in concrete phenomenal terms and not abstractly. Kenneth Caneva has explored the differences between what he calls "concretizing" and "abstracting" scientific styles among physicists, the former characteristic especially of early nineteenth-century Kantians and the latter prevalent among mid- and late-century mathematical physicists.[78]

To put the matter perhaps a bit too bluntly, Liebig's balancing act between neohumanist *Bildung* and utilitarian laboratory practice can be understood as a combination of German idealist and French empiricist philosophies. The new German culture of science that took root in the Vormärz had imported French roots, but also a strong nativist element as well. In the process of assimilating these ideas to their own agendas, Liebig and others had created a novel pedagogy with a great future.

2

Growing Up and Limbering Up

The Kingdom of Hanover

Hermann Kolbe's forefathers lived their lives in the rolling, fertile, partly wooded country around Göttingen, in Hanover's southern exclave. Sandwiched between the duchy of Braunschweig to the north, Electoral Hesse to the south and west, and Prussia to the east, this province had been ceded in the seventeenth century from Braunschweig to the country that subsequently became known as Hanover, after its largest city. In 1714 the country's ruler, Elector Georg Ludwig, also became the king of England, and for more than a century thereafter, Great Britain and Hanover had the same sovereign. During the rest of the eighteenth century, Hanover enjoyed a number of benefits from its close association with Britain. The oligarchic privy councils that effectively ran the country's affairs under an increasingly absentee succession of monarchs were relatively liberal and efficient; for instance, sound agrarian laws protected the peasants and censorship was not practiced. The political links also provided an avenue for direct cultural relations between the two countries. The creation of the University of Göttingen in 1737 is an example of the best work of the Hanoverian regime.[1]

Hanover had an ambivalent relationship with its much larger neighbor Prussia, which was becoming an important power in the eighteenth century. This ambivalence was more increased than lessened by the consanguinity of the rulers of the two countries. During the welter of the Napoleonic wars, Hanover was successively occupied by France and by Prussia, and then it was made part of a new north German kingdom. The French occupation was distasteful and vexatious. Hanoverians welcomed the restoration of the old order upon Napoleon's defeat in 1814, especially as now the country was raised to the dignity of a kingdom and it acquired additional territory from the Congress of Vienna.

But Hanover, like other German states, suffered from centrifugal and reactionary forces during the years after 1815. Hanover's chief minister after the Congress of Vienna was Count Münster, whose despotic rule from London was much resented by wide segments of the populace. As a ripple effect of the Parisian insurrection of 1830, a riot broke out in Göttingen in January 1831 that led to the fall of Count Münster. Two years later, a new and relatively liberal constitution for the kingdom was promulgated, a document that was written loosely on the English model. The University of Göttingen reflected these troubles. In the early postwar years, its enrollment topped 1500; by 1834 it had plummeted to 860.[2]

Even worse was to come. By Hanoverian law, a woman could not ascend the throne. Accordingly, the fifth son of George III, Ernst August, and not Victoria, became sovereign of Hanover in 1837, thus ever after separating the crowns of Great Britain and Hanover. Ernst August was autocratic and illiberal; one wag reported that he had committed every crime in his life except suicide.[3] One of his first official acts was to suspend Hanover's new constitution. The consequent protest of seven of Göttingen's most eminent professors, including Weber and the Grimm brothers, resulted in their dismissal. Despite expressions of outrage across Germany from citizens of a wide range of political belief, the king had the support of the Hanoverian nobility, and an appeal to the diet of the German Confederation proved fruitless. Open opposition to the regime died, and Ernst August continued his reactionary policies until the uprisings of 1848. "The decision of the Göttingen Seven," writes one historian, "propelled the German professoriate into the middle of public life. The Frankfurt Parliament of 1848 would be incomprehensible without the Göttingen protest."[4]

However, the protest and expulsion of the Göttingen Seven can also be overinterpreted. Their refusal to break their oaths by repudiating the constitution of 1833 was more a statement of conscience than a political stand. Most members of the German professoriate in the Vormärz, when exercising any political sentiments at all, stood for moderate constitutional reform; very few had sympathy for extreme democratic or republican views.[5] In this sense, as we will see, Hermann Kolbe became a typical representative of his class.

Forefathers

In German, *Kolben* means both a club or truncheon and a chemist's flask, and both of these denotations bear ironies in the case of Hermann Kolbe. The surname Kolb or Kolbe is not uncommon in central Germany; it is said that the etymology of the family name is related to the sort of person who is inclined to take up cudgels for appropriate use. Kolbe must have had this implication in mind when he designed his signet emblem in 1851: two brawny arms brandishing clubs are depicted, the one crowning the device apparently poised to strike a crushing blow on his opponent.

Kolbe's grandfather, (Johann) Georg Wilhelm Kolbe (ca. 1760–ca. 1825), was described as a preacher or minister and schoolmaster in the hamlet of Gross-Schneen, eight miles south of Göttingen, at the time of his (eldest?) son's birth—Hermann's father, Carl—in 1790. By 1808 he was schoolmaster in Grone, a settlement just outside the city ramparts. Around 1815 he became property inspector and clerk of courts in Adelebsen, a position he still held in November 1821, after which we know nothing more. It was common at this time and later for ministers to combine pastoral duties with teaching and other civil service.[6] G. W. Kolbe probably had a university education, but there is no Göttingen matriculation record, and it is not known whether he formally qualified for the pastorate.[7]

Carl Friedrich Ludwig Kolbe (18 December 1790–4 October 1870) attended the Göttingen Gymnasium, preparatory to the study of theology at the university, matriculating during the French occupation in April 1808. After completing the three-year course, Kolbe taught Latin and Greek at the monastery school in Ilfeld, near Nordhausen in Saxony. In September 1815 he received a call to lead the Lutheran congregation in Elliehausen, three miles west of Göttingen. It was there that he established his family, marrying (Dorette Caroline) Auguste Hempel (10 August 1800–10 April 1856), less than a month after her sixteenth birthday. The first of their fifteen children was (Adolph Wilhelm) Hermann Kolbe, born 27 September 1818.[8]

Auguste's father, Adolph Friedrich Hempel (1767–1834), was professor of anatomy at Göttingen. Her mother was the former Marie Catherine Louise Grabenstein, the daughter of a physician who was also for a time Bürgermeister of Göttingen. She had died by the time of her grandson Hermann's birth. Professor Hempel, as we will see, had daily contact with his adolescent grandson Hermann during the last three years of his life. Born in Neustrelitz (Mecklenburg), Hempel received his M.D. degree from Göttingen in 1789 and taught there the rest of his life. The author of a popular anatomy textbook, he was described as a straightforward rationalist in his approach toward his science, consistent with the general intellectual tone at the University of Göttingen.[9] There was still a strong current of Enlightenment rationalism in Göttingen in the nineteenth century; such speculative and metaphysical movements as Naturphilosophie, homeopathy, and phrenology never made headway there.

By all reports, Carl Kolbe was a man of energy, strong opinions, great self-confidence, and forthright manner. The words used by Hermann's biographers to describe his father's character—*gerade, unerschrocken, bestimmt, fest in sich geschlossen, ausserordentlich energisch, von festem Willen*—could as easily be applied to Hermann as well.[10] His life motto "*Tue recht, scheue niemand*" ("Do the right thing, fear no one") was carefully inculcated in his children.

The income of Lutheran pastors in nineteenth-century Germany was derived from the agricultural production of local church lands leased to tenant farmers, a sort of prebendal arrangement. The endowments of various church districts differed markedly, and so it was common for pastors to seek more lucrative positions as their careers progressed. As it happens, Elliehausen was a relatively well-endowed position, possessing not only fertile prebendal lands but also agricultural assets at the parsonage itself. In fact, the ground floor of the building provided stalls for cows and horses, part of the second story was used as a granary, and a pigpen directly adjoined the house. The pastors were expected to be farmers in their spare time; indeed, the Elliehausen pastoral farm continued in operation until 1926.

Unfortunately, a serious problem emerged soon after Carl Kolbe's arrival. The church there was barely ten years old, but it had been poorly built, indeed fraudulently, with inferior materials and without a proper foundation, and by the time of Hermann's birth, it was already dangerously and irreparably dilapidated. Failing in his attempt to raise the money needed for a new church, in December 1826 Kolbe accepted a call to the prosperous *Kirchspiel* of Stöckheim, in the Leine valley twenty miles north of Göttingen. The money for tearing down and rebuilding the Elliehausen church was raised only after Carl's brother Georg was called to be pastor there.[11]

Stöckheim was and still is a small village—there were 410 residents in 1826 and about the same number today—but the church district comprised more than 1800 people; Kolbe now presided over a handsome (and sturdy) sixty-five year old church and an ample parsonage. He needed the room. By the time of his transfer he had four young children, and his family continued to grow rapidly during the next few years. Unfortunately, to Kolbe's great distress and against his vehement objections, in the late 1830s the tenant farmers working the church lands were able to obtain the money and consistorial approval to acquire their property outright. This development suddenly and dramatically reduced his income. He began to look for yet another congregation.[12]

In September 1840 he accepted a call to the village of Lutterhausen, just outside the town of Hardegsen ten miles south of Stöckheim. Lutterhausen was even smaller than Stöckheim, but the district was prosperous, and it was here that Carl Kolbe remained for the rest of his career and the rest of his life. Weakness of old age forced him to retire in 1869, and he died in 1870. The lovely church and the spacious half-timbered parsonage, which still stand, attest to the comfortable nature of Carl Kolbe's last position. Still, Hermann's biographers—with perhaps some exaggeration—emphasize the modest character of the family's lifestyle, portraying Carl as living the life of a "simple country pastor."[13]

It is possible to say something about Carl Kolbe's religious outlook, for he published a catechetical handbook for the religious and moral instruction of children, at just about the time he needed to begin instructing Hermann and his other children. His book begins with the thesis that "we learn of the existence of God through a rational consideration of the heavens and the earth." Only after thoroughly exploring the rational basis of religious belief does he ask children to understand that "we acquire more detailed knowledge of God through the Bible." These and other passages demonstrate that Pastor Kolbe's religious opinions exhibited the moderate rationalism so characteristic of Lutheran theologians during the German Enlightenment, and especially of the anglophilic Hanoverians. One would seek in vain here for precursors of Hermann Kolbe's future fire-and-brimstone "sermons" to his fellow chemists. Pastor Kolbe's instructions are uniformly modest, gentle, and tolerant, and are informed by a positive and liberal view of God and of mankind.[14]

Determining Hermann Kolbe's religious views is not straightforward. Apparently, Kolbe intended to accede to his father's desire for him to enter the ministry, until at the age of eighteen he discovered chemistry.[15] Conversely, Edward Frankland, a good friend who spent much time with Kolbe from 1845 to 1850, described him much later as an "agnostic."[16] However, if we consider that Frankland was a "born-again" Christian during much of this period (before he began to fall into agnosticism himself), that the term *agnostic* did not even exist at that time, and that there are many demonstrable inaccuracies in Frankland's memoirs (including incorrectly labeling others as agnostic),[17] this evidence must be viewed with caution. Frankland said that he and

Kolbe had argued over religion. It could be that Frankland's early impressions were never modified by later data; from his fundamentalist perspective, sincere religious feeling tempered by rationalist moderation may have *seemed* agnostic.

An autograph memorandum in Kolbe's personnel file at Marburg states his religious confession to be "reformirt." This memo is dated 31 January 1854, which is shortly after his marriage, and it is a reasonable supposition that he converted from Lutheranism after becoming engaged. This conjecture is strengthened by the report that he initially experienced opposition from his prospective father-in-law, but that all tensions were fully resolved shortly before the wedding.[18]

Despite this (probably thoroughly formalist) conversion from the faith of his family, most evidence supports the view that Hermann remained a sincere Christian and that his religious philosophy was similar to that of his father. As professor in Leipzig, he placed the Biblical quotation "God has arranged all things by measure and number and weight" (Wisdom of Solomon 11:20) in large letters above the chart of the chemical elements at the front of his lecture theater. When Kolbe's successor Johannes Wislicenus first saw the placard, he said to his guide, "Das muss verschwinden!" When relating this story, Georg Lockemann compared the two chemists by remarking that Wislicenus, too, was the son of a minister, but a man of very different (i.e., freethinking) religious convictions—indicating that Lockemann regarded Kolbe as a traditional religionist.[19]

In an 1872 work, Kolbe argued that students of theology should be required to study chemistry and the other sciences in order to *combat* the atheistic and materialistic image of scientists and of science. This sentiment is reinforced in Kolbe's obituary of Liebig (who, Kolbe asserted, shared his views on theology and religion). The proper attitude, he felt, is not blind orthodoxy that teaches miracles and a literal interpretation of the Bible, but rather a rationalistic and science-based religious conviction that has overtones of natural theology. He wrote

> The study and recognition of the wonders of nature and of the laws by which the Creator has revealed himself to mankind in so palpable a fashion leads not, as those stupefiers of mankind would want us to believe, to atheism, but in just the opposite direction: it permits the physical as also the spiritual eye to discern the caring and cherishing hand of the Creator in thousands of features.[20]

Many passages from Kolbe's correspondence with publishers Eduard and Heinrich Vieweg also suggest a moderate rationalist Protestantism, similar to that of his father. In 1853 he recommended for publication by Vieweg a history of the early Christian church by Eduard Zeller (1814–1908), a prominent member of the Tübingen school of higher criticism and a follower of David Strauss. Zeller had recently been called to Marburg, but had been forced from the theological into the philosophical faculty by persistent charges of atheism. Kolbe explained to Vieweg what the higher criticism meant, said he agreed with Zeller's views, and commented that no one is less inclined toward atheism than Zeller, despite his record of defending conclusions that "make the hair of pious theologians stand on end."[21]

All of this suggests that Kolbe never lost the faith given to him by his father. Nonetheless, considering his upbringing as the son of a pastor and the large number of surviving personal letters that provide a window into his private life, the paucity of evidence for the character of Kolbe's religious convictions is striking. All of his biographers emphasize that chemistry was his whole life, and one can only assume that his science, simultaneously his vocation and avocation, overshadowed even his sincere religious beliefs.

Boyhood

It was in the "rustic simplicity" of Elliehausen and Stöckheim that Hermann Kolbe grew to young manhood. These villages were typical rural communities of the day; the principal occupation was growing hops, rye, oats, fruit, and stock, and the region was devoid of any significant industrialization. Indeed, all of the Germanic lands during the Vormärz were essentially rural (only about twenty percent of Germans in 1815 lived in towns or cities). Widespread industrialization did not begin until the 1840s. Moreover, partly due to the unregenerate regime of Ernst August, Hanover was more backward than most German states. It was not until the middle of the century that the power of the guilds in controlling industrial production was broken, the first railroad lines linking Göttingen and nearby cities such as Kassel and Hanover were constructed, and the kingdom of Hanover joined the Prussian—and increasingly pan-German—Zollverein. Despite their relatively modest financial circumstances and rural surroundings, however, the Kolbe family would have regarded themselves as full-fledged members of the German university-educated elite, the *Bildungsbürgertum*, with all of the attached neohumanist cultural and elevated bourgeois social implications.[22]

Indeed, the central position of Lutheran theology and the pastorate in the intellectual life of nineteenth-century Germany has often been emphasized; as Fritz Ringer recently put it, ". . . one cannot imagine the development of the German intellectual class in the nineteenth century without the pastors or their sons." They were devoted to learning and placed the highest value on humanist education. A statistical look at the social origin of German professors indicates this orientation: after sons of professors (16%), sons of ministers (15%) constituted the largest class of German academics at mid-century. The relative modesty of pastoral homes may have increased the social aspirations of ministers' sons, including a certain affinity toward the nobility.[23] Kolbe surely must have felt this pull, especially growing up as he did in the aristocratic Hanoverian kingdom and attending the nobility-dominated university of Göttingen. This affinity could only have been increased by the fact that his maternal grandparents had been two of the leading citizens of Göttingen. His best friend while attending Gymnasium there, as we will see, was the son of a nobleman.

Lockemann wrote that Hermann often fondly reminisced in latter years of his youth as "a precious time of harmless rustic freedom." Ernst von Meyer remarked that it was not the sort of upbringing that one would expect to nurture seeds of scientific greatness. His first instruction was tendered by his father, followed by a village schoolmaster, and then a capable tutor attached to the Kolbe household, whose duties increased with the growing household. It is said that Hermann derived his energetic, self-confident character from his father and a love of science from his mother. He is depicted as a happy and active child, climbing in fruit trees, pursuing gymnastic exercises in summer and ice skating in winter, preparing homemade wine from birch sap and syrup of violets, assembling beetle and butterfly collections, and assisting in household chores in the kitchen and cellar. As the eldest child, he took great pleasure in instructing his many brothers and sisters. One source of chemical interest for Hermann was a salt brine works in Sülbeck, within walking distance of Stöckheim, which he often visited; early chemical "experiments" on his mother's stove are also reported.

Kolbe's correspondence and other records have revealed the names of most of his fourteen siblings. Sister Emma died unmarried at the age of twenty-one in 1850, a loss that deeply affected Hermann. Sister "Rutsch," also unmarried, lived with Kolbe's family in Leipzig for many years. Sister Bertha, born in 1824, married a pastor in Ellierode (just two miles from Lutterhausen) named Georg Ost. Their son became Kolbe's godson and namesake; Hermann Ost studied with his uncle, had a successful chemical career at the Hanover Technische Hochschule, and became one of Kolbe's biographers.[24] Brother Carl, about fourteen years younger than Hermann, took his brother's chemistry courses at Marburg in 1851–1852. After their father's death in

1870, Hermann became the acknowledged head of the family, working hard to sort out the questions of probate and inheritance for the ten surviving siblings. Today, only three graves remain in the Lutterhausen churchyard, those of Kolbe's parents and his sister Emma. In 1878 Kolbe provided three new headstones, landscaping, and a sturdy wrought-iron fence enclosing the small plot.[25]

It is clear from later letters that Kolbe had great regard for and pride in his father. Judging by mentions of his trips in letters, his visits home were of respectable frequency, though only one visit from his father to his own home (in Marburg) is recorded. In December 1851 he personally nursed his father back from a serious illness despite the press of duties of his first semester as ordentlicher Professor in Marburg. His father accepted the role of godfather to Kolbe's first male child. In 1861 and again in 1865, Kolbe traveled to Lutterhausen for the celebrations of his father's fiftieth anniversaries of his first teaching post and his first pastoral call, respectively.[26] When he was named a foreign member of the Swedish Royal Academy of Science, Kolbe wrote Eduard Vieweg asking for a copy of the newspaper in which the news was announced so that he could send it to his father.[27] In contrast, it is remarkable that I have found but one reference to his mother in the 850 surviving letters from his pen.

Gymnasium and University

In 1831 Hermann entered the Göttingen Gymnasium, living at first with his maternal grandfather Professor Hempel at Lange Geismarstrasse 230, a short walk from the Gymnasium on Wilhelmsplatz in the heart of the old city. After Hempel's death (1834), he lived in the Gymnasium complex itself in the home of the Gymnasium's director, the philologist G. F. Grotefend, who, curiously, had had Wöhler as a pupil in Frankfurt twenty years earlier. In both of these residences, Kolbe was exposed to cultured academic households. A possibly even stronger influence was a close friendship with a classmate, a member of the Hanoverian nobility named von dem Knesebeck. Knesebeck was acquainted with the young Privatdozent at the University of Göttingen, Robert Bunsen, who informally instructed him in chemistry. Knesebeck and Bunsen both grew up in Göttingen, and one can imagine that their families may have been acquainted. Knesebeck constructed a small laboratory in his father's garden house, where he shared chemical arcana with the young Hermann. Kolbe later recollected that these events in the summer of 1837 induced him to give up his ambition to follow his father into the ministry, as his father wished, and turn to the academic study of chemistry. Unfortunately, his friendship with his schoolmate was of short duration. One day in class Knesebeck became ill; Hermann took him to his room, only to have his friend die in his arms. An overdose of opium was blamed, but whether intentional (he had a troubled relationship with his stepmother) or accidental was not determined.[28]

Hermann took a second-class Abitur in April 1838. Unfortunately, the Gymnasium's records from this period have not survived; consistent with his overall score, he is said to have been a good but not exceptional student. Ost wrote

> His friends from his years at school depict him as a boy who was eager to learn, who thoroughly studied whatever had once attracted his attention. . . . Kolbe did not possess the capability of learning effortlessly; rather, his natural gift consisted in the drive to direct himself resolutely toward fixed goals, and to immerse himself to the depths in whatever subject he attacked.[29]

Kolbe matriculated at the University of Göttingen for summer semester 1838, now intent upon a career as an academic chemist. He took two semesters of physics with J. B. Listing (Weber's successor), three of mineralogy and geology with J. F. L. Hausmann, two of mathematics with Georg Ulrich, and a course in metaphysics from J. F. Herbart. He resided at Burgstrasse 332, across from the Gymnasium. Kolbe's leaving certificate from the university attests that he studied all these subjects with "exceptional diligence," his only black mark being a four-day incarceration in the student jail on account of having insulted an unnamed personage.[30]

Of greatest importance, Kolbe studied chemistry with Friedrich Wöhler. He attended Wöhler's practicum during each of the eight semesters he spent at Göttingen, and during three of these, it was his only academic occupation. We have examined Wöhler's background and early career in chapter 1. Here it is relevant to underline the attractiveness of Wöhler's teaching and his wholehearted—and single-minded—commitment to his science. A fundamentally nonpolitical man, the traumatic episode of Ernst August's revocation of Hanover's constitution and the storm over the Göttingen Seven, which occurred eighteen months after Wöhler's arrival and six months before Kolbe's matriculation, scarcely seemed to touch his consciousness.[31]

As noted in chapter 1, Kolbe was only the third Göttingen student to register for chemistry after Wöhler's arrival there (the first two are known only from their matriculation entries). It was precisely at this time that both Wöhler and Liebig were beginning to attract comparatively large numbers of students to their lectures and especially their labs. Most of these were pharmacy and medical students, but the chemistry contingent, initially quite small, increased rapidly during the early 1840s in both universities. This constituted the first sizable contingent of serious chemistry students in history.

Kolbe's progress, however, seems to have been slower than some of his younger compatriots, for it was not until summer semester 1840 that Kolbe was allowed to perform real research. By this time, a real cohort of research students had emerged. Kolbe's compatriots included Carl Voelckel and Georg Schnedermann (Wilhelm Knop arrived in 1841), all of whom were to begin successful, though admittedly modest, academic careers before Kolbe's first professorial call. Kolbe's first research assignment, given to him by Wöhler, was to study a reaction first published by Döbereiner eighteen years earlier, namely, the preparation of formic acid by oxidizing larger organic molecules. Kolbe distilled starch and alcohol with pyrolusite and sulfuric acid and isolated ethyl formate from the reaction mixture. Neither Kolbe nor Wöhler continued this line of investigation. Wöhler published this note under his name alone, giving Kolbe credit within the article.[32]

Kolbe's second assignment was the analysis of fusel oil residue in grain alcohol. Wöhler described his results in a letter to Berzelius—at variance with the only previous analysis of the material—as "striking". Berzelius mentioned the result in his *Jahresbericht* for 1842, naming Kolbe explicitly. The paper was subsequently published in Liebig's *Annalen*—Kolbe's first scientific publication. Kolbe related in later years how Wöhler subjected this first manuscript to a stringent linguistic critique, forcing him to eliminate excess verbiage and to describe the factual details in a direct and clear fashion.[33]

About the time this paper was submitted, Kolbe suffered an attack of jaundice, and he spent the winter of 1841–1842 at home in Lutterhausen. By the spring of 1842 he was sufficiently recovered to begin his dissertation work with Wöhler. This work had been well begun but was by no means completed when Wöhler seized upon an opportunity for Kolbe. Robert Bunsen, Wöhler's successor in Kassel, had been called to Marburg in 1839. Wöhler had sent Voelckel to be Bunsen's *Assistent* late in 1841, but after one year Voelckel accepted a call to the cantonal school in Solothurn, Switzerland. So now Wöhler recommended his promising student Kolbe, not yet Ph.D., as Assistent for Bunsen, a position boasting the not exactly munificent salary of 200 thalers per year.[34]

Kolbe spent three years with Bunsen and derived great profit from his contact with the only slightly older man. Indeed, Bunsen became Kolbe's *Doktorvater*, Kolbe officially qualifying for the doctorate on 23 October 1843. Referring to the late 1840s, Tyndall reminisced

> Bunsen was a man of fine presence, tall, handsome, courteous, and without a trace of affectation or pedantry. He merged himself in his subject: his exposition was lucid, and his language pure; he spoke with the clear Hanoverian accent which is so pleasant to English ears; he was every inch a gentleman. After some experience of my own, I still look back on Bunsen as the nearest approach to my ideal of a university teacher.[35]

The circumstances that Bunsen, like Kolbe, was a native Göttinger, and that Kolbe's first introduction to chemical operations had come through Knesebeck second-hand from Bunsen, could only have increased Kolbe's regard. In later years, his relationship with Wöhler cooled somewhat (as happened with most of Kolbe's friendships) but never his ties to Bunsen. It may seem anomalous that the kind and gentlemanly Bunsen never lost his affection for his obstreperous student. This circumstance may be partly explained by the fact that in 1844 Kolbe saved Bunsen's life by carrying him unconscious out of a laboratory filled with carbon monoxide.[36]

Heinrich Debus, who studied with Bunsen from 1845 to 1847 and then served as his assistant until 1851, has provided a detailed description of Bunsen's lectures and practicum in Marburg. A set of student lecture notes from 1850 has also survived.[37] Bunsen was clearly one of the century's best lecturers on chemistry, and his numerous and apposite illustrative experiments never failed. He especially loved physical and inorganic chemistry, and he emphasized precise quantitative measurements in all operations. The small amount of theory included in his courses was thoroughly in the spirit of Berzelius' electrochemical dualism. His most popular course was his famous *Publikum* on electrochemistry, where every seat was always occupied.

In the early 1840s Bunsen taught about five to ten Praktikanten at a time, although there were sixteen by winter semester 1845/46 and occasionally over twenty in the late 1840s. Most were pharmacy or medical students who only worked the standard eight hours per week; normally four to six were advanced workers, mostly chemistry majors, who had permission to work all day every day in the lab. The practicum course started with two or three weeks on the nature of flame and blowpipe analysis. The remainder of the semester was devoted to wet qualitative analysis, for which each student analyzed (at his own pace) progressively more challenging unknowns. For four to six weeks of the analytical portion, Bunsen provided most of the supervision, but as the students gained knowledge and confidence he turned much of the burden over to his assistant. However, he was always present in the laboratory—supervising or doing his own research—to answer any student's question.

Kolbe's duties as Assistent included instruction and supervision of the Praktikanten in the laboratory and assisting Bunsen with preparations for his lecture demonstrations. Kolbe also had time and use of the lab for his own research. Why he never sought to qualify as Privatdozent is not clear; certainly his research was of the requisite quality and quantity for the necessary thesis (the *Habilitationsschrift*). A final occupation of his first years in Marburg was the preparation of a German translation of the first volume of Gerritt Mulder's *Proeve eener algemeene physiologische scheikunde* (Rotterdam, 1844). Wöhler had no doubt recommended him for this purpose to the Braunschweig publisher Eduard Vieweg, whom he had gotten to know through their mutual friend Liebig. The first seventeen letters by Kolbe to Vieweg, of what would become a total of 524 extending over forty years, concern this translation.[38]

Wöhler and Bunsen, Kolbe's two direct mentors, had much in common, and their influence on Kolbe was strong and unmistakable. Both chemists were enormously prolific and, moreover, extraordinarily skilled, inventive, and precise in laboratory operations. Wöhler, thoroughly schooled by the master Berzelius, must have often repeated to his own students his teacher's well-remembered injunction against "geschwind aber schlecht" (speedy but poor) experiments, and he provided his students innumerable models of research following this ideal. Bunsen, a virtuoso if there ever was one, was also famous for the care and accuracy of his work. In the 1830s he worked out methods for the analysis of mixed gases that were far superior to those previously used and that were only slowly spread from his laboratory. In 1841 he developed a much improved carbon–zinc battery that could be used for electrolysis experiments as well as more practical applications. Kolbe was to make immediate and highly productive use of both of these innovations.

Wöhler and Bunsen were also alike in their brilliant teaching abilities, their predilection for experimental investigations, and their habitual avoidance of theory. This is not to suggest that they always ignored the theoretical implications of their studies, nor to deny that many of their papers were theoretically important. Wöhler's work on urea and benzoyl derivatives and Bunsen's work on cacodyl are examples of theoretically rich papers. But, significantly, both scientists left the discipline of organic chemistry just when it began to explode theoretically in the early 1840s—and just when Kolbe arrived on the scene—and neither of them ever returned to the field in the succeeding decades.[39] After 1843 Bunsen excluded organic chemistry ever more effectively from both his teaching and his research. That Kolbe, the student of both of these men, became a theoretically inclined organic chemist is curious but not truly anomalous. After all, both were still doing organic chemistry when Kolbe was their student, and the strong theoretical orientation of Wöhler's teacher Berzelius and Wöhler's best friend Liebig were clearly of decisive influence on Kolbe's development.

There were also a variety of personal bonds between Wöhler and Bunsen. Since their first meeting in Giessen in 1832, their lives became curiously entwined. Stromeyer was Bunsen's *Doktorvater* and Wöhler's *Doktorgrossvater* (through Leopold Gmelin); the decisive scientific influence for both Wöhler and Bunsen, however, was exerted by Berzelius. As Privatdozent in Göttingen, Bunsen succeeded Stromeyer unofficially and temporarily until Wöhler replaced him; Bunsen then took over Wöhler's position in Kassel. Three years later, he succeeded Wöhler's first chemistry professor in Marburg, Wurzer. When Bunsen started his practicum in 1840, he closely imitated that which Wöhler had begun in Göttingen, even to the same eight hours of the week. Finally, Wöhler provided Bunsen with two of his first Assistenten, Voelckel and Kolbe. Their commonalities in personality, temperament, and scientific style resulted in strong bonds of mutual regard.

In conclusion, Kolbe must have felt it extremely natural and comfortable to work in Marburg after his education in Göttingen. Certainly he was given no reason to doubt the basis of the science of chemistry in Berzelian electrochemical dualism. This psychological and theoretical commitment was only increased by the research he carried out at both universities.

Doctoral Work

For his *Doktorarbeit*, Wöhler assigned Kolbe the task of studying the action of chlorine on carbon disulfide. This proved to be a crucial choice because it led Kolbe inexorably toward an examination of the link between organic and inorganic chemistry, a topic that was at the heart of many of the theoretical debates then raging in Europe. Carbon disulfide (modern CS_2) was a familiar substance at that time, but it was commercially unavailable, so Kolbe had to prepare it in Wöhler's lab directly from the elements. As Wöhler reported to Berzelius in a letter of 26 July 1842, it initially appeared that the products of high-temperature chlorination in a porcelain-filled tube were sulfur dichloride and a new substance that Wöhler formulated as CS+Ȼl. Berzelius considered this a "very interesting compound" in his preferred "four-volume" (i.e., doubled) formulation of it, $CS_2+C\text{Ȼ}l_2$, precisely because of its position on the organic–inorganic interface.[40] (The barred symbols represent Berzelian "double atoms," so that $\text{Ȼ}l = Cl_2$.)

Within two weeks after writing to Berzelius, Wöhler and Kolbe had determined that no novel compound was formed after all, the second product being merely a mixture of sulfur dichloride and the previously known *Kohlensuperchlorid* (literally, carbon superchloride, known today as carbon tetrachloride). Still, the reaction was significant as a smooth route to preparing large quantities of the latter material, much superior to Regnault's inefficient chlorination of chloroform. Moreover, Wöhler and Kolbe found that a new compound of carbon, chlorine, and sulfur was in fact formed when the same reaction was carried out at room temperature, though at the time of writing (September 1842) they had not yet determined its formula.[41]

Berzelius wrote that both results were "extremely important" because they represented the formation of organic products from purely inorganic reactants, and he offered advice on how it might still be possible to obtain the product that Wöhler and Kolbe had thought they had isolated in July. He concluded by once again stressing the theoretical importance of these kinds of compounds and exhorting Wöhler to look at a few additional related compounds, including one he and Alexandre Marcet had discovered three decades earlier, the substance now known as trichloromethylsulfonyl chloride (formulated in Berzelian terms as $C\not{C}l_2+SO_2$).[42]

Wöhler and Kolbe followed Berzelius' prescriptions and found that the compound formed by room-temperature chlorination was indeed the half-chlorinated carbon disulfide. In the paper as it appeared in Liebig's *Annalen*[43]—curiously, published under the single name "Heinrich Kolbe"—the new substance was formulated first as $CSCl_2$ (Wöhler's preference). Kolbe then noted that the formula "probably" should be doubled to CCl_4+CS_2 (Berzelius preference).

The Decline of Dualistic Organic Chemistry

Behind Kolbe's two mentors, Wöhler and Bunsen, lay the dominating spirit of Berzelius, whose reign as virtual king of chemical theoreticians was only beginning to be challenged while Kolbe was studying at Göttingen. Wöhler was Berzelius' official (and extraordinarily industrious) representative in Germany. The correspondence of the two chemists over two dozen years occupies nearly 1500 pages in Wallach's massive edition; the warmth of their scientific and personal relationships emerges clearly from virtually every page. Wöhler's own theoretical convictions, such as they were, nearly always came directly from the *Jahresberichte* and the *Lehrbuch* of Berzelius.

We have seen how Berzelius offered advice to Wöhler on both the experimental and interpretational aspects of Kolbe's first extended investigation; Wöhler certainly shared these suggestions with his student. Although no letters between Berzelius and Kolbe have survived, it is known that they had begun to correspond on an occasional basis at least by the summer of 1843.[44] It is clear from both correspondence and his reports in the *Jahresberichte* that Berzelius became ever more excited about Kolbe's research.[45] For his part, Kolbe revered Berzelius; he kept one highly complimentary letter from the Swede (dated 3 August 1844) for the rest of his life as a "talisman" against the seductiveness of fraudulent hypotheses.[46]

The bedrock of Berzelian chemistry, dating from Berzelius' first important paper (1803), was the conviction that chemical affinity was reducible to polar forces between electrically dissimilar molecular components. The system built up over the years by Berzelius and his school was thus called electrochemical dualism. Applied to inorganic combinations, this system became widely accepted in the 1810s and 1820s.

Applied to the emerging field of organic chemistry in the 1830s, the theory led to the formulation of various organic radicals that were thought to function integrally and electropositively, adding to negative components as metallic elements do in inorganic compounds. Whether electronegative oxygen could enter into an organic hydrocarbon radical was a disputed point right from the beginning of the older radical theory. Berzelius expressed strong doubts about the reasonableness of oxygenated radicals, preferring conceptually clean, theoretically consistent, purely electropositive hydrocarbon radicals. Liebig took the more empirical viewpoint that at least some of the oxygen associated with many organic radicals seemed to appear in entire series of compounds of the radical and so ought to be considered as an integral part of the radical.[47]

Upsetting the coherence of dualistic organic chemistry was the phenomenon of chlorine substitution, wherein highly negative chlorine appeared to replace highly positive hydrogen in organic compounds, without major alteration of the properties of the compound. Since a cardinal thesis of dualistic organic theory was the direct dependence of chemical properties on electrical characteristics of molecular components, this appeared distressingly anomalous. Adumbrated by Liebig's mentor Gay-Lussac, then by Liebig's and Berzelius' archrival Dumas, organic chlorination reactions were first thoroughly explored by Dumas' student Auguste Laurent beginning in 1830. Laurent developed a nonelectrochemical theory of organic reactions based on chlorine substitutions which he initially called the theory of fundamental and derived radicals and later the nucleus theory. According to this theory, fundamental radicals could be transformed into derived radicals either by substitution within the radical or by addition or elimination of atoms outside the radical. The most important factor for Laurent was not the identity of an atom but its position. The electrochemically opposite substances hydrogen and chlorine could play the same chemical role inside a radical; conversely, a chlorine atom inside or outside a radical would exhibit different chemical properties. Similarly, oxygen could replace hydrogen inside the radical with no great alteration of properties, but oxygen introduced outside the radical would make a neutral substance acidic.

Dumas at first attempted to salvage dualism in the face of Laurent's discoveries. Briefly allied with Liebig at the end of 1837, they published a joint manifesto that advocated a complete analogy between inorganic elemental radicals and organic compound radicals. Attacked by Berzelius as having been corrupted by Laurentian theory, Dumas indignantly refuted the charge. But in August 1838 Dumas discovered chloroacetic acid, and he noted to his surprise that the replacement of three-fourths of the electropositive hydrogen content of acetic acid by highly electronegative chlorine had little real effect on its properties.

Dumas promptly abandoned dualistic organic chemistry, formulating a new theory loosely based on Laurent's work. According to Dumas' "type theory," there are series of organic compounds each of which must be considered to be based on a single "type" formed from the same number of chemical equivalents combined in the same way. Contrary to dualistic organic theory, substitution reactions—even between electrochemical extremes, such as chlorine for hydrogen in chloroacetic acid—cannot alter the chemical type, hence cannot alter the fundamental chemical properties. In a modification of this theory (1840), Dumas conceded that substitution sometimes does alter the fundamental chemical properties without changing the total number or apparent mode of arrangement of the chemical equivalents. In such instances, a new *chemical* type is created, but without altering the "molecular or mechanical" type. Only addition or elimination of atoms could create a new mechanical type.

Despite his joint declaration with Dumas, Liebig had also been traveling away from the Berzelian orthodoxy. One of his greatest experimental and theoretical masterpieces, the joint work with Wöhler on benzoyl derivatives (1832), had suggested to him that oxygen must be considered as an integral part of certain organic radicals and that the hydrogen atom of benzaldehyde (benzoyl hydride) is replaceable by oxygen, chlorine, or other electrochemically dissimilar substances. Thus the door to apostasy was opened for Liebig.

For Berzelius, all acid–base reactions resulted from electrochemical addition of a negative anhydrous (usually oxy-) acid to a positive base. Just as potassium sulfate, for example, was regarded as the result of the combination of sulfuric acid (anhydride) and potash:

$$SO_3 + KO \rightarrow SO_3 \cdot KO$$

so potassium acetate was regarded as the combination of acetic acid (anhydride) and potash:

$$C_4H_6O_3 + KO \rightarrow C_4H_6O_3 \cdot KO$$

It must be noted that Berzelius did not regard vapor densities of compounds as indicative of molecular size, and so there was no problem with formulating acetate as a "four-volume" molecule. Indeed, only a four-volume (and not a modern two-volume) formula succeeds in representing acetic acid *anhydride* without fractional oxygen atoms. He also used an atomic weight for potassium that was twice what later chemists adopted, hence he required only half the number of potassium symbols in the formula as compared to what we have become used to. Looking at the issue in the reverse sense, we note that the modern two-volume formula for acetic acid is $C_2H_4O_2$; doubling the formula and then subtracting a molecule of water (H_2O) yields the Berzelian formula, which can be depicted as coupled to a (modern) potassium oxide component, K_2O. Although Berzelius was led to these ideas by apparently consistent application of his basic assumptions, and although they were widely accepted for many years, the result of these manipulations was that inorganic and organic chemistry became based on two different fiducial standards: two- and four-volume formulas, respectively. This conflict was to create serious difficulties for Berzelian chemistry during the 1840s and led to its destruction in the 1850s.

Initially, Berzelius preferred to consider the acid, as in the equation cited above, as the trioxide of the acetyl radical, C_4H_6. We will see in the next chapter how he shifted his position after 1838 to view the oxygen content as combined with half the carbon of the molecule, the other half being a hydrocarbon moiety: $C_2H_6 \cdot C_2O_3$. Since C_2O_3 is the (two-volume) formula for oxalic acid (anhydride), all carboxylic acids could subsequently be formulated as homologous hydrocarbons coupled with oxalic acid.

From earlier work by Davy and Dulong and his own benzoyl radical paper, Liebig developed a new theory of acid–base reactions that depended not on dualistic addition of bases to acids but rather on substitution by bases for a replaceable hydrogen atom of acids. The nonelectrochemical and substitutionist implications of this theory appeared to ally Liebig with Laurent, a point stressed by Berzelius in correspondence with his recalcitrant German friend. In fact, Liebig's organic hydracid theory led directly to a tragic polemic between them, at first waged quietly and privately, then increasingly publicly. The differences separating Berzelius and Liebig were quite real: Liebig had essentially abandoned dualistic precepts in organic chemistry, though he also had no patience for the French type theorists. The kindly Wöhler, who dearly loved both men, was caught in the middle. Unable to understand such passion over matters of theory, Wöhler tried for years without success to effect a reconciliation.

The wear on Liebig's nerves and emotions was also substantial, although not sufficient to overcome his firm (Berzelius would say obstinate) convictions. Worn down to the point of physical and psychical illness both by his arguments with Berzelius and by an increasingly vitriolic series of priority disputes with French substitutionists such as Dumas, Laurent, Malaguti, and Persoz, Liebig ostentatiously declared to Wöhler in early 1840 his abandonment of all chemical theory and his resolution henceforth to devote himself to practical pursuits such as agricultural and physiological chemsitry. Liebig was not alone in feeling an uncomfortable sense of disorientation and dismay at the rapid evolution of theory. J. F. W. Johnston described the science in 1840 as being "unhinged . . . tottering and disjointed." He called for a return to the comfortable Berzelian orthodoxy, condemned the "rage for *the new*" sweeping the Continent, and labeled Liebig as well as Dumas "chemical chartists."[48]

Liebig's abandonment of theory was not mere rhetoric. After 1839, Liebig abandoned his own hydracid theory and really ceased to participate in the organic-chemical theoretical dialectic. One indication of his disgust was his unauthorized decision in February 1840 (which was during Kolbe's Göttingen years) to publish a whimsical and hilarious French-language lampoon of the type theory, which had been written on a lark by Wöhler.[49] The putative author, S. C. H. Windler (*Schwindler* = swindler), claimed to have succeeded in gradually replacing all the atoms in copper acetate by chlorine, producing a material composed entirely of chlorine but retaining the properties of copper acetate. The paper portrayed with some justice, but also with malice, the extent to which Dumas, Laurent, and others were inclined to exaggeration. Whether due to the strain of the disputes, or to the Schwindler critique, or to the attainment of a level of professional success that reduced the hunger for acclaim, Dumas also retreated after

1840 from a leading theoretical role. Since many of the remaining leaders of German chemistry, including Wöhler, Bunsen, Gmelin, Rose, Mitscherlich, Will, and Kopp, were not inclined toward theory, and the great Berzelius increasingly appeared to the chemical world to be old-fashioned, something of a vacuum in German organic theory emerged in the 1840s—a void into which Kolbe eagerly stepped.

3

A Journeyman Chemist

The Copula Theory

Kolbe must have been an avid observer of the disputes raging in organic chemistry during his student years in Göttingen. Laurent's nucleus theory, Dumas's type theory, Liebig's hydracid theory, Berzelius's defenses of dualism, the angry self-justificatory letters by the last three of these chemists published almost biweekly for a time in the *Comptes rendus*, and the Schwindler parody, not to mention the lively private correspondence that Wöhler conducted with his feuding intimates Berzelius and Liebig, all occurred while Kolbe was studying with Wöhler.

This was also the time when Berzelius was fashioning the theory of copulated or conjugated compounds, his chief line of defense in response to the arguments of the French substitutionists.[1] Since 1837 Berzelius had been noting the discovery of a number of organic derivatives of inorganic acids and bases (especially alkyl sulfates and sulfonic acids, and ammines) that retained typical properties of the original acid or base, especially an unchanged saturation capacity. In 1839 Gerhardt coined the term *copulé* to refer to such combinations, the organic adduct being the *copule*. Berzelius then expanded this concept to explain the near identity in properties of acetic and chloroacetic acid. In the latter substance, the "sesquichloride of carbon," or C_2Cl_6—the copula—was combined with "oxalic acid," or $C_2O_3{\cdot}H_2O$, without affecting the fundamental (acidic) properties of the latter moiety. When Dumas' assistant Louis Melsens found a method to reduce the chlorinated derivative back to the starting material, Berzelius responded that acetic acid itself must be a "copulated" compound:

$$\underset{\text{acetic acid}}{C_2H_6 \cdot C_2O_3 \cdot H_2O} \underset{\longrightarrow}{\overset{\longleftarrow}{\text{or}}} \underset{\text{chloroacetic acid}}{C_2Cl_6 \cdot C_2O_3 \cdot H_2O}$$

Berzelius then developed a general theory of copulated or conjugated compounds. He supposed conjugation to be a nonelectrochemical form of chemical combination, which was why the saturation capacity of the acid remained unaltered. The hydrogen of the organic copula could be substituted by chlorine or other atoms, Berzelius conceded, but since this substitution occurred only in the passive, chemically unimportant copula, it was not surprising that the fundamental chemical properties of the compound did not change. For every substance substituted by chlorine, Berzelius now hypothesized a copula structure, always placing the chlorine in the copula. The theory provided a way for Berzelius to concede the increasingly unavoidable fact of chlorine substitution while maintaining the basis of dualistic organic chemistry. But what seemed to Berzelius to be a flexible modification showing the strength of dualism, seemed instead in the eyes of his opponents to be an unacceptably ad hoc retreat for a dying theory. In his posthumous chemical manifesto, the embittered and dying Laurent blasted the theory:

> A word let fall from the pen of Gerhardt, was thus transformed into a luminous idea for dualism. From this time everything was copulated. Acetic, formic, butyric, margaric, &c., acids,—alcohols, ethers, amides, anilides, all became copulated bodies. So that to make acetanilide, for example, they no longer employed acetic acid and aniline, but they recopulated a copulated oxalic acid with a copulated ammonia. I am inventing nothing—altering nothing. Is it my fault if, while writing history, I appear to be composing a romance? What then is a copula? A copula is an imaginary body, the presence of which disguises all the chemical properties of the compounds with which it is united. . . . The dishonesty is flagrant.[2]

But the copula theory proved to have a scientifically fruitful life. Just before Kolbe's arrival in Marburg, Bunsen provided a compelling argument in favor of Berzelius' theory by isolating the cacodyl radical and a large number of its derivatives, all of which also fit into the emerging copula theory. The acidity of oxygenated arsenic ("arsenious acid," formulated as AsO) did not seem at all affected by the addition of methyl radicals (producing "cacodyl oxide," formulated as

C_2H_6AsO). Cacodyl oxide could be reduced to cacodyl, the first unquestionably organic radical ever isolated and the first organometallic compound ever synthesized. Berzelius was thrilled at this new support for his ideas. "This is a triumphal chariot, which will overrun and smash [Dumas'] flimsy theoretical barricades," he exulted to Wöhler.[3]

The foundations of the copula theory would have been severely weakened, however, had Gerhardt's and Laurent's arguments been accepted, namely, that all compounds must be related to the same number of gaseous volumes (two), that hydrated acids are not oxides combined with water but rather are radicals with replaceable hydrogen, and that certain acids are polybasic. For instance, in response to Berzelius' suggestion that acetic acid consists of oxalic acid, $C_2O_3{\cdot}H_2O$, paired with the conjugate radical methyl, C_2H_6, Gerhardt and Laurent argued that the true formulas for the two components are half of those indicated, that there is no oxalic in acetic acid, that the true oxalic acid is dibasic rather than monobasic, and that it cannot be separated into an oxide and a molecule of water. They had similar problems with cacodyl oxide and other copula formulas.

Despite these objections, in its initial historical context Berzelius' copula theory was a successful theory that was both empirically supported and heuristically fruitful. Indeed, at the time of its formal introduction in 1839–1841, Gerhardt had not yet even fully formulated the arguments just rehearsed. Moreover, one could imagine a variety of potential compromise positions between the Berzelians and the French reformers if the two sides had been willing to look at the matter from their opponent's point of view. Despite the tendency of modern chemists to view Gerhardt's and Laurent's theories as correct, it might be said that there was as much truth to the copula formulas as there was to the French nucleus and type formulas.

Berzelius was excited by the subject of Kolbe's first research, for this was the same sort of work that had just yielded the wonderful fruit of cacodyl derivatives. This point was not lost on Kolbe, who from autumn 1842 had become a student of the chemist whose cacodyl series was still appearing in Liebig's *Annalen*. Kolbe's first discovery in Marburg was that moist rather than dry chlorine reacting with carbon disulfide resulted in an oxygenated and rearranged product, namely, one predicted earlier by Berzelius now known as trichloromethylsulfonyl chloride. Alkaline hydrolysis produced the corresponding sulfonic acid:

$$\mathrm{CS^2 + \not{C}l^2 + \not{H}O \xrightarrow{[O]} C\not{C}l^2{,}SO^2 \xrightarrow[\text{chlorination}]{\text{hydrolysis,}} \not{H}O + C^2\not{C}l^3{,}S^2O^5}$$

This last substance Kolbe named *Chlorkohlenunterschwefelsäure*, or dithionic acid (i.e., hyposulfuric acid anhydride) copulated with *Kohlensuperchlorür* (i.e., carbon "protochloride," or fully chlorinated ethane).[4]

It will be useful for what follows to understand this compound and its formulation from both a modern and a Berzelian–Kolbean perspective. The first point to note is that dualists always worked schematically with the anhydrides of acids, whether or not they had ever been isolated. For dithionic acid (*Unterschwefelsäure*), the schematic transformation from modern (hydrous) to Berzelian anhydride composition is $H_2S_2O_6 - H_2O = S_2O_5$. Second, vapor densities were not regarded as a consistent indication of molecular size, so that formulas could be halved or doubled as required by theory. *Most* organic formulas were doubled with respect to inorganic formulas—that is, four-volume organic formulas were usually standard. Thus, whenever dithionic acid, including its associated water molecule ($S_2O_5+H_2O$), occurs in a dualistic organic formula, it is equivalent to the modern sulfonic acid moiety: $-SO_3H$. Finally, the copula—carbon protochloride—could be considered either a two-volume representation of the fully chlorinated ethane discovered by Faraday (= modern C_2Cl_6) or a four-volume representation of the fully chlorinated methyl radical (= modern $-CCl_3$). The first was initially preferred by Berzelius since the copula could then be represented as a known isolated compound. After methyl was also discovered as a (seemingly) stable isolated substance, however, Kolbe preferred the second interpretation. Hence, what is seen by modern chemistry as the combination of two monovalent moieties, sulfonic acid and methyl, was viewed by copula theorists as the nonelectrochemical "coupling" of two stable, theoretically isolable molecules, dithionic acid and carbon protochloride.

In this same brief notice of 1844, Kolbe revealed that the carbon superchloride (CCl_4) formed by chlorination of carbon disulfide could be partially transformed by passage through a hot reduction tube into another of Faraday's chlorocarbons, CCl_2, today known as tetrachloroethylene. The latter compound could be further chlorinated, or chlorinated and oxidized to chloroacetic acid, depending on conditions and how many molecules were thought to take part in the reactions. In Kolbe's terms (using unbarred atoms), these reactions are

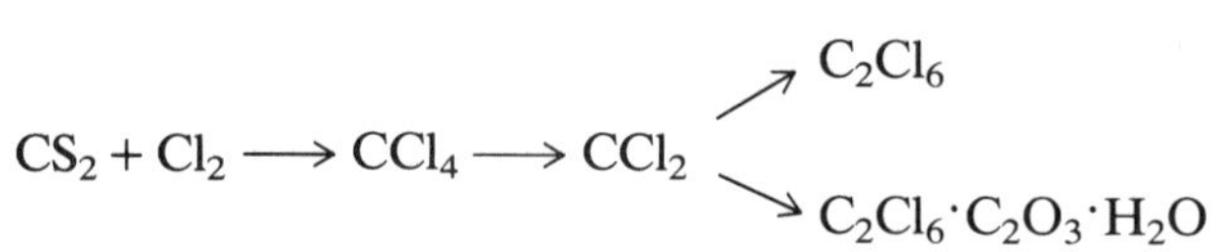

In 1845 Kolbe published his first extended paper, summarizing his work with these compounds over the previous three years, revealing a number of new derivatives, and interpreting all of them as Berzelian copulas.[5] He showed how one could reduce the trichloro sulfonic acid derivative in a stepwise fashion to dichloromethyl-, monochloromethyl-, and fully reduced methylsulfonic acid using either chemically or electrically produced nascent hydrogen. This represented his first application of Bunsen's powerful new battery, and he sagely predicted that it would doubtless prove an important instrument for the future study of chemical constitutions. The paper was fairly bursting with new compounds, new reactions, and new methods—a bravura performance.

But the paper is at least as notable for its theoretical content as for its experimental novelties. Kolbe argued that his new derivatives of dithionic acid formed a perfect analogy to the chlorinated acetic acids of the French chemists, but that the analogy only made sense when formulated as copula compounds and not as "types." The sulfur acids could be chlorinated without significantly affecting their properties only because the methyl that actually undergoes substitution is the passive copula, in precisely the same way that the hydrocarbon copula of acetic acid can be chlorinated without much affecting the "oxalic acid" to which it was coupled. In an ironic twist of language, he argued that the copulated dithionic acids could stand as "prototypes" for all copulas.[6] Furthermore, by starting with a chlorocarbon and ending with the synthesis of chloroacetic acid, Kolbe thought he had proven the preexistence of the former in the latter, confirming Berzelius' prediction.

This new research, Kolbe averred, thus transformed copulas from "the realm of mere hypothesis to a high degree of probability." Of course, the work also further broadened the field of chlorine substitution and could, like A. W. Hofmann's 1843 papers on chlorinated anilines, be viewed as supporting the "newer substitution theory" of the French chemists. This would be a hasty conclusion, he cautioned. Indeed, one could hardly imagine two more different chemical "types" than chloroform and carbon disulfide, yet Regnault and Kolbe had independently shown that each reacts with chlorine to yield an identical product, carbon superchloride. Contrary to Dumas, chemical types were not conserved. Hofmann's research only showed that aniline, too, was a copulated compound.[7]

Kolbe also followed up on Berzelius' comment to Wöhler about Kolbe's compounds occupying the border region between organic and inorganic compounds. As we have seen, one of the new reactions

Kolbe discovered was the transformation of what is now called tetrachloroethylene into chloroacetic (i.e., trichloroacetic) acid. Because he had prepared the former substance (like all of his chlorinated carbon compounds) from carbon disulfide and ultimately from inorganic carbon and sulfur, and because the product of the reaction was reducible by electrolytic hydrogen to acetic acid, here was a "nearly direct" synthesis of vinegar from the inorganic elements. These and the other reactions Kolbe had discovered showed a "continuous chain" from inorganic to organic compounds, so much so that any distinct boundary between them had disappeared. Kolbe waxed eloquent: as one can now synthesize vinegar, it may prove possible in the future to reduce acetic acid to alcohol, then to synthesize sugars and starches, in which case an immense new field for our endeavors will open up.[8]

This was the first significant organic synthesis since Wöhler's classic urea paper of 1828,[9] and the first total synthesis in history. The previous paragraph shows that Kolbe was well aware of its importance. It is also said that Kolbe was the first here to use the word *Synthese* in its modern chemical sense. In many respects, his accomplishment was even more dramatic than Wöhler's, since Wöhler had begun with starting materials derived from living creatures, whereas Kolbe had started with carbon and sulfur, whose ties with life were distant and indirect at best. To be sure, Kolbe's acetic acid synthesis did not destroy vitalism any more than his mentor's urea had, but it was an important benchmark on the route to modern synthetic organic methods.

This, Kolbe's first major paper, established patterns he would follow for the rest of his life. Above all, the article reveals his intense interest in chemical theory and, in particular, in theories concerning the inner constitutions of the molecules of organic compounds (especially organic acids). Kolbe's starting point was Berzelian radical theories. Under attack by the French substitutionists, Berzelius had devised a modification of his theory, copulas, that could countenance the distasteful new phenomenon. However, unlike Berzelius, Kolbe embraced chlorine substitution with enthusiasm and never exhibited any discomfort with the notion of electropositive radicals containing electronegative elements. Like Liebig's, his dualism was moderated by the undeniable fact of chlorine substitution. But in all essential respects, he was from first to last a committed, indeed zealous, Berzelian. For Kolbe, radicals were not mere schematic or conventional entities, but real, preexisting parts of molecules. The distinguishing characteristic of his entire career was the discernment and (wherever possible) physical isolation of these radicals.

As might be expected, Berzelius himself was delighted and highly impressed by Kolbe's work; Kolbe treasured the Swede's letter of congratulations of 3 August 1844 to the end of his life. Berzelius described Kolbe's experiments and interpretations in glowing terms in his *Jahresberichte* and praised the significance of the work in letters to both Wöhler and Bunsen. He averred publicly that Kolbe had provided "as complete a proof as is possible in chemistry" for the existence of copulas.[10] The following year (1845) he met Kolbe, for the first and only time, during his last visit to Germany.

Berzelius even used Kolbe's disproof of Dumas' types as the final decisive argument in a paper entitled "Views Regarding Organic Composition," which he submitted in February 1846 to the Swedish Academy of Sciences and subsequently published in Poggendorff's *Annalen der Physik*—the last important article of his life. Berzelius thought that Kolbe's work provided "positive proof" for the existence of copulas; it "completes the refutation of the substitution theories and the imagination-game of chemical types." He regarded as quite novel and striking the stepwise replacement of chlorine by hydrogen and vice versa, with no significant change in the properties of the compound. In short, it was the best work he had seen since Bunsen's cacodyl series.[11]

One can imagine the effect such praise from the venerated master must have had on the young Kolbe. Indeed, much of the substance and rhetoric of Berzelius' last article can be discerned in often little-altered form in many of Kolbe's own later diatribes. Berzelius here repeated (no fewer than three times, and once in italics) that the only sure guide to theories of organic compounds is the dualistic theory of *inorganic* chemistry. He also rebuked the French chemists, especially Dumas, Laurent, and Gerhardt, for letting their imaginations dictate "Phantasiebilder" that have little or no empirical support or connection to the existing theoretical structure of the science. Determining constitutions of organic compounds is both the most important and the most difficult of all goals in chemistry, he averred; all the more vital that it be done with caution, care, and circumspection.

Assistant to Playfair

In 1845 an opportunity arose for a temporary foreign position for Kolbe. The British government had just established a Museum of Economic Geology near St. James' Park in London and had hired Lyon Playfair, a former student of Liebig, as the museum's organic chemist. One important assignment given Playfair was the analysis of mixtures of naturally occurring hydrocarbons, required in connection with a Parliamentary Commission on Explosions in Coal Mines. The acknowledged master of such analytical methods was Bunsen, to whom Playfair turned for advice. Bunsen thought to send Kolbe as assistant to

Playfair and persuaded Kolbe to accept the position. This was to be Kolbe's only extended trip of his life outside the European continent.

Kolbe entered into his duties in London in October 1845, the same month that another assistant, Edward Frankland, arrived at the museum and the same month that August Wilhelm Hofmann arrived from Bonn as Professor of Chemistry at the new Royal College of Chemistry. Hofmann later related that he met Kolbe soon after their arrival at a meeting of the Chemical Society. He had been given a large official residence rent-free, and he invited Kolbe to live with him there. They became intimate friends.[12]

Frankland was seven years younger than Kolbe and not very knowledgeable in chemistry, especially regarding the latest experimental methods then current in Germany, having come directly from an unhappy pharmaceutical apprenticeship. When Hofmann married in August 1846, Kolbe established lodgings on Belvedere Road, near Frankland's residence on Doris Street. Frankland later recollected

> At this time Kolbe could speak only a few words of English, but we arranged to give each other lessons in German & English and we met, for this purpose on two evenings in the week at his lodgings. . . . He made rapid progress and was soon able to speak with facility. It was not long after this intercourse became established between us, before he began to explain to me his great interest in organic chemistry.[13]

Frankland reported that Kolbe had a "supreme contempt" for inorganic analysis, of the type that Frankland had been hired to perform, as it was of "little or no theoretical interest."[14] He was soon infected by his friend's enthusiasm for experimental and theoretical organic chemistry. Kolbe instructed him both in Bunsen's "then but little known but beautiful & delicate processes of gas-analysis" and in Berzelian theory.[15] In addition to Hofmann and Frankland, Kolbe became acquainted with most of the London chemical community, and it is said he made a particularly strong impression on Thomas Graham and Michael Faraday.[16]

Frankland described the "profound impression upon all of us" made from the first by Kolbe's exemplary care and skill in laboratory operations. "He never grudged any amount of trouble in fitting up apparatus or performing an operation, if a greater amount of accuracy could thereby be secured." After Kolbe sent a sample apparatus for explosion eudiometry, Wöhler replied with thanks, in awe of Kolbe's glassblowing skill. If all else failed, his former teacher opined, Kolbe could easily make a living as a skilled artisan.[17]

Kolbe's official duties were to analyze mixtures of gases gathered from coal mines in the hope of providing for better means of preventing explosions. He reported on this work both to Bunsen and officially to Playfair for the British government.[18] Frankland related that there was insufficient room at the museum's laboratory for these analyses, so they were all performed in Kolbe's lodgings. Playfair, as it happened, was not often present in London due to other governmental duties such as service on a commission on the potato blight and a lectureship at the military college at Addiscombe.[19] Perhaps due to this circumstance, Kolbe had considerable free time on his hands, and by December 1845 we find him using Playfair's lab for his own research purposes.[20] Toward the end of his stay in England, in the spring of 1847, he published two important papers, the second of which was a joint project with Frankland.

Kolbe knew that he of all European chemists was in a unique position to exploit a certain new field of research. He had mastered improved and as yet unpublished gas-analytical methods, as well as the mode of construction of a novel and very powerful carbon–zinc battery, both of which he had learned from Bunsen, and unlike his master, he was entranced by the theoretical goal of investigating the constitutions of organic molecules. With these tools, Kolbe set out to accomplish what decades earlier had repeatedly frustrated Berzelius himself: the electrolysis of organic acids. One of his laboratory notebooks preserved in the Deutsches Museum in Munich records experiments in this direction from 1 October 1846 to February 1847.[21] With solutions of potassium acetate, butyrate, and valerate, Kolbe made the electrolysis work, but he obtained a daunting number of solid, liquid, and gaseous products. After electrolyzing potassium valerate, for instance, he isolated, in addition to potash, hydrogen, and carbonic acid, a new saturated hydrocarbon possessing the formula C_8H_9, a new olefin C_8H_8, and an ethereal oil apparently of the formula $C_8H_9O+C_8H_9C_2O_3$. In his first paper on this subject Kolbe conceded that these were still preliminary results, but he stressed that they supported the copula formula for valeric acid, which would be $C_8H_9{\cdot}C_2O_3{\cdot}HO$.[22] Although he was not able to bring this work to a fully satisfactory conclusion while in England, this was the origin of the still useful "Kolbe electrolysis" reaction.

Frankland joined Kolbe in an attack from a different direction on the question of the constitution of organic acids. It was known that upon hydrolysis, cyanogen and benzonitrile yield oxalic acid and benzoic acid, respectively. The latter reaction, developed in 1844 by Fehling, suggested the correctness of the copula formulas for benzonitrile and benzoic acid:

$$C_{14}H_5{\cdot}C_2N \longrightarrow C_{14}H_5{\cdot}C_2O_3{\cdot}HO$$

Kolbe and Frankland determined to generalize this reaction for the simple aliphatic acids and succeeded admirably:

$$\underset{\text{methyl cyanide}}{C_2H_3{\cdot}C_2N} \longrightarrow \underset{\text{acetic acid}}{C_2H_3{\cdot}C_2O_3{\cdot}HO}$$

$$\underset{\text{ethyl cyanide}}{C_4H_5{\cdot}C_2N} \longrightarrow \underset{\text{propionic acid}}{C_4H_5{\cdot}C_2O_3{\cdot}HO}$$

$$\underset{\text{amyl cyanide}}{C_{10}H_{11}{\cdot}C_2N} \longrightarrow \underset{\text{caproic acid}}{C_{10}H_{11}{\cdot}C_2O_3{\cdot}HO}$$

In the view of the authors, this work supported the existence of copulas and, in Frankland's later assessment, established "for the first time the internal molecular structure of these acids. . . ."[23]

But in the published paper, Kolbe and Frankland expressed their views with diffidence. Well cognizant of their youth (Frankland was only twenty-one) and lack of established positions, they wisely chose to soft-pedal their novelties. Investigations of the molecular constitutions of compounds, they wrote,

> . . . are always attended with more or less danger, and those who, leaving the safer road of experiment, plunge into the depths of hypothesis, and build up theories apparently ingenious, though often untenable, frequently stumble and fall amongst a host of contradictions. It is a common error, as experience teaches, into which young chemists are very apt to fall, that, persuaded of the infallibility of their own views, and blind to well-founded objections, they endeavor to convince by quick and ready argument rather than by solid reasoning, and consequently they either offend others or feel themselves offended when contradicted.

Hence, they felt a "certain degree of timidity" in presenting these views, against those "generally received." They professed no intention of giving a "decided preference" to the ideas here defended or of forcing their opinion on others. They did, however, aver that the views were worthy of consideration.[24] One can imagine Kolbe writing these lines, worrying about being put in the same category as the French chemists and bearing Berzelius' published cautionary remarks of the previous year in mind.

In fact, Kolbe and Frankland were aware that one could explain these reactions without recourse to copulas, or even to any constitutional hypothesis, by simply using empirical formulas, as Liebig had begun to do since 1840. However, they argued that their explanation, using what we would now refer to as more resolved structural formulas, had manifold advantages. It was simpler and more consistent than empirical formulas, and it revealed direct analogies to other reactions of nitriles and cyanides otherwise hidden. Furthermore, one could then propose a single homologous series of hydrocarbon radicals as the constitutional basis not just for nitriles and acids but also for alochols and hydrides, not to mention Kolbe's own alkyl hyposulfuric acids. They even speculated on the reaction mechanism of the oxidation of ethyl alcohol to acetic acid. This was, therefore, a bold publication for two professionally unestablished academic chemists. No one else at that time was publishing this sort of experimentally grounded theoretical work in organic chemistry.

It would be hard to overemphasize the significance of the Kolbe and the Kolbe–Frankland papers. Read to the Chemical Society on the same day (19 April 1847), they represent inverse synthetic methods: a carbon atom–increasing carboxylation reaction (through the corresponding nitrile) and, apparently at least, a carbon atom–decreasing decarboxylation reaction. These reactions were the first general synthetic routes between hydrocarbons and organic acids and represent the two first great general synthetic methods ever published. Together with Kolbe's acetic acid synthesis, they were the earliest planned reactions where the carbon content of an organic molecule was deliberately altered. They were also the first synthetic reactions whose purpose was to elucidate "constitutions," or what we now refer to as chemical structures. The few pre-1847 reactions that might be considered "synthetic" were fortuitous transformations whose constitutional import was only dimly or not at all appreciated.

Back to Marburg

In the second week of May 1847, Kolbe returned to his assistantship in Marburg in the company of his new English friend. Frankland had had two job offers, one of which he declined and the other he postponed in order to study with Bunsen during that prospective summer semester. He later said that this decision "had much to do with shaping all my future life," and he always had felt grateful to Kolbe for persuading him to make it.[25] In his memoirs, Frankland described the trip, his first abroad, in captivating detail—including such important incidents as Kolbe's disgust with Frankland's negative reaction to his first taste of real German Rhine wine. Having arrived in Marburg, they rented separate rooms in the Hotel Europäischer Hof, on Elisabethstrasse across from the Chemical Institute (still today Marburg's largest hotel). The next day Kolbe introduced his friend to Bunsen, and they both began their work.[26]

In his memoirs, Frankland described the many social functions to which he was invited that summer. At the first of these he met his future wife, Sophie Fick, the sister of professor of anatomy Adolf Fick. It is amusing to read Frankland's detailed judgments of the pulchritude (or lack thereof) of a number of Marburger *Mädchen*, each carefully identified by name. Another time he accompanied Kolbe home to Lutterhausen for the wedding of one of Kolbe's sisters. This, his first introduction to German domestic life, was very interesting and agreeable to him; he spent time in the garden of the house with Kolbe's many younger siblings, the youngest of whom helped Frankland with his German.[27]

But Kolbe and Frankland must have worked efficiently that summer, for in three months they published a second paper on the nitrile hydrolysis reaction and completed another project as well. They expanded the series of acids synthesized from nitriles to a total of eight (formic, acetic, propionic, butyric, valeric, caproic, benzoic, and cuminic) and made a stronger argument for the correctness of the copula formulations than they had made in London. A further support would be the reverse process, the reduction of an acid to the corresponding nitrile. They noted that Hofmann had long been attempting this reaction and had informed them by letter that he had recently succeeded—as had Dumas. They also applied Kopp's boiling point regularities to support their formulas.[28]

Kolbe's and Frankland's second joint project in the summer of 1847, actually begun already in London, involved treating ethyl cyanide with potassium in an effort to free the ethyl radical when the potassium united with the cyanide. To the delight of Kolbe and Frankland, the reaction seemed to proceed well, but to their disappointment, the product as analyzed seemed to be methyl rather than ethyl:

$$C_4H_5{\cdot}C_2N + K \longrightarrow C_2H_3 + KC_2N$$

This makes an anomalous chemical equation, and they were unable to sort satisfactorily through what turned out to be a rather messy reaction. As Frankland later determined, the product was actually ethane (dimeric methyl); the extra hydrogen must have come from water or alcohol contamination. In the 1847 context, however, the isolation of "methyl" was a very notable result, even if the details of the reaction were still cloudy.[29] Their paper was written by Kolbe and was presented in both their names in Frankland's English translation to the

Chemical Society on 7 February 1848; it was, Frankland noted, "warmly applauded" there.[30]

In early August 1847, before the end of the summer semester, Frankland was prevailed upon by his new employer at Queenwood College (Hampshire) to return to England. He spent fourteen months in this position, another fifteen months back in Germany (acquiring a Marburg Ph.D. with Bunsen followed by a semester with Liebig in Giessen), and fourteen more months at Putney College. Frankland then received a call in March 1851 to Owens College, the predecessor of the University of Manchester. During this peripatetic period, Frankland single-mindedly and quite successfully stalked the hydrocarbon radicals. Frankland later reminisced: "The isolating of the alcohol radicals was, at this time, the dream of many chemists, whilst others doubted or even denied their existence. I was also smitten with the fever and determined to try my hand at the solution of the problem."[31] Frankland apparently allowed himself some dramatic license for this statement, for in his first paper of the series, described below, he wrote (I believe accurately) that after cacodyl, no chemist seemed to be even trying to isolate the hydrocarbon radicals except Kolbe and himself.[32] No doubt this was because almost no one besides Frankland and Kolbe believed any longer in the older conception of radicals as real, preexistent parts of molecules.

Bunsen had prepared cacodyl from the reaction of cacodyl chloride with zinc. In the spring of 1848 Frankland tried heating potassium with ethyl iodide (rather than ethyl cyanide as in his first attempt), hoping to free the ethyl. He only succeeded in isolating "ethylic hydride," $C_4H_5{\cdot}H$, which appeared isomeric with his and Kolbe's "methyl." He decided he might have better luck if he tried a less active metal, and so he selected Bunsen's reagent, zinc. The experiment, performed early the following year in Marburg, was successful, and it led to the publication of a milestone paper entitled "On the Isolation of the Organic Radicals." Frankland had not only isolated ethyl, but he had also synthesized a new organometallic substance, zinc ethyl, which he formulated C_4H_5Zn. The reaction of methyl and amyl iodide with zinc produced analogous substances, including the free (apparently stable and isolated) radicals methyl and amyl. The isolation of such radicals "excludes every doubt of their actual existence, and furnishes a complete and satisfactory proof of the correctness of the theory [of the ethyl radical in alcohol and ether] propounded by Kane, Berzelius and Liebig fifteen years ago."[33]

Frankland's proof, of course, had a flaw: his (and Kolbe's) methyl, ethyl, and so on could not be distinguished from their dimers ethane, butane, and so on (as we now regard them) without applying vapor density data to achieve consistent molecular magnitudes. Nonetheless, the evidence for isolated radicals appeared to be compelling. While Frankland was working in his laboratory in December 1849, Liebig reported to Hofmann: ". . . Frankland has isolated methyl, ethyl, and this week amyl. So what was desired as the foundation for the [radical] theory is now from this side finally here."[34] One of Liebig's reasons for abandoning chemical theory in 1840 had been the uncertain and shifting status of organic radicals, but after 1849, Liebig never again doubted their reality. This return to an earlier article of faith did not, however, induce him to return to the strife of theoretical chemistry.

Frankland and Kolbe had now isolated what they considered to be four different radicals by three distinct methods, all in less than three years. Unfortunately, Berzelius, who would have been overjoyed by these events, had died in August 1848.

Vieweg Verlag and Braunschweig

Within days of Frankland's departure from Marburg in August 1847 to take up his post at Queenwood College, an opportunity arose for Kolbe as a professional writer and editor at Vieweg Verlag in Braunschweig. Friedrich Vieweg (1761–1835) had founded this publishing house in Berlin in 1786. Distressed by Prussian censorship, in 1799 he accepted the invitation of Duke Karl Wilhelm Ferdinand to move his firm to the capital of the Duchy of Braunschweig; he built an imposing neoclassical edifice on the Burgplatz to house his firm. From its founding, the house prospered from the publication of theological and literary works. Friedrich's eldest son Eduard Vieweg (1796–1869), who became a partner in the firm in 1825 and full owner upon his father's death, changed the orientation of the press toward scientific and technological subjects, chemistry in particular. Before his entry into the firm, Eduard spent three years traveling in France and Great Britain. In Paris he met the young Justus Liebig, who was then studying with Gay-Lussac, and he formed a very close lifetime friendship with him; most of Liebig's books were published by Vieweg.[35]

In 1832, Liebig and J. C. Poggendorff (soon joined by Wöhler) hatched a plan to publish a chemical dictionary-style handbook, the *Handwörterbuch der reinen und angewandten Chemie*, and persuaded Eduard Vieweg to publish it. This was to be a detailed summary of the state of the art in chemical science, eventually issued in twelve large volumes and much imitated in other countries. Subsequent editions of this monumental project continued to appear until 1930. In the fall of 1832, Liebig set to work with great energy on some of the 400 planned

"A" articles, many of which were highly important ones: *Äther*, *Äthyl*, *Aldehyd*, *Alkohol*, *Analyse*, and so on. The editors hoped to have the first volume on the market by 1834. But Liebig had more projects planned than time, and Poggendorff was dilatory; the first fascicle did not appear until late 1836 and the first complete volume not until 1842. By 1847 only two volumes had come out, covering the alphabet from A to E, and Liebig was quickly losing interest in the completion of the project. In a letter to Vieweg on 28 April 1847, he suggested that younger talents be recruited and named nine possibilities, one of whom was Kolbe. A memo by Vieweg on this letter indicates that it was his intention to ask Buff, Will, Knapp, Varrentrapp, Erdmann, and Marschner to contribute.[36]

It was at Wöhler's urging that Kolbe was asked to become the chief editor of the project. Kolbe had just completed a translation for Vieweg of Mulder's textbook of physiological chemistry, so Vieweg had some indication of his writing skills. On 6 August 1847, Vieweg wrote Kolbe and invited him to come to Braunschweig for this assignment. Kolbe accepted immediately, indicating his intention to arrive in late September.[37]

Kolbe's Braunschweig years were happy and productive ones. He formed an extremely close relationship with Eduard Vieweg, though Vieweg was his elder by twenty-one years—a relationship that remained untroubled to Vieweg's death in 1869. He also became very friendly with Vieweg's son Heinrich (1826–1890) and with Franz Varrentrapp (1815–1877), a pharmacist who had received his doctorate with Liebig, then settled in Braunschweig. Varrentrapp taught at the Anatomisch–Chirurgisches Lehranstalt and performed research for the Braunschweig Gewerbeverein, a state-supported research institution for the promotion of trades and industries. He also wrote for Vieweg on the side. When he received a call to the University of Aachen in 1868, Vieweg persuaded him to stay in Braunschweig by making him a partner in the firm. The next year Eduard Vieweg died, and the firm was directed for the following eight years by Heinrich Vieweg and Varrentrapp.[38]

While on holiday from the University of Marburg in early June 1849, Frankland visited Kolbe for three days and spent "a very pleasant evening with Kolbe at Varrentrapp's home." Frankland reported that through the Varrentrapp family Kolbe met a woman named Franziska von Spilker and became engaged. Frankland met her on this trip and commented that "fortunately for Kolbe this engagement was soon afterwards broken off," though by which party we are not told.[39]

The four years Kolbe spent in Braunschweig were a time of great unrest throughout the continent of Europe, and Kolbe was not unmoved or unaffected by political forces. The duchy of Braunschweig had close historical, political, and family ties to the electorate of Hanover, and in the eighteenth century, Braunschweig like its neighbor had enjoyed a progressive regime under the duke that brought Friedrich Vieweg from Berlin. The medical school and trade society for which Varrentrapp worked were two examples of typically Enlightenment institutions founded in Braunschweig. A third was the Collegium Carolinum—eventually renamed the Technische Hochschule—founded in Braunschweig in 1745 on English and Dutch models and intended to educate the middle class for practical services to state and society.[40]

But the religious tolerance and civil freedoms introduced in the eighteenth century were disturbed as a result of the French wars. In 1830 the mad, reactionary, and heartily despised Duke Karl was driven from the land, and the new ruler, Karl's younger brother Wilhelm (d. 1884), granted the state its first constitution. He reinstituted the previous reforms, along with farsighted economic provisions such as entry into the Prussian customs union and introduction of one of the earliest state railroad lines in Germany. Until the founding of the German empire, Braunschweig remained a model, very much in the minority, of a progressive small German polity.

Eduard Vieweg participated fully in the political events of his day, founding at various times three influential newspapers. The longest lived was the *Deutsche Reichs-Zeitung*, begun shortly after the March 1848 revolution, which agitated for the joining of the Germanic lands into a unified empire. Vieweg was by no means always well inclined toward or in agreement with the Prussian regime, but he saw in Prussia the nucleus from which a united Germany could and must grow; Austria must in all events be excluded. Whether it was from daily contact with Vieweg or from his own earlier convictions, Kolbe's views were substantially the same. Kolbe joined a *Vaterländisches Verein* (Patriots' Society) in Braunschweig, though this likely was different from the similarly named clubs of marked republican and even radical orientation that were proliferating in the Germanic lands during the 1840s. Frankland accompanied Kolbe to a meeting of this society (1 June 1849) and reported that the members were "all unanimous in accepting the Prussian Constitution."[41]

It would appear that Vieweg's and Kolbe's political sentiments were similar to those of the so-called Casino faction of the Frankfurt Parliament, usually described as center right liberals. This group, consisting mostly of professors and industrialists, were *Kleindeutsche* who sought a unified Germany under Prussian and not Austrian leadership, who wanted constitutional guarantees but also protection against the centrifugal forces represented by republicans, democrats, socialists, and anarchists. In August 1848, at the peak of the power and optimism (such as it was) of the Frankfurt Parliament, Kolbe wrote to Frankland in the only English language words to survive from his pen: "As to politics, I pay a great confidence to our present government in Frankfurt. I hope, there will be in future but one Germany and only one king, as you have but one queen."[42]

Ost reported that it was Kolbe who safely and in secret conducted the Hessian classical historian and journalist Adam Pfaff (1820–1886) from Braunschweig to Hamburg. Pfaff had made himself *persona non grata* in Hesse-Kassel by his activity as editor of the liberal *Neue Hessische Zeitung*. When the paper was suppressed during the reaction in 1850, he had fled to Braunschweig, where he was hired by Vieweg to be editor of Vieweg's *Deutsche Reichs-Zeitung*. His escape to Hamburg, and subsequently to Brussels, was necessitated by continued pursuit by the Hessian regime.[43] This adventure underlines Kolbe's (moderate) liberalism during this period and the influence of his friend and employer Eduard Vieweg.

Chemical Editor

At Vieweg Verlag, Kolbe set to work energetically at his editorial tasks. One of his earliest decisions was notational. There was at that time disagreement on certain important points, and there had been several recent international shifts in formula conventions. Above all, should one take H = 1, Ħ = 2, or H = 2, in other words, should one notate benzoyl chloride, for example, as $C_{14}H_{10}O_2Cl_2$ (Berzelius' notation until 1826 and Liebig's until 1844), $C_{14}\text{Ħ}_5O_2\text{Ȼl}$ (Berzelius' and Wöhler's notation since 1826), or $C_{14}H_5O_2Cl$ (Liebig's notation since 1844)? To Wöhler, Kolbe expressed a preference for Ħ = 2. Wöhler agreed but remarked that Kolbe had used H = 2 in his dissertation, and he noted that there are problems of consistency in the *Handwörterbuch* as a whole since the first volumes, following Liebig's preference of the 1830s and early 1840s, had used H = 1. Wöhler suggested Kolbe check with Liebig before settling the point. Kolbe did this, and Liebig agreed with the proposal to use barred letters to remove any possible ambiguity.[44] Kolbe's pattern in his own articles at this time was to use barred letters for German publications and the unbarred (H = 2) letters in England, as had become accepted there. The English system was soon thereafter generally adopted in Europe. This system is notationally equivalent to taking $H = 1$, $C = 6$, $O = 8$, $Cl = 35$, and so on, which are the conventional equivalents that became so universally popular during the late 1840s and 1850s.[45]

Kolbe discussed these and other notational issues in articles written for the *Handwörterbuch* sometime in the middle of 1848. He explained his preference for the barred symbols, while sharply rebuking the French for maintaining Dumas' hybrid system of $H = 1$, $C = 6$, and $O = 16$ with four-volume organic formulas. This was, he thought, an "irresponsible" system that only increased the existing confusion, and was defended only to attempt to maintain an "imagined law" based on "vague hypotheses."[46] This, his first strong public derogation of the French, may have been partially motivated by feelings aroused by the violence of the February revolution in Paris and the consequent upheavals in Germany of the "March days." Kolbe's moderate liberalism did not extend to toleration of insurrection or anarchy.

In this article Kolbe also discussed how the notation of chemical formulas

> . . . attains great importance by offering us a means of representing with greater sharpness and precision the different conceptions concerning the chemical constitution of a compound, solely by the various ways of grouping a few symbols, thereby simultaneously expressing a summary of ideas that can be reproduced so briefly in no other manner.[47]

So, for example, the constitution of acetic acid might be represented by the various formulas $C{+}HO$, $HO.C_4H_3,O_3$, or $\bar{H}O.(C_2\bar{H}_3)\frown C_2,O_3$. Whereas the first is little more than an empirical formula, the second accounts for the acidic and some of the electrochemical properties of the substance. The last, however, also accounts for the evidence supporting a copula formula: hydrogen of the methyl copula can be substituted by halogen without breaking up the structure of the radical, or the methyl copula can even be transferred to another radical entirely, as, for example, in the transformation of acetyl derivatives into methyl cyanide or cacodyl compounds.

Two months later, Kolbe provided even more detail on his ideas of the constitution of acetic acid. The radical of acetic acid is methyl copulated to "oxatyl," C_2, the resulting hydrocarbon radical to be called "acetyl," $(C_2\bar{H}_3)\frown C_2$. Oxatyl was "the exclusive point of attack for the affinities of those elements," such as oxygen, that formed the various compounds of acetyl. For instance, the first stage of oxidation of acetyl was aldehyde $\bar{H}O.(C_2\bar{H}_3)\frown C_2O$, the third stage being acetic acid $\bar{H}O.(C_2\bar{H}_3)\frown C_2O_3$. In other words, reagents *add* only to oxatyl, and conversely, oxatyl undergoes only *addition* reactions. *Substitution* reactions (i.e., those that occur without major alteration of chemical properties) could take place only in the copula: for example, trichloroacetyl = $(C_2Cl_3)^\frown C_2$. Organic radicals could now be subsumed into two distinct groups: hydrocarbon radicals and conjugate radicals such as those of the acetyl class, which could then form "secondary" conjugate radicals (those that have undergone substitutions).[48]

In this fashion, Kolbe had schematically resolved the acetic acid molecule one step further than ever before by distinguishing and focusing on what has become known as the carbonyl carbon atom. He had also created a theory of greater generality and wider application: radicals fully analogous to oxatyl could be formed from such elements as arsenic and sulfur, forming the constitutional basis for cacodyl and sulfonyl compounds. In this article, Kolbe thus altered his interpretation of the organic acids from formulas based on copulated oxalic acid to a more generalized concept using copulated oxide hydrates of carbon and other atoms—a progressive and fruitful shift.[49]

The buckle symbol, $\frown$, Kolbe noted, had been proposed to him by Friedrich Otto (1809–1870), a professor of technical and pharmaceutical chemistry at the Collegium Carolinum, as a convenient way to indicate the special form of chemical bonding in copulated compounds. As for periods and commas, Kolbe defined them as equivalent in denotation; from this time forward, however, he tended to use the period for indicating the attachment of "basic water" to the remainder of an organic formula and the comma for separating oxygen from hydrocarbon moieties—though he sometimes omitted it. Parentheses presumably further emphasized the integral character of certain radicals. The option of dispensing with all of these punctuation marks and relying on simple juxtaposition of grouped letters, as he did, for instance, with C and H in hydrocarbon radicals, apparently did not sufficiently satisfy Kolbe's electrochemical instincts.

In these articles, which contain his most extended descriptions of the notational conventions that he followed with minor variation for the rest of his life, Kolbe made few of these details explicit. What does emerge clearly, though, is his conviction that carefully constructed formulas are extraordinarily useful semiotic and heuristic devices for the development and communication of theoretical ideas. In fact, to a degree perhaps unmatched by any other chemist of his day, Kolbe's notation cannot be separated from his theories. This is true because every formula that Kolbe wrote was a deliberate theoretical statement, namely, an assertion regarding the constitution of the molecule being discussed.

The *Handwörterbuch* project, barely alive for so many years, took off immediately under Kolbe's leadership. He solicited scores of articles from competent authorities and rode herd on them until they delivered; he then edited their contributions. He wrote a number of significant articles himself, among them *Formeln, chemische, Formyl, Gepaarte Verbindungen, Kakodyl*, and *Kohlenwasserstoffe*, and he revised articles on *Acetyl, Aethyldithionsäure*, and *Benzoësäure*. By 1851, the third and fourth volumes were complete through letter L, as well as a supplement volume. More progress was made in four years under Kolbe's leadership than had been made in the previous fifteen years. Liebig, Wöhler, and Poggendorff remained the "*Herausgeber*," with Kolbe as "*Redakteur*."

Kolbe also accepted another literary assignment early in his tenure in Braunschweig, which would lead to an enormous literary opportunity—and burden—occupying him for the rest of his life. Someone, probably either Friedrich Otto or Eduard Vieweg himself, conceived the idea that it would be a valuable (or profitable) undertaking to publish a German version of Thomas Graham's popular English textbook of chemistry. The first German edition (1840–1843) was a simple translation by Otto. For the second edition, conceived as a substantial rewrite of the original, Otto wrote the first volume, on general, physical, and theoretical chemistry (1845–1847), and Kolbe was asked if he would edit the organic portion of the text. On Christmas day 1847 we find Bunsen replying to Kolbe's request for advice on this point; Bunsen urged him to accept Vieweg's offer, but cautioned him not to overcommit on too many projects—sage counsel![50] We can only conclude that he must have worked at least occasionally on this project during the next three and a half years, since by the time of his arrival in Marburg, he had a manuscript of at least a portion of it, which he put to good use in preparing his lectures. But the manuscript went through several incarnations before it actually began to appear in fascicles in 1854.[51] By then, Kolbe had long been referring to it as "*meine Organische*." It was no longer a translation, nor even a translated revision of Graham's work, but a detailed advanced organic chemistry text written by Kolbe *ab ovo*—although still published with a second title page giving the Graham–Otto imprimatur.

Frankland mentions, and we are told by biographers, that Kolbe had occasion to work in Varrentrapp's laboratory during his years in Braunschweig. No evidence for this is apparent from Kolbe's publications or letters, and fifteen years later he explicitly stated to Frankland that he had had no opportunity for laboratory work in Braunschweig.[52] He published only one article containing original experimental work during this period, and that work appears to have been done when Kolbe returned to Bunsen's lab in Marburg for six weeks or so in the summer recess of 1848 and again for "a week or so" in the summer recess of 1849, as Frankland reported. He returned for a third time in the summer of 1850, spending a month in Marburg, Giessen, Frankfurt, and Heidelberg.[53] Frankland related: "He was then in his prime, full of enthusiasm for organic chemistry and earnestly hoping for a professorship, which would afford him the much desired opportunity for work."[54]

The work done by Kolbe on the first of these occasions was particularly fruitful. On the first day of August 1848, Kolbe wrote Frankland from Marburg, hoping that Frankland would arrive before he had to return to Braunschweig. He was working on the electrolysis of malonic and acetic acids. Six days later, he reported to Vieweg, "My chemical work, for which Bunsen's laboratory offers in fullest measure all the necessary equipment, has already achieved the geatest success, so that within a short time I will have reached the goal I had set for myself. Bunsen himself shows the greatest interest in my experiments. . . ."[55] A paper summarizing the results of this work was sent early the following year to Liebig for the *Annalen* and to London for the *Journal of the Chemical Society*.[56] Liebig thought the work "magnificent" and agreed with Kolbe's conclusions. "It has been a long time," he wrote, "since I have read an article that has excited me as much as yours has. The thoughts are as lovely as the methods, and the development is masterly." He had dabbled along similar lines himself, he said, but without this kind of success, and was happy to leave the field entirely to Kolbe.[57]

In this paper, Kolbe succeeded in sorting out many of the difficult details of the electrolysis reaction he had been working on since London. The hydrocarbon radical from decarboxylated potassium valerate he named "valyl"; he guessed that it must be the radical of the still unknown butyl alcohol. Even more important, acetic acid, barely mentioned in the London paper, now gave clean results: hydrogen, carbon dioxide, and methyl were all among the products, and the methyl gas proved chemically identical to that which he and Frankland had prepared from ethyl cyanide. (Kolbe's "radicals" are now considered to be their respective dimers, octane and ethane.) Whereas in 1844 he had fully synthesized acetic acid, this most fundamental organic substance, he had now successfully analyzed it into its component parts. They were the radicals predicted by his Berzelian theory.

This success gave Kolbe the impetus to write a long critical review of recent research on organic radicals, together with the various theories that were then contending to explain those results. Although he was at work on this review as early as March 1850, it was not until that fall that he completed and submitted the article, both to the *Annalen* and to the *Journal of the Chemical Society*. Hofmann was

Foreign Secretary to the Chemical Society, and in a reply to Kolbe, he offered either to translate Kolbe's article or to have it translated under his direction.[58]

Kolbe began by summarizing the disputes of the 1830s and 1840s between the (principally German) Berzelians and the French substitutionists, centering on the constitution of acetic acid and its derivatives. It was now beyond question that substitution does take place without altering the fundamental chemical properties of the substance and that organic radicals are by no means inviolable, as Berzelius had initially wanted to maintain. But Kolbe insisted that this by no means destroyed, or even substantially weakened, the electrochemical theory. One need only hypothesize further structure in one's radicals to provide a fully adequate theoretical explanation for all the new reactions.

For acetic acid, this meant that it was no longer sufficient to suppose a single integral "acetyl" radical, C_4H_3, combined with three equivalents of oxygen to form the anhydrous acid, as Liebig had been doing for the past decade. Kolbe's new formula was $(C_2H_3)\frown C_2$, in which the oxatyl group C_2 "presents the exclusive point of attack for the powers of affinity of oxygen, chlorine, &c.," as described in his *Handwörterbuch* article of 1848. This in turn suggested a new mechanism for the reaction in which alcohol is oxidized to aldehyde:

$$(C_4H_5)O.HO + 2O \longrightarrow HO.(C_2H_3)\frown C_2,O + 2HO$$

Kolbe thought that the oxygen must induce the ethyl radical to split into methyl plus C_2H_2, whereupon the two hydrogen equivalents in the latter moiety are captured by the extra oxygen, while the existing oxygen in the compound remains. Oxidation of aldehyde proceeds then by the direct addition of two more oxygens to the C_2. Substitution can complicate the notation of conjugate radicals, such as sulfobenzoic acid:[59]

$$HO.\left(C_{12}\begin{Bmatrix}H_4\\SO_2\end{Bmatrix}\right)\frown C_2,O_3$$

In addition to the conjugate radicals found in copulated compounds, Kolbe named his second class of hydrocarbon radicals, possessing different chemical properties, the "ether" or "alcohol" radicals, which provided an idea of the constitutions of the homologous ethers and alcohols and their derivatives. A third class of radicals, new to this article, was the "homologizing hydrocarbon radicals," C_nH_n, as in succinic or adipic acids:

$$HO.(C_2H_2)\frown C_2,O_3 \quad \text{or} \quad HO.(C_4H_4)\frown C_2,O_3$$

For the first time, we see here Kolbe admitting the possibility of dibasic organic acids, and he conceded that it might be necessary to reformulate succinic and sulfobenzoic acids as

$$2HO.\begin{cases}(C_4H_4)\frown C_2O_3\\ C_2O_3\end{cases} \quad \text{and} \quad 2HO.\begin{cases}\left(C_{12}\begin{cases}H_4\\ SO_2\end{cases}\right)\frown C_2,O_3\\ SO_3\end{cases}$$

where the numbers of atoms are simply doubled and rearranged in the first formula and another sulfate group is added in the second.[60] A curious aspect to this particular manipulation is that the second formula does not maintain the correct atomic ratios.

It is unclear precisely how Kolbe assigned some of his structures, in particular how he knew which compound was in the ether and which in the conjugate radical series. For example, according to Kolbe's theory, methane and ethane are actually methyl hydride and ethyl hydride, whose radicals are in the ether series. The chemically similar chlorinated derivatives perchloroethylene and perchloroethane are, however, conjugate radicals:[61]

$$(C_2Cl_3)\frown C_2,Cl \quad \text{and} \quad (C_2Cl_3)\frown C_2,Cl_3$$

To be sure, Kolbe had shown six years earlier that one could transform perchloroethylene into chloroacetic acid, justifying the first formula. He had also, however, shown how to convert methyl hydride to methyl chloride to methyl cyanide to acetic acid, which ought to have suggested that the methyl in methyl hydride is also a conjugate radical.

Another weakness of Kolbe's position, and one that he recognized, was the uncertain status of electrochemical-dualist precepts in his theory. Although his organic electrolyses appeared to demonstrate anew the general validity of electrochemical ideas, his specific predictions of the outcome of the experiments not infrequently failed verification.[62] He conceded that in many organic substances, it even seems impossible to determine which elements are positive and which are negative, such as carbon and nitrogen in cyanogen or carbon and hydrogen in hydrocarbons. Perhaps hydrogen's electrochemical properties are different in organic versus inorganic compounds—after all, hydrogen certainly has different properties in the normal gaseous versus the nascent state. So might chlorine have a less negative character in organic compounds than in its natural condition. These, he admitted, must be seen as conjectures regarding yet unsolved difficulties.[63]

In any case, Kolbe had developed here a detailed electrochemical radical theory that he was proud to compare with the type theories of the French. His view was that Dumas, Laurent, and Gerhardt had rushed to discard Berzelius' theoretical edifice on the basis of a single new phenomenon—substitution. This was overly hasty, to say the least, since all one needed to do was discard the now untenable belief in the immutability of radicals in order to resurrect the time-tested Berzelian theory. He concluded

> It would be ridiculous to allow a single fact, difficult of explanation, to induce us to throw aside at once a theory which has served us for so long a period as a trustworthy guide in the difficult field of organic chemistry, and has preserved us most securely from the errors of a code of laws like that which has been laid down by Laurent and Gerhardt—unless we had some better theory to substitute for it. . . . [C]hemistry is indeed something better than a mere arithmetic problem, into which Laurent and Gerhardt endeavor to convert it.[64]

In many respects, this was an impressive performance. The article is detailed, well documented, and (aside from some nationalistic aspersions in the conclusion) a fair-minded summary of much recent research in organic chemistry. And in one respect at least, it was revolutionary. Kolbe here adumbrated for the first time a concept that would prove central in the future development of the science, what modern organic chemists refer to as *functional groups*, and he began the process of locating functionality on specific parts of the "constitutions" or structures of molecules. It was Berzelius who first suggested that acetic acid could be considered as schematically dissectible into hydrocarbon and oxycarbon moieties; it was Kolbe who generalized that notion and drew out its implications. Kolbe showed that it is at the "oxatyl" carbon (in modern vocabulary, the carbonyl carbon) where the chemical functionality of the molecule is concentrated. This thesis was amply supported by dozens of reactions of acetyl derivatives and other methyl compounds. He attempted the same thing with all of the scores of compounds discussed in his paper.

But it was precisely in this respect where the greatest weakness of the paper lies. Kolbe aggressively followed the same pattern of identifying functionality and its location within the molecule that proved so successful for acetic acid, even for those much more numerous cases where little empirical data existed from which to conclude such details. His paper is full of formulas suggesting specific details of constitution and implying predictions of chemical behavior that were unsupported or unexamined in 1850. Along the same lines, he continued the proliferation of notational symbols that were poorly defined or even undefined. Berzelius had used periods or commas, while others among his contemporaries used parentheses and brackets, but Kolbe's pen was exceedingly restless with regard to the density of such symbols, and he added single and double buckle symbols to identify conjugate and homologizing hydrocarbon radicals. This often produced quite complex formulas (some examples of which have been cited) to which might be attached various empirical interpretations or predictions. For psychological and structural reasons, he could not write a formula without deliberately suggesting these interpretations and predictions.

For instance, the implication of his formulas for dibasic sulfobenzoic or succinic acids (as cited earlier) suggests that the two acid functions of each molecule are not chemically the same. It was eventually determined that in one case, this suggestion was correct, while in the other case it was not. More damaging and more to the point, in neither case did he have any evidence on which to judge the issue. Kolbe's readers had the right to expect the verbal and symbolic distinctions between alcohol and conjugate radicals to match consistent differences between the chemical behavior of those radicals, such as their replaceability with halogen atoms, but he simply failed to pursue this evidence. As a final example, the formulas for the chloroethylenes and chloroethanes cited above imply chemical nonequivalence of both the carbon atoms and the chlorine atoms within these molecules, but there was at that time no evidence for either equivalence or nonequivalence.

The Defection of Hofmann and Frankland

Some additional problems in this paper concern Kolbe's orientation toward Hofmann's and Frankland's research on substitution reactions. As already noted, Kolbe and Hofmann were exact contemporaries, arrived in London the same month, and became very close friends during the next two years. Hofmann later reported the results of some joint work that they performed in this period.[65] At this time, Hofmann and Kolbe were of very similar minds regarding constitutional theories. Hofmann's work on aniline appeared fully to support Berzelius' copula theory, the aniline derivatives all being formulated as ammonias with the structurally unspecified hydrocarbon $C_{12}H_4$ as a copula. Consistent with the theory, all the substituted anilines retained the characteristic basic properties of ordinary aniline, analogous to free ammonia. In June 1848, two months before Berzelius' death, Hofmann published a paper extolling Berzelius and his theory and adducing all the recent powerful experimental support for copulas, including Kolbe's work on acids and his own on bases. Still, he plaintively noted, "vainly have I hoped" to split aniline directly into NH_3 and $C_{12}H_4$.[66]

But the following year he found compelling reasons to abandon the copula theory, turning to embrace a thoroughgoing substitution theory and Liebig's amidogen theory (NH_2 plus a radical). Hofmann's conversion appears to have turned on one notable failed analogy between ammonia and aniline: whereas ammonia and benzoic acid can react together losing two water molecules (four "water equivalents") to form benzonitrile, there is no corresponding reaction between *aniline* and benzoic acid. The only way theoretically to rationalize this fact was to consider aniline not as a compound containing proximate ammonia but rather as one containing proximate amidogen (NH_2) in its structure; then there would be insufficient hydrogen in aniline to yield two water molecules unless some additional hydrogen were abstracted from the hydrocarbon moiety. Thus, aniline is not $NH_3,C_{12}H_4$, but rather $NH_2,C_{12}H_5$.[67]

As trivial an adjustment as this may seem, the implications were large. Copula theorists insisted wherever possible on using isolable compounds as their proximate components and explaining their combination as nonelectrochemical addition. This joining of proximate components, they averred, had nothing to do with substitution of hydrogen. By contrast, Hofmann's new constitution for aniline did indeed suggest substitution of Laurent's phenyl radical for a hydrogen atom of ammonia, and it denied the preexistence of ammonia in aniline. In short, Hofmann was saying that aniline could not be a copulated ammonia; it was instead a *substituted* ammonia.

In this paper, Hofmann placed all the weight motivating his conversion on the nitrile-forming reaction just discussed. Nonetheless, he did mention Wurtz's discovery earlier that year of methylamine, ethylamine, and amylamine (what Gerhardt called primary amines), which appeared to be ammonia with one hydrogen substituted by organic radicals. This new development was surely a motivating factor as well. At the very end of the paper, he mentioned one more new kind of reaction that, he thought, put the matter beyond all question: his own discovery, just made, of secondary and tertiary amines, representing further substitutions of the second or third hydrogens of ammonia or of the second hydrogen of aniline, by various hydrocarbon radicals.

This paper was read to the Chemical Society on 5 November 1849. By 26 December, Hofmann had submitted to the Royal Society a detailed memoir on the subject just broached, detailing one of the most classic pieces of organic chemical research in the nineteenth century. Using alkyl iodides as reagents, Hofmann described the preparation, properties, analysis, and theoretical classification of dozens of new substituted ammonias, including mono-, di-, and tri-methyl, ethyl, and amyl amines and anilines.[68] In a letter to Kolbe written while the paper was in press he described the work and related how much he had enjoyed it: "This investigation was a great deal of fun for me, for in 6 weeks the whole matter was settled. These reactions are so precise, that not a single experiment failed. The number of bases has now become virtually unlimited."[69]

In a letter that has apparently not survived, Kolbe must have communicated to Hofmann in 1850 an outline of his ideas for the review article on organic radicals, for in the letter just cited (undated, but ca. March 1850 by context), Hofmann reacted to some of Kolbe's ideas. Kolbe must have been attempting to maintain a strict interpretation of copulas, for Hofmann, now a true substitutionist, responded:

> I cannot yet accommodate myself to your point of view, in particular it seems more complicated than mine, and moreover it does not explain why ethyl can be inserted twice into aniline and three times into ammonia. This is a fact which I, as you can see, have established by these experiments. But I will suspend judgment until I have heard all your reasons.[70]

Kolbe's other good friend from his London years was simultaneously undergoing a similar conceptual evolution. It was Frankland who in 1848 had introduced the use of alkyl iodides as reagents, and he used them with phenomenal success in the preparation of new organometallic compounds. These reactions appeared to proceed by substitution mechanisms. The radicals seemed to substitute indifferently for hydrogen, not only in the (mostly hypothetical) metal hydrides but also in Paul Thenard's alkyl phosphines and in Hofmann's and Wurtz' nitrogen bases.

At the same meeting of the Chemical Society where Hofmann first mentioned his preparation of secondary and tertiary amines from alkyl iodides and signaled his conversion from copulas to substitution, Frankland's first paper on organometallic compounds, in which he drew similar implications, was read in absentia. An alkyl iodide, Frankland wrote, was analogous to hydriodic acid or any other hydracid, and the alkyl radical can substitute for hydrogen in a variety of organic and inorganic substances. He drew a specific analogy to Wurtz', Hofmann's, and Thenard's compounds, all the result of substitution reactions. Frankland was quite serious about this matter: the same comment was repeated in a paper read to the Chemical Society on 18 February 1850.[71]

Thus, in addition to his strong statements and recently revealed evidence in favor of the original radical theory of 1832–1834, from the beginning of his second German period (October 1848 to January 1850) Frankland was also paying close attention to the newest research on substitution and fitting his results into that theoretical program as well. In a classic paper written shortly after his return to England, Frankland was explicit. One could depict metal oxides as "true molecular types" for the organometallic compounds, he stated, citing Laurent and Dumas as authorities for the concept. For the fourth time in three years, he emphasized the analogy between his new compounds and those of the type theorists: "It is obvious that the establishment of this view of the constitution of the organo-metallic bodies will remove them from the class of organic radicals, and place them in the most intimate relation with ammonia and the bases of Wurtz, Hofmann, and Paul Thenard." Frankland was clearly offering the olive branch to the French type theorists: "The formation and examination of the organo-metallic bodies promise to assist in effecting a fusion of the two theories which have so long divided the opinions of chemists, and which have too hastily been considered irreconcilable."[72]

This apparent defection from the copula theory (and implicitly at least toward the French chemists) of his two loyal friends must have given Kolbe pause, and may account for the long delay between draft and publication of his 1850 article. There is evidence in Kolbe's paper of a serious effort toward accommodation of the uncomfortable new research, but also of the construction of a seawall against the incoming tide of typist ideas. Hofmann and Wurtz were "certainly right," he said, regarding the amidogen constitution of aniline and primary amines. Furthermore, Hofmann's work did demonstrate to a certain extent the truth of the type theorists' dictum that substitution of electrochemically foreign elements (such as chlorine or bromine in aniline) does not fundamentally alter the substance's properties. But that same research also showed that the *extreme* interpretation of type theory is wrong: a steady decrease of basicity occurs as aniline becomes more halogenated, until finally it becomes virtually neutral in character. Hofmann's and Frankland's research also placed beyond any doubt the actual existence of organic radicals as components of compounds, a point that some French chemists such as Dumas and Gerhardt had at times denied.[73]

Kolbe then asked whether it was the "radical hydrogen" of ammonia (i.e., that hydrogen atom whose removal creates amidogen) which is substituted by methyl or ethyl to create methylamine or ethylamine, and he answered his query in the affirmative.[74] Thus, Kolbe assumed the chemical nonequivalence of the three hydrogens of ammonia, simply from the circumstance that in some compounds nitrogen has only two hydrogens associated with it. The fact that Hofmann had shown how to substitute all three hydrogens in succession by similar and smooth reactions did not seem to weaken this assumption.

This is an example that once more illustrates Kolbe's habit of assuming theoretical details that are unmotivated, or even subtly contradicted, by empirical information; others have been cited earlier. His mind was so intensely and habitually oriented toward chemical theory in general, and molecular constitutions in particular, that such assumptions seem to have been an integral part of his mentality. This nearly obsessive concern would continue in the future to be both his greatest strength and his greatest weakness as a scientist.

4

Gerhardt and Wurtz

Liebig and Dumas

As Hermann Kolbe's relationships with the principal French chemists during his early career were influential on his ideas and actions, both in these years and later, we must spend some time with them.[1] A summary of the early history of the theories of substitution, nuclei, and types has been provided at the end of chapter 2. The leading personalities in this story were Liebig and Dumas, who first met as young men in Paris in 1823 and who spent the next two decades jousting with each other in developing overarching concepts in organic chemistry.

It was perhaps inevitable that these men would become rivals. Both were demon workers with extraordinarily creative minds, cultivating a field that had too many mysteries and too few facts. Both had occasion to accuse the other, sometimes justly, of experimental work that was "geschwind aber schlecht." Both had occasion to accuse the other, probably also sometimes justly, of poaching results. As violent as their disputes at times became, by 1840 they found themselves not very far apart—though neither man was then willing to admit this to the other.

In his worst moments, Liebig thought of Dumas as a true charlatan or "Schwindler" who was not above using questionable tactics or sleight of hand to achieve renown and whose greatest concern was pursuit of effect, flourish, and the rhetorical turn of phrase, all for the sake of personal ambition. For his part, Dumas often viewed Liebig as a heavy-handed and hotheaded chemical empire builder. After a brief alliance at the end of 1837 and the beginning of 1838, Liebig became dissatisfied with the pact he had made with Dumas. In 1840 he made a "total break" from the Frenchman, the quarrel resulting from substitution theories and based upon some real issues along with some pure misunderstandings. Dumas was a "tightrope dancer," a "Jesuit," a "highwayman," and a "thief," like "nearly all Frenchmen."[2] To Berzelius, Liebig complained,

> These Frenchmen truly have no feeling of true honor, no sense of justice and fairness, they have for many years been occupying themselves with theoretical speculations that are useless for science, and solely to satisfy their own vanity and arrogance; they have discovered that the word Radical must be banned and must be substituted by the word Type. This is the greatest of their discoveries. Unfortunately when I step forward there is in Germany only envy and weakness, so I stand completely alone, no one who has enough power to stand up to them supports me. In short, it is a bad time and I am very unhappy, and have turned from these miserable matters to applications of chemistry to physiology, which now interest me tremendously.[3]

Unfortunately, here again Liebig collided with Dumas, as Liebig became convinced in 1842 that Dumas had stolen his original ideas on plant and animal nutrition, and the heat of discord only became more intense.[4]

Even in the midst of some of these disputes, however, Liebig was able to recognize Dumas' merits and to concede when he had been in the wrong, and when the violence of his replies sometimes had done nothing but damage.[5] On 23 April 1850, Liebig wrote his friend C. F. Kuhlmann in Lille, whom he was about to visit to help dedicate a new factory. He was very much looking forward to seeing Dumas there, as he was anxious to renew their old friendship,

> . . . since I have always very highly esteemed Herr Dumas as one of the most outstanding and ingenious men among the chemists and scientists of our day. Perhaps more than any other chemist in Europe I found myself in the position of judging and prizing the value of his work, since we very frequently encountered each other in our investigations, and have cultivated the same fields.[6]

Liebig's hopes for the encounter were realized, as he wrote to Wöhler:

> We all arrived at the same time, embraced each other, and everything was fine. Dumas was extremely cordial, and looked so young that I hardly recognized him. His wife and daughter were with him, to serve as witnesses to the plans for revenge that he had brewed. On Whitsunday the celebration was splendid and merry, the next evening a banquet, to

> which the civil and military leaders of Lille were invited. At the end of the banquet Dumas stood up, gave a long speech, flattered me with various puffery, and finally took a decoration for the legion d'honneur from his pocket, and handed it to me along with the brevet in the name of the President of the French Republic. I was unprepared and thought I would faint; but I managed a speech and received an accolade. Thus he revenged himself on me. Despite all he has a magnificent nature.

The following year Liebig dedicated a new edition of his *Chemische Briefe* to Dumas, and the two exchanged a number of warm letters until Liebig's death in 1873.[7]

It will be noted that during the entire period over which we have followed Kolbe's early career in chapters 2 and 3, Dumas and Liebig were feuding intermittently. In fact, the interval between Kolbe's first published paper and his long review article on radicals and types precisely brackets the period when Liebig's and Dumas' relationship was at its lowest ebb. Kolbe certainly imbibed an extremely negative view of Dumas from Liebig, who was one of his idols and models and whose diatribes were often openly published in the scientific literature. Berzelius, and his former student Wöhler, the author of the Schwindler satire, also had opinions of Dumas and other French chemists which were not much more positive than Liebig's. But Dumas had retreated about 1840 from a leading theoretical role, replaced by such chemists as Laurent and Gerhardt and in the 1850s by Wurtz as well. Kolbe's relationships with Gerhardt and Wurtz paralleled Liebig's relationship with Dumas, except for the lack of a final reconciliation. It was with Gerhardt and Wurtz that Kolbe felt the strongest sense of rivalry, enmity, hatred—and occasionally, even affinity, if not regard.

Gerhardt

Charles Gerhardt (1816–1856) was born in Strasbourg to a bourgeois Jewish family with both German and Alsatian roots. With the prospect of helping his father make a success of a white lead factory in which he had a major financial interest, Gerhardt was sent to the Karlsruhe Technische Hochschule in 1831, then two years later to the Leipzig Gewerbeschule. Here he lodged and studied with O. L. Erdmann and published his first scientific paper in Erdmann's (and later Kolbe's) *Journal für praktische Chemie*. During winter semester 1837/38, he studied in Liebig's lab in Giessen, thereby acquiring an influential protector. On Liebig's advice, he then traveled to Paris where he was associated with Dumas and H. Sainte-Claire Deville at the Faculté des Sciences and the Sorbonne and with A. Cahours at the Jardin des Plantes.

Liebig, who just then was becoming disillusioned with polemics over theory, sought to discourage his protégé's predilections in that direction:

> You will destroy your future and irritate everyone, like Laurent and Persoz, if you continue to make theories. . . . The Academy [of Sciences in Paris] has always reserved for itself the right of making laws in science, and it considers anyone else doing this as a thief and an assassin. . . . You always think you are on the neutral soil of Germany, but in fact you are standing on ground which contains all sorts of combustible matter.

Liebig urged Gerhardt to tie himself personally to Dumas, who has "a magnificent character" and likes to take promising young chemists under his wing. "But for the love of God, don't write any more theories except for German journals!"[8]

Even as a young man, Gerhardt was headstrong, outspoken, and dogmatic, and his relationship with Dumas was from the beginning somewhat uneven. His ardent republican political convictions did not mesh with Dumas' moderate conservatism, and his ties to Liebig created friction in a period when Dumas' relations with the Giessen school were starting to unravel. Moreover, Liebig was correct in perceiving a fundamentally positivistic flavor among the leaders of Parisian chemistry, which conflicted with Gerhardt's visceral theoretical orientation. In 1841, Gerhardt received a provisional appointment at Montpellier, which was made permanent in 1844. In a letter of gratitude to Liebig, Gerhardt attributed both this career triumph and his general success as a chemist to his teacher's influence. But no sooner had Gerhardt arrived in Montpellier than he began to complain bitterly to Liebig of provincial life and the poorly provisioned laboratory facilities.[9]

In 1842 Gerhardt published one of his greatest papers and simultaneously committed one of his greatest faux pas. He had come to the conclusion that organic chemists needed to unite their formula notation with inorganic conventions by taking all formulas at two volumes instead of the usual four. This would eliminate many dualistic formulas that posited the preexistence of water, ammonia, oxides, and so on in organic acids, bases, and salts, but since the advent of type theories such electrochemical rational formulas were no longer popular in France anyway. In the context of French chemistry of the early 1840s, a very good case could have been made in favor of such ideas. They were later viewed as constituting the essential step away from a misleading and inconsistent notation to unified, consistent, and essentially modern formulas. Although Gerhardt did not say so and may not in fact have realized it, his reform proposal was equivalent to accepting the hypotheses of Avogadro.[10]

Unfortunately, Gerhardt committed two serious errors in this paper. One mistake was a certain degree of conceptual and linguistic confusion. Even when a historian reads this paper with full knowledge of both contemporary and subsequent events in mind, it is sometimes difficult to extract Gerhardt's precise meaning. There were a few egregious lapses: one example is Gerhardt's claim that atoms, volumes, and equivalents are exact synonyms. A second paper on the same subject published the following year solved a few of these problems, but not the most serious ones.[11]

Even worse for the future of his career, Gerhardt expressed himself in socially inappropriate terms. When his paper was read at the Académie on 5 September 1842, some members of the audience were shocked to hear what seemed to be arrogant and imperious language from this young man, and they let him know in no uncertain terms. Dumas had wanted to suppress the theoretical part. V. Regnault was furious at some of his formulas being declared "wrong," and Baron L. J. Thenard said the style would not have been appropriate even for a Lavoisier. Later that month back in Montpellier, Gerhardt recounted the disaster in a letter to his friend Cahours. He said he had asked Thenard timidly later that day why he didn't like the paper. "For a moment I thought he was going to eat me," Gerhardt related. He and others agreed, Thenard had said, that the language was "not French" and "not academic" and that the style was infuriatingly imperious (such-and-such a formula is "false," the notation "must be changed," and so on). Thenard concluded by screaming "Adieu, monsieur!" at least ten times before Gerhardt got the hint and excused himself. Liebig received the same news from Gerhardt and replied, "You have Italian blood, too hot for the Parisians." Liebig noted that Gerhardt's fatal error was not to have heeded his warning about avoiding theory in Paris.[12]

In the fall of 1843 Gerhardt met Laurent, and they quickly became friends and comrades in arms. Gerhardt's judgment of Laurent in his letters to Cahours underwent a reversal from derogation to adulation. Laurent, like Gerhardt, was an uncompromising iconoclast and an ardent republican, cut off from the power structure in Paris and a professor at a provincial faculty, namely, Bordeaux; he may also have been Jewish.[13] Also like Gerhardt, Laurent was collegially closer to the Giessen circle than to Paris. He had visited Giessen for a time during the autumn of 1844 and had quite favorably impressed both Liebig and his assistant August Wilhelm Hofmann (it was under Laurent's influence that Hofmann began to study the halogenation of aniline). His angry confrontations with Dumas over priority matters during the late 1830s, which had alienated his former teacher, could not but have stirred sympathy for him in Liebig's circle.

Laurent was soon converted to Gerhardt's atomic weight reform, and in December 1844 Laurent wrote to Liebig and Hofmann to urge them to accept it as well. Unfortunately, Liebig considered the reform to be nothing more than yet another French theory, and declined; Hofmann was presumably too cautious and too professionally insecure to associate himself with the reformers. "Eh! bien, marchons seuls," Laurent wrote Gerhardt in resignation.[14]

A more serious problem arose a year later when Gerhardt infuriated Liebig by criticizing as "completely false" his work on a series of complex nitrogen compounds. Liebig promptly published a blistering diatribe entitled "Herr Gerhardt und die organische Chemie, Erste Artikel," openly accusing Gerhardt of the vilest perfidy and calling him a "highwayman."[15] He also urged Laurent to break off his "monstrous alliance" with Gerhardt. "If you associate yourself with him," Liebig wrote, "it is you who will lose because he has nothing to lose. Read carefully my article, and tell me whether this man has the truth in his soul."[16] But Laurent refused to abandon his friend and wrote him a kind letter of support and consolation. Laurent related that on his recent visit to Giessen, Liebig had apologized to him for his earlier unjustified attacks and had vowed that his next angry polemic would rest six months in his desk first! He assured Gerhardt not only that Liebig would cool off, but that he would eventually realize that the new atomic weights are correct.[17]

Gerhardt had criticized Liebig's formulas largely on the basis of the "even-number rule" developed by Laurent from certain regularities in the numbers of hydrogen atoms in organic formulas noted by Gerhardt in his 1842 article. The rule as generalized by Laurent (1845) states that although there may be any number of carbon and oxygen atoms in organic formulas when expressed in Gerhardt's two-volume convention, the sum of all the hydrogen, halogen, and nitrogen atoms must be an even number.[18] At the time of its formulation, this law was merely an empirical generalization unmotivated by theory, but it seemed to hold for every well-studied organic compound; the only exceptions were cases where there was some reason to suspect analytical inaccuracies, and a handful of contrary instances, such as Liebig's nitrogen compounds. It was little wonder that Liebig was unconvinced by this sort of reasoning.

By this time Laurent had given up his unhappy post in Bordeaux and was living in poverty in Paris. As he and Gerhardt were virtually cut off from the traditional French journals, they had started their own, entitled *Comptes rendus des travaux de chimie*. This was essentially a review journal modeled on Berzelius' *Jahresberichte*, as well as an organ for publication of their own papers.

It would seem that at least in one respect Gerhardt had learned his lesson, for after 1842 he pursued a much more openly, indeed often dogmatically, positivist course. Having long denied the preexistence of the dualists' water and oxides in acids and salts, he now denied the real existence of any and all radicals within compounds. In fact, he argued that it was beyond human capabilities to discern any information at all regarding molecular constitutions, and so he began to rely exclusively on empirical molecular formulas. In this he was resisted by his compatriot Laurent, who used constitutional theories habitually and continuously, though professing more a conventionalist than a realist philosophy of science.[19]

Perversely, this new Gerhardtian positivist–empiricist organic chemistry was rich in new doctrines and ideas, among which were the revised atomic weights and Laurent's even-number rule. The former was an early version of the reform established nearly two decades later at the Karlsruhe Congress of 1860; the latter represented the first seed of what would eventually grow into the theory of atomic valence. Gerhardt also began to develop the concept of homologous series, an idea first broached by Dumas, which would prove extremely productive and heuristically important. These ideas were treated systematically in Gerhardt's first book, *Précis de chimie organique* (Paris, 1844–1845). Although this work was not well received—not even by Laurent, who subjected it to a searching critique—and although Gerhardt himself described it later as a "horrible old book," it contains the basis of many of his contributions to organic chemistry.[20]

The way out of the morass was finally perceived by Laurent. In the summer of 1846 he wrote Gerhardt:

> I have carefully pondered equivalents. The words equivalents, atoms and volumes cannot be synonymous. . . . There are atoms, there are proportional numbers, but not equivalents. . . . Indeed, the proportional number is a number chosen arbitrarily; one can take it to agree with the atom or volume and to give the simplest notation. The *equivalent* is different: it is that quantity of a simple body which must be employed to replace another simple body and play its role.[21]

These thoughts were published that fall in a major theoretical paper that fully clarified Gerhardt's terms and straightened out the conceptual confusion in the 1842 reform proposal. "M. Gerhardt's atom," Laurent stated, "represents the smallest quantity of a simple body which can *exist in a compound*. My molecule represents the smallest quantity of a simple body which must be employed to *produce a combination*; this quantity splits in two in the act of combination." For instance, an oxygen *molecule* splits into two oxygen *atoms* to form two molecules of water; a molecule of chlorine splits into two atoms, one of which substitutes for a hydrogen atom of an organic molecule, with the other combining with the hydrogen thus split off to form a molecule of hydrochloric acid.[22]

In hindsight, Laurent's modification of Gerhardt's reform appears to establish at a stroke the basis for modern chemistry. In fact, it did nothing of the kind. In the late 1840s the Laurent–Gerhardt system was only a schematic proposal with considerable esthetic appeal, but based essentially on thoughtful rationalization rather than empirical verification. Moreover, it was in conflict at many points with well-accepted chemical theory and was still troubled by a number of anomalies. Laurent himself recognized all of these weaknesses and urged Gerhardt not to lose courage.[23] The theoretical anarchy of this period is well summarized by Laurent's ironic summary description of a course he had been invited to teach at the Sorbonne in the summer of 1847:

> *Introduction*. The entire science is placed in doubt. Atoms are perhaps divisible; dualism is attacked; nomenclature is insufficient; ditto for classifications; all compass bearings are lost, we need a guiding thread. Embarrassment of the professor who teaches a science in which he does not believe.[24]

Laurent was then suffering both extreme penury and depression. "What misery! What a bitch life is! Not a sous in my pocket!" he cried out to Gerhardt.[25]

The following year Gerhardt converted Laurent's sarcasm into a book entitled *Introduction à l'étude de la chimie par le système unitaire*, dedicated to Laurent. It is here where Gerhardt most clearly stated his conviction that molecules must be treated as unitary entities and must be described only by empirical formulas. Reactions of synthesis and analysis say nothing about the constitutions of compounds, for every chemical reaction sets up a violent motion among the atoms in the molecule that totally scrambles their arrangement. Organic chemical theories, he said, must therefore be based on taxonomic rather than constitutional ideas, on reactions rather than structures. In this he was representing a well-established French positivist tradition in chemistry that was fundamentally opposed to many German, English, and Berzelian conceptions.[26]

This book, reflecting the theoretical anarchy in chemistry, appeared the same month when political anarchy emerged in Paris, the opening wedge of the long sought republican revolution. Understandably, the fortunes of Laurent and Gerhardt suddenly appeared promising again. In March Laurent was named assayer at the national mint, where he fitted up a small laboratory; Gerhardt immediately abandoned his position in Montpellier to join him there. Arago and Quesneville, republicans, and Cahours, a liberal, greeted the revolution warmly; many who had more conservative or centrist convictions were frightened by it. Dumas appeared to change camps, writing a discourse strongly favoring the republic; the apparent hypocrisy scandalized those on the left.[27]

Gerhardt called on Dumas on 26 March, and described the encounter in a letter to his wife, still residing in Montpellier. Dumas received him coldly and was about to show him the door when Gerhardt suggested that Dumas' conduct "had been unworthy."

> At once he became timid as a lamb. So he was afraid. I changed tactics and told him that he had been wrong to abdicate the good position that he had had in science, to turn to politics and administration; that he alone in France was able to understand my ideas (mine and Laurent's) . . . Then Dumas protested of his devotion to science, vigorously denied any hostility to me, and began to weep! Was he sincere at this moment? I have no idea, but I want to believe it. He shook my hand several times and we parted perfectly well. Our conversation had lasted nearly an hour and a half, and surely no one has ever dared to tell Dumas the truth as energetically as I. . . . I am sure now that he will do anything for me to prevent me from denouncing him.[28]

There were several vacancies in Paris, and Gerhardt and others mounted a vigorous attack on the system of *cumul* that allowed prominent academics such as Gay-Lussac and Dumas to accumulate almost any number of posts simultaneously. Laurent and Gerhardt both sought to influence the provisional regime's minister of culture Hippolyte Carnot (brother of Sadi) to procure appropriate positions, with promising results. Gerhardt reported to his wife that all of his former enemies were treating him very well now. "If there is justice in the world I will succeed this time; now or never."[29]

Unfortunately for Gerhardt, Carnot was gone by July, and by the following year the republican tide was waning; all of his hopes began to evaporate. In the spring of 1849 he described himself to his wife as being in a "veritable state of fever," reading ten or a dozen newspapers every day, but lying low for the sake of his family. During the June riots he wrote: "I must tell you, you are making a coward of me! If the insurrection is vanquished this time, it will be the death of democracy in all of Europe." Vanquished it was.

Gerhardt's enemies, especially Dumas, emerged stronger than ever before. Even more tragic, from working in his cold and damp cellar laboratory at the mint—described by Gerhardt as a "veritable glacier"—Laurent contracted a severe case of tuberculosis in December 1850; he lingered on over two years before finally succumbing. His last days were bitter in the extreme.[30]

Now essentially unemployed and desperately in need of a protector, Gerhardt wrote Liebig in an attempt to patch over their quarrel:

> You know, monsieur, that basically there is no important difference between our theoretical opinions. . . . One could still today define organic chemistry as the chemistry of compound radicals; it is only a question of making very precise the meaning of the word radical, and of removing from it that absolute signification which you yourself have never assumed, and which has been interred with Berzelius.[31]

Liebig's response, although friendly, was noncommittal. To Hofmann, Liebig expressed suspicions concerning Gerhardt's sincerity and motives. In February 1851, using money lent by Gerhardt's Scots mother-in-law, he and the infirm Laurent began to accept pupils in a private "École de Chimie Pratique," and both began to write definitive treatises of chemistry.[32]

This was the low point for the reform movement of Gerhardt and Laurent. Its revival was associated with remarkable breakthrough discoveries by Gerhardt and Alexander Williamson and by the work of Wurtz, to whom we now turn.

Wurtz

(Charles) Adolphe Wurtz was born on 26 November 1817, the eldest son of Jean Jacques Wurtz and the former Sophie Kreiss. The elder Wurtz, the only child of a "simple bourgeois family" of Strasbourg, became a Lutheran pastor in Wolfisheim, a small village three miles west of the city; Sophie's father, also named Jean Jacques (Johann Jakob), was also a Lutheran pastor, at the church of St. Pierre-le-jeune in Strasbourg. It is said that Adolphe's boyhood in Wolfisheim provided him with both a robust constitution and a love of the countryside. He spoke Alsatian with his family, but learned French and Hochdeutsch in his youth (he is said to have spoken both languages without a trace of regional accent).[33]

The parallels here to Hermann Kolbe's family background and upbringing are remarkable. Born less than a year apart, both died in 1884 at the age of sixty-six. Despite their different nationalities, both were raised in German-speaking households of the same social class; both were eldest sons of Lutheran country pastors who led congregations in small towns near important university cities. Both remained sincere Protestants and became outspoken nationalists in their later years.

In 1826 Wurtz' father was called to Strasbourg as third pastor at his father-in-law's church, and Adolphe entered the Strasbourg Protestant Gymnasium. His father wanted him to enter the ministry, but Adolphe had conceived a passion for science. As a sort of compromise, Adolphe was allowed to study medicine, entering the Strasbourg Faculty of Medicine in 1835. Between 1839, when he was named the faculty's Chef des travaux de chimie, and his departure for Paris in 1844, he worked mostly in the laboratory of Amedée Cailliot (1805–1884). He also spent five months (summer semester 1842) at Liebig's laboratory in Giessen, and he always considered Liebig as one of his mentors. In the spring of 1844, he went to Paris seeking a chemical career. He started with Balard at the Faculté des Sciences, then after a year was appointed Dumas' préparateur at the Faculté de Médecine. Simultaneously he accepted the post of Chef des travaux chimiques at the École Centrale des Arts et Manufactures, founded by Dumas sixteen years earlier. He soon became Dumas' favorite and ultimately was to inherit the mantle of the master.

Of course, fellow Strasbourger Charles Gerhardt was also a presence in Wurtz' early years. Most biographers note, correctly, that Gerhardt and Wurtz were schoolfellows at the Strasbourg Gymnasium and that Wurtz translated Gerhardt's first book, the *Précis* (1844–1845), into German. The conclusion has often been drawn that Wurtz came to Paris already in the Gerhardt–Laurent orbit, against the established ideas of such figures as Liebig and Dumas.[34]

The situation was quite the opposite. Although Gerhardt was only fifteen months older than Wurtz, he was two Gymnasium classes in advance and left a year earlier than was normal, when Wurtz was but thirteen years old; they appear not to have become acquainted then.[35] Moreover, Wurtz did not offer to translate Gerhardt's book from a position as disciple. Rather, on a trip home in 1842, Gerhardt inquired of Cailliot regarding an appropriate bilingual Alsatian chemist; Cailliot recommended Wurtz, who was happy to do the work both for the chemical interest and for the money involved.[36] Gerhardt was in Montpellier for the first four years after Wurtz arrived in Paris, and even after Gerhardt returned to Paris in 1848, Wurtz saw little of him. Wurtz scarcely knew Laurent personally.[37] During Gerhardt's lifetime, Wurtz never adopted the new chemical "equivalents" championed in the *Pré-*

cis. Finally, at the time Gerhardt began work on the *Précis* and hired Wurtz as translator (1842–1843), he was still on good terms with Liebig. He was rightly seen as a disciple of both Liebig and Dumas and had not yet joined forces with Laurent.

By agreeing to translate Gerhardt's *Précis*, Wurtz was not declaring his allegiance to Gerhardtian ideas, and those ideas had not yet been labeled as heretical. Rather, he was accepting an assignment that might well lead to professional advancement, as well as providing much needed income. By the time Wurtz finished his translation in 1846, open warfare had broken out between the Gerhardt–Laurent alliance and the uneasy Dumas–Liebig axis, but Wurtz could not have foreseen this situation four years earlier. In short, all the evidence suggests that in the 1840s and early 1850s, Wurtz was a loyal adherent of the established chemical theories of Liebig and especially of his patron Dumas.

Wurtz' early work illustrates many of these ideas. In Liebig's lab he investigated hypophosphorous acid, presenting the results in terms of Liebig's hydracid theory (and also citing Liebig's predecessors Davy and Dulong).[38] Wurtz' first papers after his association with Dumas exhibit a distaste for dualist ideas and a concern for merging organic and inorganic chemistry under unitary assumptions. They also contain references to Dumas' "edifice" analogy for molecular structure (despite conventional empiricist protestations). Wurtz even proposed PCl_3 as the "type" of most phosphorus compounds—choosing a word that had been used in 1839 and 1840 by Liebig as well as Dumas.[39]

As we have seen, from about 1840 both Liebig and Dumas simultaneously and apparently independently retreated from the new directions they had been heading. Both began to pursue more practical and purportedly less theoretical subjects such as agricultural and physiological chemistry. Liebig turned away from his unitary hydracid formulations toward more dualistic oxide formulas. In 1844, Liebig shifted from four-volume Berzelian atomic weights, which he had been using for fourteen years, to Gmelin's four-volume conventional equivalents. In approximate modern terms, this was equivalent to moving from doubled molecular formulas to formulas with doubled numbers of even-valence atoms but correct numbers of odd-valence atoms. This shift was connected to the other changes since the new formulations were regarded as perfectly empirical, in contrast to Berzelius' theoretically derived atomic weights.

For several years, Wurtz mirrored the ambivalence of Liebig and Dumas. On the one hand, he continued to advocate substitutable radicals and the unification of organic and inorganic formulas. On the other hand, he followed Liebig's "empiricist" atomic weight shift of

1844 and continued to use many of the formula conventions attacked by Gerhardt and Laurent. For instance, his novel substance "ammoniacal cyanic ether" (modern ethyl isocyanate, C_2H_5NCO) was written C^2AzO,C^4H^5O,AzH^3, a formulation that reflects an implicit dualism as well as incompatibility with Gerhardt's new atomic weights and with his constitutional ideas.[40] This example is not an arbitrary one: alkaline hydrolysis of this substance led to Wurtz' first major discovery, ethylamine, in 1849. It was the first primary amine, indeed the first amine of any kind to be recognized, and the progenitor of an immense number of synthetic organic bases.[41]

Hofmann provided an interesting gloss on this event. Hofmann and Wurtz had become good friends during Wurtz' stay in Giessen in 1842, where Hofmann was assistant to Liebig. From 1847 they had the opportunity to see each other regularly, for Wurtz had frequent occasion to visit his sister and her husband in England, and Hofmann, who was professor at the Royal College of Chemistry in London from 1845, found himself often in Paris.[42] Moreover, they had a major chemical interest in common, namely, organic bases. Hofmann related that Wurtz had missed the discovery of ethylamine for quite some time since he was assuming that hydrolysis of the ethyl isocyanate ought to proceed analogously to the recently published Frankland–Kolbe hydrolysis of ethyl cyanide, that is, to yield propionic acid and ammonia. Thus, reasoning on the basis of the copula theory, Wurtz assumed that the ammoniacal gas released in the reaction was simply ammonia, and he was long mystified at being unable to isolate any sort of oxidized hydrocarbon in solution. It was only when this gas happened to ignite from a fortuitous flame that light shone in Wurtz' mind, as at his laboratory bench: the gas was in fact the chief product of the reaction, an ethyl-substituted ammonia. Wurtz quickly published the result, commenting that ammonia now could be regarded as the "type" for perhaps all organic bases. Wurtz was only repeating the very words of Liebig, who nine years earlier had predicted the existence and even the properties of ethylamine.[43]

Hofmann, who had been working on substituted anilines, felt he had been forestalled; in a letter to his mentor, he remarked on Liebig's clairvoyant prediction. In his reply Liebig also expressed some chagrin, for he had read Wurtz' paper without even remembering his own earlier remarks, and it was only Hofmann's letter that had brought them to mind (a revealing indication of his disinclination toward chemical theory after 1840). "I was so enchanted by Wurtz' work," he added, "that I wrote him and congratulated him on such lovely discoveries. . . ." He went on to say that Hofmann's studies on aniline had been "very interesting . . . every new compound is the first member of a new series of homologous compounds, and the thought from which the compounds arose is like a seed corn, which bears its fruits in the minds of others doing similar work."[44] Late that year (1849), Hofmann wrote Liebig about his latest syntheses, the details of which were soon to appear in a major paper published in the *Philosophical Transactions*: secondary and tertiary amines. "These are incredibly remarkable things you have discovered"; Liebig responded, "we are all ecstatic over them. . . . These are very valuable experiments and fully conducive to lead to a definitive view [of the constitution of the organic bases]. . . . This is not a small jump out in front of the French, although I do not mean thereby to include our friend Herr Wurtz."[45] But it was indeed Wurtz' turn to feel forestalled.

The competition between Hofmann and Wurtz in no way affected their friendship. Indeed, in March or early April 1850, Hofmann visited Wurtz in Paris; he not only had a delightful time seeing the sights with his comrade, but described the trip in amazing detail in his biography of Wurtz many years later. During this trip, Baron Thenard hosted a dinner party in Hofmann's honor; his son Paul introduced the company to Hofmann with the words, "Voilà tous les jeunes chimistes de Paris, *moins les deux*." Hofmann related this incident in a letter to Liebig, commenting that "this is characteristic of the general mood that has gradually formed in France," that is, against Gerhardt and Laurent.[46]

The relationship between Gerhardt and Wurtz was never more than coolly correct. Gerhardt felt a degree of resentment, bordering on animosity, that the younger Wurtz was so favored by the powerful Dumas: Wurtz was made *agrégé* (loosely, assistant professor) at the Faculté de Médecine in 1847 and then took over Dumas' lectures on organic chemistry two years later, simultaneously continuing his duties at the École Centrale, all while Gerhardt was without any official position. Particularly galling was Dumas' almost fulsome praise in an official report on Wurtz' papers on amines, when in fact Gerhardt felt that Wurtz had used some of his own work without acknowledgment.[47] "Quelle chance il a, çe garçon," he later exclaimed to a correspondent.[48]

In a major summary of the chemistry of the new aliphatic amines that appeared at the end of 1850, Wurtz tried to apportion proper credit to his predecessors and competitors, especially Hofmann and Gerhardt—which he had certainly not previously attempted. In a theoretical conclusion he argued, as Hofmann had earlier that year, that his and his colleagues' results could most easily be explained by expanding Liebig's amidogen theory to include secondary and tertiary amines, in which alkyl radicals such as methyl, ethyl, propyl, and butyl substitute for hydrogen atoms in ammonia. He also argued that these results were fully compatible with, indeed provide further support for, Kolbe's idea that the homologous organic acids were alkyl radicals substituted for the nonbasic hydrogen of the simplest acid (although that series progenitor he took, following Gerhardt, to be formic rather than oxalic acid). He also strongly affirmed, with Kolbe, that the elements that form organic molecules are not thrown "pell-mell, without order, without arrangement, without predisposition" (as Gerhardt had sometimes implied), but rather are arrayed in a definite and ordinarily stable structure. He felt that determining those molecular constitutions was one of the most important goals of chemical theory.[49]

The early 1850s were a busy time for Wurtz. He relinquished his position at the École Centrale to open a private chemical laboratory school and to take a position at the short-lived Institut Agronomique in Versailles. The lab school folded in 1851, and the Institut was closed the next year. In the meantime, Wurtz courted and wed Constance Opperman, the daughter of a wealthy Paris banker. At about this time (the beginning of 1852), Dumas decided to take advantage of Wurtz' linguistic skills and his German and English connections by inviting him to prepare abstracts of selected foreign articles for every monthly issue of the *Annales de chimie*. Finally, in 1853 Dumas retired from teaching in the Faculté de Médecine; Orfila's death at almost precisely the same time (12 March) gave Wurtz two chairs in the Faculté, along with the financial and career security he had been seeking.

The Breakthrough of Gerhardt's Reform

By this time, the Gerhardt–Laurent reform had begun to make headway in the chemical world. Although most French chemists, including all the influential ones, continued publicly to oppose Gerhardt and Laurent, they did begin to receive some private sympathy. As early as 1846 Thenard was treating Gerhardt kindly again, and Dumas began quietly to support Laurent in 1847. In Great Britain and the United States a few disciples began to appear.[50]

It was Williamson's publications on etherification and the water type in 1850–1851 that marked the critical break for the type theory and for its French partisans. His novel syntheses of asymmetric ethers (such as methyl ethyl ether) provided the first compelling experimental support for Laurent's and Gerhardt's conceptions of the relative constitutions of those theoretically most central substances alcohol, ether, and acetic acid. More indirectly, he argued for the truth of the entire new French system, including the atomic weight reform. "In your formulas," Williamson wrote Gerhardt, "I see the future."[51]

On the heels of this work, and taking advantage of the recent research of Hofmann, Wurtz, and others, from 1851 Gerhardt developed his modified type theory. In this theory, four inorganic substances (water, ammonia, hydrogen, and hydrogen chloride) served as models for large series of organic derivatives. Ethers, acids, acid anhydrides, aldehydes, and ketones were all schematically derived from water by replacement of one or both hydrogen atoms; similarly, amines and amides were related to ammonia through substitution of hydrogen by radicals, alkyl derivatives were derived from the "hydrogen type," and chlorides from the "hydrogen chloride type."[52]

In 1852 Gerhardt succeeded in producing asymmetric organic acid anhydrides, providing another compelling argument in favor of the view that organic acids must be seen as analogous to water, but containing no water per se even in their hydrated forms. In brief, the dualists assumed organic acid formulas twice as large as Gerhardt's and Laurent's so that every acid could be formulated to have two replaceable hydrogens, namely, the hydrogens of the associated water molecule. Extract the water molecule, the dualists argued, and you produce the acid anhydride. What Gerhardt succeeded in doing, in a complete analogy to Williamson's asymmetric ethers, was to create structurally asymmetric or mixed acid anhydrides, such as acetic–benzoic anhydride. The dualists could write acceptable formulas representing the new reaction, but such an equation predicted two products, a mixture of acetic anhydride and benzoic anhydride. The actual production of a single asymmetric product could only be accounted for by the smaller, non-dualistic formulas for the acids. Another way to put the matter is that the only way to force monobasic acids such as acetic and benzoic acids to form anhydrides is to join two molecules, since the water that must be removed requires two hydrogen atoms from two molecules; the production of asymmetric anhydrides made it inescapably clear that two *different* molecules must be involved. Thus, there is no preformed water in acids, merely replaceable hydrogen atoms.

Gerhardt knew immediately that he had a winner. He was so excited that his letter announcing the discovery to his student Gustave Chancel is barely legible. In May and June 1852 he read two major papers on acid anhydrides to the Académie.[53] The discovery was received with a great deal more immediate acclaim than Williamson's had been, despite the fact that Williamson had predicted Gerhardt's reaction the previous year and that Gerhardt's theoretical argument was identical to Williamson's (Gerhardt did not mention the Englishman's name). Indeed, Williamson's work seems to have been little known, or at least little noticed, on the European continent until 1853.

Even Gerhardt's most implacable adversaries were impressed. After his second memoir was read to the Académie, he wrote Chancel, "You could not imagine how everyone has changed toward me: Regnault shook my hand, Dumas nearly hugged me, even Fremy paid me the sweetest compliments . . . What a success!" Dumas wrote a somewhat restrained but strongly favorable report on Gerhardt's paper for the Académie. Thenard became an enthusiastic advocate for Gerhardt's career.[54] The following year Gerhardt published a large review article, carefully crafted to be "academic" in tone, although, he thought, "terribly revolutionary."[55]

For once, others agreed with Gerhardt. Chancel congratulated Gerhardt, predicting a "great sensation" in Germany and England, along with many new converts—perhaps even Liebig himself. Indeed, Liebig soon wrote Gerhardt, praising the discovery as "one of the most brilliant of recent times," the (Williamsonian) theoretical argument being "as simple as it is elegant." "It is very strange," he concluded, "that these two theories, although completely opposed, are now unified into a single one which explains all the phenomena in the two senses."[56] Liebig was but echoing the words in Gerhardt's letter to him two years earlier, urging their reconciliation. In October 1853 Gerhardt visited Giessen and was treated very warmly by Liebig.[57]

As Chancel predicted, during the next two or three years the Gerhardtian system began to attract converts, especially in Germany—a story that is explored in greater detail in chapter 6. A few Russian and a few more English and American disciples began to appear. In France, Malaguti, Quesneville, Cahours, Pelouze, and Gerhardt's students Chancel and Chiozza had adopted elements of the reform even before the acid anhydride work.

Important for this turning of the tide was not only Gerhardt's acid anhydride work, but also his massive and magisterial *Traité de chimie organique*. It was begun in 1851, and by August of the following year, Gerhardt had written close to 600 printed pages worth of manuscript for the first volume, but with no publisher in prospect. It was reading this manuscript in 1851–1852 and conversing with its author that converted Kekulé to Gerhardt's views. On 28 November 1852, Gerhardt wrote Chancel: "Grreat news! I have sold my organic chemistry to Didot for the trifling sum of ten thousand francs." Gerhardt related that Didot had inquired whether Gerhardt might be willing to complete the last unfinished edition of Berzelius' text. He had replied that the book was already out of date, even though the last volume had only appeared two years earlier, but coincidentally, he had the manuscript of a text himself. They agreed on publishing Gerhardt's manuscript under the title "Traité de chimie organique *de* Berzelius entièrement refondu et enrichi des descouvertes récentes *par* C. Gerhardt," with the spine reading "C. Gerhardt: Traité de chimie organique de Berzelius." Gerhardt added that Laurent had joked that if Berzelius had known this would happen, he would have died a week earlier.[58]

The book was published in fascicles starting in June 1853 and was exceedingly and justly popular. Liebig himself heartily praised the work and told Gerhardt on a visit to Giessen in 1854 that the book would have ten times as many German readers as French.[59] Gerhardt worked on it at a furious pace; by February 1856 he finished the manuscript, and the final fascicle appeared in print by May.[60] In five years, he had written a four-volume work of nearly 4000 printed pages. The book provided a detailed description of the entire science of organic chemistry of his day, along with a defense of his ideas. Ironically, however, all but the last volume utilized the older chemical equivalents, for Gerhardt had for once adopted the pragmatic strategy of presenting the material in the most acceptable possible form.[61]

In his preface, dated June 1853, Gerhardt advertised his avoidance of all molecular speculation as a fruitful methodology that he had been employing for the past ten years (since the *Précis*). He justified his return to the older Liebig–Gmelin equivalents as a way of showing the world how irrational they are. He would trust to time, he said, the consummation of the reform that his colleagues had not yet generally adopted. But at the end of the preface, he hinted that the time may now have arrived, following the work of Williamson, Hofmann, Frankland, Wurtz, and others, to begin building more than just a conventional system of chemical theory.[62]

Finally, Gerhardt had achieved the career success he had been seeking, though unfortunately not in Paris. In 1855, through Thenard's influence, Gerhardt was offered and accepted two professorships at his hometown university, Strasbourg. In April 1856 he was named Corresponding Member of the Académie, a cherished honor. However, in August of the same year, soon after the publication of the final theoretical volume of his *Traité* and just when it appeared Gerhardt might finally be able to enjoy the acclaim and financial security he had long wanted, he died of a sudden fever.

The Conversion of Wurtz

Meanwhile, Wurtz had reached a decisive turning point in his theoretical commitments. On 18 July 1853, Gerhardt read to the Académie what Wurtz subsequently characterized as "one of his most beautiful papers," demonstrating that not only could alkyl radicals substitute more than once in ammonia, but acetyl and other acyl radicals could do the same, producing secondary amides. What was particularly remarkable about these compounds was that the basic property of the ammonia was not simply suppressed, but rather was eliminated entirely; the new substances were *acidic*, even though derived from the "type" of ammonia.[63] Whereas Dumas in the older type theory had declared arrangement (or "constitution") to be the principal determinant of chemical properties, Gerhardt was arguing a thesis of dualistic chemical theory, that the *kinds* of atoms in the molecule determine properties; or, more precisely, Gerhardt had taken the flexible position that both factors were decisive.

Two weeks later, Wurtz provided an elaboration of Gerhardt's idea.[64] He found that his isocyanates react with acetic anhydride to yield a tertiary amide containing one ethyl and two acetyl radicals united by a nitrogen atom, but he expressed the results in terms of the water rather than the ammonia type. He then added, "I note here that the relations that exist between water and substances derived from the water type are expressed in a neater and simpler manner with the aid of the equivalents adopted by M. Gerhardt than by employing the notation ordinarily used."[65] Moreover, he found the opportunity to state that he regarded ether as containing two (rather than one) ethyl radicals substituting the hydrogen of water (rather than adding to oxygen), a hallmark of Laurent's and Gerhardt's new ideas. Hitherto he had been faithful to Liebig's and Dumas' conviction that alcohol was nothing more than a hydrated form of ether. However, although here clearly signaling his new view that major elements of the Laurent–Gerhardt system were superior to the older one, he failed to use the new "equivalents" in this paper.

We possess an eyewitness report of the personal encounters between Wurtz and Gerhardt that summer. A student of Wurtz reminisced:

> There [in Wurtz's laboratory] lively, passionate scientific discussions were held between the two young masters. Gerhardt, of an outspoken, brusque, and violent character, and Wurtz, fiery, but glib and subtle, argued over atoms and molecules. At times Gerhardt, whose conviction was profound, became impatient at the resistance of his contradictor, and I have seen him break in his fingers the piece of chalk with which he had just traced the formulas on the blackboard.[66]

Gerhardt responded formally at the Académie to Wurtz' paper. He declared that it made little difference to him whether the new amides were portrayed as derived from the water or the ammonia type, for all of his formulas were only *synoptic*—general, flexible, and empirical and designed to do no more than summarize the experiments. Wurtz, however (he said), was still under the thrall of Dumas' types. According to that concept, the chemical properties of the derivatives had to be directly related to the properties of the progenitor, so it made no sense to classify acidic derivatives under a basic type. Gerhardt concluded that the issue came down to Wurtz' treatment of formulas as hypothesized *before* experiments and used as a guide in interpreting them, a procedure that once more revealed Dumas' influence.[67]

Two weeks later (on 29 August 1853), Wurtz responded in turn. Yes, their viewpoints regarding the constitution of the amides were essentially the same since they were assuming the same groupings of radicals. Yes, his attitude toward formulas was different from Gerhardt's. His were *constitutional* in orientation, possessing a "true molecular signification," indicating, inter alia, "the arrangement of the simple or compound molecules," by which he meant, in modern vocabulary, the arrangement of atoms and radicals within the molecule. But no, he did not regard types as "purely mechanical and inert" with respect to properties; the type imprints "un cachet particulier" on all its derivatives, hence the inappropriateness of deriving acidic secondary amides from the basic ammonia type.[68] In all of this Wurtz was thereby pleading guilty, and proudly so, to Gerhardt's charge that Wurtz was still loyal to Dumas' type theory. It was only in his first systematic theoretical treatise a decade later that Wurtz finally averred, fully in the spirit of Gerhardt, that for the sake of simplicity, compound amides should be regarded as ammonia derivatives even though they do not resemble amines chemically.[69]

It would seem that it was only during 1853 that Wurtz became aware of Williamson's dramatic experimental work on the ethers and his theoretical development of the water type, for his first mention either of Williamson's name or of the water type occurred in his first paper of this series. Williamson had published his three seminal papers on the water type in 1850 and 1851, but only the first had appeared in a French translation and that was not in one of the principal French chemical journals. Wurtz may have first come across Williamson's work in connection with his duties as foreign correspondent for the *Annales de chimie* (after January 1852), through his visits to England and his friendship with a mainstay of the London chemists, Hofmann, or through his discussions that summer with Gerhardt. Once familiar with the work, however, he quickly assimilated it to his own concerns, as documented in the series of papers just discussed. It must have been in the summer or autumn of 1853 that he wrote Williamson to ask him to prepare a résumé in French of all three articles for publication in the *Annales*. The sixteen-page résumé appeared in the January 1854 issue accompanied by editorial notes by Wurtz explaining the background and circumstances.[70]

Wurtz' high opinion of Williamson's work is revealed in a letter of 18 April 1854.

> I would be delighted, my dear Williamson, to send you occasional summaries of my investigations; I will do this each time I present something to the Institute and even before I have fully completed my researches. I hope thereby to be in the position soon of proving to you my good will!

He then praised Williamson's most recent publication, on sulfonyl chloride, and offered to publish it in the *Annales de chimie*. He added, "I must tell you that your article on etherification has created a sensation" in Paris.[71]

Clearly it had created a sensation with Wurtz, at least. Williamson had developed a strongly realist and constitutionalist program in connection with the water type; his formulas, he stated carefully, depict "an actual image of what we rationally suppose to be the arrangement of constituent atoms in a compound." In his third paper (1851), he had even suggested a mechanistic interpretation of Wurtz' 1849 isocyanate hydrolysis.[72] Wurtz had always signaled a similar constitutionalist orientation, but until this point he had not found the ideas of Laurent and Gerhardt compelling. Wurtz' reorientation dated from his polemic with Gerhardt in August 1853 and his approximately simultaneous discovery of Williamson—not to mention his promotions. But clearly he much preferred Williamson's realist interpretation of the new chemistry to Gerhardt's positivist one.

Wurtz stated this reorientation unambiguously in subsequent papers. In later historical accounts, he repeatedly accorded Williamson principal authorship for the newer theory of types—even after he became an ardent champion of Gerhardt.[73] In 1855 he published two remarkable papers, both of which show the strong (and explicitly acknowledged) influence of Williamson. Inspired by Williamson's development of the multiple water type, Wurtz showed how a triple water model could well account for the reactions of glycerin, with the glyceryl radical forming a bond ("lien") between the three water molecules.[74] Later that year he revealed what history knows as the "Wurtz reaction," using it to make a theoretical argument —explicitly modeled after Williamson's of 1850—to provide what he regarded as "decisive" and "conclusive proof" that the interpretation of the new school regarding the isolated hydrocarbon "radicals" was superior to Kolbe's and Frankland's explanation. At the end of this paper, he provided a brief but reasonably complete summary of the principal points basic to the new chemistry, unequivocally declaring his adherence to it.[75]

Wurtz' public conversion, we have seen, appears to date from the summer of 1853, the same point when he achieved financial and personal security. Was there a connection? Is it not possible that he was in fact converted much earlier, refraining from declaring himself for fear of alienating Dumas and others during a period when he was still trying to achieve the difficult task of establishing a Parisian career? Although this thesis is attractive, there are difficulties with it. If he were a secret partisan of Laurent and Gerhardt in 1851 and 1852, he would not have gone out of his way during those years to posit rational formulas directly contradicting the new chemistry, such as water = HO, alcohol = C^4H^5O,HO, and "sulfobutylate de potasse"[76] =

$$\left.\begin{matrix}\mathrm{KO} \\ \mathrm{C^8H^9O}\end{matrix}\right\} 2\mathrm{SO^3}.$$

It would have been a simple matter to write formulas that were consistent both with Dumas' and with Gerhardt's and Laurent's ideas. Such empiricist formulas were in fact quite prevalent then and would have excited no comment at all. In such a way he could easily have preserved both his orthodoxy and his flexibility for the future revelations of his present convictions.

If Wurtz had been a closet reformer, he also would not have pursued an oral and written polemic with Gerhardt in the summer of 1853, or at least not with the form and content that we see. Finally, we would not see evidence of a *gradual* conversion that summer, such as his retention of elements of Dumas' theory. The evidence suggests that he was won to the new views not exclusively or even principally by patronage, peer or career pressures, but by evidence and argument—even though it is true that he may well have felt freer to be convinced by those arguments after the possible career penalties had been removed.

Another sign of the gradual character of his conversion is the curious fact that he continued to refrain from actually using the new "equivalents;" he adopted them in his papers only from the beginning of 1859. Old habits die hard; moreover, he may still have felt a certain loyalty to his patron Dumas that inhibited him, even after he had achieved financial and career security. Another factor may have been what I would characterize as a certain visceral conservatism in Wurtz' character, a trait that can be discerned at many points in the preceding narrative.

Finally, it should be noted that the atomic weight reform of the 1850s and 1860s differed in character from previous seemingly similar shifts. A number of different conventional systems had been proposed and used during the previous decades: Dalton's, Davy's, and Thomson's early weights; Wollaston's "equivalents"; the French system initiated by Gay-Lussac and Dumas; Berzelius' atomic weights, represented in two principal modifications (before and after 1826); the proposals of Gerhardt and Laurent, from 1842; and finally, the reform championed by Gmelin and Liebig from about 1845, to which most European chemists pledged fealty. The older chemists had lived through most of these battles and must have been growing tired of the constant changes. Every shift made major portions of familiar pedagogy obsolete, and each time the formulas for most substances had to be recast and committed to memory anew. The latest (Gmelin–Liebig) reform was only a few years old, within memory of even the youngest members of the community, when men such as Williamson and Kekulé began to use the Gerhardt–Laurent two-volume formulas. As noted earlier, even Gerhardt himself had used conventional equivalents throughout most of his *Traité*, for he was afraid that otherwise the book would suffer poor sales. Indeed, the guiding thought behind the Gmelin–Liebig reform was that the new equivalents were regarded as the most empirical of all possibilities, possessing explicitly conventional status. Since they were fully independent of any theory, they could be used indefinitely: the final and ultimate atomic weight system. In short, resistance by the community to yet another shift was natural and understandable.[77]

Furthermore, advocates of the newest reform were making an important new claim, namely, an ontological one. Despite the positivistic form of Gerhardt's theories, he had no hesitation in declaring the old formulas pure and simply "false" and his the only correct ones. Laurent and Williamson took a much more strongly and explicitly realist stance toward the new weights and formulas, with Williamson providing compelling chemical arguments for their truth. Kekulé made a similar explicitly ontological claim in 1854: "It is not merely a difference in notation, but rather an actual fact" that the formula for water is H_2O.[78] So one may sympathize with Wurtz in pondering his decision whether to sign on to the new reform: such a move would signal not merely an opinion that the new weights and formulas were more convenient, but that they were *true*. One would want to be very certain before making such a claim. In 1853 he would only go so far as to say that the new formulas were "simpler and neater," not that they were the only correct ones.

Wurtz appears to have become fully convinced of the actual truth of the Gerhardt–Laurent weights and formulas as a result of the work done during an immensely productive period of his life, 1856 to 1858. During these three years, he published eighteen papers on glycol and its many derivatives and on organic acids, diacids, and hydroxyacids.

He was also influenced by Kekulé's and Couper's 1857–1858 publications on structure theory. Kekulé regarded himself in some measure as a disciple of Wurtz, giving Wurtz significant credit for leading ideas in his major paper of 1858. Couper was working in Wurtz' laboratory when he published a substantially similar paper nearly coincidentally with Kekulé. Wurtz was also a major influence on the third principal founder of structure theory, A. M. Butlerov. In the autumn of 1858, Wurtz finally decided to discard equivalents; in all his papers from 1859, he adopted the new atomic weights and formulas using a notational convention devised by Williamson.[79]

It was just before this final shift by Wurtz that he expressed feelings of isolation and insecurity to his old teacher Liebig. At the last election to membership in the Académie, Fremy had been chosen, and Wurtz conceded that he was not undeserving; but even Berthelot had received more votes that he. Balard had voted for Deville, and Dumas had not supported him. This was an unfortunate omen, indicating that "for my future nothing is secure." He felt more appreciated in Germany than in his homeland.[80] It is curious that this pessimistic letter was written in February 1858, just *before* Wurtz decided to shift to the new atomic weights and to carry out an extended public campaign for the new chemistry; it also suggests that Wurtz did not yet feel professionally secure. Nothing could have been better calculated to increase his feelings of insecurity and isolation than to begin such a campaign for views that he knew would be unpopular with important people. That he went right ahead with it suggests that he was convinced that the new chemistry was true and would ultimately prevail. He must have been heartened by the obvious successes of the Gerhardt–Laurent reforms in Germany. Until the victory in France was won, however, he was bound to create a professionally uncomfortable life for himself. He could not have known in 1858 that at his death twenty-six years later the reforms would still not have fully succeeded in his native France.

5

Early Years in Marburg

The Call to Marburg

At mid-century, when he published his review article on theoretical organic chemistry, Kolbe was thirty-two years old and had been seeking to enter an academic career for several years. The usual route to this goal included publication of substantial research, which he had done, but also teaching as Privatdozent or ausserordentlicher Professor, which he had not. Kolbe was not alone in his quest. The decade that had just passed was one of those periods during which almost no movement was made in the German chemical professoriate that would allow opportunities for younger scholars. This was the context in which Hofmann, failing opportunity in Germany, accepted the call to the Royal College of Chemistry in London.

Moreover, this slow job market came at a time when the founders of the later German dominance in chemistry, such as Bunsen, Wöhler, and especially Liebig, were producing substantial numbers of highly qualified and highly motivated students. Complaining of his heavy teaching responsibilities, Wöhler wrote Liebig in 1851:

> You are the one who is really to blame, by raising chemistry to its great reputation through your achievements and writings, that we must slave as we do, since now the whole world wants to do chemistry. But the damage you have inflicted must be borne.

At the same time, Heinrich Debus informed his Doktorvater that he had just accepted a post at Queenwood College. "It would not have been easy," Bunsen replied, "for you to have given me greater joy than by this news . . . I have often thought of you recently, not without concern, in this time of ever increasing overcrowding in our field . . ."[1]

But it was just at this time that the death of N. W. Fischer at Breslau and the nearly simultaneous retirements of Leopold Gmelin at Heidelberg and J. N. Fuchs at Munich provided one of the first great chains of professorial successions in the field. After Fischer's death (on 19 August 1850), the Prussian authorities were able to lure Bunsen from Marburg to Breslau by promising him a new laboratory; as Bunsen learned from his Berliner colleagues, Kolbe would have received the call had he declined.[2] Upon Gmelin's retirement, an inquiry went first to Liebig, who considered the possibility for a time but then declined.[3] The choice fell once more on Bunsen, who, with the promise of another new lab in his pocket, transferred to Heidelberg in 1852. Bunsen's replacement at Breslau was C. J. Löwig, who arrived from the Zurich Technische Hochschule in 1853; Löwig's successor at Zurich was Wöhler's former assistant Georg Staedeler. Fuchs' retirement in Munich led to another call for Liebig, which this time was accepted (1852); Liebig's successor in Giessen was Heinrich Will.

Although he remained in London, Hofmann was involved with many of these negotiations. While Liebig was deliberating over Heidelberg, he wrote Hofmann to say that should he (Liebig) decide to decline, he would recommend Hofmann in his place; were he to accept, he intended to recommend Hofmann as his successor in Giessen. Hofmann replied confidentially that he would indeed like to return to Germany and would be inclined to accept a call to Heidelberg, but not to Giessen.[4] Hofmann nearly did get the nod from Heidelberg, but it appears his conditions may have been too expensive for the Badische authorities.[5] A few months later, Liebig informed Hofmann both that Bunsen had the call and that he had accepted Munich as "an honorable *retraite*" from Giessen. He added, "If you are not called to Giessen, the Darmstädters are asses."[6] He also tried to exert influence for Hofmann in the Breslau deliberations of 1852—an undistinguished city, to be sure, but "the *bridge* to Berlin."[7] Hofmann's unwillingness to be a candidate for either Breslau or Giessen was motivated by personal and political reasons, and probably also by the unappealing prospect of exchanging a lucrative position in a world capital for a less remunerative position in a provincial town.[8]

Kolbe was growing ever more impatient for a call, and his two chief promoters, Wöhler and Bunsen, well knew it.[9] After losing out to Bunsen in the Breslau succession, Kolbe set his sights and hopes on Marburg. In January 1851, Prorektor Nasse inquired of Hofmann whether he would be inclined to accept a call to Marburg. Hofmann's reply was long, gracious, and frank, but it was clearly negative in tone.[10] Bunsen's opinions regarding other possible candidates for his own succession were then officially solicited.

Bunsen was prepared. He had obtained letters of recommendation for Kolbe from Liebig and Wöhler, assembled them together with two letters he had received five years earlier from Berzelius praising Kolbe, and submitted all four along with his own cover letter. Bunsen, Liebig, and Wöhler were all in agreement: with the single exception of Hofmann, who probably would not accept the call, Kolbe was clearly the best choice among the younger chemists. All praised his ability, accomplishments, and reputation. Liebig considered his papers of the late 1840s to be "masterpieces" and noted that only "external circumstances" had conspired to prevent him from accomplishing even more.[11] Six days after Bunsen submitted these letters to the philosophical faculty, the faculty voted to recommend Kolbe. The following month (March 1851) Kolbe duly received the call to Marburg as ordentlicher Professor and Director of the Chemical Institute, and the contract was signed in April.[12]

Settling In

Daniel Hassenpflug, the Kurhessian interior minister, obviously knew that he was in a position of strength in his contract negotiations with Kolbe; his ministry reduced the laboratory budget from 1000 to 700 thalers and offered Kolbe only half of Bunsen's salary, 600 rather than 1200 thalers. This was only three-fourths the salary Wöhler had been given at the Kassel Gewerbeschule twenty years earlier and was also less than Bunsen had accepted for an Extraordinarius position at Marburg in 1839. Desperate as he was, Kolbe had little choice but to accept the poor conditions. He moved from Braunschweig to Marburg on 7 May, took the oath of office on the 9th, and on the following day the Chemical Institute was ceremoniously opened and delivered to him.[13] Two days later, homesick for Braunschweig, he wrote to Vieweg saying that he planned to begin his lectures in a week; the laboratory was in a "desolate" condition, and it would take that long to get it even roughly into shape.[14] Although the lab was very familiar to Kolbe—he had worked there off and on for the past nine years—most of the inventory had been Bunsen's personal property, which he had understandably taken with him to Breslau.

Bunsen's laboratory was the first satisfactory chemical teaching and research facility at the University of Marburg, and it had been operating for ten years at the time of Kolbe's arrival. In 1825 a small university laboratory had displaced a Freemason's lodge in the former refectory and dormitory of the thirteenth-century monastery known as the Deutsches Haus, which to this day stands beside the early Gothic church of St. Elizabeth, on the Lahn River at the foot of the Schlossberg below Marburg Castle. When in 1841 Bunsen was promoted to ordentlicher Professor and was transferred from the medical faculty to the philosophical faculty, he expanded the facility and began instructing a small number of advanced students there. The meager laboratory budget, initially 600 thalers per year, was insufficient even for current expenses, not to mention the capital expense of stocking with apparatus, reagents, and preparations, so that Bunsen had been forced to spend his own money for this purpose. The servant's salary also came from Bunsen's pocket. By mid-century the lab was outmoded and in no way comparable to those at Giessen or Göttingen. In 1849 Tyndall referred to the "scoundrel-like appearance" of the place, while praising its master to the sky. Other university occupants in the Deutsches Haus were pharmacy, zoology, midwifery, and lying-in, as well as the court treasury office. The only thing to be said for the ancient building is that the six-foot-thick walls provided excellent temperature stability for eudiometric measurements![15]

The entire university was in an undernourished and unhealthy condition. Marburg had a long and proud history—it was the first university to be founded as Protestant—but on Kolbe's arrival, there were only 31 full professors and 262 students, making it one of the smallest universities in Germany. Eight years later the enrollment had fallen to 216. Both university and town were narrow-minded and provincial. Frankland was the first Englishman ever to take a Ph.D. in Marburg, and he averred that in Marburg Englishmen were "considered to be more or less mad." His future wife Sophie Fick he guessed to be the only woman in town who could speak English.[16] Frederick Guthrie, who studied with Kolbe from 1854 to 1855, wrote home to a friend soon after his arrival of the "very sleepy hollowism which prevails in this dreary valley of desolation;" he referred to himself, perhaps not entirely tongue in cheek, as "the one civilized inhabitant of Marburg."[17] But the troubles in the university were due to more than provincialism; they were also the result of political problems in Hesse-Kassel, problems that had reached a state of crisis when Kolbe arrived in Marburg.

After 1815 Hesse-Kassel retained its status, though now meaningless, as an electorate of the Holy Roman Empire (hence its synonym Electoral Hesse or *Kurhessen*), under the successive reigns of their Royal Highnesses Wilhelm I, Wilhelm II, and Friedrich Wilhelm. All three were equally arbitrary and avaricious, and inveterate enemies of constitutional and industrial reform. They enforced strict religious orthodoxy, refused to countenance new industries or factories (fearing nuclei of social unrest), engaged in a devastating tariff war with Prussia, and delayed the development of railroads. Ripple effects from the July revolution of 1830 in France provided only a brief respite; from 1832 Friedrich Wilhelm found a henchman in Interior Minister Hassenpflug. Hassenpflug's administration was so hated that puns on his name soon proliferated: "Hessenfluch," "Hessens Hass und Fluch," or Bismarck's priceless coinage "Kassenfluch in Hurhessen."[18] Hassenpflug finally managed to alienate even the elector, who discharged him in 1837.

The revolution of 1848 forced Friedrich Wilhelm to agree to thoroughgoing reforms, but they did not last. With the failure of the revolution and the onset of reaction, Hassenpflug was brought back to lead the administration. He and his sovereign were so heartily despised that this proved difficult. In September 1850 Friedrich Wilhelm dissolved the uncooperative diet and put the country under martial law, but he had no army, as virtually the entire Hessian officer corps promptly resigned. He and Hassenpflug then fled to Frankfurt to seek the aid of the German federal diet, which proved more cooperative. On 1 November, a Bavarian–Austrian force occupied the country. The Prussians, who relied on free transit through Hesse-Kassel for access to its Rhine provinces, were disconcerted, and war between Prussia and Austria threatened. But Prussia was then in no position to fight, and it was forced to yield to the federally sponsored occupation. A new constitution was promulgated in the spring of 1852, one which imposed few restrictions on the elector. The reactionary regime survived until the resurgence of Prussia in the 1860s.[19]

This political atmosphere had made it very easy for Bunsen to decide to leave Marburg, and other colleagues were leaving as well. Kolbe was the first professor to be hired after Hassenpflug's return; since he had never been Privatdozent, many on the faculty assumed he was unqualified and had been selected for political reasons. It was not even clear that a scholar who had never habilitated was legally entitled to a full professorship. Even worse, Kolbe was replacing the university's most distinguished professor. So Kolbe's arrival was greeted with a good deal of mistrust on the part of his colleagues; as he later put it, his position was seriously undermined by partly malicious and partly thoughtless gossip. But by November 1851 he reported to Vieweg that the mistrust had entirely disappeared.[20]

Every semester Kolbe lectured on experimental (inorganic) chemistry six days a week at nine o'clock and taught an eight-hour practicum during the same hours as Bunsen and Wöhler. Advanced Praktikanten worked mornings from eight to one (including Saturdays) and afternoons two to five, in summers until six. He was given one assistant, with a salary of 200 thalers. Kolbe was also expected to teach a Publikum every semester, lecturing once a week for two hours; these courses embraced such topics as stoichiometry, eudiometry, introduction to organic chemistry, chemistry of daily life, and "discussions on chemical subjects." From winter semester 1855/56, he began to alternate organic with inorganic experimental chemistry, teaching the former every winter semester on Monday, Tuesday, Thursday, and Friday at nine.[21]

Six weeks into his first semester Kolbe wrote Vieweg that he was teaching inorganic experimental chemistry and a Publikum on theoretical organic chemistry, and using his partially written manuscript of the Otto–Graham text for the latter.

> I'm having particular fun with this course, and I think I can conclude from the constant interest which my students show for this subject, which is normally so dry since it cannot be illustrated by experiments, that I have pursued the correct path.[22]

He reported that nearly all of the thirty students who began the course were still attending classes, which he thought was unusual. During Kolbe's first three semesters, such positive indications appeared to augur well for the future: he continued to attract reasonable numbers of students, around thirty for his lectures and close twenty Praktikanten. Even better, about a half-dozen of the latter were foreigners.[23]

What soon became clear, however, is that these indications of a healthy institute had more to do with the reputation that Bunsen had established than with anything Kolbe had done. The average numbers in all of Kolbe's classes dropped by about a third during the early 1850s, and the supply of foreigners—the cherished symbol of a reputable program—dried up. Between 1853 and 1861, he had at most one or two Praktikanten from abroad at a time, and for nine semesters none at all. The number of chemistry majors fell precipitously; for instance, between 1857 and 1861 only one new major enrolled. In some semesters he had trouble reaching a total of twenty students in all of his classes combined. This dismal situation was partly caused by the political climate in the town and university, which made study there less attractive; Kolbe also blamed his miserly administration. Equally relevant, however, was Kolbe's own failure to mount a highly visible research program during his first eight years in Marburg. As early as 1853 Kolbe could detect the trend. Not only was enrollment bad, but good faculty were accepting calls elsewhere. Although the university had had prospects for becoming a good place to study science, he wrote Vieweg, he now doubted that this would happen.[24]

In all, Kolbe taught around 240 Praktikanten during his twenty-nine semesters at Marburg, of whom 216 names can be documented. Eighteen of these men took Ph.D.s with Kolbe, and another ten were authors of papers from the lab and/or were postdoctoral guest workers. Thirty-three were foreign: ten Russians, eight Englishmen, five Swiss, three Scots, three Americans, an Irishman, a Bolivian, a New Zealander, and a Dane.[25] Due to the vagaries of surviving data, a reliable analysis of the fields of study of the Praktikanten is possible only for six semesters early in Kolbe's stay (winter semester 1851/52 to winter 1852/53, and winter 1854/55 to winter 1855/56). Aggregate numbers from these six semesters show that no less than sixty-four percent of the lab workers gave "natural sciences" as their university subject, and about sixteen percent indicated "chemistry." Only ten percent indicated "pharmacy;" unlike most other German states at this time, Hesse-Kassel required pharmacists to be trained in a separate university institute for pharmacy. The remaining ten percent of Kolbe's students were in miscellaneous fields, mostly medicine or philosophy. The vast majority of Kolbe's lab students were not studying to become academic chemists, but rather would seek careers in such fields as chemical industry, mineralogy, mining, metallurgy, forestry, agriculture, horticulture, pharmacy, medicine, teaching, and civil service.

Unsystematic data from Kolbe's late Marburg period suggest that these approximate proportions continued, except that as Kolbe's reputation rose, the percentage of chemistry students increased at the expense of the percentage of students of natural sciences. What is striking about all of these numbers is the very high percentage of science students and low numbers of medical students, which contrasts sharply with previous assumptions and calculations.[26] Assuming that there are no serious statistical anomalies here (e.g., large numbers of medical students giving "natural sciences" as their field), we can only conclude that the Marburg medical students preferred to take courses in pharmacy and pharmaceutical chemistry rather than the strongly theoretical organic chemistry of Kolbe. If this is the case, it would help to explain Kolbe's chronically low student numbers and his enmity with the director of the Pharmaceutical Institute, Constantin Zwenger.

Kolbe suffered personally from the paucity of students, as his income (above his inadequate salary, which remained unchanged for ten years) was partly dependent on enrollment. He received 2 louis d'or (about 11 thalers) from each student in his beginner's lab course and 3 louis d'or each from the all-day workers. Starting in winter semester 1861/62, he raised his fee for the all-day practicum to 4 louis d'or

(22 thalers). Each Praktikant also had to pay for his own chemicals. Auditors for his lectures paid 6 thalers for the four-hour and 8 thalers for the six-hour courses. Kolbe often remitted these fees for certain students.[27] The Publika were, of course, always free. As a consequence, Kolbe earned only about 300 thalers per semester from his students in the slower years of the 1850s.[28]

It is possible to arrive at an approximate figure for Kolbe's total income during this period. He was paid by Vieweg for his textbook writing at the rate of 3 louis d'or per sheet (sixteen printed pages);[29] he produced an average of around seven sheets' worth of manuscript per year in the mid-1850s, for about 115 thalers per year. His writing for the *Handwörterbuch* was less prolific and perhaps less highly paid, though figures are lacking. Since very few students were promoting, he earned little from examination fees, and there is no evidence that he did any consulting work. Thus, counting all sources of income, Kolbe earned about 1400 thalers per year, equivalent then to about U.S. $1000 or £200.[30]

It was not much on which to raise a family in a bourgeois manner. In 1859 he told Vieweg that despite working from morning to evening in the laboratory with his students, his family was living practically hand to mouth, making at most a tenth of the income (from student fees) that Will in Giessen enjoyed.[31] On New Years' Eve of 1860 he complained of the impossibility of maintaining a standard of living "consistent with one's class," for which he would need nearly twice the salary. Moreover, although housing in small-town Marburg was slightly less expensive than in the cities, food and other commodities cost just as much as in Frankfurt or Kassel, a result of the arrival of the railroads.[32] Kolbe repeatedly asked Vieweg for advances and loans, until Vieweg finally rebelled; he sent the requested money, but pointedly indicated how much in arrears Kolbe had gone. The severely chastened Kolbe wrote an abject reply, asking for forgiveness and explaining his habitually poor money management, but he did not stop his practice of regularly prying advances from his friend.[33]

As far as salary alone is concerned, the comparisons shown in Table 1 will place Kolbe's situation in perspective.[34] Kolbe's salary was clearly on the low end of the scale for the 1850s, when most of his peers were earning over 1000 thalers, and he may have had the lowest actual income of all. Will's salary was even lower than Kolbe's, but Liebig had built up the Giessen enrollments to the point that Will could make a very good living even with a poor salary. As for Liebig's fabulous new salary in Munich, it was clearly based not only on his eminence but also on his condition that he teach no Praktikum, so that he needed to replace that lost income. From 1865 on, salaries for

Table 1. Salary Comparisons

Chemist	*Place*	*Year*	*Salary* In Each Country's Currency	*Salary* In Kurhessian Thalers
Will	Giessen	1853	600 florins	350
Kolbe	Marburg	1851–1861	600 thalers	600
Kopp	Giessen	1853	1200 florins	700
ord. Professoren	Prussia	1834	850–1200 thalers	850–1200
Bunsen	Marburg	1849–1851	1200 thalers	1200
Thomson	Glasgow	1840s	200 pounds	1350
Kekulé	Ghent	1858–1867	6000 francs	1500
Liebig	Giessen	1841–1852	3200 florins	1850
Horsford	Cambridge	1854	1800 dollars	2500
Hofmann	London	1845	400 pounds	2700
Liebig	Munich	1852	5000 thalers	5000

German Ordinarien in chemistry roughly doubled, to the 1500–4000 thaler range.

Kolbe's problems during the 1850s were partly a product of the attitudes of his administration and partly due to his notable lack of tact or subtlety. For instance, Kolbe's response to his penurious laboratory budget was simply to overspend; on a budget of 700 thalers per year, he was already 800 thalers in the red by 1853. The administration's solution was to amortize the debt by reducing the budget by 150 thalers for each of the following five years. This was "bread and water" in Kolbe's view and not at all in the spirit of how the Chemical Institute had been treated during the Bunsen era. The lab had been virtually empty of apparatus and supplies on his arrival, and the overspending had been necessary as capital expenses; in the future, he assured his ministry, the 700 thaler budget would suffice, but by no means could he survive on 550. His protests fell on deaf ears.

In June 1854 the administration proposed to defray expenses by making each Praktikant pay 10 thalers per semester directly to the university. Kolbe cried foul again, arguing that his students already had to pay anywhere from 5 to 20 thalers or more per semester for their chemicals and supplies, plus his honorarium, making their total costs anywhere from 17 to 40 thalers or more. (The institute's annual budget sufficed only for such continuing expenses as heating and lighting, apparatus, instruments, and the cost of lecuture experiments.) Other universities were much cheaper, Kolbe argued, and he asserted that the move would hurt enrollments even further. A compromise position was adopted: each student would pay a pro rata contribution at the end of the semester to eliminate whatever deficit had accrued. Kolbe complained bitterly, at one point even threatening to close the lab to students.

Finally, in 1860 the budget was raised modestly to 800 thalers, but Kolbe was to be held personally responsible for any deficits. He now argued that 800 thalers was no longer sufficient. Chemistry was unfairly treated in comparison to medicine and horticulture; the budget for anatomy alone, he said, was 1400 thalers. By 1862, Kolbe's rising star gave him leverage; he overspent every year, daring the administration with impunity to follow through on its threats. After accepting the call to Leipzig in 1865, Kolbe tried to get back at his enemies in the university by publicizing his disputes, but his request to publish his complaints was denied, and his internal report was buried in the university archives.[35]

In fact, Kolbe's assertions were often exaggerated or inaccurate. His budget for the lab was not out of line with what other German universities were providing at the time. As for fees charged to Praktikanten, Göttingen may have been somewhat cheaper, but Giessen was a bit more expensive. The price that students paid to study in Marburg was probably fairly typical of the day. Furthermore, Germany was in general financially very attractive compared to the relatively more expensive universities in England and America.[36]

The real problem was that Kolbe's models for the function of an academic laboratory, namely, as serving both scientific pedagogy and scientific research and as an integral part of the university, were Liebig's lab in Giessen and Wöhler's in Göttingen. Liebig had gradually succeeded in persuading the authorities of Hesse-Darmstadt to support his lab adequately by means of an annual budget; Wöhler had arrived at what was then probably the only reasonably endowed lab in Germany. This would prove to be the model for the future, a model that was already in evidence by the 1840s. Bunsen, in contrast, had at least while in Marburg followed the older pattern of stocking the lab out of his own pocket and using it virtually as private property, even though he had a budget for apparatus and glassware. This approach was a residue of the era when many chemistry professors ran private pharmaceutical institutes on the side, as Liebig had between 1826 and 1835.[37] Since Hassenpflug may have shared this assumption, it is little wonder that his ministry did not understand or appreciate Kolbe's constant requests.

Similarly, Kolbe had adopted Liebig's attitude toward the place of the study of chemistry in higher education. Natural science, it was thought, was a vital element of liberal education, highly valuable even for future lawyers, ministers, and civil servants, in that one learns thereby to think clearly and logically. A thorough grounding in scientific theory was a necessary foundation not just for future academics but also for future practical chemists, since applications come not from random trial and error but from deep understanding. Finally, laboratory experience was vital to learn the science properly at all. These attitudes were in conflict with the older image of chemistry as mere industrial soap boiling and drug compounding, a view still held by many of Kolbe's colleagues and superiors. Kolbe particularly resented the middle-level committees, composed of faculty colleagues largely from law and the humanities, that oversaw their charges with a combination of ignorance and malice and treated him like a schoolboy who had exceeded his allowance.

But Kolbe made a pest of himself over small issues as well as large ones, and his truculence gradually isolated him from his faculty colleagues as well as from the administration. He was hot-tempered, self-righteous, and more than a little arrogant. His violent disagreements with the mathematician Friedrich Stegmann drew an official rebuke by the prorector, as did his purchase with university funds of an expensive bust of Liebig for his lecture room. The liberal theologian Eduard Zeller, with whom Kolbe was good friends in 1853, later derailed a possible call to Tübingen when he apparently described Kolbe to the Württemburg authorities as a rude and insufferable man. Kolbe's relationship with the director of the Pharmaceutical Institute, Constantin Zwenger, was worst of all; he repeatedly accused Zwenger of undermining his position and poisoning his friendships. The details of these entanglements, which became heated from the summer of 1857 on, cannot be discerned from surviving documents, and it is unclear to what extent Kolbe was personally to blame for the problems.[38] But despite these disagreements, he seems to have had at least a few good collegial friends.[39]

Home Life

Kolbe's domestic life must have served as a psychological anchor during these troubled times. Already by September 1852 he had successfully courted his future wife, Charlotte von Bardeleben. Then twenty years old, Charlotte was the youngest daughter of General-Major Wilhelm von Bardeleben, a fifty-five-year-old officer in the Hessian army, a decorated veteran of the Napoleonic wars, and then commander of the Marburg garrison. Her mother was Johanna née

Holzförster, the daughter of a physician.[40] The Bardelebens appear to have been reasonably prosperous, though by no means wealthy, and provided a small dowry.[41] After an initially difficult time with his future in-laws—they had had little time to get to know him before the engagement—Kolbe was soon well reconciled with the family.[42] It was probably at this time that Kolbe converted from the Lutheran to the Reformed Church. The wedding took place on 10 May 1853, after which the couple spent three weeks on a honeymoon at the Vieweg home in Braunschweig.[43] Kolbe's letters to friends exude the purest bliss in connection with his marriage, both at this time and ever after. Shortly before the wedding he wrote Vieweg praising his future wife's domestic and economic virtues and looking forward to making good use of the new "peace and spare time" that "a married man" naturally enjoys. He later admitted to having been a very bad housekeeper before his marriage.[44]

Charlotte Kolbe was devoted to her husband and children, as well as to her parents, both of whom died in 1859. Not much detail about her is revealed in Kolbe's correspondence. The only direct testimony about her was from her second cousin (and eventual son-in-law), Ernst von Meyer, who described her as a woman of rare kindheartedness, possessing a true philosophy of life and an "exquisite inclination toward genuine cheerfulness," in sum, a "figure of light."[45] Her health was not robust. She escaped a typhus epidemic in the winter of 1855–56, but nearly died when it struck again in the summer of 1861; it took her almost a year to fully recover. A nearly fatal uterine infection in the spring of 1865 similarly required several months' convalescence.[46] She also suffered periodically from serious respiratory and liver complaints.

Their first child, a daughter born in early May 1854, died three days later of a spinal hemorrhage. Kolbe, and more especially his wife, were heartbroken.[47] Guthrie gives a brief but engaging portrait of their relationship as viewed by Kolbe's students that summer: "He (Kolbe) is terribly wrapped up in his wife and comes into the lab. with as serene and demure a face as if we hadn't seen her kiss him a moment before at the window."[48] On 27 September 1855, Kolbe's thirty-seventh birthday, a healthy boy was born whom they named Carl, after his paternal grandfather. Both grandfathers as well as Vieweg came to Marburg for the christening in October.[49] The next child, Johanna, named after her maternal grandmother, was born on 18 February 1857.[50] Carl was eventually to study chemistry with his father, and Johanna was to marry one of her father's students, Ernst von Meyer. Three of the remaining five of Charlotte's pregnancies resulted in miscarriages, one of which resulted in the serious infection mentioned above. Two more Kolbe daughters, Maria and Elisabeth, were born in June 1860 and January 1868, respectively.[51]

According to the testimony cited below, Kolbe and his wife socialized actively early in their marriage (Frankland's reminiscences show how many opportunities there were for socials, parties, balls, and such in Marburg). However, after Johanna was born they withdrew into a very small circle. They both hated parties, Kolbe once confessed to Vieweg.[52] In response to Vieweg's surmise in the fall of 1859 that Kolbe's social life was interfering with his timely production of manuscript, Kolbe answered

> For two years I have been living like a real hermit, and have been (not to brag about myself) quite industrious. To be sure, at the beginning of my marriage I had a rather heavy social schedule, but for a long time it has been limited to a few families and solid men, who really provide intellectual stimulation, even if not in my field. From personal experience I know that there is a danger in the circumstance that in such a small town as Marburg so many people know you, and must know you; I need only open my eyes to see how many scholars are thereby destroyed. I have long since vanquished the power of these relationships, and have shaken off all obligatory social intercourse. My wife and I both feel very happy doing so.[53]

This is the most extensive description of their restricted social circle, but there are other similar passages in Kolbe's correspondence after 1856.[54] The Kolbes only resumed socializing actively around 1869.

Like his wife, Kolbe did not enjoy vigorous health. In 1841 he suffered from jaundice, and Frankland reported that in London he was frequently ill, complained of a bad heart, and asserted that he would not live long.[55] He was subject to frequent bouts of bad colds and influenza, each usually resulting in a week's confinement in bed; for instance, in one fifteen-month period (December 1853 to March 1855) he reported five such illnesses to Vieweg.[56] His letters reveal frequent worry about these illnesses, although one also contains the observation that he is a "fatalist" regarding serious diseases such as typhus.[57] In two letters from the 1870s he expressed the hope of living at least to the mid-1890s; he fell a decade short of this goal.[58]

A period of very serious health problems began early in 1857. On 24 March he wrote Vieweg

> What must you think of me, that you haven't heard anything at all from me for so long.—I don't know what it is with me, I am not my old self any more. Sick since January, unable to work, for the last three weeks confined to bed; only a couple of days ago was I able to leave my bed,

> and still feel very weak, so that writing really exhausts me. This time a gastric catarrhal fever afflicted me; the doctor thinks it is very much connected with my rheumatism of last summer. The intolerable catarrh has now eased somewhat, so that now at least I have sighted land.[59]

But it must have been within a day or two after writing this that he was struck with a high fever, which left him in bed for weeks with excruciating rheumatism in his joints. In the middle of April he was still strongly affected, barely able to write. A second attack in late April was even worse, confining him to bed for two weeks with such violent pain he could scarcely move a muscle; a third attack in May laid him up another week.[60]

There followed a long period of slow convalescence, with thin water gruels, sweat cures, and Spanish fly plasters. He was able to begin his lectures again by early June, but delivered them all summer in his slippers due to the pain and found them exhausting. A half-hour's work at his desk often put him into such a state of perspiration that a change of clothes was necessary.[61] A number of small relapses occurred that summer, but by the end of August he could spend a couple hours at a time writing. In autumn he continued to improve; however, attacks of influenza in December 1857 and February 1858 further complicated matters. In May 1858 he was struck once more with the same rheumatic fever. This time he decided to try a mineral-bath cure to alleviate the rheumatic symptoms, although he was skeptical of its medical efficacy. To his surprise, two weeks into a four-week stay in Wiesbaden he felt enormously improved; the only annoying aspect was the enforced idleness that his physician had imposed. After his return to Marburg he reported feeling "newly reborn."[62]

Nonetheless, he was never thereafter able to free himself fully from the effects of rheumatism. After a severe recurrence in the summer of 1861 he decided to try Bad Nauheim, closer to Marburg and cheaper than Wiesbaden.[63] He continued to return to Nauheim nearly every year, even after his transfer to Leipzig, though he gradually transferred his allegiance to more distant destinations such as Marienbad in Bohemia, Sassnitz on the Baltic Sea in Prussia, and Gersau on the Lake of Lucerne in Switzerland. As his finances continued to improve while his health slowly got worse, he would regularly visit such resorts twice or more in the same year.

There were psychological and emotional symptoms of his illnesses, as well; conversely, as Kolbe himself conjectured, his physical symptoms may have been exacerbated by his emotional problems. In describing to Vieweg in October 1859 Zwenger's past "Machinationen" against him, Kolbe said the dispute made him long to leave Marburg but that it had not upset his work. His illness, however, had been more problematical:

> To be sure, I was earlier disturbed in so far as my illness was the sole consequence of my intense emotional agitation [die alleinige Folge der heftigen Gemüthsbewegungen] and of my anger over this stroke of fate. But I have long since conquered this.[64]

Some passages in letters from the following year also suggest emotional distress. In April the "unremitting east wind" was "getting on everyone's nerves"; his enemies were whispering in Liebig's ear against him, he conjectured both to Vieweg and to Liebig himself.[65] Zeller had conspired against him, he thought, to deny him the Tübingen call (this seems to have been true).[66] In October he complained of an "irritated-nervous state" (nervös-gereizte Stimmung) and a bad mood or depression (Verstimmung) lasting for weeks, which made him incapable of any serious work. He blamed this mood on the Hessian ministry, which made him personally liable for deficits in his laboratory budget that year. This decree from the administration was a "bolt from the blue" that had him "beside himself" (ausser Fassung) for weeks. His only hope, he said, was that the "bigoted reactionary" government would not last much longer.[67]

In July 1861 he again complained of a "deep depression" (tiefe Verstimmung), lasting "some months," which made him unable to do any serious mental work, connected with physical exhaustion from lectures and laboratory supervision. He thought this exhaustion had led to symptoms of his old rheumatism that were beginning to reappear. "I myself believe," he added, "that my current indisposition is not purely physical, but is also connected with severe depression."[68] In March 1862 he was again struck with physical exhaustion, which for two weeks prevented him from working; he thought it may have been an effect of the bad laboratory air.[69] Over the years, he complained periodically of "nervous exhaustion" and often blamed the lab work that forced him continually to breathe noxious vapors.[70]

Kolbe's writings after 1870 gave many contemporary and later observers the idea that he may have been mentally ill, suffering symptoms of paranoia, delusions of grandeur, and inappropriate rage. Whether this behavior was connected to his health problems of the late 1850s and early 1860s remains uncertain. What seems beyond question, though, is that beginning in 1857 Kolbe had a series of severe illnesses, a number of rancorous collegial disputes, and a sharply reduced social life. The death of his mother in 1856 and the birth of Johanna early in 1857 may have been psychological factors in this change. Kolbe drew emotional sustenance from his family, his students, and a small circle of correspondents including above all Vieweg, but also Bunsen, Liebig, Varrentrapp, and a few others. This pattern did not appreciably change for the rest of his life.

Classroom and Laboratory

Kolbe's teaching was from the first very successful. His letters suggest that he put a great deal of energy and thought into his lectures and practica, and the students seem to have responded very favorably. From as early as November 1851, Kolbe often spoke with pride and confidence of his students. Given one assistantship by the university, Kolbe filled the position during his first few years in Marburg with undistinguished chemists; later followed the much more capable Rudolf Schmitt (1857–1861) and Eduard Lautemann (1861–1865). Schmitt subsequently made a career at the Dresden Polytechnikum. Praktikanten and guest workers from the Marburg period included Adolph Claus, Edmund Drechsel, Georg Fischer, Wilhelm Gerland, Carl Graebe, Peter Griess, Ludwig Mond, Carl Ulrich, Jacob Volhard, and Julius Ziegler; foreign students included Frederick Guthrie, Alexander Crum Brown, E. T. Chapman, N. A. Menshutkin, Maxwell Simpson, Francis Wrightson, and A. M. Zaitsev. The paucity of students in the early and mid-1850s was a source of frustration for Kolbe, who had more ideas for research projects than he had skilled hands available to him.[71]

Guthrie gave his friend Henry Roscoe an intriguing picture of what it was like working with Kolbe in 1854 (he noted that he was the only organiker in the lab at the time and was working on electrolysis experiments):

> I am obliged somewhat to take the law into my own hands, and stick to one thing 'till I've got a distinct result or nonresult, for my much honoured professor comes almost every morning with "Herr G. es ist mir eingefallen dass Sie so u. so ein Versuch machen können," etc and the most provoking thing is that the "Versuch" is often so tempting that it requires no small amount of moral courage to keep in the straight and narrow path.[72]

In 1855 Kolbe described Guthrie as a "well-educated man . . . very industrious . . . and extremely nice," with some fine chemical research already to his credit. Guthrie went on to become Frankland's assistant at Owens College, eventually gaining a professorship at what became Imperial College in London. He became a physicist and had a distinguished teaching career.[73]

In summer semester 1862, Carl Graebe also studied with Kolbe and described his impressions and his schedule in letters to his parents:

> I like Professor Kolbe very much indeed. Only the laboratory equipment is unfortunately deficient. . . . I rise at six, read a little, and about seven-thirty walk to the laboratory, where I remain until dinnertime, with the exception of two days, when I ride from 11 to 12. In the afternoon I am in the laboratory from 2 till 5 or more often 6. Then I bathe and take a walk; by around 9 I am usually home, so that I use my lamp more than I expected. Kolbe pleases me ever better; he gives a magnificent lecture.[74]

Jacob Volhard, a Giessen Ph.D. who worked with Kolbe that same summer, described the lab as an ordinary-sized, low-ceilinged room (it was about 25 by 35 feet), crammed with twelve to fifteen workers, each with his own charcoal fire as well as alcohol lamp. It was intolerably hot, crowded, and confined, usually filled with noxious vapors and possessing no ventilation, running water, or gas. And yet, Volhard added, he and all his former lab mates looked back on their Marburg years with nothing but fondness.[75]

Henry Armstrong, who studied with Kolbe in Leipzig in 1867–1870,[76] regarded him as "a true chemist if there ever was one" and never thought of him "in terms other than those of admiration and affection," although Armstrong did concede that he later became "peculiar" and "little short of a monomaniac." He was "almost bourgeois in appearance, a typical professor of the old school, though with a wonderful sparkle of intelligence in his eyes and a most endearing personality when you learnt to know him—not the ogre he has been painted." Armstrong also characterized him as "a puritan and an enemy of insincerity," honest and forthright in the greatest degree, and subsequently much misunderstood and unjustly neglected.[77] Ernst von Meyer emphasized that it was necessary to be in Kolbe's inner circle of friends in order to see his warmth, amiability, and humor.

> In personal intercourse with him the sharpness typical of his critical writings was entirely absent. A social man, inclined toward merry jokes, fond of witty rejoinders and word plays, he was entirely different from the impressions of those who viewed him from afar.

The violence of his opinions, Meyer concluded, was due entirely to the fact that he was "a fanatic for the truth" and always insisted on speaking his mind no matter whose toes he was stepping on.[78] Ost, also a student, relative, and obituarist, chose almost the same words: "The image of Kolbe impressed on his immediate surroundings was substantially different from the appearance from afar. Kolbe was a complete gentleman, and moreover of exceptional amiability and benevolence."[79]

Kolbe was an enthusiastic follower of Liebig's style of instruction, but he faulted Liebig for not going far enough and not remaining completely consistent with his guiding ideas. He agreed with Liebig that chemistry must be taught the same to all students; that it must be theoretically based but also firmly grounded in experimental practice; that the overall goal must be to teach the student to "think chemically"; that the laboratory must be as much a pedagogical as a research institution; and that advanced students must be considered vital scholarly collaborators. Consistent with their *Lernfreiheit*, students often took a variety of courses in their first two or three semesters before specializing exclusively in chemistry.[80]

Kolbe thought it best to have beginners complete the lecture course in experimental chemistry before entering the lab, although he never insisted on this in practice.[81] "The heart of chemical instruction," he wrote, "is not, as in many other disciplines, in the lectures, but rather in practical work in the laboratory." Learning chemistry solely from lectures and books is like teaching a blind man about colors or studying swimming on dry land, he thought. From books, one can learn one's chemical ABCs, even to spell and to write, but to speak, think, or philosophize chemically requires immersion in *phenomena*. Such emphasis, common to many others in addition to Liebig and Kolbe in the nineteenth century,[82] is derived in part from a phenomenalist and sense–intuitionist perspective, as discussed in chapter 1. It is also derived from the accurate perception that scientific work is partly craft-based, tightly packed with all manner of tacit knowledge that can only be learned from exemplars and practice.[83]

We know more about the details of Kolbe's practica than about his lectures, for which we possess no documentation. The first semester in lab was devoted to qualitative analysis. Starting with nonmetals, inorganic acids and their salts, the students progressed through some 110 analyses of compounds and mixtures, concluding with complex and rare metallic compounds. The laboratory course met four days a week, two hours a day, and was overseen jointly by Kolbe and his assistant. Unlike in Giessen, however, no lab manual was used. Instead, close personal supervision demonstrated to each Praktikant the methods of careful observing, recording, and understanding of every reaction, by which he was gradually led essentially to write his own manual. Kolbe and his assistant would carefully examine and critique each student's annotations, allowing nothing to remain unclear. The goal was total immersion in phenomena, so that the student truly absorbed into mind and soul the essence of every observation, event, and operation. Like

Bunsen (and not like Liebig), Kolbe spent much time with rank beginners, letting his assistants spend more of their time with more advanced students.

Kolbe thought this combination of intensive personal supervision and self-instruction was twice as efficient as the Giessen pattern, so that by the middle of the second semester of lab work, he regarded most students as ready for the next stage. After about one semester of quantitative gravimetric analysis, the student was put to work for a few weeks making organic preparations, which served the triple purpose of introducing the student to more complicated operations and apparatus, acquainting him at least with parts of organic chemistry, and providing the more advanced students, assistant, and director with starting materials for their research. Finally, the student was taught organic elemental analysis.

At this point, typically after three semesters, the student was assigned the task of studying a classic chemical paper chosen by the professor and was directed to repeat the work or to make a small variation in it. Kolbe thought that this was the best way to induct students into proper scientific method and writing as practiced by the masters of the field. Kolbe would then suggest a small original project, often motivated by his current theoretical interests, that the student could work through essentially independently. From this point on, the student was invited to make his own way through the scientific thickets. Depending on a variety of circumstances, Kolbe might or might not continue closely to direct the work of the student. But in any case, having passed through a two-year chemical novitiate, the student was regarded and treated as an independent worker and was invited to spend about forty-five hours per week in the lab rather than eight. Any publications from the student's work appeared under the student's own name, or as a collaboration if warranted. Early in Kolbe's Marburg period, about a third of the Praktikanten—a small handful at best—were advanced students. By the end of his stay, when he had become famous, more like two-thirds to three-fourths of the workers (ten to twenty) were full-timers. A number arrived in Marburg already with advanced training, even Ph.D.s.

Kolbe thought that it was vital that every chemistry student go through this same apprenticeship, whether they were planning to be "Chemiker vom Fach" or were studying an ancillary discipline such as agriculture, pharmacy, metallurgy, or chemical engineering. Those who followed his course had learned to think chemically, and those who could think like chemists, he believed, would have the mental flexibility to solve efficiently a wide variety of unpredictable problems encountered in their future professions, while those who had followed the prevailing "cookbook" pedagogy would not have the proper independent and scientific training to succeed. Such sentiments had already been expressed by Liebig in his critiques of Prussian and Austrian chemistry, and Liebig's influence is evident in much of the description just concluded. In one important respect Kolbe had gone beyond Liebig, namely, in his emphasis on personal instruction and on having the Praktikant gradually write his own laboratory manual on the basis of his own personal observations. It is probable that the idea for this pedagogical innovation came from Bunsen, who always preferred that students make their own observations rather than look them up in a book.

Kolbe was proud of the excellent esprit de corps and exceptional diligence of his students. This seems not to have been an idle boast, if his Marburg lab was anything like that in Leipzig. Concerning the late 1860s, Henry Armstrong reminisced

> Kolbe's laboratory, in those days, afforded wonderful opportunities. About a dozen of us were doing advanced work, in preparation for the Degree—seeking independence. Each had his *Arbeit*—his definite problem—in view, as his chief aim in life: we were all proud of being called on to show that we could do something. This was the distinctive feature of the German system. At most two or three had themes from the Professor—the rest were carrying out ideas of their own; the work was, therefore, varied. Whatever suggestion we made to Kolbe, he never discouraged us; his habit was to grasp the lapels of his coat, then to reply: "Try it, try it." We disputed with him constantly before the blackboard, often for hours together, nearly always taking exception to his theoretical views—but without his being offended. And we constantly compared notes together. Each of us, therefore, was interested in the solution of a whole series of problems.[84]

Graebe's and Volhard's reports from 1862 (cited previously) are consistent with this picture. Guthrie's report from 1854 suggests stronger guidance, even a degree of control, from Kolbe, but it is not inconsistent with these other testimonies. Similarly, V. V. Markovnikov, who studied with Kolbe in Leipzig in 1866–1867, wrote to A. M. Butlerov, then professor at Kazan,

> Kolbe was especially attentive to all the workers, and, contrary to expectations, he did not behave like a commanding general at all. On the contrary, he was very glad to discuss and argue, and I have already succeeded several times in locking horns with him. He himself often approached and asked me how one should understand formulas which were not written by his method.[85]

Alexander Crum Brown also noted that as a student of Kolbe's in 1862 he was "always able to explain any theoretical views of mine to him by first translating them. . . into his language, but I am quite sure he would fail to recognize his own ideas if translated into our language."[86]

Kolbe emphasized that "in my laboratory I allow everyone who maintains his chemical convictions—which I respect—to continue working in his own direction, and am pleased whenever he obtains the expected results."[87] When he finally acquired a large laboratory in 1868, he had no trouble filling it to overflowing with admiring students. However intolerant Kolbe may have been to the differing views of colleagues at other universities, he was always indulgent and broad-minded with his own students.

Textbook Author

Kolbe's poor research productivity during his early years in Marburg had several causes: the paucity of advanced students; the poor condition of the lab and its inadequate budget; the labor of preparing lectures for the first time; and the various difficulties, both personal and health-related, that he and his wife suffered. In addition, his work for Vieweg on the *Handwörterbuch der reinen und angewandten Chemie* continued through his early years in Marburg. This work provided some much-needed income, but consumed a great deal of his valuable time in an activity that was not very useful for his scholarly reputation. With the press of other business in the first six years of his professorship, Kolbe was increasingly dilatory with his editorial tasks. In the summer of 1857, Vieweg suggested—partly at the suggestion of Liebig, who was then unhappy with Kolbe's polemics and one-sided approach[88]—that he turn over many of his responsibilities to Hermann von Fehling, professor in Stuttgart. Kolbe agreed.[89] Kolbe had completed the editing of volumes three through six (letters F through R) and a supplement volume. Fehling took over from volume seven and completed the entire work in nine volumes by 1864; he also immediately began a second edition. In 1871, a *Neues Handwörterbuch der Chemie*, which was published until well into the twentieth century, continued the work. Kolbe was very distressed to see these later editions adopt structure-theoretical ideas.

It was not difficult for Vieweg to persuade Kolbe to give up control over his beloved *Handwörterbuch*, partly because a new brainchild had been displacing it: his textbook of organic chemistry. Recall that at the end of 1847 he had accepted the assignment to write the organic part of the third Graham–Otto edition, but five years later he had apparently not been able to proceed very far in his task. However, in a letter to Vieweg of 24 January 1853 he reported being hard at work and was optimistic at making sufficient progress to allow the printing to start by the first of May.

There are, of course, many reasons to write textbooks, and Kolbe was doubtless sensitive to the various opportunities. Organic chemistry around 1850 was obviously theoretically labile, and no full modern texts existed. This meant that the first to enter the field with a satisfactory treatment would sell well. The competitor texts, which were some of the first textbooks of organic chemistry, were either out of date (Liebig and Löwig) or too short (Schlossberger and Strecker); he did not then know about Gerhardt's *Traité*, which was on the verge of appearing.[90] Perhaps even more important, the lability of the field meant that the first successful textbook author had a real chance to establish his theoretical views as the pedagogical standard for the discipline, in the manner that Berzelius had done a quarter century earlier.

In addition to these pecuniary and programmatic motivations, a third factor may have been present. Kolbe had not yet worked out all of the details of his incipient system; the inconsistencies that we observed in his 1850 theoretical paper indicate as much. Writing a textbook can provide an author with the opportunity to view the entire discipline from the ground up, and working through all the details can materially assist in developing a finely articulated theoretical system. Whether or not Kolbe consciously intended his endeavor to serve this function, we will see that in writing the first volume, he followed a meandering journey through a theoretically rich chemical landscape, finally ending up in a position that gave him a powerful tool for exploring unknown scientific territory.

In January 1853 Kolbe projected his *Lehrbuch* at two volumes totaling about 1300 pages, and he wanted to produce them quickly, presumably within a couple of years.[91] His estimates of size and schedule proved hopelessly unrealistic—it would be a quarter century and over 3300 pages of text in four volumes before the first edition was completed. In February he wrote with satisfaction: "The Organic is flourishing and growing. There is no lack of heat and fire from my side." But in March he reported difficulties. He was having to rewrite various sections of the manuscript begun in Braunschweig due to various newly published theories based on recent discoveries. He thought that the two years of teaching he had completed in Marburg had been necessary preparation for this task, and his own ideas had thereby evolved. May came and went, as did the entire summer, without Vieweg receiving manuscript to begin the printing, though Kolbe continued to emphasize the need for his work to appear as soon as possible.[92]

Finally, in the middle of October Kolbe was able to send manuscript for the first two or three sheets, though not enough for the first fascicle. (Like many publishers of his day, Vieweg put out longer works in parts or fascicles, often consisting of six printer's sheets or ninety-six pages each, which were retailed separately and eventually bound by booksellers or by customers.) By then, unfortunately, the field was beginning to get crowded. In September Vieweg had sent Kolbe the first published fascicle of a new organic chemistry, Theodor Gerding's edition of William Gregory's text.[93] Kolbe was able to calm his and Vieweg's fears about this book—it was outdated and poorly translated, and would not sell.

> Not from that side is competition for my Organic to be feared, but rather from the German edition of Gerhardt's new chemistry. Probably in order to recommend himself in Germany, since as I understand he is seeking a position at a German university, partly also no doubt for the money, Gerhardt has with this work undertaken a continuation of Berzelius' chemistry (what Berzelius wrote about organic chemistry in his last edition is null and void), and, with incomprehensible—I would call it Jewish—self-abnegation, has disregarded his own views as defended in his chemical papers, in preference to the diametrically contrary principles of Berzelius, and has cozied up to Berzelius' thoughts and conceptions, as far as this is still possible now.[94]

Kolbe was principally referring to the fact that in his *Traité*, Gerhardt had used the older conventional equivalents, though in his preface still professing loyalty to his new atomic weights. To Kolbe, this was an insidious and cunning trick that just might work, by blinding and seducing readers.

Even worse from the competitive angle, Gerhardt's *Traité* appeared, to Kolbe's surprise, to be "a good book." Hence, it now became vital to get the first fascicle of his text out as soon as possible. Because his duties as editor of the *Handwörterbuch* were still quite burdensome, he asked for a reprieve until he could finish the manuscript for the first fascicle. Moreover, the writing on the textbook was no longer easy for Kolbe. "I've had a hard time with it," he wrote Vieweg. "You see, research by several chemists of the very greatest scientific importance, especially by Gerhardt and Williamson, has appeared in recent months, which threatens to destroy the entire existing structure of currently accepted doctrines in chemistry." Kolbe reported having spent most of August and September studying these papers and devising responses; he had had to revise much of the already written manuscript in light of the new work. Otto had expressed a fear that Kolbe's volume would end up being too Gerhardtian to fit properly with the Berzelian inorganic portion of the Graham–Otto project, but Kolbe reassured Vieweg on that point.

> The result of my multi-week, if I might call it philosophical investigation at my writing desk, is this, that I believe I have found the strongest proofs against Gerhardt's theories, though many parts of the latter must essentially influence the writing of an organic chemistry. Accordingly, I hope that Otto will be satisfied with my views and you with my work.[95]

Kolbe continued to work away at the all-important introductory portion. He revealed to Vieweg one source of his distress at Gerhardt's pretensions to be continuing Berzelius when he confessed that he, too, wished to be viewed in precisely that sense. "Don't make fun of the juxtaposition," he wrote, "when I say that it is unfortunate that Berzelius is no longer here to stand up to the extravagances of foreigners in theoretical chemistry; that is why I have undertaken to become a critic."[96] On 1 March 1854, Kolbe reported additional progress in his struggle, namely, his recent formulation of a "fundamental proof of the untenability of the newest French–English hypotheses on the constitution of the most important organic compounds." He said he was fortunate to have had time to incorporate it into the existing manuscript. He and Vieweg agreed on a double fascicle of 11 sheets (176 pages) as the first published installment; it appeared in bookstores in June 1854.[97]

In early April he sent Vieweg a draft of the foreword. He there declared his intention of following the Berzelian model, disclaimed any connection (other than the rubric under which he wrote) with Graham's organic chemistry, then concluded with some polemics. With a few significant exceptions, he wrote, it had been the French chemists who, on the basis of a few isolated observations and no true scientific foundation, had propounded flimsy hypotheses with an air of infallibility, all to destroy the existing theoretical structure of chemistry. By contrast, "I declare myself as an adherent of the conservative party of our science," which demands critical examination of new theories before allowing them entrance. He ended with a sarcastic taunt:

> A certain self-discipline is necessary in the writing of a textbook not to give preference to hypotheses that one believes to be probably correct, but which still require further confirmation. I have striven not to fall into this error; but likewise I have not been able to bring myself, as in a recent work, to place in the foreground views which are considered false even by the author himself.[98]

Vieweg rightly pointed out that this reproach was by no means clearly aimed, and he did not know to whom it referred. It could even be taken to refer to Liebig himself, since within the last decade he, too, had altered his atomic weights and retreated from various theoretical positions!

Kolbe responded that he intended to single out Gerhardt, who he thought had adopted Berzelian coloration only to make money in Germany, thereby committing "perfidy against himself and risking his entire scientific reputation."[99] He still hesitated to name Gerhardt openly for fear of being seen as slamming a competitor. But neither Vieweg nor Bunsen had objections to that idea, so he eventually decided to insert Gerhardt's name into the last sentence just cited. Upon further thought, he decided he need not fear competition, for the book proved not to be as Berzelian as he had feared and would not fool the German public.[100]

As Kolbe continued to receive installments of the German edition of Gerhardt's book, his estimation of it, already surprisingly positive, continued to rise. His favorable opinion is surprising not only because Gerhardt was a competitor whose theories he considered incorrect, but because the man was also both French and Jewish. We have already noted a few Francophobic slurs published by Kolbe as early as 1848 and an unfavorable reference to Gerhardt's religious background to a private correspondent. After 1870, Kolbe's prejudices intensified greatly and were expressed in an increasingly public fashion; his tirades eventually affected the course of his career. We will return to this subject later.

As far as Gerhardt's *Traité* was concerned in the context of 1854, Kolbe still despised the theories, the style, and the "truly French" organization ("pretentious and absurd"), but he admired the detail and care with which the work was written, and Gerhardt's enormous productivity.

> To be sure I cannot work half as fast as Gerhardt, whose capacity for work I cannot believe. . . . I could imagine an author capable of working with such steam power, if the work were hasty and superficial. But one cannot accuse Gerhardt of this fault. . . . Gerhardt's citations are strikingly accurate, as in general the book is written with exceptional diligence.[101]

From the start, he found much there that he could use for his own text.[102] One letter to Vieweg suggests that he followed Gerhardt's treatment very closely indeed; he quickly added that when doing so, he was careful always to check Gerhardt's literature references, for he did not want to experience the "Spectakel" Gerhardt would make were he ever to carry an error over from Gerhardt's book into his own.[103] He was very open to Vieweg in admitting his new respect for Gerhardt;[104] a quarter century later he even opined that this was the only good book to have come out of France in the last thirty years.[105]

The first installment of Kolbe's textbook, consisting largely of introductory and critical matters, got mixed reviews. Erdmann was furious at some Kolbean criticism directed at him (Erdmann had been drifting in a Gerhardtian direction), and he wrote Kolbe an angry letter. Liebig also wrote Kolbe, mixing compliments with the complaint that there were too many polemics in it; in a letter to Vieweg shortly thereafter, Liebig characterized Kolbe as "one-sided."[106] Vieweg's letters to Kolbe have not survived, but it appears that Vieweg also tried to exert a moderating influence on Kolbe's fiery prose, for Kolbe appears somewhat defensive and conciliatory in subsequent letters on the question of polemics. But a man of Kolbe's self-confidence, conviction, and emotion could not be reined in indefinitely.

6

Confronting the Reform Movement

The "quiet revolution" of the 1850s transformed the theoretical foundations of chemistry and provided the basis for its explosive growth in intellectual and technological applications during the second half of the century. It was a complicated affair that had many interconnecting aspects.

1. First was the *decline of electrochemical dualism*. By the mid-1840s, virtually everyone had lost faith in at least the strong program of dualism, and many French chemists opposed it ardently. It had been weakened by innumerable examples of substitution reactions in organic chemistry, accumulating since the 1820s, few of which appeared to agree with dualistic principles. The gradual loss of what had been such an important and productive principle was distressing and disorienting to many. The general retreat from pursuit of theory by such leading figures as Liebig and Dumas was related to this perceived theoretical chaos.
2. A direct and obvious consequence was the *rise of a unitary view of molecules*. If molecules are not dissectible into paired, electrically opposed components (classical radicals) as demanded by dualism, some drew the conclusion that they must instead be holistic structures whose internal details cannot be determined by chemical reactions at all. This view, implied by some of Dumas' rhetoric in his type theory papers of 1838–1840 and pursued in more than rhetorical fashion by other Frenchmen,

such as Gaudin, Baudrimont, and Gerhardt,[1] appeared from the German perspective to be the overriding French orthodoxy of the 1840s and 1850s.

3. This unitarian viewpoint, however, was simultaneously opposed, also in France, by *new (substitutionist) arrangement theorists*. Dumas, who during the heyday of the old radical theory had practiced the art of constructing "rational formulas" with the best of the German Berzelians, did not really abandon the quest for molecular constitutions when he embraced type theory, as his defense of "mechanical types" and his use of other structuralist metaphors show.[2] The very notion of a "type" and the centrality of substitution suggested that the arrangement of the atoms within a molecule must be just as important as their identities. The structuralist program was thus an obvious concomitant of type theory. These ideas were intimately related to Laurent's nucleus theory; Laurent frequently remonstrated with Gerhardt regarding the latter's apparently absolute rejection of arrangement theories.[3] Those pursuing typist arrangement theories were not so different in principle, then, from those Germans attempting to rescue the rational formulas of dualism by means of an increasingly obsolescent copula theory.

4. The decline of copulas was directly connected to the *emergence of a new type theory*. The natural outgrowth of all of the previous trends (except the radical unitarian doctrine), this theory developed suddenly in 1849–1850. Wurtz' synthesis of primary amines, Hofmann's synthesis of secondary and tertiary amines, and Frankland's work on zinc alkyls, all published in 1849, were each viewed by their authors essentially as substitutions of organic radicals for the hydrogen atoms of inorganic hydrides (in the cases just mentioned, ammonia and the hypothetical metal hydrides). Paul Thenard's work of 1847 on the methylation of phosphine (PH_3) was now interpreted by the new type theorists in similar terms, though Thenard had not initially done so himself.[4] The following year (1850), Alexander Williamson provided a fourth new type and the most important one of all, the water type. In this work, Williamson not only ensured the success of the new type theory but also effectively promoted two central reforms championed earlier by Laurent and Gerhardt.

5. One of these reforms was the *establishment of consistent molecular magnitudes*. Organic chemists had never regarded vapor densities as consistently determinative of molecular size,

partly because to do so would have created a number of apparent anomalies. To maintain elements of dualistic theory, it had proven necessary to use "four-volume" organic formulas double the size of chemically comparable inorganic ones, and most chemists subscribed to these formulations. For largely schematic and esthetic reasons, hence ones that few of their contemporaries found compelling, Gerhardt from 1842 and Laurent (with greater consistency and cogency) from 1845 had advocated taking all formulas, organic as well as inorganic, at the same number of gaseous volumes, preferably two.

6. A largely separate but equally critical issue was *atomic weight reform*. By 1850 this came down essentially to the issue of whether or not to double the atomic weights of carbon, oxygen, and sulfur, that is, to halve the usual number of such atoms in organic formulas. The smaller weights, implying, for example, a formula for water of HO (or H_2O_2) and a formula for the ethyl radical of C_4H_5, had become nearly universally accepted.

Williamson's Asymmetric Synthesis Argument

Williamson's work proved to be the pivot on which turned the ultimate breakthrough of the movement as a whole, because an effective experimental argument that he devised in 1850 and that saw various applications in the next five years essentially settled the molecular magnitude issue, provided compelling grounds for adopting the newer type theory and the atomic weight reform, and further weakened the vestiges of dualism enshrined in the copula theory. During his student years in the 1840s in Heidelberg, Giessen, and Paris, Williamson (1824–1904) was a student of Gmelin and Liebig, and he became the first major advocate of Laurent's and Gerhardt's chemistry.[5] In 1850, newly installed as professor of analytical chemistry at University College London, he decided to try to settle the important issue of relative molecular magnitudes of alcohol (in modern terms, CH_3CH_2OH or Et-OH) and ether ($CH_3CH_2OCH_2CH_3$ or Et-O-Et).[6]

From the 1820s through the 1840s, Liebig and most other chemists had assumed that alcohol was simply the hydrate of ether, since acid dehydration of alcohol yields ether—this despite the fact that the vapor density of ether is nearly twice that of alcohol.

$$\text{For Liebig: } \underset{\text{alcohol}}{C_4H_{10}O.H_2O} \longrightarrow \underset{\text{ether}}{C_4H_{10}O} \text{ plus } H_2O$$

Laurent and Gerhardt, however, argued from 1846 on that alcohol is a monoethyl and ether is a diethyl water derivative. Reacting Frankland's reagent ethyl iodide together with a substance first prepared by Liebig, potassium ethoxide, Williamson succeeded in substituting ethyl for the replaceable hydrogen of alcohol, resulting in a novel ether synthesis that was elegant, smooth, low-temperature, and high-yield. In Williamson's terms,

$$\begin{aligned} \mathrm{EtOH} + \mathrm{K} &\longrightarrow \mathrm{EtOK} \\ \mathrm{EtOK} + \mathrm{EtI} &\longrightarrow \mathrm{Et{-}O{-}Et\ plus\ KI} \\ \text{Overall: } 2\,\mathrm{EtOH} &\longrightarrow \mathrm{Et{-}O{-}Et\ plus\ H_2O} \end{aligned}$$

This reaction could still be explained by the older theory if one assumed that the potassium ethoxide and ethyl iodide were each separately transformed, yielding *two* ether molecules rather than one. But an additional advantage of the Williamson reaction was its flexibility in producing new "tailored" ethers, and an ingenious modification by Williamson cut off any retreat. By reacting *methyl* iodide with potassium ethoxide, the two theories would now predict two distinct results. The Gerhardt–Laurent (Williamson's) theory required the production of a single, homogeneous product, an asymmetric ethyl methyl ether. Liebig's theory would require a *mixture* of products to result from the reaction, the symmetrical (normal) ethyl ether, derived from the ethoxide, and a new symmetrical methyl ether, derived from the iodide.

$$\begin{aligned} &\text{Liebig:} && \mathrm{C_4H_{10}O.KO + C_2H_6I_2 \longrightarrow C_4H_{10}O + C_2H_6O} \\ &\text{Williamson:} && \mathrm{EtOK + MeI \longrightarrow Et{-}O{-}Me} \end{aligned}$$

The consequent production of a single ethyl methyl ether settled the issue, at least for Williamson. As Gerhardt and Laurent had been urging, the ether molecule must be essentially twice the size of the alcohol molecule, not simply a dehydrated version of it; more generally, there is no pre-formed water in alcohols or acids, nor pre-formed oxides in salts. The Gerhardt–Laurent molecular magnitudes reform appeared to be established at a stroke, and dualistic formulations were further weakened, if not destroyed. Many chemists agreed with Williamson, as we shall see in the next section.

The atomic weight reform required another kind of argument. The focus of the ether molecule was the oxygen, which Williamson saw no reason to subdivide. A single oxygen atom, $O = 16$, could then serve as the material link or bond between two ethyl radicals in ether, an ethyl and a hydrogen in alcohol, or two hydrogens in water. The argument was cogent, but not as compelling as that for the structure of ether; it involved an appeal both to Ockham's razor and to structuralist instincts.[7]

The best evidence for the strong impact of Williamson's asymmetric synthesis argument is its widespread deployment in the five years after 1850. In 1851 Williamson used it again himself, in conjunction with a messy ketone reaction, to advocate Gerhardt's and Laurent's formula for acetone—and solving one of their troubling anomalies in the process. As we saw in chapter 4, in the spring of 1852 Gerhardt synthesized organic acid anhydrides by a reaction route analogous to Williamson's (and already predicted by Williamson the year before). He also applied Williamson's *logic* as well, by producing *asymmetrical* anhydrides (such as acetic–benzoic anhydride) that could be explained only by the new chemistry. Gerhardt thought this accomplishment was "terribly revolutionary," the best paper of his life. Indeed, it convinced a wide variety of skeptics both that monobasic organic acids contain no water even in their "hydrated" form and that the water molecule must have two hydrogen atoms. This was the event that propelled Gerhardt from his pariah status to career success—if unfortunately only briefly, as he died in 1856.

The next chapter will show how in 1854 William Odling and August Kekulé, then living in England, independently applied versions of Williamson's argument. Also in 1854 Williamson himself used it once more. He succeeded in chlorinating sulfuric acid in two separate stages, producing in the first reaction a substance now known as chlorosulfonic acid, $HO.SO_2.Cl$. If one were to reason on the basis of the conventional formula for sulfuric acid, $HO.SO_3$, the halfway chlorination ought to have produced at best a mixture of $HO.SO_3$ and $Cl.SO_3$, and not a homogeneous asymmetric product. It was clear to Williamson from this reaction that there is no preformed water even in dibasic inorganic acids. Again, the reaction had indicated the inconsistent molecular magnitudes of the hitherto prevailing views: the conventional sulfuric acid formula $HO.SO_3$, like the HO formula for water, needed to be doubled to compare consistently with the usual organic formulas (in more unitary, nondualistic terms, $H_2S_2O_8$), and then the larger Gerhardtian atomic weights had to be applied to oxygen and sulfur (yielding H_2SO_4). Viewed in this light, his argument was that, in sulfuric acid, the sulfuryl radical SO_2 is the material link between HO and Cl in chlorosulfonic acid and between two HOs in sulfuric acid, just as oxygen is the linking element in ether and in water.[8]

Assumptions of relative molecular sizes and atomic weights were also crucial in interpreting the new radicals created by Kolbe and Frankland during the years from 1847 to 1850. The radicals formed in these reactions are now thought to combine together as they are formed, to create dimers:

$$2\ CH_3CO_2K \longrightarrow CH_3{-}CH_3 + 2\ CO_2 + 2\ K \quad \text{(Kolbe electrolysis)}$$
$$2\ C_2H_5I + Zn \longrightarrow C_2H_5{-}C_2H_5 + ZnI_2 \quad \text{(Frankland reaction)}$$

Nothing in their reactions of formation, however, indicates this dimerization, and Kolbe and Frankland wrote versions of the simpler equations as follows:

$$CH_3CO_2K \longrightarrow CH_3 + CO_2 + K$$
$$C_2H_5I + Zn \longrightarrow C_2H_5 + ZnI$$

These implied production of the actual monomeric radicals. Although Gerhardt and Laurent immediately argued for the dimer formulas of the new substances, it is understandable that Kolbe and Frankland, and initially most other chemists, were not persuaded of the truth of a more complex and less obvious reaction route—especially because the central argument that the Frenchmen used was their mysterious and unmotivated even-number rule.[9]

But by March 1850, newly converted from copula to type formulations, Hofmann became convinced that Gerhardt and Laurent were right.[10] He attempted to settle the issue by an experiment designed to decarboxylate valeric acid pyrolytically, but the results of the experiment were messy and unhelpful. Still, he appended a thorough brief for the thesis that the "radicals" were dimeric molecules. They had none of the properties that one would expect from true radicals, such as extreme reactivity, or additive reactions with oxygen or halogens. That they had proven so difficult to extract from their compounds suggested all the more strongly that they ought, if they were true monomeric radicals, to have properties similar to the highly active "radicals" (i.e., metals) extracted by Davy from soda and potash. On the contrary, the new hydrocarbons had all of the indifferent paraffinic properties of the predicted homologues of marsh gas.

Another anomaly was that the differences between the boiling points of three adjacent homologous "radicals" were found in each case to be 47°C. This was about twice the fairly consistent boiling point differences between nearly all adjacent homologs, as found by Hermann Kopp. The anomaly would vanish if the formulas for the "radicals" were doubled and intermediate unknown hydrocarbons were interpolated. Again, whereas amylene ($C_{10}H_{10}$) and amyl hydride ($C_{10}H_{12}$) boil at 39°C and 30°C, respectively, "amyl" (Frankland's and Kolbe's $C_{10}H_{11}$) boils at 155°C, which did not seem reasonable to

Hofmann; a doubled formula would again restore consistency in the place of anomaly. Finally, the "radicals" all had doubled vapor densities compared to similar olefins and hydrides.

In short, both physical and chemical properties suggested that these molecules were twice the size of what Frankland and Kolbe thought. Hofmann conceded that this meant supposing an *in situ* and *in statu nascenti* dimerization reaction, for which there was no direct evidence. He pointed out, however, that *in situ* polymerization or condensation reactions were already known and accepted, such as the formation of acetone from acetates and mesitylene from acetone.

In December 1850 Benjamin C. Brodie, Jr., adduced additional arguments in favor of the dimer formulas.[11] At the end of his paper, he suggested a crucial experiment modeled on Williamson's recently published asymmetric synthesis argument. Frankland had reacted ethyl iodide with zinc to produce what he thought was ethyl and what Brodie and Hofmann were interpreting as ethyl–ethyl. If one instead used *mixtures* such as ethyl iodide and amyl iodide for the reaction, one might be able to produce asymmetric radicals, in this instance ethyl–amyl. The formation of such an asymmetric radical as a product would demonstrate that dimerization was occurring, since according to Frankland's conception one would expect to produce only mixtures of symmetrical radicals, ethyl and amyl. But Brodie reported that neither he nor Hofmann had had success with the reaction.

That success was achieved by Wurtz in 1855.[12] In place of Frankland's, Brodie's, and Hofmann's zinc (Frankland, presumably following an earlier suggestion of Liebig's, had also tried potassium), Wurtz got the reaction to work using sodium, which he said combined the advantages of acting less energetically and being cheaper than potassium. Following the Williamsonian strategy indicated by Brodie and Hofmann,[13] Wurtz produced the "mixed" radicals ethyl–butyl, ethyl–amyl, buty–amyl, butyl–caproyl, and methyl–caproyl. The argument from this evidence for the larger dimeric formulas for the Kolbe–Frankland "radicals," averred Wurtz, was "of the same order and just as conclusive" as Williamson's for the larger ether formula. One of Hofmann's predicted interpolated compounds was also a product of this research, namely, butyl–amyl (in Wurtz' four-volume formulas, $C_8H_9.C_{10}H_{11}$), which was half-way between "butyl" (Kolbe's "valyl," in Wurtz' formulation $C_8H_9.C_8H_9$) and "amyl" ($C_{10}H_{11}.C_{10}H_{11}$). The boiling point, Wurtz emphasized, just matched Hofmann's prediction. Wurtz even succeeded in producing similar asymmetric radicals from Kolbe electrolyses.

Dimerization introduced a new complexity into the explanation of the reactions, but Wurtz, again following Brodie and Hofmann, cited other accepted instances of this kind of reaction. He was even able to make reference to a project he had carried out under Liebig in 1842 involving the reaction of hydrochloric acid and copper hydride with release of hydrogen gas, which supported Dumas' 1828 speculation on the "dimerization" of hydrogen atoms to form hydrogen molecules.[14] Wurtz also pointed out that this was one of the theses defended by Laurent when, in 1846, he clarified Gerhardt's ideas on atoms and molecules. He mentioned that it had also been proposed and defended more than forty years earlier by Ampère.

The "Wurtz reaction" is still taught in organic chemistry classes today, but it is a messy reaction producing problematical product mixtures, and it is no longer synthetically useful. This is all the more reason for admiring Wurtz' achievement in unraveling the complex chemistry and properly analyzing the liquid and gaseous products, then drawing the important theoretical conclusions—even if the latter were not original to him. He concluded the paper with an advocacy brief for the new chemistry. He used the new atomic weights in this section, "for greater simplicity,"[15] but in general continued to employ the older conventional equivalents for another four years.

It is interesting that all of the protagonists in the reform movement sketched here expressed themselves after the new type theory emerged in terms of compatibility and complementarity rather than conflict. In 1850, Frankland, Hofmann, Wurtz, and Williamson all implied independently that their work on organometallics, complex amines, primary amines, and ether, respectively, suggested a consolidation of radical and substitution theories.[16] The same year Gerhardt used similar language in attempting to reconcile with Liebig, and Liebig returned the rhetoric two years later.[17] Also in 1852, Frankland stated unequivocally that his and others' recent research promised "to assist in effecting a fusion of the two theories which have so long divided the opinions of chemists, and which have too hastily been considered irreconcilable."[18]

Finally, in 1855 Wurtz put forth an extended argument having to do with "condensed" or multiple types. He pointed out that although the old radical theory used a binary conception of constitution with addition mechanisms, the new chemistry used a similar binary constitutional orientation with substitution mechanisms. Thus, "far from being contradictory, they complement one another."[19] These visions of continuity and complementarity were clear-sighted and not simply an instance of soft-pedaling revolution. The newer type theory was a natural and logical outgrowth of the development of the science to 1849; it incorporated much of the older theories, while at the same time justifying many of the new ideas of Laurent and Gerhardt.

Attack and Counterattack: Kolbe Versus Williamson

As we saw at the end of chapter 5, Kolbe was taken by surprise by Gerhardt's and Williamson's papers of the early 1850s and he recognized their importance. Kolbe was far from alone in seeing the compelling character of the new research, and he became increasingly alarmed by the rapid gains that Gerhardt seemed to be making. In France, establishment figures such as Fremy, Regnault, and especially Dumas and Thenard ceased quarreling with Gerhardt and began to support his academic career. Not only did Gerhardt's students such as Chancel and Chiozza adopt elements of his system but also Wurtz, Cahours, Pelouze, Malaguti, and Quesneville. A few Russians and Americans, as well as such Britishers as Odling, Brodie, Henry Roscoe, and Williamson, became converts.

Of even greater concern to Kolbe, there were obvious signs that Gerhardt's views were gaining currency among younger and mid-career German chemists as well. Leopold Gmelin had adopted Laurent's classification scheme in 1848, and Liebig had had high praise for Gerhardt's acid anhydride research. Worse, Kolbe's good friends Hofmann and (German-educated) Frankland had clearly been moving toward the "typist" camp since 1849. Hofmann privately opined at the time of the acid anhydride work that this sort of argument "removes the last support for the idea that [alcohols and acids] contain water." Later he characterized the argument as "irresistibly convincing."[20] A new generation of German chemists—including Kekulé, Karl Weltzien, Emil Erlenmeyer, Heinrich Limpricht, Ludwig Carius, Lothar Meyer, Adolf Baeyer, and Leopold von Pebal—were declaring themselves as converts during the early to mid-1850s.

Kolbe could take little comfort that the *éminences gris* of German chemistry had not signed on, for he well knew that they were uninterested in the theoretical dialectic in general. "What opinion Liebig now has on the subject [of Gerhardt's work] I do not know," he wrote Vieweg. "I suspect none at all, like Wöhler, since neither one seems to have much interest for such matters."[21] The same was true for Bunsen, as Kolbe well recognized.[22] One member of the "classical" generation who did convert to Gerhardt's system, if not to his notation, was Otto Erdmann, Gerhardt's first chemistry teacher and founder of the *Journal für praktische Chemie*.[23] In sum, Gerhardt's views were making headway in the early 1850s in major chemical centers such as Heidelberg and Göttingen and were establishing outposts all over the Germanic lands.

In addition, virtually the entire chemical faculty at Giessen became

Gerhardtians. Heinrich Will, Liebig's successor, is discussed shortly. Adolf Strecker had been successively student, assistant, and Privatdozent under Liebig before his call to Christiania (Oslo) in 1851 and had written Kolbe's favorite short organic chemistry textbook. Strecker had always been favorably inclined toward Gerhardt's and Laurent's research, and by 1854 he was clearly a convert; at the Karlsruhe Congress he was quite vocal in his defense of the new chemistry.[24] He was a talented chemist, much respected in his day. Finally, it took little reading between the lines of the annual *Jahresbericht der Chemie* to perceive that its editor Hermann Kopp, Ordinarius for theoretical chemistry at Giessen after Liebig's departure, had by about 1854 become attracted by the French ideas. The sober and cautious Kopp, seven years older than Williamson, later recalled the latter's work as providing "brilliant confirmation" of the new chemistry, "demonstrating beyond question" the Gerhardtian constitutions and quickly garnering willing converts.[25] The defections of Strecker and Kopp were particularly bad blows for Kolbe. They were both intimate friends of his (*Duzfreunde*), and Kolbe had always admired their careful conservative approaches.

Kolbe worried about how to respond to the new research, especially because he was writing the critical theoretical sections of his textbook in 1853, just as the tide appeared to be turning against him. During the spring, summer, and early fall of that year, Kolbe worked furiously to devise counterarguments to the Williamson–Gerhardt theory. When he thought he had the desired disproofs, he incorporated them into his first textbook installment, which appeared in June 1854. But he was so concerned at the signs of revolution in the air that he decided to hurry his critiques into print as articles at the earliest possible moment.

The first of these was an attempt to disprove Williamson's theory that acetic acid consists of Williamson's "othyl" radical C_2H_3O united to oxygen, the latter atom also linked to a hydrogen atom.[26] Erroneously translating his own conventional equivalents into Williamson's atomic weights by doubling the number of his hydrogen atoms (rather than using the correct transformation algorithm of halving the numbers of his carbon and oxygen atoms), Kolbe concluded that Williamson's acetic acid formula implied the presence of *two* methyl groups. If that formula were true, then one of the methyls could in principle be replaced by another radical, ethyl for instance. But experiment failed to achieve this end, so Williamson's theory was at least weakened, if not directly falsified.

This line of thought demonstrates both that Kolbe fully grasped the import and compelling nature of the asymmetric synthesis argument in particular—he was after all trying to turn Williamson's own powerful gun upon him—but also that he had only a superficial understanding of the new chemistry in more general terms. Two years earlier, upon publication of the German translation of Williamson's third ether paper, Liebig had even provided his readers with two versions of the correct transformation algorithm, but Kolbe must have misunderstood it.[27] Having rejected the atomic weight reform years earlier as a passing French fancy and supremely confident of the truth of the older system, he had simply never successfully worked through the details of the arguments.

Kolbe's attempted refutation was actually performed and written by an English student of Kolbe's, Francis Wrightson, who is otherwise little known; Kolbe later affirmed that the reasoning was his. The research was complete by about June 1853 and was published in the August issue of the *Philosophical Magazine*. The following month a reply by Williamson appeared. Without explaining the precise character of the fallacy committed by Wrightson, Williamson contented himself with demonstrating that "the result which he failed in obtaining is incompatible with the othyle theory of which he conceived it a consequence, and the result he obtained is decidedly confirmatory of the theory which he expected it to upset."[28] Wrightson's confused and ungracious rebuttal shows that he had not understood Williamson's reply[29]—nor apparently did his teacher, who repeated the same arguments the following year.

In the July and August 1853 issues of the *Annalen der Chemie*, a two-part German translation of Gerhardt's detailed memoir on acid anhydrides appeared, and in September a German translation of his and Chiozza's work on amides was published. It was to these articles in particular that Kolbe must have been referring when he wrote Vieweg of the "research of the very greatest scientific importance . . . which threatens to overturn" the entire existing theoretical foundation of chemistry. By the time he wrote this letter (on 16 October 1853), he thought he had hit upon "the strongest proofs against Gerhardt's theory" without having had to enter the laboratory.[30] But that fall, Wrightson having left Marburg, Kolbe put another English student, Frederick Guthrie, to work on additional organic electrolyses to refute Williamson. By late fall he had written a detailed critique, which he read to the Marburg Naturforschende Gesellschaft and sent to Liebig for the *Annalen* and to Hofmann for the *Journal of the Chemical Society*.[31]

Kolbe brought a number of points to bear against Williamson (and secondarily against Gerhardt). Williamson had taken no notice, it seemed, of his proof that methyl exists in acetic acid and its derivatives. Synthetic methods of the sort Williamson had used are untrustworthy indicators of constitutions, Kolbe felt, less certain at least than analyses. In addition, Wrightson's evidence, he thought, had not been effectively countered in Williamson's rebuttal. As for the new asymmetric compounds, nothing prevented one from assuming that the new methyl ethyl ether, for instance, was a combination of ethyl oxide (ethyl ether) and methyl oxide (methyl ether), i.e., $C_4H_5O.C_2H_3O$. To the possible objection that such electrochemically similar molecules would have no reason to cohere, Kolbe responded that many other similar instances were known, such as compounds of chlorine and iodine. Another point of difference between the two theories was that Kolbe's acetic anhydride formula posited three methyl hydrogens, whereas Gerhardt's and Williamson's had six. Were the latter correct, acetic anhydride with an odd number of chlorine atoms would be possible; conversely, such compounds would be excluded by Kolbe's formula because they could not be represented without use of non-integral atomic coefficients. But no such compounds were known to exist.

Finally, Kolbe regarded the electrolysis of acetic acid as irrefutably falsifying Williamson's theory. According to Williamson, the constitution of acetic acid was analogous to water, with the othyl radical taking the place of one of water's hydrogen atoms. Accordingly, electrolysis of the acid ought to proceed analogously to that of water, with othyl and potassium (or at least their decomposition products) appearing at the negative electrode and oxygen emerging at the positive pole. In fact, nothing but hydrogen is found on the negative side, with methyl and carbonic acid appearing at the positive. Especially for this reason, Kolbe concluded that Williamson's theory was "easily refuted."

Kolbe's paper was read at a meeting of the Chemical Society on 20 February 1854; there was an element of drama as Williamson had prepared a reply to be read immediately following. Henry Watts wrote to H. E. Roscoe at Heidelberg shortly before the event, very much looking forward to the "jolly row" in store.[32] Williamson wrote the same correspondent

> Of course I will answer [Kolbe's paper], though from what I have seen of its contents there seems little chance of my converting him to more rational views on the subject, as he does not enter into or understand the point of view opposed to his view.[33]

Williamson's reply was a polemical masterpiece, in which carefully reasoned refutations alternate with less substantive forms of verbal pyrotechnics.[34] In response to Kolbe's methodological critique of synthesis as a valid means of determining constitutions, Williamson attacked Kolbe's favorite approach, analysis, at the same time blasting his use of undefined and empirically illfounded symbolic distinctions:

> It would be just as reasonable to describe an oak-tree as composed of the blocks and chips and shavings to which it may be reduced by the hatchet, as by Dr. Kolbe's formula to describe acetic acid as containing the products which may be obtained from it by destructive influences. A Kolbe botanist would say that half the chips are united with some of the blocks by the force *parenthesis*; the other half joined to this group in a different way, described by a *buckle*; shavings stuck on to these in a third manner, *comma*; and finally, a compound of shavings and blocks united together by a fourth force, *juxtaposition*, is joined on to the main body by a fifth force, *full stop*. The general use of unmeaning signs has become so habitual to Dr. Kolbe, that whenever anything has to be explained, he performs the task to his own satisfaction by inventing a sign for its unknown cause. . . . Signs and words are doubtless indispensable means for the expression of facts or thoughts; but Dr. Kolbe uses them instead of facts, and as a substitute for ideas.[35]

Williamson averred that in admitting the existence of the new compound ethyl methyl ether, however he may wish to formulate it, Kolbe had thereby conceded the essential point at issue, the doubled size of ethers relative to alcohols. (This is an example of rhetorical sleight of hand, as Williamson must have been aware that Kolbe intended his argument about the combination of ethyl and methyl "oxides" to apply only to the asymmetric compounds, not to ordinary ether itself.) He reaffirmed his response to Wrightson, expressing satisfaction at being permitted now to direct his rebuttal to the real source of that fallacious reasoning. Finally, he pointed out that the crucial experiment that Kolbe had conceded would win the day for the Williamson–Gerhardt theory—production of an acetic acid anhydride with an odd number of chlorine atoms—was easily produced from the reaction of trichloroacetyl chloride with sodium acetate.

Williamson's mastery of both the rhetorical and substantive forms of scientific discourse is highlighted by comparing his reply to Kolbe with Gerhardt's response.[36] Published in the same issue of the *Annalen* immediately preceding Williamson's, Gerhardt could only sputter with indignation that Kolbe had attacked "Williamson's" theory, when the ideas that the "overly sensitive" Kolbe opposed so vehemently were his (Gerhardt's) alone. The matter was serious because, by appearing in Liebig's journal, the German public might infer that Kolbe's views were shared by the master (here Liebig added a footnote cautioning that "This should not be assumed"[37]). Gerhardt's outburst once more indicates his tragic flaws: arrogance, bad grace, and lack of tact. His priority claim was not unreasonable, but here it served only to alienate his friends. And he did not bother to attempt to refute Kolbe's arguments rationally.

While all these events were taking up Kolbe's attention, his friend Edward Frankland was also directing a friendly but devastating critique at the copula theory. Frankland's most famous memoir, which contains a limited statement of the concept of valence, was published in the *Philosophical Transactions* in 1852, but in German not until March 1853, when presumably Kolbe first read it.[38] In this work, Frankland argued not only that the reactions of organometallic compounds showed that metal atoms have a maximum combining capacity with other atoms or radicals. He also suggested that the only proper way to envision the reactions is to depict organic radicals as replacing inorganic atoms such as oxygen, rather than simply adding to the metal without altering its chemistry, as in the copula theory.

For instance, according to the copula theory, it should have been easy to oxidize or halogenate the highly electropositive zinc methyl, but all such attempts failed. The only way further combination predicted by the theory could take place was if the organic copulas became detached from the metal atom. In this way, Frankland demonstrated that metal atoms were very much chemically altered by their "copulation" and that their combining capacity was indeed reduced—moreover, precisely by the number of copulas. This applied even to the prototype of copulated radicals, cacodylic acid, in which arsenic was connected to three oxygen atoms and could not be further oxidized to the five-oxygen analog of uncopulated arsenic acid. In Frankland's new type-theoretical interpretation, this was because two methyl groups had taken the place of two oxygens. In short, the postulates and predictions of the copula theory had been systematically contradicted at every turn. Only an application of type theory could save the phenomena.

Accommodation and Defiance

Little wonder, then, that a note of panic appeared in Kolbe's October 1853 letter to Vieweg. Having adjusted his theories and devised what seemed to him to be decisive refutations, Kolbe wrote the critique discussed above, and incorporated it into the first installment of his textbook, which appeared in June 1854. In his introduction, he commented that a number of empirically shallow theories, proposed mostly by French chemists, threatened to destroy the true basis of organic chemistry, the electrochemical radical theory. It was therefore necessary to begin his text by treating these ideas in detail.

The task facing organic chemists today, Kolbe averred, is the elucidation of chemical constitutions by discerning the "rational compositions" of molecules, that is, the manner in which their proximate components are combined. Kolbe had nothing but contempt for those empiricists whose love of experimentation (*Experimentirkunst*) was greater than that for true scientific research and who only sought to produce new substances as fast as possible without investigating their constitutions. He stated that we will "never" be able to gain a "clear view" of the *spatial* arrangements of the atoms in a molecule; but we are now and will be in the future able to discern the "functions" that atoms and proximate components perform with respect to each other.[39]

The only sure guide to rational constitutions, Kolbe continued, is Berzelius' electrochemical radical theory, founded on the conviction that organic chemical theory must be based on analogies to inorganic chemistry. There are problems with organic electrochemical theory, Kolbe conceded. Above all, we often cannot isolate the components of organic molecules, and even when we can, the determination of their electrochemical properties is problematic. But we should maintain a sense of perspective and not abandon an old, trusted, and useful theory at the first sign of difficulty. Have confidence in the flexibility of a good theory, he counseled, and trust to the future to work through the anomalies.[40]

Consciously or not, Kolbe illustrated this flexibility by the adjustments that he continued to make in his own version of the theory. In the 1840s he had viewed oxalic acid, $HO.C_2O_3$, as the progenitor of all organic acids. Beginning in 1848 he had preferred to view acids in a more general fashion, as deriving from an ultimate resolved radical C_2 to which one could *add* methyl to form the more proximate acetic acid radical $(C_2H_3)\frown C_2$, whose hydrated trioxide was the acid itself. Now in 1854 he viewed the methyl group as *substituting* for the hydrogen of "formyl," $H\frown C_2$; this was very close to Wurtz' theory of 1850. Since Gerhardt had shown in 1853 that his oxygen-containing acetyl radical (methyl carbonyl) can substitute repeatedly for the hydrogens of ammonia, Kolbe no longer doubted that radicals can contain oxygen. Consequently, he now felt compelled to distinguish between a more resolved radical $(C_2H_3)\frown C_2$ and a more proximate radical $(C_2H_3)\frown C_2O_2$ in acetic acid. Similarly, benzoyl compounds all contain an oxygenated radical $(C_{12}H_5)\frown C_2O_2$. Kolbe noted that this matched Liebig's and Wöhler's initial formulation of the benzoyl series in 1832, despite Berzelius' distress at oxygen inside a radical.[41]

Thus, whereas in 1850 he had conceded to type theory the thesis that chlorine and other atoms may substitute in organic radicals without necessarily causing major alterations of properties, he was now expanding this concession to include all-important oxygen. Moreover, he now viewed *hydrogen* substitution as the schematic means of constructing all organic acids and acid derivatives from formyl. Other systems were analogous: amines were copulated nitrogens produced by substitution of hydrogens by hydrocarbon radicals, and so on. The fact that halogen substitution does not much alter chemical properties of hydrocarbon radicals proves that "the grouping of their proximate components remains the same" and that their "types" are unaltered. This is true in some cases for inorganic substituting groups as well, such as amido, nitro, and sulfuryl groups.[42] Dumas could not have argued with greater force or clarity for the assumptions of type theory than Kolbe did here.

Despite these concessions, however, Kolbe remained resolutely faithful to dualistic organic chemistry. Although halogen substitution for hydrogen cannot be explained by means of electrochemical precepts, that does not mean that the precepts are wrong. Chlorine, perhaps even hydrogen, might undergo an alteration of electrochemical properties when it enters organic molecules. After all, elemental phosphorus has several chemically different states, and nascent hydrogen is very different from the stable gaseous element. Thus, he believed it would be "frivolous" to abandon dualism because of the single anomaly of substitution.[43]

Kolbe also retained the copula theory as the last major element of dualism left in organic theory. Copulas, he continued to affirm, have little influence on properties of the prototype acid. For example, formic, acetic, and propionic acids are quite similar despite the presence of hydrogen, methyl, and ethyl as copulas, respectively. Accepting Frankland's demolition of the copula theory would require conceding that atoms have fixed combining capacities, irrespective of the electrochemical properties of the adducts—a challenge directed to the very heart of dualism.

But Frankland's evidence was hard to deny. Although stressing that he did not agree with Frankland's conclusion, Kolbe admitted that there did seem to be a general law at work. One might be able to account for it, he thought, by a qualitative electrochemical argument relating to the varying affinities for oxygen of different elements. Since positive elements such as potassium and calcium combine with oxygen in only a small number of proportions, whereas negative elements such as halogens, sulfur, and nitrogen combine in a much larger number, perhaps the addition of alkyl groups to, say, nitrogen makes the nitrogen less negative, hence less capable of combining with additional oxygen. Thus, the total number of atoms or radicals combined with nitrogen remains roughly the same. Even so, he noted that his tentative suggestion "in no way explains" the extreme regularity or exactness of the relationship revealed by Frankland.[44]

Although often sounding here much like a "typist" himself, in a historical section he poured scorn over the "games with formulas" that

> . . . for many years the best chemists of France have been playing, with unbelievable self-deception. Rarely indeed has a new theory been introduced into science with such ostentation and such great confidence in its infallibility, as Dumas' type theory. Seldom has unscientific behavior and the desire immediately to derive general laws of nature from isolated facts been more severely punished as in this case. Of the numerous laws with which the type and substitution theories wanted to enrich science, one after another had to be retracted, and finally nothing more remained than the naked fact that in many organic compounds hydrogen can be substituted by chlorine without essentially altering the molecular grouping of the atoms or in many cases the characteristic properties of the compound.[45]

Kolbe noted the "remarkable fact" that it was almost exclusively German scientists who had developed the radical theory, and he ventured the opinion that the French had opposed the theory "because it had not developed on French soil." Laurent's theory of fundamental and derived nuclei, which some had considered a sort of radical theory, was nothing of the kind to Kolbe; it was a "painting of the imagination," reminiscent of Naturphilosophie. An example of Laurent's penchant for "paper-and-pencil laws of nature" was his notion that oxygen substituting within the nucleus does not alter the characteristic neutrality of organic nuclei (to form aldehydes, ketones, ethers, etc.), whereas if one, two, or three oxygen atoms are added to the exterior of the nucleus they create mono-, di-, or tribasic acids, respectively.[46]

Some, he noted, had criticized the radical theory for hypothesizing fictitious entities, just as he was criticizing Laurent's nuclei. These cavils no longer had any force since he and Frankland had isolated a number of hydrocarbon radicals. Others had suggested that methyl, ethyl, etc. were not real (monomeric) radicals but rather dimeric molecules. To the extent that this argument was based on the application of Laurent's even-number rule, it was just another example to Kolbe of a fictitious French law. Hofmann's claim that one ought to have expected extreme chemical lability from hydrocarbon radicals is also flawed; after all, the radicals are homologous with hydrogen, which is stable and fairly unreactive except regarding oxidation. Finally, the evidence from boiling points could also be countered. Kopp's laws were known to have many exceptions, such as comparing H–OH with

CH_3–OH, and it could well be that the new radicals represented further exceptions. (This despite the fact that in 1847 Kolbe himself had used boiling points to argue for his interpretation of the constitutions of the fatty acids.) In Kolbe's view, the case for monomer versus dimer formulas was simply unproved, and in such instances one must prefer the simpler hypothesis.[47]

At the end of the general section of his textbook, Kolbe printed a truncated version of his critique of the Williamson–Gerhardt ether theory. He chose to omit the Wrightson argument and the methodological critique of synthesis as providing access to constitutions, although this was probably not in reaction to Williamson's reply, which he had presumably not yet seen. However, the week after his critique and Williamson's rebuttal were read in London, Guthrie's experiments in Marburg cast further light on the matter, and Kolbe just had time to incorporate the consequent changes into his manuscript. If, as Williamson had argued, potassium ethoxide has a constitution analogous to water, then in electrolysis the constituents occupying the places of the hydrogen—potassium and ethyl—should share the latter's electrochemical behavior and emerge at the negative electrode. As Guthrie found that only hydrogen appeared there, the theory appeared to be weakened.[48]

Kolbe, however, put the matter more strongly. To Vieweg he declared Guthrie's evidence a "fundamental proof of the untenability of [Williamson's] hypotheses," and in his text he stated that this "fundamental experiment . . . completely refutes" the French–English theory.[49] Words such as "proof" and "refutation" leave no space for doubt or debate. It is more than a little ironic to see such strong wording coming from Kolbe, who here and elsewhere openly conceded the anomalous character of many electrolyses and the unpredictability of the electrochemical properties of organic compounds and their components. Furthermore, Heinrich Will was soon to note that Kolbe failed to see that his own formulas were fully as susceptible to his own critique as Williamson's. But even without such internal contradictions, the investment of perfect certainty in a purely analogical argument suggests methodological problems.

The Battle Lost

There is considerable evidence here of defensiveness, along with often emotional and even shrill counterattacks. Despite Kolbe's brief sense of triumph with Guthrie's electrolyses, there continued to be good reason for his discomfort. Four months after the premiere of Kolbe's textbook and a month after the appearance in German of Williamson's and Gerhardt's rebuttals, Heinrich Will published a devastating critique of Kolbe's views. Will (1812–1890) had been a Liebig student in the late 1830s, subsequently becoming Privatdozent and first assistant in Giessen. When Liebig opened a branch laboratory for beginners in 1843, Will was placed in charge. As a consequence, he taught large numbers of students, many of whom achieved prominence in the next decade. Many of those who are counted (and counted themselves) as Liebig students actually were taught mostly by Will, including Williamson, Hofmann, Strecker, Kekulé, and Erlenmeyer.

Although Will had a strongly empirical approach, it is interesting to note that the topic of his disputation for Habilitation (1844) was the thesis that no compounds exist that have an odd number of equivalents of carbon—a cardinal thesis of Gerhardt's chemistry.[50] A fellow student and confidant of Kekulé's during the years 1849–1851 reported that Will's lectures opened many windows.

> Later our discussions, both in private and in the larger circle of our friends, moved increasingly into the purely theoretical realm. At that time in our circle the perception was already stirring, partly still unconsciously, that the strict radical theory was not the sole all-redeeming dogma of chemistry. The seed of this perception, I believe, was to be found in Will's lectures on organic chemistry.[51]

Will succeeded Liebig as ordentlicher Professor when Liebig transferred to Munich in 1852. By 1853, according to his student Volhard, Will was giving a "clear, convincing, indeed enthusiastic portrayal" of Gerhardt's system in his lectures, to the point that Volhard and his comrades came to swear on type theory and honor Gerhardt as the reformer of organic chemistry.[52] The next year Will published his commentary "On the Theory of the Constitution of Organic Compounds" in the *Annalen*.[53]

Will argued that Gerhardt's atomic weight reform must be adopted simply because it was becoming ever clearer that formulas written in conventional equivalents *always* have even numbers of carbon and oxygen atoms. There was thus no empirical justification for retaining C_2 in preference to C (i.e., $C = 6$ in preference to $C = 12$). Furthermore, the asymmetric synthesis argument as applied by Williamson and Gerhardt, and the new bases of Wurtz and Hofmann, had placed the molecular magnitude issue (the doubled sizes of ethers and acid anhydrides with respect to alcohols and acids) beyond any question.

Unlike Williamson, Will understood that Kolbe's depiction of ethyl methyl ether as a compound of ethyl oxide with methyl oxide was not intended to apply to ordinary ether. But if the asymmetric Williamson reaction links these two compounds together so firmly, then it must also link ethyl oxide with itself in the symmetrical reaction, which is the equivalent of Williamson's ether formula. The latter makes much more chemical sense, Will argued, for a dimerized ethyl oxide has no analog in either organic or inorganic chemistry.

As for Kolbe's crucial electrolysis argument, Will suggested that it would have some force if directed against the existence of such oxygenated radicals as Williamson's othyl or ethoxide. But since Kolbe had now conceded the existence of oxygenated radicals, his and Williamson's formulas were equivalent. Translating Williamson's formula for acetic acid into Kolbe's notation, the two were

$$C_2H_3\frown C_2O_2,O.HO \quad \text{and} \quad \left.\begin{matrix} C_2H_3C_2O_2 \\ H \end{matrix}\right\} O_2$$

These formulas depict the same atoms combined in the same way. The only significant difference between them is that Kolbe suggested a chemical distinction between the two extra-radical oxygen atoms, a distinction, Will noted, that has no empirical basis. But even conceding this distinction, any electrochemical argument that applies to one of these formulas must perforce apply to the other. If Kolbe had disproved Williamson's formula, he had simultaneously disproved his own.

The issue here as elsewhere, Will concluded, was that inorganic chemistry could no longer remain the analogical guide to organic chemistry. The reform of organic chemistry, begun by Gerhardt, provided the entrée into the chemistry of the future, and it could no longer be in doubt that his reform would win the day. Will, the model of the mid-career establishment German chemist and the newly appointed successor to Liebig as ordentlicher Professor at Giessen, was declaring himself for the young (French and English) Turks.

Much later, in a biography of Will, Hofmann wrote of this period of rapid gains for the new views: "With some, this transition happened silently, they slipped as it were right into the new theory; in fact there were some who the night before had been implacable opponents of Gerhardt and Laurent, and awoke the next morning completely converted." In his biography of Wurtz, Hofmann chose similar language to describe the period: "The time had arrived in which, one after another, the most ardent opponents—often overnight, and without providing any reasons whatever for their conversions—made their pronouncements."[54] Hofmann later affirmed that a "revolution" had taken place during the 1850s—and he was not the only one to apply this word.[55] The evidence presented here suggests that this transition came earlier, faster, and more completely in Germany than has hitherto been thought. Most active and theoretically oriented German chemists had converted several years before the Karlsruhe Congress of 1860, the event usually described as the turning point.

Wurtz's discovery in 1855 of the asymmetrical or "mixed" radicals was the final blow for Kolbe. Kolbe may have been able to satisfy himself that ethyl oxide and methyl oxide are sufficiently distinct electrochemically to provide a basis for combination, but it was difficult to imagine such a distinction between, say, the eight-carbon butyl and the ten-carbon amyl, especially considering the absence of any heteroatoms and the extreme stability of their combination. Butyl–amyl's exact fulfillment of Hofmann's boiling point prediction—based on Kopp's laws—was another problem for Kolbe. It was fortunate that the imminent publication of the fascicle of volume six of the *Handwörterbuch der reinen und angewandten Chemie* that contained the beginning of letter "R" gave Kolbe the impetus to write a summary article on "radicals," further modifying his views. He wrote the article in late December 1855, immediately after a summary of Wurtz' paper appeared in Liebig's *Annalen*.[56]

The piece contains a strong affirmation of the complete substitutability of the hydrogen of organic radicals by virtually any element, even electrochemical opposites, sometimes with minimal alteration of properties. How this undeniable truth can be reconciled with electrochemistry "remains for the moment still an open question," but by no means falsifies that theory. He portrayed the issue of monomeric versus dimeric radicals likewise as open, and rehearsed all his previous rebuttals to Gerhardt, Laurent, Hofmann, and Brodie.

However, Wurtz' new research was "without a doubt . . . of much greater significance for this question" than the earlier boiling point argument. The vapor density and boiling point of the mixed radical butyl–amyl, for instance, fit right between butyl and amyl, and it just matched Hofmann's prediction, which strongly suggested that the symmetrical radicals were butyl–butyl and amyl–amyl. But after fairly summarizing this evidence, Kolbe still demurred: "It remains for the moment undecided and questionable whether these facts have sufficient probative force" to compel a doubling of the radical formulas. After proclaiming Wurtz' evidence highly important, he did not even attempt a reasoned refutation, but rather simply judged the issue still open.

Frankland later stated that through correspondence during the year 1856 he was able to persuade his friend Kolbe to retract the critique of the law of maximum combining capacities that had appeared in his textbook in June 1854.[57] Unfortunately, none of these letters have been found. It would seem, however, that at least the beginning of the correspondence on this subject may have been earlier than Frankland indicated, for this article reveals that Kolbe's conversion dates from the end of 1855. Significantly, Kolbe now not only accepted Frankland's law for metals, nitrogen, phosphorus, antimony, and arsenic but expanded it to include carbon as well. If one takes as one's type carbonic acid, formulated as $2HO.C_2O_4$, and assumes that it has a maximum combining capacity of four, then replacement of one oxygen atom by methyl results in acetic acid, $HO.(C_2H_3)C_2,O_3$. (In Berzelian fashion, the half-molecules of water HO were considered external and not essential to the composition of the acid. One needed simply to omit an HO group in moving from a dibasic to a monobasic acid.) A second methyl in the place of oxygen would yield the neutral substance acetone, $(C_2H_3)_2C_2O_2$ (omitting the second half-molecule of water).[58]

Kolbe's "carbonic acid theory," as it evolved into its final formulation by 1860, is described in detail in chapter 8. Here it suffices to note that it was essentially a "newer type theory" very similar to, and clearly influenced by, those of Williamson, Gerhardt, Hofmann, Frankland, and Wurtz. Kolbe later conceded that he had become a type theorist (but always insisted that his types were sharply distinct from those of Gerhardt and his followers). He also eventually conceded that he had been wrong when in 1854 he attacked Williamson's interpretation of his ether synthesis.[59]

7

Kekulé, Wurtz, and the Rise of Structure Theory

The fundamental axiom of structure theory is the idea of valence, and the foundation of valence is the nonequality of an element's chemical equivalent and atomic weight. We say the chemical equivalent of oxygen is eight, since oxygen combines with the lightest element, hydrogen, in a gravimetric ratio of eight to one. Because the atomic weight of oxygen is sixteen, its valence is two (that is, it combines with two atoms of hydrogen). By 1853, Williamson had introduced a modification of a Berzelian notation that expressed this idea: if one places a bar through the letter that signifies an atom, it can indicate that this letter (or atom) stands for two conventional equivalents (Ɵ = 16 and Ȼ = 12).[1] The convention spread among the new type theorists in the 1850s and was officially adopted at the Karlsruhe Congress in 1860. It persisted for a few more years, until most chemists (outside France) had adopted the new weights, whereupon the bars were dropped.

The concept of valence thus depends directly upon the ideas of chemical equivalence and atomic weight. Accordingly, valence ideas could not have been (and indeed were not) formulated with consistency until this terminology and the ideas behind it were introduced; conversely, once the foundation was there, valence concepts were quickly developed. Furthermore, once valence regularities were perceived, structural ideas soon followed. This partially explains the complex series of changes that rapidly overtook theoretical chemistry in the 1850s and 1860s, and the many bruised feelings and priority disputes that broke out at that time.

The French Connection

The deep background of valence was a long tradition of speculations regarding submolecular and even subatomic structures. Examples from the early nineteenth century include the provocative ideas of Avogadro, Prout, Ampère, and Gaudin, which were familiar to most chemists of the middle decades of the century and have been studied by modern scholars. It is well known that Dumas was fascinated by some of these writers, and his explicitly philosophical writings of the 1820s and 1830s, as well as some of his scientific papers, suggest that he took the idea of submolecularity very seriously as a way to explain some of the more puzzling common phenomena of chemistry. Nor was Dumas speaking for himself alone when in 1836 he affirmed "that the chemists'. . . *atoms* are nothing but molecular groups."[2] What he meant, in modern vocabulary, is that elementary molecules normally consist of more than one atom. But Dumas always let this suggestion remain vague and undeveloped, for the details seemed inaccessible to experiment.

The proximate background to the rise of valence involved the work of Gerhardt and Laurent in the years after 1842. In 1846 Laurent provided consistent definitions of equivalents, atoms, and molecules, defended two-volume formulas, and applied this axiomatic base to chemical theory. His calculational control for the two-volume criterion was the even-number rule.

Laurent divided all elements into two classes: *monasides* or *monads*, such as carbon, oxygen, sulfur, and selenium, and *dyodides* or *dyads*, such as hydrogen, nitrogen, chlorine and other halogens, phosphorus, arsenic, sodium, and silver. The even-number rule states that the number of atoms of all dyads in a molecule must be even. If it was not, the formula was incorrect and had to be doubled or otherwise modified. Following his rule, Laurent could tell at a glance the falsity of formulas such as those for water = HO, cyanogen = CN, and ethyl = C_2H_5. He explained the rule by supposing that dyadic elements exist in the form of "binary" molecules such that each atom must always be joined to a complement (a *moitié complémentaire*). In contrast, monads can exist either singly or in pairs. Thus, dyads always appear in twos, while monads may appear in any integral number.[3]

But Laurent encountered difficulty in applying his principles to certain metals. "It was then that I asked myself," he wrote soon before his untimely death in the spring of 1853, "whether the atoms of chemists might not be divisible." He then echoed the very words of Dumas in supposing "that the chemical atoms are but molecular groups, composed of a certain number of minuter atoms."[4] This hypothesis removed the anomalies. For instance, by hypothesizing two sorts of iron atoms of different sizes—composed of different numbers of identical iron *sub*atoms—both the ferric and ferrous series of compounds would conform to the even-number rule. The hypothesis would also account for systematic chemical differences between the ferric and ferrous series and could easily be applied to other metals such as copper, mercury, and platinum.[5]

There is no indication in this passage as to when Laurent arrived at his idea of subatomic particles, but as early as 1843 he was already applying it to certain elements. For instance, if the atoms of manganese metal are composed of groups of twenty-four "atomes plus petits," then the three series of salts of manganese could be represented as compounds of three different smaller conglomerations of manganese subatoms. Gerhardt adopted Laurent's notational convention, if not his physical hypothesis, for his *Traité de chimie organique*.[6]

Laurent's proposals of physical hypotheses to explain the empirical phenomena covered by the even-number rule and the varying valence of certain metals broached the idea that subatomic structure might explain the varying capacities of different atoms to unite chemically with atoms of other elements. The same concept was hidden within his insistence (here following his friend Gerhardt) that atomic weight formulas HCl, H_2O, NH_3, CH_4, and so on are actually more empirical and more consistent than the equivalent weight formulas H_2Cl_2, H_2O_2, N_2H_6, C_2H_4, and so on. Indeed, by the time of his death he had come to the conclusion that so-called "equivalents" as defined and used by chemists since Wollaston and Davy were nothing less than a rival set of atomic weights, no more empirical and no less hypothetical than Berzelius' atomic weights. This was a philosophically correct conclusion that was inadequately heeded both by nineteenth-century chemists and by twentieth-century chemical historians.[7]

Even before Laurent's death, Alexander Williamson explored the implications of Laurent's work by arguing for the "bibasic" character of various radicals and, at least by implication, the oxygen atom. Shortly thereafter, William Odling, an associate of Williamson, wrote of the "replaceable, or representative, or substitution value" of the atoms of a number of elements. From a different starting point, but also consciously following substitutionists such as Dumas, Laurent, Hofmann, and Wurtz, Edward Frankland provided the first explicit and reasonably general statement of valence regularities, even though he formulated the argument in equivalents.[8] But it was Adolphe Wurtz who proposed a clear—indeed, virtually the only—physical hypothesis that could account for *why* atoms have "polybasic" character, or "substitution values," or "saturation capacities," to use the language of Williamson, Odling, and Frankland.

Wurtz' valence hypothesis emerged in the context of his work in following up Williamson's research on polybasic radicals. Upon the appearance of an article wherein Williamson described the formation of trinitroglycerin and triethoxymethane and used formulas suggesting that both were combinations of tribasic radicals, Wurtz prepared a French abstract of the paper for the *Annales de chimie*, and published a related article of his own in the same issue (April 1855), immediately following Williamson's.[9] Developing Williamson's argument more fully than Williamson had done, Wurtz depicted glycerin as a tribasic radical schematically derived from the propyl radical by abstraction of two more hydrogen atoms. The tribasic glyceryl radical, derived from a triple water type, can form a bond (*lien*) tying the three hydroxyl groups together. He also reinterpreted Berthelot's description of mono-, di-, and triglycerides as analogous to mono-, di-, and tribasic acids. Wurtz argued that the correct analogy was to mono-, di-, and trisalts of a tribasic acid.

As he later described these events, this work led him to wonder if it might be possible to test the truth of all of these ideas by synthesizing the intermediate link between tribasic *glyc*erin and monobasic alcoh*ol*, namely, a dibasic (or "diatomic," as he now began to call it) substance he chose to name "glyc-ol." On 24 March 1856, Wurtz acetylated and then hydrolyzed ethylene iodide and isolated the expected product, the first dialcohol.[10] He had opened up a huge new field, and he pursued it aggressively, preparing in the following years an impressive array of polyfunctional organic alcohols and acids.

Wurtz' valence hypothesis was founded on an atomic analogy to the polyfunctional radicals he had begun to introduce and study. It was proposed in an offhand fashion at the end of his July 1855 paper on mixed radicals, in the context of a general defense of the reform movement of Laurent, Gerhardt, and Williamson. He thought that Williamson's and Odling's single, multiple, and mixed types could be explained by considering all of them as formed from multiply condensed hydrogen. Water, for instance, was nothing but a double hydrogen type, with the dibasic oxygen atom replacing two of the four hydrogen atoms. Ammonia was three hydrogen molecules with one atom from each molecule replaced by a single tribasic nitrogen atom. A double water type, as in Williamson's sulfuric acid formula, was quadruply condensed hydrogen, and so on.[11]

Citing the reformers as well as earlier work by Dumas, Ampère, and himself on the compound nature of hydrogen and other elemental molecules, he adopted the symbols previously used by Odling to indicate "basicity": oxygen is dibasic, O″, and nitrogen is tribasic, Az‴. A footnote reads: "This property of nitrogen can be explained by supposing that it is itself formed from 3 juxtaposed and inseparable atoms." Thus, Az‴ could be written as az^3, ammonia as H^3az^3, phosphorus (P‴) as p^3, and phosphorus trichloride as Cl^3p^3. "But as this notation rests on considerations which are not susceptible of a rigorous demonstration, I renounce them for the moment."[12]

Wurtz was suggesting that polyvalent atoms are polyvalent because they are accretions of monovalent equivalents; nitrogen has a valence of three and an atomic weight of fourteen because it consists of three equivalents adhering together in some fashion, each equivalent weighing $14/3 = 4.67$ and each capable of forming a single bond to another atom, such as hydrogen. He never developed this intriguing speculation to any degree of detail, though he subsequently referred to it many times, calling the particles p or az "little atoms" or "subatoms."[13] He said he relinquished the idea after the discovery of the variability of valence; it was hard to imagine how the phosphorus atom, for instance, could be alternately p^3 and p^5.

Although in this paper Wurtz adopted Laurent's notational convention and elements of his theory of subatoms, he was here applying the notion differently than Laurent had. This was an original hypothesis, the first of its kind, designed to explain the law of atomic valence—neither of which, theory or law, Laurent had developed. Wurtz' idea provided a conventient visualization of both valence and equivalence, since the valence is equal to the number of subatoms in an atom and the equivalent is the weight of a subatom. In the 1860s the shorthand definition of equivalent became "atomic weight divided by valence," a simple but imprecise and misleading formula still often used by writers of elementary chemistry textbooks.[14]

Wurtz' subatomic speculation was founded on an analogy between atoms and molecules. Just as Williamson, Odling, and he himself had already shown that there were "polybasic"—or as Wurtz began to call them in 1856, "polyatomic"—radicals, Wurtz thought (following Odling and implicitly Williamson as well) that the atoms themselves must also be polybasic or polyatomic. The word *atomicity* had already been introduced by Gaudin, with a denotation similar to the twentieth-century one, namely, referring to the number of atoms in an elementary molecule. Wurtz borrowed the term to denote what we now refer to as valence, and he may well have wanted it to carry a similar implication to Gaudin's usage: the "triatomic" (trivalent) nitrogen atom is itself a kind of "molecule" consisting of three monovalent subatoms.

Despite the offhand character of Wurtz' proposal, colleagues took notice. In the 1860s the subatomic speculation was adopted and defended by such chemists as Alexander Crum Brown, Emil Erlenmeyer, Charles Delavaud, and Christian Blomstrand, and in the 1850s it may have been influential for Archibald Scott Couper and August Kekulé as well (as we will see in the next section). A notational convention introduced in 1859 by Kekulé, which was probably based on Wurtz' idea, was adopted in several variations during the 1860s by Wurtz, Alfred Naquet, Joseph Wilbrand, Pierre Havrez, George Carey Foster, and Blomstrand.[15] But after 1869 neither the notation nor the theory was seriously defended.

The Education of August Kekulé

Friedrich August Kekulé was born to a respected family of civil servants in Darmstadt, Liebig's hometown, on 7 September 1829.[16] His father, Ludwig Karl Emil Kekulé, was a senior military advisor to the Grand Duke of Hesse, and the family lived in prosperous bourgeois style during August's youth. August distinguished himself in the Darmstadt Gymnasium, especially in the study of languages (English, Italian, French, and Latin), in chemistry, in mathematics, and in drawing. He had a nearly photographic memory, which manifested itself from his youth and served him well into old age. Recovering from weak health as a child, by adolescence August was handsome, tall, strong, and athletic, an enthusiastic gymnast and dancer. It is said that as a young bachelor he was highly regarded by the opposite sex and in demand as a dance partner. In his last year at Gymnasium, Kekulé's father died, which reduced the family's income to a fraction of what it had been. As his father had wished, Kekulé determined to study architecture, a field with considerable financial promise, at the University of Giessen. In his second semester there (the summer of 1848), he was "seduced" by Liebig's course on experimental chemistry into changing fields. Because chemistry was still at that time a discipline with poor prospects—there were but four positions for chemists in the entire state—his mother was not happy at this decision, but August was firm in his resolve.

Although Kekulé always considered himself a student of Liebig, from whom he heard lectures on introductory, inorganic, and agricultural chemistry, his training in laboratory operations was at the hand of ausserordentlicher Professor Heinrich Will, and his introduction to organic chemistry was due jointly to Will and Adolf Strecker, then Privatdozent. As we have seen, Will and Strecker would soon become two of the earliest from the German chemical establishment to subscribe to the reform movement. His closest friend among the chemistry students, Reinhold Hoffmann, later wrote of his frequent evening discussions with Kekulé, wherein Kekulé played the role of the professor and he that of the "shy and amazed pupil." Kekulé, he said, was already a walking library of chemical literature—more reliable even than the *Jahresbericht*, it was said—and "already exhibited a desire to trace the sources of knowledge." This critical tendency, Hoffmann thought, was due primarily to Will's lectures.

> In contrast to Liebig, who often enjoyed saying of rather obscure areas, "Gentlemen, this we know with absolute certainty," Will was fond of suggesting, even for uncontested issues regarding the constitution of organic compounds, that the matter could be considered from another point of view, from which were often discovered relationships that would otherwise have remained unremarked and unknown.[17]

This was particularly true regarding organic radicals, Hoffmann reported. Liebig's certainty regarding the existence of radicals is understandable, and Will's apparent skepticism is curious, considering that this was precisely when Frankland seemed to be isolating one radical after another in Liebig's laboratory during the fall of 1849. Between 1843 and 1852, however, Will and his novice Praktikanten occupied a branch lab physically separate from Liebig's main lab (on the Seltersweg, now Frankfurter Strasse). Will's relationship to Frankland is unknown, and there is no evidence that Kekulé and Frankland became acquainted at this time. Kekulé entered Liebig's lab only in winter semester 1850/51.

Kekulé's older half-brother Karl (Charles) had emigrated to London and by this time had become a prosperous grain merchant. He offered to finance a *Wanderjahr* for his younger brother. Kekulé pondered the choice between Berlin or Paris, and asked Liebig his opinion. "Go to Paris," counseled Liebig, who had done the same in his youth. "You will enlarge your horizons, learn a new language, and become familiar with a world capital. But you will not learn chemistry there." In May 1851, on the journey to Paris, Kekulé saw Gerhardt's *Introduction à l'étude de la chimie* in the window of a Frankfurt bookshop, purchased it, and read it while crossing the French countryside.

In Paris, Kekulé heard lectures by Dumas, Cahours, Regnault, and Wurtz, and spent a good deal of time with the latter. Soon after his arrival, while strolling along a boulevard, Kekulé saw an advertisement for Gerhardt's private school and he signed his name in the list (Laurent was already an invalid, and Kekulé never met him). His first meeting with Gerhardt, Kekulé later recalled, lasted from noon to midnight, and they saw each other twice a week the entire year Kekulé was in Paris. Gerhardt offered Kekulé an assistantship, with the apparently generous remuneration of half the fees of all his Praktikanten. Asked the obvious question, Gerhardt responded that there was then but one Praktikant, and so Kekulé declined the offer. But he read Gerhardt's partially written *Traité* in manuscript, and he gained enormously from his frequent and friendly contacts with the older Frenchman.

In early March 1852, Kekulé returned to Darmstadt to be with his mother, who died the following month. After passing his Ph.D. exam in Giessen on the basis of a project carried out under Will's direction, Kekulé accepted a position as assistant in Adolph von Planta's private lab at Reichenau castle near Chur, Switzerland. The theoretically and physically quiet fifteen months he spent there, analyzing alkaloids amid the glorious Alpine scenery, gave him ample free time, he later reflected, to work through the controversial and confusing ideas he had been exposed to in Paris.

In the fall of 1853, Kekulé was offered another assistantship through Liebig's offices, this time with John Stenhouse, a former Liebig student, at St. Bartholomew's Hospital in London. Kekulé was inclined to refuse the offer, for he considered Stenhouse a "Schmierchemiker," but fate intervened. By chance, Bunsen came to Chur to his brother-in-law's estate, and Kekulé met him for the first time there. Bunsen persuaded Kekulé to accept the offer, saying that he would certainly learn a new language, but no chemistry. Bunsen was, of course, just as wrong as Liebig had been.

Kekulé arrived in London in the last week of 1853. Life there was made considerably more pleasant by the presence of two other young Germans: Hugo Müller, an assistant of Warren De la Rue, and Kekulé's old student comrade in Giessen, Reinhold Hoffmann, assistant to Williamson. Hoffmann, who shared a flat with Kekulé, later recalled

> If Kekulé's professional activity had been dull with Planta, it now became even drier in Stenhouse's laboratory. . . . Kekulé sought and found refreshment and relaxation with Williamson at University College, where he frequently picked me up on his way home in the evenings. If Odling happened to be there as well, the friendly battle over the relationships of the atoms and radicals among each other, and to their commanders-in-chief Gerhardt, Laurent, Kolbe, etc., was soon under way, and the blows fell thick as hail.[18]

Hoffmann further related that Kekulé gradually became acquainted with A. W. Hofmann, Thomas Graham, and Edward Frankland, who occasionally visited London from Manchester. Long, casual conversations among Kekulé, Müller, and Hoffmann always turned to current questions of chemical theory and frequently placed long-held views in doubt.[19]

Kekulé and Williamson made a most favorable impression on each other. Kekulé later recalled

> Chemistry then stood at a turning point; the theory of polybasic radicals was emerging. The ingenious Odling also consorted with Williamson. Williamson insisted on clear formulas, without Kolbe's commas and buckles or Gerhardt's brackets. That was an excellent schooling for me, generating an independent spirit.[20]

In a Latin curriculum vitae for the University of Heidelberg (1856), Kekulé wrote, "I must not fail to make mention of Williamson, that wisest of men and most learned of philosophers, who was not my teacher but my friend, and to whom I owe so much."[21] Soon after his arrival, he wrote to Planta: "I'll tell you about Williamson some other time, for once you start on him you're not soon finished, he has too many sides to be able to be characterized in a few words."[22]

Kekulé summarized his intellectual heritage in the following fashion:

> I had had the opportunity in Paris to become acquainted with the yet unpublished views of Gerhardt, and now the good fortune was allotted me, in lively and friendly intercourse with Williamson, to become thoroughly familiar with the patterns of thought of this philosophical intellect. Originally a pupil of Liebig, I had become a student of Dumas, Gerhardt, and Williamson; I no longer belonged to any school.[23]

Kekulé could not have chosen a better time to arrive in London. Within two months of his arrival, Odling had published his article on "substitution values" of atoms and "mixed types," Williamson had published on chlorosulfonic acid, and the "jolly row" between Williamson and Kolbe had been played out before the Chemical Society. There can be no question as to where Kekulé's sympathies stood. During this period early in 1854, Kekulé conceived the idea of sulfurating the organic acids, whereby theoretical conclusions similar to those of Williamson and Odling could be drawn.

Kekulé's sulfurations succeeded beautifully, producing thiacetic acid and related compounds, and he drew the expected conclusions. Published in both the *Proceedings of the Royal Society* and in Liebig's *Annalen*, the paper contained the following Williamsonian statement:

> . . . the new (Gerhardtian) formula notation is truly a better expression of the facts than the notation that has hitherto been customary. . . . It is

> not merely a difference in notation, but rather an actual fact, that . . . sulfur, like oxygen itself, is *dibasic*, so that one atom is equivalent to two atoms of chlorine.[24]

In this statement Kekulé went even further than the letter, if not the spirit, of Williamson's theory, for up to this time Williamson had never explicitly claimed polybasicity in an *atom*. Kekulé later based a priority claim for valence on precisely this point—conveniently forgetting about Frankland's paper of 1852 and Odling's of a few months earlier.

Kekulé emerged here as an advocate of the reform movement in chemistry. He argued for the new chemistry not just with his German friends and in the company of Williamson and Odling but also with his colleagues in Stenhouse's lab. He sought, unsuccessfully, to become the German translator of both Gerhardt's *Traité* and Laurent's *Méthode*. His habilitation colloquium in Heidelberg (15 March 1856) was a defense of Gerhardt's views.

Kekulé was still in London when Odling published "On the Constitution of the Hydro-carbons" (March 1855) and when Wurtz published on mixed radicals (July 1855). Odling's article[25] used the new weights and focused on the hydrocarbons, rejecting any absolute significance to the concept of radicals and introducing a "marsh gas type" (CH_4) along with Gerhardt's hydrogen, hydrogen chloride, water, and ammonia types. Wurtz' paper defended the reform movement and contained his subatomic speculation discussed previously.

In a famous story, Kekulé later recalled a dream he had one summer evening on the top of a London omnibus, wherein atoms formed lines and chains in a giddy dance and whirled before his eyes—a dream that marked the birth of the structure theory, in Kekulé's view, since that night he wrote down the forms he had seen and later published ideas based on them.[26] The event, if true, probably dates from the summer of 1855, since Kekulé mentioned that an evening conversation with Hugo Müller had led to the revelation; Kekulé's roommate and best friend Reinhold Hoffmann, who departed London in August 1854, is not named and so presumably was not then in London. The dream story dovetails nicely with the presumed proximate influence of Odling's and especially of Wurtz' 1855 papers. The evidence for Wurtz' influence[27] includes the following:

1. Kekulé's curious sausage-shaped atomic symbols, used by Kekulé until 1866, provide a literal image of what Wurtz was suggesting. Carbon, for instance, was represented as four monovalent spheres linked together by a sort of hoodlike "sausage-casing." These "sausage formulas" may be renditions of Kekulé's dream vision.

2. Kekulé first used the latter symbols shortly after the putative dream took place (in manuscript and classroom by 1856 or 1857, published by 1859).
3. Kekulé published an explanation (1867) that "polyvalent atoms, with respect to their chemical value [valence], can be viewed in a sense as a conglomeration of several monovalent atoms [subatoms],"[28] which is precisely Wurtz' idea.
4. Upon publishing his structure theory in 1857–1858, Kekulé stated that, in addition to the efforts of Gerhardt, Williamson, and Odling, who had contributed to his structural ideas, "the development of my views first became possible" by "reading between the lines" of Wurtz' "classic researches."[29]

My own conjectural scenario has Kekulé pondering Wurtz' article in the latest issue of the *Annales* just arrived from Paris[30] that evening in the summer of 1855, and with a mind appropriately focused on hydrocarbons by Odling's recent paper. Departing London shortly thereafter for home and frantically seeking an academic position, Kekulé had no immediate chance to write up his ideas for publication. But the opportunity was not long in coming.

Kekulé on the Nature of Carbon

Kekulé spent the fall of 1855 in Darmstadt pondering his future. Brother Charles had offered to set him up in London as a practical chemist, but Kekulé had regarded that as the most miserable sort of "chemical woodchopping" and "bootpolishing." On a six-week journey, he traveled to Giessen, Marburg, Frankfurt, Wiesbaden, Bonn, Berlin, and Heidelberg to explore the possibility of habilitation, speaking personally with Will, Kopp, Kolbe, Fresenius, Böttger, Baumert, Mitscherlich, Rose, Bunsen, and others. Kekulé decided that Heidelberg offered the best opportunity for an organic theorist since the dominant figure, Bunsen, was an inorganic and physical–chemical empiricist. Ever since Liebig's transfer to Munich and effective retirement from the laboratory, Bunsen's had become the most famous teaching lab, which meant a large number of potential students, hence income. After lengthy negotiations, Charles Kekulé was finally persuaded, against his better judgment, to advance his brother enough money to enable him to attempt "a stupid scholarly career."[31]

In January 1856 Kekulé moved to Heidelberg and two months later had obtained the necessary permissions and passed the examination for the *venia legendi*. Bunsen was not in the habit of providing laboratory facilities for Heidelberg Privatdozenten who were not also his assistants. Kekulé was therefore forced to find facilities for himself, and his relationship with Bunsen always remained cool, although each recognized the other's merits. He rented a house on the Hauptstrasse, set up his residence as well as a laboratory and lecture room there, hung out his shingle, and began to take on paying customers for summer semester 1856. His first Praktikant was his old friend Reinhold Hoffmann, his second a promising young man named Adolf Baeyer (who had also studied directly with Bunsen). With Kekulé working alongside them, the small lab was filled to capacity.

Kekulé's rent was reduced the following year by the offer of Emil Erlenmeyer to share the expenses and use of his lecture room, which was considerably more spacious than the laboratory, as it could accommodate ten chairs. Erlenmeyer, a pharmacist who had also earned a chemistry Ph.D. from Liebig in Giessen, had been living in Heidelberg since 1855 doing consulting for artificial fertilizer manufacturers and pursuing, with mixed success, some entrepreneurial activity. Using an advance from his father-in-law for support, he habilitated in 1857 and thereafter entered into an academic career.[32] Kekulé, Baeyer, and Erlenmeyer became good friends and learned much from one another.

According to his later autobiographical sketch, Kekulé wrote down a version of his structure theory shortly after his arrival in Heidelberg (it must have been the spring or summer of 1856), and showed it to two close friends (Baeyer and Erlenmeyer?), both of whom expressed doubts. Deciding that either the time or the theory was not yet ripe, he laid the manuscript away in a drawer. Apparently, he decided it was wiser to concentrate on producing a number of small but respectable empirical studies before coming forward with a broad-ranging theoretical interpretation of organic chemistry.[33]

But Kekulé was incapable of producing empirical studies that did not directly relate to his theoretical obsessions, and the first work to emerge from his new laboratory illustrates this. Sent to Liebig at the end of December 1856, Kekulé's paper attempted to settle the controversial topic of the constitution of the fulminates. Kekulé's analyses and specific constitutional proposal in this paper are less important than the fact that he took the opportunity to defend a marsh gas type that could be used to represent many small organic molecules. He carefully added that he was using the word and concept *type* not in Gerhardt's positivistic "synoptic" sense but rather in the sense of Dumas' *mechanical* types. He thus emphasized the structuralist connotations of the newer type theory in the tradition of Williamson. The same explicit reference to mechanical, that is, structural, types occurs in Baeyer's first paper from Kekulé's lab.[34]

In the spring and summer of 1857, papers appeared by an ausserordentlicher Professor at Göttingen, Heinrich Limpricht (a former assistant of Wöhler), and two of his assistants, Louis von Uslar (a former assistant of Kolbe) and Otto Mendius.[35] Two years earlier, Limpricht had been the first in Germany to publish a textbook based on Gerhardt's type formulas, but in Kekulé's opinion he had not properly understood the true sense and implications of the newer type theory. These two papers gave Kekulé the opportunity he had been looking for to publish his structural ideas, for Limpricht had opened himself to criticism in several ways, and such a critique provided the rhetorical launching platform for more general considerations.

Kekulé's specific criticisms[36] focused on the concept of copulated compounds, on Gerhardt's "basicity law" to which Limpricht made repeated reference, and on Limpricht's retention of the oxygen equivalent $O = 8$ within the Williamson–Gerhardt water type. Kekulé was philosophically troubled by Limpricht's use of types in a purely schematic rather than realist sense. This is what he meant when he wrote, seven months later, that "some chemists have fully adopted the external form of the newer type theory, while either misunderstanding or interpreting differently the underlying idea."[37]

Kekulé not only included Limpricht and his students in this group, but also Gerhardt and Odling. All these chemists had often proposed formulas that were impossible to represent consistently by the "theory of polyatomic radicals" as Kekulé conceived it.[38] In 1854 Williamson had made a similar criticism of one of Gerhardt's formulas, and Kekulé was following his friend's lead here; he also thought of Dumas and Wurtz as defenders of this realist version of type theory. The proper criterion for the newer types was whether the formulas could be constructed according to the accepted atomicities of the radicals and elements composing the compound, linked together in atomic–molecular chains. A consistent theory could be constructed following this criterion, Kekulé averred, and that theory would make the hypothesis of copulated compounds unnecessary.

His first sketch of this theory—which we know today as structure theory—is an intentionally incomplete version, dated 15 August 1857 and published in the November issue of Liebig's *Annalen*.[39] Kekulé here attempted to demolish the surviving remnants of copula theory and at the same time provide the first *systematic* exposition of the "theory of polyatomic radicals," whose origin and proper interpretation he ascribed mostly to Williamson. Molecules are "contiguous juxtapositions" (*Aneinanderlagerungen*) of atoms, he stated, constructed according to atomic valences, which he called "atomicities" or "affinity units."[40] He thus traced the atomicities of radicals to the atomicities of the constituent atoms, which were monobasic as in hydrogen and halogen, dibasic as in oxygen and sulfur, or tribasic as in nitrogen and phosphorus. A footnote[41] asserts that carbon is "tetrabasic or tetratomic," as in CH_4 or CCl_4. He defined a *radical* as simply that portion of a molecule that persists unchanged in any given reaction; thus different radicals may be assumed in the same molecule, depending on the reaction being considered. They are a theoretical expedient only, nonexistent in and of themselves.

Limpricht published a reply to this paper, not so much denying the truth of Kekulé's assertions but suggesting that Kekulé had distorted his views in opposing them.[42] Limpricht's reply in turn provided the inducement for Kekulé to write a more complete version of his theory, dated 16 March 1858, which was published two months later and grandly entitled "On the Constitution and Metamorphoses of Chemical Compounds and on the Chemical Nature of Carbon."[43] After a polemical introduction in which Kekulé elaborated and clarified his earlier criticisms, he boldly declared

> I consider it necessary, and given the present state of chemical knowledge in many cases possible, to go back to the elements themselves of which compounds are composed, in order to account for the properties of chemical substances. I no longer consider the principal task of the times to be the determination of atomic groupings, which due to certain properties can be considered as radicals, in this way assigning compounds to a few types which thus have scarcely any more significance than pattern-formulas. I believe on the contrary that we must extend our considerations to the constitutions of the radicals themselves, that we must determine the relations of the radicals among each other, and that we must derive the nature of the radicals as well as that of their compounds from the nature of the elements. My earlier considerations on the nature of the elements and the basicity of the atoms serve as the point of departure for these views.[44]

Now, as he had failed to do earlier, he explored the implications of considering carbon as a tetratomic element. Just as examples were known in which atoms of a single element could link to each other—especially in the elemental gases hydrogen, oxygen, nitrogen, and chlorine—carbon could reasonably be supposed to do the same. For compounds containing two carbon atoms, which possess between them eight "affinity units," two units, one on each atom, must be used to link them together. Hence, six affinities remain for attachment to other atoms, and C_2 is indeed a "hexatomic radical" as in, for example, the ethyl series. Every subsequent carbon atom added to the "skeleton" (this neologism first occurs in this article) requires two valences to be incorporated into the chain and so adds a total of two free valences to bond hydrogen or other atoms.[45]

Of course, other bonding patterns exist as well. Hydrogen atoms may be only *indirectly* bonded to the carbon group, that is, they may be held only through the intermediacy of a polyatomic atom such as oxygen or nitrogen, which is itself bonded to the carbon chain—such as in alcohols. Alternatively, more than one carbon group may be indirectly held together through the intermediacy of such a polyatomic atom—such as in ethers. In either case, those atoms that are indirectly or incompletely bonded to carbon, that is, those atoms in which not all of their affinity units (valences) are bonded to one and the same carbon group, Kekulé began to call "typical" atoms (that is, atoms of the type). Typical atoms in this sense thus included the replaceable hydrogen atoms of alcohols, acids, and amines, the oxygen atoms of alcohols, ethers, acids, and esters (except the carbonyl oxygens), and the nitrogen atoms of amines. Kekulé later defined a *radical* as a group consisting of the basic hydrocarbon skeleton plus all the atoms directly and completely bonded to it, i.e., excluding only the typical atoms.[46]

This definition of radicals is an extension of Liebig's and Laurent's idea of distinguishing atoms that are "inside or outside of the radical," the distinction having been provided by the type notation. For many years Kekulé used this more theoretical definition for taxonomic purposes, while simultaneously applying his more empirical and flexible definition stated in the previous paragraph. But the former definition is only slightly less general than the latter, for atoms "of the radical" are only occasionally attacked, which is to say that most chemical reactions exclusively involve atoms "of the type," which compose what modern chemists refer to as functional groups.

Given such an ambitious program, the epistemological question of the determinability of atomic arrangements within molecules through the study of chemical reactions was a point that Kekulé needed to address directly. Kekulé was a student of the structural agnostic Gerhardt and of the cautious realist Williamson (who had studied directly under Auguste Comte, the father of positivism). Consequently, he was sensitive to the need for epistemic caution. He emphasized that the determination of the actual physical arrangements of atoms was as yet impossible. The chemist needs to exercise extreme care in drawing conclusions about "constitutions" from the study of chemical reactions since reactions necessarily produce intermolecular and sometimes also intramolecular rearrangements. "Rational formulas" are nothing but "reaction formulas," he proclaimed, and can be nothing more in the current state of the science.[47]

But Kekulé believed, pace Gerhardt, that the chemist is by no means powerless to gain information on constitutions, especially if one proceeds while simultaneously renouncing the attempt to determine actual spatial atomic arrangements. At the end of his 1858 paper he proposed three axioms (above and beyond simple application of the valence rules) that he believed were justified to be used in determination of constitutions. First, homologous series are to be constructed schematically by using the "simplest arrangement" of carbon atoms, in which each carbon atom in a chain is connected to each of its neighbors by a single affinity unit (Kekulé essentially ignored olefinic and aromatic compounds in this article). Second, in reactions that proceed without loss of carbon, it is to be assumed that the constitution of the carbon skeleton of the reactant is preserved in the product. Finally, when the carbon skeleton is broken up, aspects of its original constitution may be deduced from the constitutions of the products.[48]

The last words of Kekulé's article were a formalist obeisance to positivist ideals, a safe cautionary note for a young man attempting against bad odds to make a career—moreover, a statement that appears to have been closely modeled on the final paragraph of Wurtz' July 1855 paper. Kekulé wrote

> Finally, I think I should emphasize that I myself place only a subordinate value on considerations of this sort. But since, in the total absence of exact-scientific principles in chemistry, one must for the moment be content with conceptions based on probabilities and convenience, it seemed appropriate to communicate these views, since, it appears to me, they give a simple and rather general expression especially for the most recent discoveries, and therefore their application may perhaps facilitate the discovery of new facts.[49]

Of course, Kekulé did succeed in his career, his first call coming just four months after his second structure theory paper appeared. When Jean Servais Stas traveled to Germany to seek a successor to D. J. B. Mareska at the University of Ghent, his attention was first drawn to Limpricht. However, both Liebig and Bunsen pointed out to him that Kekulé was far more cosmopolitan, would have no trouble in lecturing in French, and had already earned a sterling teaching reputation. (When Privatdozent Ludwig Carius took over the organic chemistry course after Kekulé's departure from Heidelberg, he was shocked to see Kekulé's eighty auditors gradually dwindle to six!)[50] Impecunious, energetic, and ambitious, Kekulé was overjoyed at the call, even though the exigencies of becoming a Belgian scholar were daunting, and the Ghent laboratory proved to be stripped bare upon his arrival in the middle of November 1858.

A letter from Kekulé to Erlenmeyer gives a good impression of his activities during his first semester in Ghent, chief of which was the writing of a textbook of organic chemistry:

> I have again been working regularly until one or two o'clock, often, as I did in Heidelberg, until 11:30 or 12 on my textbook and only then beginning to piece together a lecture. . . . In the morning at 9 I go to the laboratory, lecture three times a week from 10 to 11:30 . . . at 12 a frugal bite to eat in the lab, at 4:30 dinner—the high point of the day, and my only pleasure in life other than cigars (genuine Havannas, I can afford them now!)[51]—then a strong cup of coffee, and around 6 or 6:30 back to my digs, first for digestion and then for work. My lifestyle is simple. No diversions except Sundays, when I eat around one and then take my required constitutional walk. No longer do I go to theater; concerts and balls are neglected; no beer; no social visits. If with such a lifestyle I fail to achieve the dignity of "scholar" it is not my fault; of "man" there is little enough left over.[52]

In his old age, Kekulé recalled that Liebig told him that he who does not ruin his health with overwork would never make it in academic chemistry. Kekulé averred that for years he followed this advice to the letter.

Kekulé did of course achieve "scholarship," one sign of which was the appearance of his textbook. Like Kolbe's, it was published in fascicles, the first (240 pages) appearing in June 1859; after 1867, when it was about two-thirds completed, Kekulé ceased writing. Also like Kolbe's, the first fascicle contained an interesting and perceptive, if one-sided, history of recent chemical theory. It also contained a longer revised version of Kekulé's theory of the atomicity of the elements (structure theory) and unveiled in published form Kekulé's idiosyncratic sausage formulas.

This work enjoyed tremendous success, far outstripping Kolbe's textbook. Not only did the fascicles appear faster, but it was both modern and imaginative in its outlook, and chemists took notice. It proved an effective means of publicizing and propagating the views of the newer French–English school to new audiences, especially to German students. After 1865, it also helped quickly to establish Kekulé's benzene theory. When the first fascicle appeared, Rudolf Fittig said he read it with such excitement that he could hardly put it down. His testimony is significant, not least because he was a student of Wöhler and Limpricht at Göttingen and so had no direct or indirect personal connection with Kekulé. He was referring to this period when he wrote, toward the end of his life, that "in a certain sense all of us were Kekulé's students."[53] This was not true for Kolbe, however, who later commented that he barely glanced through it since, as he said, he was sure he could learn nothing from it.[54]

Structure Theory and the Problem of Independent Codiscovery

Even setting aside the problem of Kolbe's and Frankland's contributions to the development of structure theory, the sequence of events leading to Kekulé's 1857–1858 papers is complex. In both of these papers, Kekulé tried to make clear his intellectual debts to his predecessors, especially Gerhardt, Williamson, Odling, and Wurtz, and he disclaimed any great originality. There is no reason to read any exaggerated modesty into his protestations, for the theory of polyatomic radicals and the valence ideas being discussed by these and other chemists did indeed provide at least the implicit basis for most of the content of Kekulé's articles. What Kekulé had done that was new was to provide the first really systematic exposition of these ideas and to explore some novel implications—especially that of the combination of carbon atoms with one another, within the restrictions of tetravalence.

The latter proposition cannot be underestimated; it is the foundation of structural organic chemistry. And yet we can find intimations of even this suggestion in the work of many of the above-named predecessors of valence theory. Moreover, less than a month after Kekulé's second paper was published, Archibald Scott Couper independently proposed a theory that was identical to Kekulé's in all of its essentials, including carbon self-linking.[55] Couper's claim to be an independent codiscoverer was valid, but his work both before and after 1858 was virtually without influence on the science of his day. Thus, it is not further discussed here, other than to remark that a case can be made that Wurtz' subatomic speculation may have influenced the genesis of his, as well as Kekulé's, theory.[56] Frankland and Kolbe, of course, also had claims to some elements of Kekulé's theory, and their story is treated in the next chapter.

Williamson, a model of gentlemanly reserve, never urged his priority for polyatomic radicals and valence theory beyond what Gerhardt, Wurtz, and Kekulé granted him, even though many historians now see him as the real genius behind the entire reform movement culminating in the Karlsruhe Congress. As for William Odling, he had been at the end of 1853 the second after Frankland to attempt a general statement on atomic valence and the first to do so using atomic weights—at the same time developing the concept of "mixed types." In 1855 he had also clearly laid out the marsh gas type. He tactfully pointed out these contributions in an article published shortly after Kekulé's second structure theory paper. He remarked, "Kekulé has recently published a paper in which these views are brought forward more definitely, and with an amount of illustration heretofore not possible."[57] Any incipient priority fight between Kekulé and Odling essentially died at this time. Kekulé's own view, which as far as can be determined remained unexpressed directly to either Williamson or Odling, was that he was the first to apply the Englishmen's ideas systematically to *atoms* rather than radicals.[58]

Wurtz reviewed Kekulé's structure theory paper in the first issue of his *Répertoire de chimie pure*, a sort of French version of Berzelius' *Jahresbericht* and an official organ of the new Société Chimique de Paris. In general quite complimentary, Wurtz approved of Kekulé's call to go "back to the elements" by constructing molecules according to atomicities of the atoms, analogous to the way the theory of polyatomic radicals had been developed. He pointedly noted, however, that others before Kekulé had already gone back to the elements: Williamson, Odling, and he himself, in 1855. Moreover, "I do not think I deceive myself," he wrote, "in believing that my recent work on the syntheses of polyatomic alcohols has provided this idea with the experimental confirmation which it lacked." He cited with approval Kekulé's comment that his (Wurtz') ideas had formed Kekulé's "point de départ."[59] Just after Wurtz' review was published, Wurtz wrote to Kekulé congratulating him on his recent call. He also sent a copy of the first issue of the *Répertoire* and asked if he would consent to being the correspondent for Belgium. In a letter dated 15 February 1859 Kekulé gladly consented, and used the opportunity to make an argument for his own priority. (He thought enough of the matter to write an exact copy of the letter for his own files; it is only this copy that has survived.)

Kekulé pointed out that in his thiacetic acid paper he had explicitly drawn attention to the monatomic character of hydrogen and chlorine and to the diatomic nature of oxygen and sulfur. This was more than a year before Wurtz' 1855 paper appeared. Kekulé continued

> Your lovely investigations on the glycols are even more recent; so they can have served just as little as a point of departure for my views, and indeed I only remarked that the *further development* of my views only became possible through your experiments. As for the views of William-

> son and Odling in the papers you cited, I gladly acknowledge, and did so in my publication, that the ingenious considerations of these chemists formed the point of departure for my views. But whether the idea of the basicity of the atoms *as we presently conceive it* is expressed "clearly" in these papers is less certain. In fact it doesn't seem so to me; I think that if one goes back that far, then no doubt Laurent and Gerhardt could be considered to be the originators of this idea.[60]

Wurtz replied to Kekulé: "I assure you that had I known of those passages in your [1854] memoir I would not have written the little note to which you allude." He promised to rectify the error publicly. In his next published article, he added a footnote wherein he expressed regret at having unfairly omitted citation of Kekulé's 1854 paper.[61]

About a month later, during Easter vacation of 1859, Kekulé had the opportunity to visit Wurtz in Paris. He described his trip to Erlenmeyer and raved about Wurtz: "Capital fellow!—it is not possible that two people could agree more closely with the general conception of a science than we two."[62] In 1865 he counseled his student Ladenburg to continue his education in Paris with Berthelot, since "Wurtz is the same as I."[63]

Shortly thereafter, the first fascicle of Kekulé's textbook appeared, and one of the first colleagues to whom he sent a complimentary copy was Wurtz. In the accompanying letter, Kekulé asked Wurtz, "if you should happen to have a free moment," for his opinion of the theory contained therein. Kekulé wrote

> I think you will not consider it mere customary flattery when I tell you that I lay more value on your judgment than on that of most modern chemists. In principle I will enter no priority claim for theoretical views as long as this can possibly be avoided; I will also not, à la Couper, blow my own horn, trumpeting my views as a "new chemical theory"; I will leave it calmly to time and to the legal sense of others to demonstrate whether in the development of these views anything reverts to me, and how much. If I now request your opinion on this, it is because I believe that the viewpoint which the majority of chemists in the future will have can be little different from the viewpoint that you have now.[64]

In this fascicle Kekulé had dealt with Wurtz' research in a complimentary fashion. Wurtz had been the first, Kekulé affirmed, to have correctly interpreted Berthelot's work on the glycerides, and his "ingenious conceptions and brilliant discoveries" of the di- and trialcohols had subsequently served to develop the theory of polyatomic radicals.[65] Wurtz replied to Kekulé's letter

> I willingly recognize that you have gone further than your predecessors in seeking to consider the fundamental properties and basicity of the elements in the conceptions regarding the constitution of compounds and chemical formulas. I consider your idea of the polyatomic nature of carbon as both sound and fertile, susceptible of important development. I must also thank you for the role you ascribed to me in your chapter on polyatomic radicals. I would not wish to claim more that what you have attributed to me.[66]

But in the same letter, Wurtz expressed frustration that his work, appreciated in Germany and England, was being neglected in his own country. This frustration explains why, despite his agreeable concessions to Kekulé, during the next few years he repeatedly (though carefully and with proper appreciation for the work of others) urged his own priority in certain respects.[67] In brief, Wurtz thought he deserved credit in two senses. The subatomic speculations in his 1855 article, and the associated exposition of "multiply condensed hydrogen" as the schematic basis for multiple and polyatomic types, provided a general theoretical foundation for valence that had not hitherto existed. This contribution was historically significant and heuristically valuable even if the theoretical details were still speculative, indeed, even if they were false.

More importantly, Wurtz thought that his elucidation of the glyceryl radical and his discovery of glycols offered the first real experimental confirmation of the existence of polyatomic radicals and that it properly illustrated their significance and fruitfulness. He was not the first to introduce the idea of polyatomic radicals, he was happy to concede, but his work provided the completion and perfection of the theory explaining them. Moreover, Wurtz believed that his proof of the existence of polyatomic radicals constituted a solid argument for the existence of polyatomic elements.

But these were relatively small merits compared to the great synthesis and advancement made by Kekulé, and he always granted Kekulé full credit for the application of valence ideas to carbon itself (even urging the superiority of Kekulé's claim to that of his own student, Couper). There is no question that Wurtz recognized the fundamental breakthrough achieved by Kekulé, and he no doubt felt chagrined that he had not gotten there first. In the fall of 1858, presumably under the influence of Kekulé's articles, he finally decided to adopt the atomic weight notation of Gerhardt, Laurent, Williamson, Odling, and Kekulé.

Wurtz' first paper using the new formulas—the same paper in which he corrected his earlier omission of Kekulé's 1854 valence argument for oxygen and sulfur—was a long review article on glycols, in which a great deal about structure theory appears. Wurtz stated that Williamson's work had had "a large part in the development of this theory," but that Kekulé's "extremely important theoretical article" of 1858 had put the pieces together, literally as well as figuratively. The principle behind the theory was to place the idea of bonds (*liens*) between polyatomic elements and radicals in the foreground. The new theory not only unified radical and type theories, but it also provided the basis for demonstrating the unity of inorganic and organic chemistry. According to Wurtz, organic radicals "offer more points of attack" than inorganic radicals, which lends a greater "fragility" to "the molecular edifice," otherwise they are the same. "The ideas, the hypothesis if you will, which express the molecular structure of the first, are exactly applicable to the second."[68]

Although since 1853 he had openly defended the reformed chemistry without adopting the new formulas, from this time onward Wurtz exhibited total commitment to the reforms and a missionary zeal for propagating the work of Laurent and Gerhardt, both of whom had since died before the age of forty-five. In 1860 he was a principal organizer, along with Kekulé and Weltzien, of the Karlsruhe Congress, whose subtext was reform. He also presented a formal lecture before the Société Chimique on his discovery of the glycols, using it as a pretext for evangelism. In 1862 he delivered an "Éloge" in memory of both Laurent and Gerhardt to the Société des Amis des Sciences,[69] again taking the role of advocate as well as obituarist. In the same year he spoke before the Chemical Society of London "On Oxide of Ethylene" and its theoretical implications.

In 1863 he presented a book-length series of lectures to the Société Chimique covering all of chemical theory, which was subsequently printed in both French and English. In 1864 he lectured at the Collège de France with similar intent and results. In 1866 he began publishing his *Leçons élémentaire de chimie moderne*, using the reformed weights. Finally, in 1868 Wurtz began to publish a detailed dictionary of chemistry similar to the *Handwörterbuch der reinen und angewandten Chemie*, the first fascicles of which consisted of a history of chemical theory. Both the history and the dictionary itself proved to be clever means of propagating the new doctrines. However, despite all these efforts, many in France continued to use the older equivalents long after they were rejected elsewhere. Wurtz tried one last time, writing a historical precis of the atomic theory in 1879. But it was only after his death in 1884 that the ultimate victory for atomic weights in France was won.[70]

One more claimant to structure-theoretical ideas, Aleksandr

Mikhailovich Butlerov, requires discussion. A student of Nikolai Zinin and Karl Klaus, Butlerov had already been teaching for nine years at the University of Kazan when he took a nine-month trip to western Europe in 1857–1858. He met with Kekulé twice during this period and spent five months in Wurtz' laboratory, at just the time when Couper was formulating his ideas on structure theory. He was already then theoretically active, as demonstrated by a presentation he gave to the Société Chimique in February 1858 and a commentary on Couper's theory published in Liebig's *Annalen* later that year. At this time, Butlerov's thinking on the constitutions of organic compounds was similar not only to Couper's but also to Kekulé's, Frankland's, and Kolbe's. However, he took exception to Couper's too absolute formulations and to his complete rejection of type theory in both its older (Dumas') and newer (Gerhardt's) forms. He said he much preferred Kekulé's version of structure theory, which in any case had appeared slightly earlier than Couper's.[71]

On a second trip to western Europe in 1861, Butlerov gave what proved to be an influential paper at the Naturforscherversammlung in Speyer. He now disavowed his earlier rejection of Couper, affirming with the Scotsman that both type theories must be completely discarded. He criticized Kekulé and other structuralists for retaining vestiges of types that had done nothing but introduce confusion and inconsistencies. What was necessary at this point in the science was to apply the "theory of chemical structure" in a thoroughly consistent fashion. "Chemical structure," in Butlerov's definition, can be considered to be equivalent to an application of Kekulé's "atomicity of the elements," with an added epistemological caution that actual physical constitutions (spatial arrangements) were as yet inaccessible to chemists, who could only determine *apparent* structures from purely *chemical* phenomena. Kekulé, Kolbe, Wurtz, and others had been making a similar caution, but it was Butlerov's vocabulary that caught on.[72]

Soviet historians of chemistry during the Stalin and Khrushchev periods attempt to show that Butlerov's concept was unique, fruitful, and modern, hence that he was the founder of the "classical theory of chemical structure." This thesis is vitiated by the fact that Butlerov himself never made such a claim, merely asserting (vocally and repeatedly) that from the beginning he had called for a clear and consistent application of the theory, that he had been one of the most successful of the first generation of structural chemists, and that his contributions had often been neglected. All of these claims were accurate. In chapter 10 we will return to Butlerov and discuss his relationship with Kolbe.

All of the nuanced and often overly subtle priority skirmishes concerning the origins of structure theory highlight the complexity of this period of rapid theoretical evolution. The speed of development and the near coincidence of multiple contributions is partially explained by the circumstances outlined at the beginning of this chapter. Valence and structure were certainly in the air breathed by chemists of the 1850s, a point noted by Kekulé in his autobiographical sketch. One critical thesis, however, remained unelaborated until Kekulé's structure theory papers appeared, a thesis upon which all of structural chemistry depended: the self-linking of carbon atoms. Why was this one crucial idea so relatively slow to emerge when it seems to be a simple application of valence?

The answer apparently is that electrochemical dualism was not completely dead, even among the "typists" of the 1850s, and for good reason. Among the followers of Laurent and Gerhardt, it may have become accepted that atoms of hydrogen, oxygen, nitrogen, and chlorine form elemental molecules by combining with themselves, but this phenomenon was troublesome since it remained theoretically inexplicable. A further expansion of the concept to carbon in nonelemental molecules must have stuck in the craw of even the most modern reformers. After all, it is an understandable and even necessary tendency for scientists to attempt to explain the microworld by referring to the macroworld, hence to invoke macroscopic forces such as gravity, or more commonly, coulombic attraction and repulsion, to explain chemical forces.

Although some typists (Brodie and Couper, for example) attempted to develop electrochemical theories that might account for at least some cases of self-combinations of atoms, these attempts never achieved much success.[73] Indeed, one reason for the general retreat into positivism of European chemists around 1840 had been the perceived general failure of electrochemical theory and the absence of any fundamental analog to replace it. Little had changed during the following two decades. The key to the discovery of carbon self-linking—and hence, the key to structure theory—had to be a certain willingness to suspend belief in *any* fundamental explanatory analogy to chemical affinity and simply to trust the direction that the leading systematists of chemical reactions were heading. In other words, it required a certain empiricism, or even positivism, of outlook, and Kekulé clearly possessed such a vision.

Ironically, though, it was precisely because of Kekulé's realist tendencies, in the form of a structuralist conviction of the ontological reality of chemical atoms and the physical linking function that they can perform (a conviction derived from Williamson and Wurtz), that he was able to conceive of a theory of chemical *structure*. Kekulé himself thought that this realist tendency could be traced back to his architectural training, which had engendered in him an "irresistible necessity for visualizability (*Anschaulichkeit*)" and had led him to become the architect of molecular rather than of wood, brick, or stone edifices.[74]

8

Carbonic Acid and Natural Types

While Kekulé was developing his ideas on carbon tetravalence and the use of that concept in formulating molecular skeletons from valence-linked carbon atoms, Kolbe and Frankland were traveling a partially independent path that led to similar ideas. Influenced by electrochemical instincts, these chemists did not develop the second element of structure theory (carbon linking) in the manner that Kekulé did, and so their vision of a general theory of organic "constitutions" looks somewhat different. Both the similarities and the differences between the Kekulé and the Frankland–Kolbe versions of what came to be known as structure theory provide insight into the development of this fundamental set of ideas.

Frankland and the Origins of the Carbonic Acid Theory

Edward Frankland's personal odyssey following the period of his direct association with Kolbe comprised a stay at Giessen, then professorial appointments at Putney College, Owens College (later the University of Manchester), and St. Bartholomew's Hospital, London, before he became Faraday's successor at the Royal Institution in 1863.[1] During this period, his scientific work exhibits a curious blend of old and new, of elements of the reform movement commingled with vestiges of Berzelian electrochemistry.

Frankland's first paper after leaving Marburg and the Kolbe–Bunsen orbit (read to the Chemical Society in November 1849) announced his discovery of the zinc alkyls, and he placed this work explicitly and repeatedly in the type theory tradition of Dumas, Gerhardt, Wurtz, Paul Thenard, and Hofmann.[2] He was convinced, and rightly so, that he had discovered an invaluable tool by which he could effect substitutions that were in a sense inverse of the traditional ones. Rather than replacing hydrogen atoms with halogen atoms, the intensely electropositive zinc methyl and zinc ethyl might be used to replace halogen or oxygen atoms with methyl or ethyl radicals, in a manner analogous to the way metallic zinc replaced iodine in the formation of zinc alkyls. He laid out this research plan first in an unobtrusive passage in his long paper published in the spring of 1852.[3] As early as 1851, as he later told Kolbe, he sought to realize this plan by trying to react zinc ethyl with phosgene or carbon tetrachloride.[4] But these experiments were not successful. The first chemist self-consciously to succeed in employing metals to alkylate organic compounds was Wurtz (1855), following a route suggested by Hofmann and Brodie (1850), who were themselves influenced by the work of both Williamson and Frankland.[5]

The theoretical conclusion of Frankland's 1852 paper stated a restricted form of the regularity since known as valence. Valence was intrinsically supportive of types and subversive to electrochemical theory since, in Frankland's words, the "combining power" of an atom "is always satisfied by the same number" of other atoms, independent of electrochemical properties. But Frankland also pointed out that the extreme version of type theory, which attempted to account for chemical properties *solely* on the basis of the arrangement of the atoms in the molecule (that is, by assignment of compounds to immutable types) is likewise untenable. For example, it is impossible to account for the difference in properties between antimony trioxide and antimony trimethyl, or between arsenic acid (AsO_5) and cacodylic acid (AsO_3Me_2), without considering the electrochemical properties of the components. Gerhardt came to a similar conclusion the following year: in replacing the hydrogen atoms of ammonia successively by three acetyl groups, a dramatic alteration of properties was apparent.[6]

Due to the difficulties attached to working with the particular substance Frankland had chosen, he continued to have little success in pursuing the promising program outlined in his 1852 paper. A full three years later he reported his subsequent work on zinc ethyl to the Royal Society and to the British Association for the Advancement of Science, but he still indicated only in outline form his future intent to attempt systematic alkylations of organic halides and oxides by means of zinc methyl and zinc ethyl. Kolbe mentioned in a footnote a few years later that Frankland had reported to him in a letter from this period that he had synthesized two mercaptans by reacting zinc alkyls with carbon disulfide. For some reason, Frankland never published this result; he did, however, later affirm that he had tried to alkylate carbonic acid with zinc alkyls.[7]

In a paper of June 1856 Frankland announced the research program for a fourth time, but once again publicly temporized.[8] He fully recognized the high promise of this plan; if he were successful, it would enable him to "ascend the homologous series of organic bodies," thereby attaining "clearer views of [their] rational constitution."[9] In both his collected papers and his autobiography he said that he knew that it was "an obvious and easy step" to apply valence ideas to carbon and that during this period he continued to try without success to react zinc alkyls with carbonyl chloride and carbonic acid. He noted that James Wanklyn achieved the experimental success, and Kekulé the theoretical success which he had been seeking.[10] In summary, from 1851 on Frankland consistently attempted to develop reactions that could exemplify valence theory as applied to carbon.

By 1855 he had drawn even closer to the thinking of the "typists," which is indicated by his noting a variety of new electrochemical anomalies, citing with approval an 1850 paper by Brodie, and suggesting that zinc ethyl "appears to belong to the so-called water type, and to consist of two volumes of ethyl and one volume of zinc vapour . . ." In 1857 both Kekulé and Baeyer cited this statement as further empirical evidence for "diatomic" elements, in this case zinc. But Frankland continued to formulate zinc ethyl as ZnC_4H_5 (i.e., a half-molecule of the compound according to the reformed chemistry), and there is no indication that he had become convinced of Laurent's and Gerhardt's central thesis that formulas should consistently express volume relationships.[11] Frankland, like his friend Kolbe, remained convinced

> . . . that, although all the electro-chemical theories hitherto proposed were far from satisfactory, yet, that amongst the factors of chemical action, the electrical character of elements could not be denied a place, without ignoring and leaving unexplained some of the most remarkable of chemical phenomena.[12]

Kolbe remained a more thoroughgoing electrochemist longer than did Frankland. Chapter 6 treated Kolbe's tentative rejection of Frankland's valence thesis in the first fascicle of his textbook (published in June 1854 concurrently with Frankland's Friday Evening Discourse at the Royal Institution just cited), on the basis of its apparent violation of electrochemical tenets. Frankland and Kolbe later stated, separately and publicly, that through correspondence during the year 1856 Frankland succeeded in persuading Kolbe of the truth of the "law of maximum combining capacities."[13] Unfortunately, this correspondence has not survived.

There is evidence that Kolbe was at least partially converted to Frankland's way of thinking as early as the end of 1855. In the article "Radicals" for the *Handwörterbuch*, completed by the first week of 1856, Kolbe discussed Frankland's law of combining capacities, which he now seemed to accept, as applied to carbon. He formulated acetic acid as "methyl carbonic acid," wherein one methyl radical substitutes for one of the two replaceable oxygens of the "dibasic" carbonic acid:

$2HO.C_2O_4$	$HO.(C_2H_3)C_2,O_3$
carbonic acid	"methyl carbonic acid" (acetic acid)

He then offered the "conjecture" that the C_2 of carbonic acid might be able to suffer a further substitution of methyl for oxygen, which would result in a compound that "would have the composition of acetone." The article shows signs of persistence of earlier ideas, as the concept of copulas was still strongly represented.[14]

Kolbe developed this approach further in a short sole-authored theoretical article in the *Annalen der Chemie*, written in December 1856 and published in March 1857, which he ever afterward considered to be the first statement of his "carbonic acid theory." In addition to deriving the homologous fatty acids and acetone schematically by substitution of the two replaceable oxygens of carbonic acid with hydrocarbon radicals, he noted that aldehydes could also be related to the same type by replacing one of the methyl groups of acetone by a hydrogen atom. A final speculation suggests the possibility that substitution of a third oxygen of carbonic acid by alcohol radicals might provide the rational composition of the homologous alcohols and ethers.[15]

When this paper appeared, Frankland thought that Kolbe had violated a private trust by publishing in his name alone. Considering the circumstances just described and the discussion that follows, it is likely that Frankland had suggested the leading idea of this theory in a letter to Kolbe in the summer or autumn of 1856, in the course of marshaling all possible arguments to persuade his friend of the truth of the valence law. Given the content of Kolbe's article on radicals written at the beginning of that year, it is clear that his mind must have been prepared for conversion to the general case, but it also appears that he had not yet forsaken copulas.

The story cannot be fully sorted out from existing documents, but several hitherto unpublished letters shed some light on the matter.

After Kolbe's paper appeared, Frankland refused to write to his friend for five years.[16] When Kolbe finally succeeded in prevailing on Frankland to tell him what had precipitated the break, Kolbe wrote two long letters of exoneration.[17] We have Frankland's response to the first of these. "I am of the opinion," Kolbe wrote to Frankland,

> that the foundation of the recent development of chemistry, and especially of our carbonic acid theory, is the truth, first expressed and brought to the attention of chemists by you, that the affinity of elementary radicals is always satisfied by the same number of atoms, regardless of their chemical character. From the instant I correctly understood the sense of your perception it was immediately clear to me that the organic dithionic acids are nothing more than sulfuric acid in which one of the six oxygen atoms is substituted by an alcohol radical, and just as quickly both of us became convinced that fatty acids, acetones, and aldehydes must arise through replacement of oxygen atoms in carbonic acid by alcohol radicals. We immediately agreed to work in collaboration to confirm this hypothesis experimentally by means of your zinc ethyl . . . You undertook to treat carbon [tetra]chloride and carbon [di]sulfide, and I phosgene and carbonic acid, with zinc ethyl.[18]

Kolbe reminded Frankland that after he had come to realize the truth of the law, he and Frankland had successfully worked through all their remaining differences and formulated the carbonic acid theory. Kolbe had written Frankland urging an immediate joint theoretical article outlining the rationale for the experimental program. Frankland had demurred, Kolbe reminded him, saying it was not customary in England to publish purely theoretical papers. But a paper by Piria shortly thereafter (October 1856)[19] made clear that they were close to being forestalled. Thus, in December Kolbe wrote out the eight-page theoretical paper and omitted Frankland's name (he continued), out of respect for Frankland's wishes. He did, however, send the manuscript to his friend, with the hope, he said, that Frankland might still be induced to append his name to it. He quoted directly from Frankland's reply of 18 January 1857—which does not appear to have survived—in which Frankland stated that he "sufficiently agree[d] with what you have said," that their views were "fundamentally the same," and that it was "the highest time, that something should appear from us on the subject."

Frankland appears to have intended these phrases as a clear request for joint authorship,[20] but apparently Kolbe did not interpret them as such. In any case, Kolbe had already made a firm decision in favor of single authorship more than two weeks *before* he received Frankland's reply (but he did not tell Frankland this!), for on 4 January 1857 we find Kolbe writing Vieweg:

> . . . it is a joy to be able to say to you that I have now come to a clear understanding of the important theoretical questions with which I have so long been continuously occupied, and which are so important for the presentation of the subject in my Organic [textbook]. I have just sent off a paper concerning this for publication in the *Annalen*, and am now proceeding with redoubled pleasure and energy on the continuation of my Organic, which, if I stay healthy, should move rapidly forward.[21]

The first few pages of the published paper use the pronoun "ich" and refer to Frankland by name. Kolbe then changed to "wir," doubtless intending thereby to refer to both of them, but most readers probably inferred the editorial "we" and concluded that inconsistent editing accounted for the change to the plural pronoun.

The matter was of the highest importance to Frankland. Characteristically modest and extremely reticent about priority claims for his whole life, he thought enough about the theory and his co-authorship of it to sever relations with a good friend to whom he owed a great deal:

> . . . what I complain of is, that owing to your having in this last memoir [of 1860] ignored my participation in the views of the previous memoir [of 1857] you have practically claimed & obtained the exclusive authorship of the Carbonic Acid & carbonic oxide types. I do not believe that you intended this, but such has been the result, &, as I attach much value to the theory, you will easily conceive that the present position of the matter is unsatisfactory to me & I should wish some way to be devised for setting it right.

Frankland suggested the means of "setting it right" might be a paper by Kolbe alone or by both of them, regarding the historical development of the theory.[22]

Kolbe tried to satisfy Frankland by writing an historical introduction to a volume of his collected papers published three years later. He declared there that the 1857 paper had been joint with Frankland; moreover, "Frankland had a large share, indeed much larger than is generally known, in the formulation of this hypothesis [of deriving organic compounds from carbonic acid]." The germ of the idea lay in Frankland's theory of atomicity of elements. Kolbe wrote:

> The unclear conception of the mode in which the so-called copulas were chemically combined was a great weakness in [my] hypothesis of copulated radicals. It is Frankland's merit to have been the first to throw light upon this, and thereby to have completely done away with the idea of copulation, by recognizing the fact that the various elements possess definite saturation capacities (atomicities).[23]

Although their correspondence was reestablished after Kolbe's self-exoneration, their relationship was never really warm thereafter, Frankland describing it as "more a scientific than a social friendship."[24] Frankland always considered Kolbe's 1857 paper as co-authored; it appears that way in the edition of his collected papers, Frankland noting editorially that his name had been omitted by "inadvertence."[25] Kolbe's son-in-law Ernst von Meyer agreed in his history of chemistry that this article was "joint" with Frankland.[26]

The affair was no less important to Kolbe, who really did begin thinking of the paper as his own, as is clear from his letter to Vieweg of 4 January 1857 and passages in his papers and textbook after 1857. His monographic paper of 1860 refers to the 1857 paper unambiguously as his alone. This was what was so galling to Frankland and may have led the latter, as we shall see, pointedly to ignore Kolbe in his own review article of 1860. Despite his clear public declaration of 1865, Kolbe was still referring to the paper as his in 1881.[27] Kolbe, however, did recognize Frankland's crucial role, publicly as well as privately, in the creation of the carbonic acid theory. Even if Frankland's name did not appear in the byline, Kolbe gave Frankland extensive credit *within* the 1857 paper and was absolutely clear in his letters to Frankland concerning his conviction of Frankland's necessary role.

Kolbe was fully cognizant of the importance of the idea sketched in the paper. Within days of its publication, he proclaimed to Vieweg a "small revolution" in organic chemistry, and if anything, he was understating the matter.[28] However, with the theoretical details and research prospectus now public, Kolbe initially had just as little success at obtaining experimental verification as Frankland. Not only were organozinc compounds difficult to work with, but Kolbe became seriously ill shortly after sending his paper off at the beginning of January 1857, and remained a semi-invalid for seventeen months. Wurtz had already forestalled one element of Frankland's program by "ascending the homologous series of organic bodies" using alkyl iodides and a metal (sodium, not zinc).

Early in 1858 news came of an even more direct application of the theory. James Wanklyn, Frankland's assistant at Owens College in 1856 and a fellow Lancastrian, went the next year to Heidelberg on Frankland's recommendation to study with Bunsen, and there attempted to synthesize new organometallic compounds, especially sodium and potassium alkyls. He succeeded in this goal, although he was unable to isolate the compounds in pure form. Even more interesting, he found that the "affinity" of sodium ethyl was so powerful that it attacked carbonic acid (anhydride) itself and transformed it into sodium propionate.[29] Here was a dramatic realization of the carbonic acid theory, for it appeared to represent a direct substitution of ethyl into carbonic acid. On 8 March 1858, Frankland wrote a gracious letter congratulating his former assistant.

> I am delighted to hear of your success with the Propionic Acid experiment. It is to me as you may imagine of peculiar interest & importance on account of its affording such a beautiful & satisfactory proof of the correctness of the views . . . expressed by Dr. Kolbe & myself in Liebig's *Annalen* more than a year ago.[30]

Kolbe was less happy. He had just succeeded in obtaining propionic acid, unfortunately in only trace amounts, by digesting potassium carbonate with zinc ethyl in ether, but he had not had time to publish when he got the news of Wanklyn's discovery.[31] He immediately wrote to Bunsen, accusing Wanklyn of treachery in poaching on his and Frankland's scientific territory. Bunsen replied with a friendly but strongly worded letter defending Wanklyn's integrity, and he enclosed Frankland's letter to Wanklyn as further evidence.[32] Kolbe let the matter rest, but he was not convinced. In his view, Wanklyn had "skimmed the cream" off of his and Frankland's discovery.[33] Frankland may also have had later second thoughts, for after another clash in 1867 he consistently opposed Wanklyn's further professional advancement, and with considerable success.[34]

Already aggrieved by Kolbe's action of a year earlier, Wanklyn's paper was a second blow for Frankland (a third, if Wurtz' 1855 research is to be counted). Frankland decided at this time to lay out his own ideas more explicitly. At a Friday Evening Discourse on 28 May 1858, Frankland presented in type formulas the elements of the carbonic acid theory, which he stated to his audience—as he had to Wanklyn—that he had published jointly with Kolbe in the *Annalen* article of March 1857. One and two methyl radicals substituted for one and two oxyen atoms of carbonic acid (C_2O_4) yield acetic acid and acetone, respectively, and a methyl and a hydrogen atom yield aldehyde. Furthermore, two methyls substituted for two hydrogens of methyl hydride (marsh gas) produce the new substance methyl-ethyl, and one methyl and one ethyl radical substituted into marsh gas result in the substance he himself had first isolated by chemical means in 1849—"ethyl" gas. He thereby illustrated what he had meant to say in published papers since 1852 by his intention to "ascend the series of homologous organic bodies," although he was still lacking the experiments to demonstrate the idea.[35] He sent Kolbe an offprint of this paper.

Once again, this paper represents an intimate amalgamation of old and new. The type formulas, substitution mechanisms, and hydrocarbon formulations are entirely in the spirit of the reformed chemistry, even if Frankland still retained the double carbon atom. He now even seemed to be explicitly conceding that ethyl gas was twice the size of the ethyl radical. However, the carbonic acid formulations could only be understood by reference to Berzelian dualistic formulas, with one and two unspecified water molecules (actually half-molecules, OH, where O = 8) understood to be present for monobasic and dibasic acids, respectively. The schematic clarity of matching four oxygens, rather than two, in carbonic acid, to the four hydrogens of marsh gas must have seemed to be decisive in Frankland's mind, counterbalancing what must have seemed to others a greater inconsistency. His amphibious theoretical position may have been as important as the experimental difficulties in holding Frankland back from rapid exploitation of the new ideas.

Frankland read a long review article "On Organo-Metallic Bodies" to the Chemical Society on 7 June 1860, a section of which further illustrates the application of his 1852 valence law to carbon compounds. For the fifth time, he indicated his intention to substitute alkyl radicals for chlorine in alkyl chlorides; experiments, he said once more, were in progress. But he noted that Wanklyn had already shown experimentally the tenability of the theory as applied to carbonic acid, and the theory could apply as well to aldehydes, alcohols, glycol, and glycerin. He did not even mention Kolbe. He wrote

> It would be greatly easy to extend this view of the constitution of organic carbon compounds; but the above examples are sufficient to indicate its general application somewhat more fully than I have previously done [here he cited Kolbe's *Annalen* paper—as if it were his own—and his 1858 Royal Institution talk], and more than this is not desirable until the hypothesis has been further supported by experimental results.

Curiously, he included neither this nor the 1858 Royal Institution article in his collected papers.[36]

Frankland's formulas suggest that he accepted the molecular magnitudes argument from the Williamson asymmetric syntheses, for he again depicted ethyl as a dimer, water as $O_2\left\{\begin{matrix}H\\H\end{matrix}\right.$, and zinc ethyl as $Zn_2\left\{\begin{matrix}C_4H_5\\C_4H_5\end{matrix}\right.$. He even attempted unsuccessfully to demonstrate the truth of the last formula by the Williamsonian strategy of preparing ethyl methyl zinc.[37] But he also did not seem to understand all of the details of the reformed magnitudes, for his propyl ether still contained only three double carbons rather than six. That he also had not accepted the atomic weight reform is clear, not from his use of double carbon atoms (for the reformers themselves sometimes used these for various reasons), but from the fact that he thought nothing of separating the individual equivalents of oxygen or zinc in certain formulas. For Frankland, these equivalents must have been the same as chemical atoms. Thus, despite his status as one of the discoverers of the law of valence, he had not yet made a fully consistent distinction between the concepts of valence and equivalence, which the reformers were so careful to specify and which was so fundamental to their movement. This failure was a result of his continued adherence to the use, and ontological atomistic interpretation, of conventional equivalents, which tend to mask the phenomenon of valence.

Kolbe's Development of the Theory

In the carbonic acid theory, Kolbe had essentially come to accept Frankland's type-theoretical formulation of organic compounds. His reorientation toward types corresponded to a peaking in his feeling of respect toward Gerhardt. In chapter 6 it was noted that Kolbe regarded Gerhardt's *Traité* as a very good book in many respects; he used it heavily in writing his own book and was astounded at Gerhardt's rate of production of high-quality manuscript. Shortly before Gerhardt's death, he wrote Vieweg, "I will no longer oppose Gerhardt. I fully recognize his accomplishments; his corrupt theoretical speculations will soon be outlived, and he will be the first to ignore them." And again, in 1858, he stated that he now gave Gerhardt "my full recognition of his services to organic chemistry."[38] Having fully embraced types in the publication of his carbonic acid manifesto, Kolbe set about to treat the theory systematically throughout his textbook.

Consistency was a serious problem. He had already been working on the book for several years, and by the time of his conversion to Frankland's views, he had published five fascicles in three major installments. Kolbe suffered from the circumstance that he was producing a detailed and strongly theory-based survey of the field at a time when the theory was evolving with astonishing speed. Consequently, each installment of the book contained different and sometimes inconsistent theoretical details. That he himself was so theoretically inventive, and honest with his readers about the ideas and hypotheses that he was using, exacerbated the problem of internal consistency. As he wrote Vieweg about the time that the first volume was completed (1859),

> In the last four years organic chemistry has found itself in a stage of development in which only just now a crisis has arisen. In composing my textbook, I had three possible paths before me. I could write the text without any particular system as a foundation and without having any firm theoretical viewpoint, which is the situation in which Wöhler and Liebig find themselves; or I could follow Gerhardt's swindle, which even the best minds have done . . . or, finally, I needed to follow my own path, which I must tread all the more cautiously and circumspectly, since everyone is watching me extremely closely in order to find and breach every weak point in my system. . . . Be assured, dear Vieweg, that that which has hitherto failed to recommend my book, that it did not follow the usual accepted patterns and took up an independent position, will be viewed very soon as an advantage, and will transform its current enemies into friends. In this regard I look forward to the near future, confident of victory.[39]

In the summer or autumn of 1857, feeling the need to treat the carbonic acid theory as soon as possible in his text, he wrote a section entitled "Theoretical Views Concerning the Composition of Alcohols, Acetones, Aldehydes and Related Acids," and he fit it into the fourth installment, double fascicle 6/7, somewhat out of place.[40] Production of this manuscript was agonizingly slow, however, due to his recurrent rheumatic fever. Vieweg was losing patience; the book was not selling well for several reasons: its slow appearance, its idiosyncratic and polemical characteristics, and its insecure and shifting theoretical base. "I assure you," Kolbe responded, "that I have for a long time been in the clear regarding all theoretical problems whose solution was earlier so important."

> And do not think that in the description of organic compounds I see myself in strict opposition to the various prevailing viewpoints. To be sure, I place my views, which I regard as more correct, in the foreground, and in doing so I believe I bring much that is good and new; but I consider also the viewpoints of others as well. Aside from a few chapters, no one will find anything in my book to be partisan. On the contrary, I am the conservative, who only rejects eccentricities and abnormalities.[41]

The installment was finally published in December 1857.[42] Kolbe was extremely anxious to see it in print. As he wrote Vieweg on 20 December,

> The double fascicle appearing now contains a great deal that is new, and I think that several chapters will capture the interest of readers, especially the small section "theoretical views on the constitution of alcohols etc."

> Even if many chemists will not yet at present accommodate themselves to these views, I am absolutely convinced that the theory will gain rather general currency in the very near future.[43]

The theoretical section of the fascicle announced some dramatic departures for Kolbe. No longer did he use Otto's buckle symbol to indicate conjugation, nor any symbol at all, although he still employed the term. No longer did he make any meaningful distinction between his two classes of organic radicals—the ether or alcohol radicals in hydrocarbons (and their derivatives) and the conjugate radicals in acids (and their derivatives). He permitted himself the use of "the currently popular expression" of "types" and described both carbonic oxide C_2O_2 and carbonic acid $(C_2O_2)O_2$ as "diatomic" (zweiatomig)—a term that had been in use for only eighteen months among type theorists such as Gerhardt, Wurtz, and Kekulé. Alcohols were now formulated in an analogous fashion to aldehydes, acetones, and acids:

$HO.\left.\begin{matrix}C_2H_3\\H_2\end{matrix}\right\}C_2O$	$\left.\begin{matrix}C_2H_3\\H\end{matrix}\right\}C_2O_2$	$\left.\begin{matrix}C_2H_3\\C_2H_3\end{matrix}\right\}C_2O_2$	$HO.(C_2H_3)C_2O_3$
alcohol	aldehyde	acetone	acetic acid

Hydrogens and methyls simply substitute for oxygens in acetic acid, and ultimately, oxygens in carbonic acid, to produce all of these compounds. Carbonic acid is diatomic because two oxygens are more loosely bound and more readily replaced than the other two, just as one of the two oxygens may be readily replaced in acetic acid (as in the formation of amides or thiacetic acid). Somewhat inconsistently, he also referred to carbonic *oxide* as diatomic because two atoms (e.g. of oxygen or of chlorine) can add to it.

Kolbe attempted to show that his theory had heuristic power by making several predictions. There is no reason why methyl radicals should not be induced to continue substituting for the two remaining hydrogens of ethyl alcohol, he stated, so the following compounds should be possible:

$HO.\left.\begin{matrix}C_2H_3\\H_2\end{matrix}\right\}C_2O$	$HO.\left.\begin{matrix}C_2H_3\\C_2H_3\\H\end{matrix}\right\}C_2O$	$HO.(C_2H_3)_3C_2O$
methyldihydro-carbon oxide hydrate (ethyl alcohol)	dimethylhydro-carbon oxide hydrate (isopropyl alcohol)	trimethyl-carbon oxide hydrate (*t*-butylalcohol)

If these latter two substances were produced, they would have to be called "pseudoalcohols," since Kolbe predicted from the theory that they would not share the property common to all other known (aliphatic) alcohols of being oxidizable to aldehydes and acids. The first would be isomeric with propyl alcohol (from propionic acid), and would oxidize only to acetone. The second would be an isomer of butyl alcohol (from butyric acid), and would suffer no oxidation at all (without decomposition). These predictions were derived by implicitly applying the concept of tetratomicity to the carbonyl carbon group (C_2), although that idea was neither defined nor named here.

Other predictions concerned the recent discovery of glycol, which Wurtz had proclaimed to be a dialcohol. Kolbe objected to Wurtz' categorization since Wurtz had failed to demonstrate the oxidation of glycol to an aldehyde or acid—again, the distinguishing characteristic of alcohols. Kolbe thought that partial oxidation of glycol would not in fact yield an aldehyde; rather, either acetic anhydride or a homolog of glycerin would result:

$C_4H_4O_2.2HO$	$C_4H_3O_3$	$C_4H_3O_3.3HO$
glycol	acetic anhydride	glycerin homolog

This prediction was based on Kolbe's supposition that glycol was merely a hydrated form of aldehyde, $C_4H_4O_2$. He conjectured that it ought to be possible to produce glycol directly, by applying hydrating agents to aldehyde. In this section he did not fully justify his preference for considering glycol as an oxide rather than a dialcohol. His real reasons were detailed only in the next installment, published a year later, and are discussed below.

Glycerin is also no alcohol, according to Kolbe, but rather the trihydrate of a tribasic oxide, $C_6H_5O_3.3HO$. Acrolein is not the aldehyde of glycerin but of allyl alcohol. Oddly, Kolbe continued, the basic oxide of glycerin, $C_6H_5O_3$, has the same composition as propionic acid. Although, like all organic acids, propionic acid has one oxygen atom bound more loosely than the other two, perhaps glycerin possesses the same hydrocarbon radical but with the three oxygens bound in a different fashion, that is, in a chemically equivalent way. This would explain why propionic acid is monobasic, while glycerin is tribasic.

Before his next installment appeared, he needed to take account of all the new research that had appeared in the eventful years 1857 and 1858, and once more this occasioned some important modifications in his ideas. About the middle of June 1858 he returned from a Wiesbaden cure, full of energy and feeling "newly reborn" after his long and serious illness.[44] After the summer semester ended, he was able to make rapid progress. Indeed, as early as 11 August 1858, he presented a paper to a local scientific society that contained a precis of the theoretical content of the *Lehrbuch* installment that he was then only starting to compose.[45] Kolbe persuaded Vieweg to cancel sheet number 42 (the last sheet of fascicle 7, already published), so that changes could be made in it. The double fascicle 8/9, including a revised sheet 42,[46] was complete in manuscript form by the end of 1858 and was published very early in 1859.[47]

Chief among the new published work that he needed to deal with was that of Wurtz, Debus, and Kekulé. Wurtz' further work with glycol derivatives and other polyfunctional alcohols and acids was especially important. Wurtz was then in his prime, publishing during this period important articles at the rate of one every six weeks or so, filled with fascinating new substances and all used to develop the theory of polyatomic radicals. Debus had succeeded in oxidizing alcohol into a double aldehyde (glyoxal), an aldehyde–acid (glyoxylic acid), and a hydroxy–acid (glycolic acid). Wurtz had shown that glycol could be oxidized to glycolic and oxalic acids, which constituted Kolbe's own criterion of alcohol character and thus should have satisfied him that glycol was a dialcohol.[48] Since glycol was derived from ethylene, a reaction route from ethylene to oxalic acid was now apparent; Debus' reactions showed that alcohol, too, could be transformed into oxalic acid. Because all agreed that ethylene and alcohol had two carbon atoms (four carbon equivalents), it could hardly be doubted any longer that oxalic acid also had two carbon atoms rather than one and therefore was dibasic.

Kolbe now accepted this conclusion as experimentally demonstrated, but he still resisted characterizing glycol as a dialcohol just because it could be oxidized to an acid. Glycol does not share other (unnamed) properties of alcohols; moreover, there are theoretical reasons for formulating glycol as an oxide hydrate rather than as a dialcohol. He had decided that ethylene must be carbonyl (C_2) united with both a methyl group (C_2H_3) and a hydrogen atom, since it is derived from the carbonic oxide type and so their constitutions must be analogous. The reason ethylene is analogous to carbonic oxide is that both compounds undergo similar addition reactions, such as with chlorine. Because the two chlorine atoms both add to the same carbonyl radical (carbon atom) in carbonic oxide (there is only one), they must also add to a single carbonyl radical in ethylene. So ethylene must be methy carbonyl hydride. Since glycol is derived chemically from ethylene, it must be ethylene oxide hydrate:

$$2HO.\left.\begin{matrix}C_2H_3\\H\end{matrix}\right\}C_2O_2$$

Oxidation of this substance is not analogous to oxidation of ethyl alcohol to acetic acid, Kolbe thought, since it has only one lone hydrogen replaceable by oxygen, while ethyl alcohol must oxidize both its radical hydrogens to form acetic acid. Wurtz' glycol when oxidized should yield a homolog of glycerin ($3HO.C_4H_5C_2O_3$), namely, $3HO.(C_2H_3)C_2,O_3$. The *true* alcohol derived from glycolic acid has not yet been synthesized, Kolbe argued; it would have the composition

$$HO.\left.\begin{matrix}\left(C_2\left\{\begin{matrix}H_2\\HO_2\end{matrix}\right.\right)\\H_2\end{matrix}\right\}C_2O$$

If this hypothetical substance were to be oxidized under mild conditions, it would yield a novel aldehyde isomeric with acetic acid. Kolbe thought that unhydrated ethylene oxide, not yet isolated, must be an isomer of normal aldehyde (and not, as he had suggested earlier, identical to it).[49] The difference in their natures is a consequence of a difference "in the molecular grouping of their components, in other words, in their constitutions."[50]

In effect, Kolbe was suggesting that there were two possible isomers possessing the elemental composition of glycol. One was Wurtz' compound, an oxide; the dialcohol was certainly possible, but it had not yet been synthesized. Of course, Wurtz had succeeded in oxidizing glycol to glycolic acid, but this offered no proof that these two compounds had analogous constitutions. Rather, a rearrangement must take place during the oxidation, in which a hydrogen of the methyl group is replaced by "hydrogen peroxide," HO_2 (a modern hydroxyl group), while the carbonyl carbon is further oxidized. Kolbe cited Kekulé's recent paper, in which Kekulé had described hydrolyzing monochloroacetic acid to glycolic acid, to demonstrate that glycolic acid was simply acetic acid with a hydrogen peroxide group substituted for hydrogen, and not the acid derivative of glycol. He thereby summarily excluded the possiblity of Wurtz' point of view, namely, that it was both of these at once.[51]

As problematic as the supposition of all these hypothetical isomeric compounds was, Kolbe seemed untroubled. At least there was a certain consistency here. He asserted that Wurtz' derivation of glycol from ethylene also falsified one of his (Kolbe's) ideas, given emphasis earlier in his textbook and defended since 1850, namely, that ethylene derivatives are all derived from the vinyl radical, C_4H_5, wherein the four carbon atoms are chemically equivalent. Instead, ethylene is a methyl group united along with a hydrogen atom to the double carbon atom, "carbonyl," whose direct oxidation (addition of two atoms of oxygen to carbonyl) results in aldehyde. Accordingly, there ought to be two distinct monochloroethylenes resulting from substitution of either the single hydrogen atom or one of the three methyl hydrogens.[52]

In this installment of his textbook, Kolbe found a place for a second explicitly theoretical chapter, entitled "Theoretical Considerations Regarding the Saturation Capacity of Simple and Compund Radicals."[53] It began: "Our views regarding the mode of composition of the elements in organic chemistry have received a significant expansion by the perception that, in addition to the customary monatomic radicals, there exist also diatomic, triatomic, and tetratomic radicals." Monatomic radicals include methyl, ethyl, and acetyl; diatomic radicals include methylene, ethylene, and carbonic oxide. Triatomic radicals include formyl (methine) and the hydrocarbon radical of acetic acid, and finally, "carbonyl" itself, C_2, is tetratomic. All of these facts can be derived from the last. Adding methyl to carbonyl reduces its atomicity to three, and three additional oxygens create acetic anhydride. Replacing the oxygen of acetic acid by alkyls or hydrogens creates aldehydes, acetones, and other acids. Carbonic acid, marsh gas, or carbon tetrachloride represent the satisfaction of atomicities of carbonyl by four atoms of oxygen, hydrogen, or chlorine.[54]

Kolbe announced an important new chemical law in this context: the basicity of an acid, whether organic or inorganic, is always equal to the number of oxygen atoms outside the radical. In the following series, the radicals between the square brackets were regarded as stable groupings whose oxygen content did not affect basicity.

$HO.[NO_4],O$	nitric acid
$2HO.[S_2O_4],O_2$	sulfuric acid
$HO.(C_2H_3)[C_2O_2],O$	acetic acid
$2HO.[C_2O_2],O_2$	carbonic acid

What mattered was the number of oxygen atoms outside these radicals, always matched by an equal number of water molecules (or, according to Wurtz' and Kekulé's notation, half-molecules) on the other side. Thus, Kolbe regarded nitric acid as the hydrated oxide of the monatomic radical NO_4, acetic acid as the hydrated methyl derivative of the diatomic carbonic oxide radical C_2O_2, and carbonic acid as the dihydrated dioxide of the same radical. The oxygen within the radical could be replaced without altering the basicity, as in the conversion of acetic acid to alcohol (which still forms salts with strong bases). If extra-radical oxygen atoms are replaced, however, the basicity is reduced, as in the conversion of acetic acid to neutral amides.[55]

Finally, Kolbe argued that dibasic organic acids are naturally derived from a *double* carbonic acid type, with succinic acid being

$$2HO.(C_4H_4)''\left\{\begin{matrix}C_2O_2\\C_2O_2\end{matrix}\right\}O_2$$

The C_4H_4 has a methyl carbonyl hydride composition like ethylene. Organic sulfonic acids, based on the dibasic sulfuric acid type, must have anlogous constitutions, and Kolbe devoted a substantial section of the chapter to these compounds. They also can form dibasic acids upon a double type; for example, Buckton and Hofmann's new "disulfometholic acid" was $2HO.(C_2H_2)''(S_2O_4)_2''O_2$. It ought to be possible, thought Kolbe, to add the elements of water HO to this acid by heating it in a sealed tube, thereby transforming it into a new tribasic acid, $3HO.(C_2H_3)(S_2O_4)_2''O_3$.[56]

From October 1858, when Kolbe was nearing completion of the writing of this installment, until early January 1859 when it was in press and about to appear, he stressed in every letter to Vieweg the novelty, originality, and importance of these ideas. The installment was the "most difficult" of all to write, but its significance would soon be recognized; he had intentionally delayed publishing the material elsewhere to save it for the book. Still, he recognized explicitly that these theoretical advances were only made possible by incorporating the "striking" results of a number of important articles that had appeared in the journal literature during the past half year.[57]

It is hard to escape the conclusion that Kolbe must have been influenced by Kekulé's structure theory articles, which appeared in Liebig's *Annalen* in November 1857 and May 1858. There is no question that he always read each issue of the *Annalen* carefully. He cited in this installment another article of Kekulé's (on monochloroacetic acid) appearing in the March issue as supporting his viewpoint, and in 1860 he cited Kekulé's first structure theory paper—but only to criticize Kekulé's excessive use of mixed-type formulas.[58] Moreover, nothing indicates (and much suggests the contrary) that he routinely ignored type theory papers, even if he treated some of them dismissively and did not always understand them fully. By now, in any case, he was regularly using many other type theorists' work—indeed, he had become a type theorist himself. As was the case in Kekulé's structure theory papers, Kolbe laid stress on new research regarding the aliphatic sulfonic acids, behind much of which lay the work of Kolbe's close friends

Hofmann and Strecker and Kolbe's former assistant von Uslar. A principal theme of Kekulé's two papers had been to destroy the copula theory, and for the first time Kolbe eliminated the word *copula* from his vocabulary. The development of the idea of atomicity in fascicle 8/9 has much in common with Kekulé's concept, including Kolbe's first explicit statement of the "tetratomic" character of carbon.

Moreover, Kekulé's definition of an organic radical as essentially the hydrocarbon skeleton of a molecule and all heteroatoms fully bound to it finds a parallel in Kolbe's new basicity law. Indeed, it is fascicle 8/9 that first announces Kolbe's conversion to the general concept of polybasic radicals and acids, a cardinal thesis of the reformed chemistry. Williamson had been the first to argue that the basicity of a radical confers an equal basicity on an acid derived from it, such that the dibasic sulfuryl, carbonic oxide, and oxalic radicals all result in dibasic acids. Kekulé had transformed this idea, using Liebig's and Laurent's vocabulary and his own structure theory, into the phrase "inside or outside the radical," which Kolbe followed. And, in a final irony, Kolbe's basicity law was essentially that proposed by Laurent in 1836, a law that Kolbe had particularly singled out for sarcasm and scorn in the first installment of his textbook!

The first volume of Kolbe's textbook required just one more fascicle for completion, along with an addendum updating the work from the journal literature that had appeared during the five years it had taken for the volume to be published. It was finished by the end of July 1859 and finally appeared that autumn. At this point he decided to write a summary of the modified carbonic acid theory for publication in the *Annalen*, since the number of readers of his textbook was very small compared to the readership of Liebig's journal. Few in the German chemical community were even aware, much less approved of, the details of his theory before 1860, and he lamented this feeling of isolation to his friend Vieweg. Once the paper was shipped off, he wrote with badly mixed metaphors that he was "now looking forward, sitting fast in my saddle, to the coming storms."[59]

The paper was entitled "On the Natural Connection of Organic and Inorganic Compounds, the Scientific Basis of a Natural Classification of Organic Chemical Bodies." To Vieweg he often referred to it as his "confession of faith" (Glaubensbekenntnis). Richard Anschütz referred to it as a kind of counterpart or rejoinder (Gegenstück) to Kekulé's structure theory paper of 1858,[60] and it seems likely that Kolbe thought of it this way as well. Dated 27 September 1859, it appeared in the March 1860 issue of the *Annalen*. The four types of Gerhardt's theory had been used to try to unify organic and inorganic compounds,

Kolbe related, but the relationships they revealed were unnatural, artificial, and superficial, leading to an "empty game with formulas" and a "dead schematicism"; "nature does not limit herself to variations on four themes." A few formulas from Kekulé's structure theory paper of 1857 were cited as exemplifying this game-playing approach. However, Kolbe's theory—as the title proclaimed—offered a "natural" and "scientific" basis for such a unification. The fundamental axiom was "*organic chemical bodies are invariably derivatives of inorganic compounds, and have been formed from the latter, some in a direct fashion, by wonderfully simple substitution processes*," namely, substitution in carbonic acid.[61] Deriving organic compounds from this single progenitor is not only simpler and more general than Gerhardt's approach, it also focuses on the most important substance for plant physiology and biosynthesis. In this way, the chemist's scheme nicely parallels nature's.

The ideas in this article are the same as in fascicle 8/9 of his textbook, but they are treated here with substantially greater cogency and consistency. Kolbe wisely pruned most of the hypotheses and predicted isomers that had appeared in his text, while highlighting the ones that he obviously thought the most of, especially the prediction of isopropyl and tertiary butyl alcohols. The organization is clear and systematic, and the exposition magisterial.

He also discussed two additional important and theoretically troublesome organic diacids. Kolbe suggested that malic and tartaric acids are succinic acids with one and two hydrogen atoms, respectively, of the hydrocarbon radical replaced by hydrogen peroxide, HO_2. In a sense, this substitution is simply an insertion of oxygen O_2 into the radical, hence the names *Oxybernsteinsäure* and *Dioxybernsteinsäure* for malic and tartaric acids. (Following Kolbe, German organic chemical nomenclature to this day preserves the prefix *Oxy-* to indicate hydroxy substitution.) They remain dibasic acids because this oxygen is adding inside the radical, and according to Kolbe's basicity law, this has no effect on basicity. If any of these diacids were to decarboxylate (lose carbonic acid C_2O_4), however, the molecule would lose one of the two dibasic radicals C_2O_2 upon which the double type was formed. A third oxygen would then be lost from outside the radical, reducing the basicity to one, and the fourth oxygen to make up C_2O_4 would come from one of the two attached water molecules. The hydrogen thus set free would enter the hydrocarbon radical.[62]

The paper just described constitutes Kolbe's virtually final formulation of the theory of organic compounds. After more than a decade of wandering in the theoretical wastes, Kolbe felt he had arrived in the promised land of chemical truth. To the end of his life he regarded the ideas in this paper as both necessary and sufficient to eventually expose to the light of day all the exotic creatures in the tropical jungle of organic chemistry.

The Response

On the day the article was officially published, Kolbe wrote Vieweg, concerned about the general course of battle in his war with the reformers. The majority of chemists were now following Gerhardt—even his good friend, the otherwise sober and reasonable Strecker—and Kolbe was isolated in his opposition.

> In my view this whole movement has already outlived its usefulness. Such mindless ideas as are to be found in Weltzien are its dying echoes. The adherents of Gerhardt's school in Germany as in France have shot their bolt. They are riding to death the poor old nag that carried Gerhardt forward for a time, without producing anything other than variations on the old theme. Not even the entire great Göttingen school has educated anything approaching a capable and intelligent chemist. Limpricht has created a spectacle, but produced not a single piece of work that shows anything like ingenuity or sharpness of intellect. Fully as unthinking a man, but correspondingly foolish a chatterer, is Kekulé. In a word, the whole mindless business, on which I expressed my judgment in the beginning and concluding sentence of the article just now appearing, is bankrupt and has been outlived. What people of this kind are doing is not science, but a game.— But I see I am letting myself go on too long. The future will tell.[63]

Considering these views, it must have been extraordinarily discouraging to Kolbe to see that the initial response of many of his colleagues was to place Kolbe himself in the midst of the "whole mindless business" of newer type theory and polyatomic radicals. Wurtz was the only member of the reformist camp to reply in print. In a review of the article in his *Répertoire de chimie pure*, he wrote: "Monsieur Kolbe has so fully adopted the fundamental idea of types that not only does he want to multiply them, but even, with Gerhardt, to assume condensed types, as are represented by [his] molecules of carbonic acid." The carbonic acid type, Wurtz affirmed, is nothing more than the water type with diatomic carbonic oxide functioning as the oxygen atom, as Williamson had formulated it in 1851.

Wurtz thought Kolbe's argument that Gerhardt's four types are artificial was not a trivial one; nature should indeed not be limited in this way. But what Kolbe did not see, Wurtz argued, was that a more general principle lurks behind these types, namely, successive degrees of condensation of matter. Hydrogen, water, ammonia, and marsh gas represent hydrogen once, twice, thrice, and four times condensed, with the oxygen, nitrogen, and carbon atoms representing two, three, and four atoms of hydrogen, respectively, in one unit of action. This is what he had meant to express in his subatomic speculation of 1855. Wurtz thought that the idea of types ought to be replaced by the idea of the atomicity of the elements, which is the foundation of the phenomena that types attempted to express. "Here is a clear, simple, and general principle, which deserves for this reason to be placed at the base of a system of chemistry." Kolbe's approach was fundamentally the same, Wurtz urged, whether he was aware of this or not. "If, then, the ideas this chemist presents are novel, the innovation resides rather in the form than in the essence, and I believe that I have shown that even the form is not fortunate; in truth, he combats Gerhardt's types by counterfeiting them."[64]

Kolbe's friend Hermann Kopp saw the matter similarly, although he phrased his criticism more kindly. Reviewing the 1860 paper in the *Jahresbericht*, Kopp wrote: "Kolbe still argues against relating organic compounds in general to the hydrogen, water, and ammonia types, as also against the recent assumption of mixed types. But he reveals himself as a de facto adherent of the 'type theory' by conceiving organic compounds as derivatives of inorganic compounds." Kopp also pointed out that Kolbe had used multiple types to formulate diacids, and suggested the influence of Gerhardt. Three years earlier Kopp had also referred to the Gerhardtian and Williamsonian elements in the Kolbe–Frankland carbonic acid paper.[65]

However, Kolbe's paper impressed one very important reader: Justus Liebig. Liebig wrote Kolbe on 3 April 1860 to express his "great satisfaction" with the paper; the schematic derivation of malic and tartaric acids from succinic acid was "the triumph of your theory."[66] Liebig himself never penetrated the sense of newer type and structural formulas. To Wöhler he admitted his disinterest in modern chemical theory and complained about the thoughtless manipulation of formulas that was going on around him. Wöhler's opinion was similar.[67] In the 1860s Kolbe's and Frankland's articles were some of the few that excited Liebig. He told each of them separately that only in their work could he see the guiding influence of "a large scientific idea" directing the experimental program. To Frankland he wrote: "What Wöhler and I saw thirty years ago in dreams, that is, in our imagination, you are now on the road to realizing . . ." To Kolbe he sympathized regarding the "Formelspielerei" of many who were not on the correct path.[68] Ironically, by the time Liebig was writing this letter Frankland had become a full-fledged member of the structuralist school, whose manipulation of formulas was entirely in their spirit, while Kolbe was making a sharp distinction between his and nearly everyone else's formulas.

Liebig's support fortified Kolbe's spirits at a time when it appeared all the world was converting to the reform. Led by Kekulé, a small cabal including Weltzien, Baeyer, Roscoe, and Wurtz planned an international conference whose ulterior motive was to attempt a world standardization of the Gerhardt–Laurent atomic weights and molecular formulas.[69] The need for such a meeting was by no means uniformly conceded. Some, such as Leopold von Pebal who had been a confirmed Gerhardtian for years, thought such a degree of unity already prevailed that it was not necessary. He expressed this sentiment in a letter to Roscoe, adding "Kolbe will hardly be converted!"[70] Others, including some reformers themselves such as Williamson and Brodie, thought it was unseemly to attempt, or even to give the impression of attempting, to legislate scientific conviction. Several, including Erdmann and Kopp, espoused such sentiments at the meeting itself.[71]

Lothar Meyer, a leader of the reform movement after 1860 and a good friend of Pebal and Roscoe from their Heidelberg period, occupied an ambivalent and interesting position before the meeting. He was still studying with Bunsen in Heidelberg when Kekulé arrived there early in 1856, and Kekulé's eager advocacy for Gerhardt's type theory made a strong impression on Meyer, according to his later reminiscences. "How very well I remember even today," he recalled,

> . . . the debates lasting hours and days, in which he won ground step by step. The authority of the accepted dualistic theory and the decided aversion of our honored master to get involved with the new business with formulas [Formelkram] explains why we gradually came over to the other side only after energetic resistance.[72]

But in Meyer's case, it would appear, this did not happen before the Karlsruhe Congress. In July 1860 he wrote Roscoe saying he felt obliged to come to the "idiotic church counsel, to propose the election of an infallible formula-pope," but feared what the edicts of the "Committee of Public Safety" at such a "National Convention" might prove to be.

> The good Gerhardt, if he only knew what kind of scandal is being made of his innocent types! . . . These "types" are a dangerous toy for fools; but the expression "type *theory*" is an insult to science, which recognizes a theory of gravitation, of light, of capillarity, etc., but which can never agree to having a notation called a "theory."[73]

As a matter of form, Weltzien wrote Kolbe to ask for his support. Kolbe replied that since the science was so dominated at the moment by the Gerhardt–Williamson–Kekulé–Wurtz school such a conference would do more harm than good. Any resolutions taken at the meeting would have to be overturned upon the fall of the reform movement, an event Kolbe expected to see "in the very near future."[74] Kolbe had every intention of skipping the meeting, but at the last minute he let himself be persuaded by Fehling to go for one day of the three-day conference. It was, he wrote Vieweg, a real "farce," just as he predicted it would be.

> Weltzien let himself be used by Kekulé, to give the latter the opportunity to pass himself off to the assembled chemists as the great chemist of the future. Instead of some kind of result or agreement (the most important questions were never even raised), only a general dissatisfaction and disappointment was achieved. Kekulé gained precisely the opposite of what he wanted; he made a real fool of himself and turned everyone against him.[75]

This was, of course, a jaundiced point of view, but the organizers themselves were likewise somewhat disappointed by the absence of concrete results. However, the conference had long-term effects that were not visible in the fall of 1860. Lothar Meyer, who had been so cynical in July, was profoundly affected, not so much by the events themselves in Karlsruhe, but by having been handed an offprint of Cannizzaro's *Sunto di un corso di filosofia chimica*. He read it on his journey from Karlsruhe and "repeatedly" after his arrival home. He "was astonished by the clearness that the brochure provided concerning the most important points at issue. It was like blinders being removed from my eyes; doubts disappeared, and a quiet feeling of certainty replaced them."[76] The Russian chemist Mendeleev related a similar story regarding Cannizzaro's influence.[77]

The general historical impression that Cannizzaro was the decisive personality in Karlsruhe and that his brochure helped achieve completion of the reform is probably accurate. The work also appears to have had a strong effect on Frankland. In assessing the relative contributions of various chemists to the formulation of structure theory, he gave Cannizzaro a higher place than Kekulé, for it was Cannizzaro, he said, who established the atomic weights that made structure theory viable—a point made by Meyer as well.[78] In any event, Frankland's publications show that in the months before Karlsruhe he had still not accepted some important elements of the reform, whereas by 1862 he was fully in the modernist camp. Wurtz and Hofmann, who had been reformers for years, only began to use the two-volume atomic weights in their published papers around the time of the conference, and this may also have had some impact on Frankland's thinking.

But it would be erroneous to conclude that Kolbe was the only hold-out after Karlsruhe. The elder spokesmen, especially Liebig, Wöhler, Bunsen, and Dumas, continued to prefer the apparently more empirical conventional equivalents. This was partially because they held themselves aloof from the theoretical trends of the day, as they themselves, as well as their ardent defenders such as Kolbe, often admitted. But exceptions existed even among the younger active chemists. Among the French, Marcellin Berthelot was just as contemptuous of the reform movement as Kolbe was. The empirically minded Friedrich Beilstein, another friend of Kekulé's from his Heidelberg period, accepted the new atomic weights, but was disgusted by the facile manipulation of structural formulas.[79] Even Kekulé complained to Meyer, shortly after the Karlsruhe Conference, of "Constitutions-formel-spieler."[80] Indeed, there were many who were happy multiplying possible formulas far beyond necessity—Kolbe himself was susceptible to this charge. In fact, it was surely to Kolbe that Kekulé was referring in the phrase just cited.

Was Kolbe a Type Theorist?

To comprehend the character of Kolbe's route to his mature theory, it is useful to recapitulate the four stages of his passage in the period from 1856 to 1860. In his article on radicals completed by the beginning of 1856, Kolbe appears to have accepted Frankland's valence regularity, from an application of which he suggested that methyl could substitute once or (perhaps) even twice for oxygen atoms of carbonic acid, to form acetic acid and (perhaps) acetone. By the end of that year, following correspondence with Frankland on the subject, he further generalized the scheme in a theoretical paper for the *Annalen*, now suggesting, with Frankland, that hydrogen as well as alkyl radicals could replace oxygen in carbonic acid; the experimental demonstration was, however, still lacking. Evidence from correspondence suggests that only then was Kolbe beginning to realize the far-reaching significance of this new idea.

Kolbe entered into the third stage during the summer or fall of 1857. By this time he fully accepted both the vocabulary and the general validity of type theory; he abandoned copulas, conjugate radicals, and buckles, and asserted the "diatomic" nature of both carbonic oxide and carbonic acid. His prediction of the existence and properties of isopropyl and *t*-butyl alcohol emerged from these considerations. The fourth and final stage was achieved by the summer of 1858, when he first proclaimed the "tetratomic" nature of the carbonyl radical (C_2) and for the first time accepted polybasicity as a general phenomenon in acids. Once he had crossed this threshold he was able to devise a general theory that was substantially similar to that of the type theorists. But there were substantial differences as well.

One element in Kolbe's failure to find full accommodation with the reformers was the extremely literal sense in which he interpreted chemical formulas, his as well as those of other chemists, and his inability to see what was intended by others' notation. One obvious instance was his continued retention of conventional equivalents, which created difficulties in communication. During the 1860s, most German chemists began to use the newer weights. As early as 1862 Kolbe was in a distinct minority, and he must have felt the peer pressure in many ways. It was not only his ingrained conservatism that held him back from conversion, but also his discovery of the "basicity law," which states that the basicity of a polybasic acid is equal to the number of oxygen equivalents thought to be located outside the radical (and always matched by an equal number of "molecules" of water, HO, on the other side of the formula).[81]

Kolbe was not the only one to write aggressive critiques in the pages of his textbook. In fascicles of his own textbook published in 1861 and 1864, Kekulé criticized Kolbe's formulas. One problem, Kekulé argued, was Kolbe's inconsistent notation for alcohols. In the formulas for hydroxy acids, Kolbe wrote alcoholic groups as HO_2, whereas alcohols themselves were notated in the same manner as acids, namely, O.OH. More fundamentally, in Kolbe's basicity law an increment of *two* oxygen equivalents is always necessary when increasing basicity. This itself should have led Kolbe to double his equivalent for oxygen; in Kekulé's view, the law was "explained by the circumstance that two *half* atoms of oxygen are required to form a whole atom." If the oxygen equivalent were doubled to halve the number of atoms in the formula, then marsh gas rather than carbonic acid must instead be used as the carbonaceous type, "whereby this viewpoint would coincide with that developed in this textbook."[82]

Kolbe, Frankland, and Butlerov all pointed out later that each "O" in formulas written in conventional equivalents could be considered to represent not an atom but an affinity unit, thus remaining consistent with valence theory.[83] Such arguments were later used by Kolbe and Frankland to maintain that they had been the earliest to formulate carbon tetravalence. In an unpublished paper written in 1883, Kekulé rejected this claim. In Kolbe's theory, alkyl radicals or hydrogen schematically and successively substitute for the oxygen of carbonic acid, $2HO.C_2O_4$. Kekulé wrote

> Probably no one has really understood these formulas. It is not at all clear why the attached water equivalents (i.e., half water molecules) vanish as the oxygen atoms (i.e., half oxygen atoms) are replaced. The whole derivation is only possible if, proceeding from the hypothetical hydrated carbonic acid, hydroxyls or water residues are replaced by hydrogen or radicals. In any case the entire procedure has nothing at all to do with the tetravalence of carbon.

Kekulé also pointed out that even if Kolbe's argument were accepted, his own structure theory articles were published earlier than the installments of Kolbe's book in which the tetravalence of the carbonyl radical C_2 was first advanced.[84]

In the most general case, Kolbe interpreted the goal of any chemical formula, once properly formulated, to be the ultimate expression of the chemical properties of the substance. If a new reaction could not be expressed by the formula, Kolbe usually regarded it as a serious anomaly, often rejecting the formula and the idea behind it, even if it was his own. Conversely, he rarely hesitated to predict unknown reactions or to posit unknown isomers as logical consequences of his formulas. This was one reason why the details of his theoretical system kept shifting throughout the 1850s. The emerging structuralist camp, following a perception of Gerhardt, began to see that such conflicts were not so much anomalies as expressions of the incompleteness of most formulas; only a fully resolved structural formula, and one that was adequately understood through a multitude of reactions—such as the formulas for acetic acid or triethylamine—could be judged as ultimate. By "fully resolved" the structuralists meant showing each atom in the molecule and how it was connected by valence bonds to every other atom. This was an essential point, first made in an explicit fashion by Kekulé in 1858.

For example, Kolbe thought that the chemical analogies of ethylene with carbonic oxide (which both undergo addition reactions with chlorine or oxygen) meant that the atoms adding to ethylene had to be assumed to be adding to a single carbon site, as they necessarily did with carbonic oxide. Consequently, both ethylene and its chemical derivative glycol had to have methyl groups. Like the reformist type theorists, Kolbe was convinced (inter alia by Kekulé's monochloroacetic acid hydrolysis) that glycolic acid was a substitution product of acetic acid and so did not have a methyl group in its constitution. Thus, he concluded that glycol and glycolic acid could not have analogous constitutions. This assumption compelled him to propose several yet undiscovered substances: an oxidation product of glycol only isomeric with glycolic acid; a reduction product of glycolic acid only isomeric with glycol; an intermediate oxidation product of this last compound only isomeric with acetic acid; and an unhydrated ethylene oxide only isomeric with ordinary aldehyde. Kolbe must have considered the hypothesis of all these unknown compounds to be a less serious price to pay than relinquishing the analogy of ethylene and carbonic oxide.

Kolbe shared with Frankland three variances from the Williamson–Kekulé–Wurtz school. One was the belief that valence could vary (e.g., that ethylene and carbonic oxide both possess a *di*valent carbon). A second was the ontological commitment to conventional equivalents, by Frankland until 1862 and by Kolbe until 1868. A fully resolved formula in which the focus of resolution is equivalents did not and probably could not yield structure theory in its most fruitful form. Even with atomic weights there proved to be enough anomalies to structure theory tenets (such as varying valences in oxygen and sulfur, multiple bonds in olefins, aromaticity, and so on) to require a laborious sorting-out process through the 1860s and 1870s. With conventional equivalents, it was barely possible to perceive valence itself (as exemplified by Kolbe's intellectual path, though Frankland managed it in 1852), much less the most fruitful kind of structure theory. To mention just one example, it was the commitment to equivalents that allowed Kolbe to separate his water "molecules" from the rest of the formulas.

Finally, one can perceive a persistence of instinctive electrochemical dualism, in Frankland until 1860 and in Kolbe to the end of his life. It was this vestigial dualism that prevented Kolbe from accepting the crucial chain-forming ability of carbon. It is difficult enough to understand polyvalence, not to mention chain-forming, under dualistic assumptions. Even after formulating his carbonic acid theory in its final form (by 1860), Kolbe was quite clear that he still regarded coulombic forces as the explanation for chemical affinity. Carbon atoms did not "link" together, and they had no "bonds" at their disposal. Rather, organic molecules were composed of proximate pieces (radicals) that substituted electrochemically for one another, or for certain atoms (such as oxygen in carbonic acid).

To use modern language to describe Kolbe's conception, chemical affinity is always isotropic, like all other (physical) forces. The methyl radical did not have a "bond" or a "hook" or any other form of localized or spatially oriented valence force. It substituted for another atom or radical *as a whole*, acting isotropically and coulombically. He could not explain these coulombic interactions in detail, and he knew it. No matter. He was absolutely convinced that chain formation was impossible because it seemed to exclude the only correct basis of chemical affinity—electricity. Thus, it happened that he was to develop a theory that was largely equivalent to structure theory but which denied that critical idea. That was his carbonic acid theory.

One more philosophical underpinning of Kolbe's theory requires mention. After 1859 Kolbe conceded his critics' charge that he, too, was a type theorist. Indeed, he often admitted the early fruitfulness of Gerhardt's theory. The certain indication that adherents of the theory had overreached themselves, Kolbe thought, came when they began to propose multiple and mixed types that had no material connection with the original type. How could carbonic acid, for instance, be formulated on the water type when it is the carbonic oxide moiety, not oxygen as in water, that is the dibasic entity? By deriving most organic compounds directly and materially from carbonic acid, Kolbe was proposing "natural" and "real" types, in contrast to the merely "formal," indeed "fictitious" types of the "modern" type theorists. The merely formal is not real, he felt, therefore the structural chemists were just playing superficial games on paper.[85]

The curious aspect of this line of argument is that it has very much in common with Dumas' original concept of types as both mechanical and material progenitors of their derivatives. As Dumas had done earlier, Kolbe was asserting the ontological significance of the type for its progeny. Or as Wurtz put it before he converted to the Gerhardtian view, the type imprints a "cachet" on all of its derivatives. A consequence of this view was that formulas must be taken very seriously and very literally. The type formula specifies one atom, outside the bracket, that is chemically as well as visually central. A derivative belongs to one type or to another, but not to more than one.

In departing from this point of view, Gerhardt had begun to edge toward a position according to which, at least in principle, all the atoms in a molecule could be accorded the same importance. Different reactions reveal different aspects about the molecule, each of which can be expressed by a different formula and all of which are valid. This establishes a sort of chemical "democracy" among the atoms in a molecule, in contrast to the hierarchical structure of the early type theory—and of Kolbe's types. Multiple and mixed types were anathema to Kolbe precisely because they represented a movement toward this democratic, nonhierarchical view.

It was most natural for Gerhardt to combine this view with a radically empiricist position, a position that Kolbe as an instinctive theorist detested. Here again Kolbe found himself in the same league with Dumas (although Dumas' theoretical and realist orientation was often hedged and even disguised by empiricist protestations). For Williamson, Kekulé, Wurtz, and other theorists in the reformist school, a combination of Gerhardt's relativist view of the atoms in a molecule and

Dumas' realist approach toward "mechanical types" together provided the cognitive route to structure theory. It was also necessary that the emergent structuralists cease to worry excessively about what precisely constituted valence and simply abandon familiar macroscopic physical analogies.

These considerations explain Kolbe's apparently inconsistent position. Gerhardt's radical empiricism clearly was not science to him. Nor was the overly schematic, formalist, conventionalist direction of the newer type theory (otherwise known as structure theory). Schematism is not realism, Kolbe insisted, and conventionalism is a farce; at best, such workers were playing superficial games, at worst they were destroying the science by accepting false speculations as truth. Moreover, abandonment of physical analogy to accept chain formation was in a sense a pragmatic and empirical response to an anomaly. Kolbe could not accept such a move because he was a passionate theoretician and a man of firm principles (although his opponents would have expressed this characteristic as obstinance).

The situation can be expressed somewhat facetiously (and in a way that Kolbe himself would find objectionable, due to his burning hatred for Dumas) as follows. Kolbe, a conservative to the bottom of his soul, retained elements of Berzelian chemistry as long as possible, and dualist vestiges survived in his system until his death. But when, during the period 1853–1860, he was forced to concede many critical points and become a type theorist himself, he remained true to his conservatism by, in effect, choosing the oldest form of the theory (Dumas') and then refusing to modify it for the last quarter century of his life.

9

The Great Break

Characterization and Causes

The chapter title refers to a sharp change in the fortunes of Kolbe's personal research and that of his school, as well as to a parallel change in the chemical community as a whole. Economists refer to an inflection point in a country's per capita economic output as the "take-off" that marks the beginning of an industrial revolution. Following this model, academic chemistry, led especially by the German *organikers*, achieved a kind of take-off during the 1850s and 1860s. Thereafter, the study of chemistry was transformed from a small affair conducted by elite scholars with an equally elite student clientele to the kind of routine mass education that is familiar in the modern world. A concomitant of this transformation was the relatively sudden recognition by political leaders of the potential applicability of chemical knowledge.

Kolbe's career followed this pattern, but with an uncharacteristically sharp inflection at the year 1858. First, let us look in table 2 at some rough-and-ready numerical measures of the size and research productivity of the Marburg research group, comparing Kolbe's first seven and a half years (from his arrival in May 1851 until late 1858) with the following year and a half (from the beginning of 1859 until mid-1860). To provide further reference points for comparison, data are also drawn from Bunsen's lab at Marburg during the five years before Kolbe's arrival (1846–1851) and average enrollment figures for the university as a whole.[1]

Table 2. Research Productivity in Marburg

	Bunsen 1846–1851	*Kolbe 1851–1858*	*Kolbe 1859–1860*
Student numbers			
Ave. Praktikanten in lab	22	14	12
Ave. attendance in chem. lecture	25	21	13
Ave. enrolled as chemists	13	4	0–1
Ave. total at University	263	243	233
Papers published			
By lab director	10	4	6
Collaboratively	1	0	4
By students alone	16	10	9
Total	27	14	19
Aggregate intensive measures			
Papers per year by lab director	2	0.5	4
Total papers per year	5.4	1.9	12
Papers per Praktikant per year[2]	0.25	0.14	1

During the fourteen years examined here, there was a gradual decline both in the prestige of the university and in its overall enrollment. Nonetheless, compared to the other changes being measured, the decline is small enough—about ten percent—that this factor can be neglected. Obviously, throughout the entire period, chemistry at Marburg became less popular, at least as indicated by the sizes of classes and numbers of majors. A factor independent of intrinsic merit operating at least in the 1850s must have been the circumstance that Kolbe was relatively young and still little known in comparison to his predecessor.

As far as research productivity is concerned, Bunsen's later years in Marburg must be regarded as quite successful, considering the time and place. Such was not the case with Kolbe's early Marburg period. In rough terms, he published personally a fourth as often as Bunsen had; his lab as a whole produced a third the annual number of papers; and even after taking into account his smaller number of Praktikanten, per capita productivity was half that of Bunsen's. His own research and that of his school was, to put it bluntly, moribund.

This situation was transformed starting at the end of 1858. In the ensuing eighteen months, his personal productivity was eight times what it had been, and that of his research group increased by a factor of seven. The transition between these periods was razor sharp. Kolbe's last paper before the transition point was his theoretical prospectus, joint with Frankland, dated December 1856. In the fall of 1858 his lab was suddenly bursting with activity (if not with students), and in the last week of that year and the first week of 1859 he wrote or edited seven contributions by him and/or his students; they were published as a group in the March 1859 issue of the *Annalen*.[3] In the course of 1859 he wrote five new solo papers and another one with a student, and he edited three more by students. Three additional papers came out of his lab by mid-1860. Although I have not tabulated it, in the next five years—his last in Marburg—he managed to maintain close to the same impressive level of productivity of this remarkable eighteen-month period. This activity made him internationally famous; by the mid-1860s he was generally regarded as (perhaps after Hofmann) the most eminent German chemist of his generation. It led to his call to Leipzig in 1865 and to Bonn in 1866 (the latter of which was refused).

Even superficial examination of the numbers in table 2 suffices to show that the explosion was not facilitated by any increase in numbers of students at his disposal; in fact, the numbers continued to decline. He reached simultaneously a low point in numbers and a high point in excitement in March 1860, when he wrote to Vieweg about how his theory had opened inviting prospects of discovery. "If only I had more hands," he cried, "that is, more capable students, who could help me to exploit this treasure trove before others use it." That semester he had but six auditors in his lecture, thirteen at various levels of competence in his practicum, and no chemistry majors at all. The university enrollment had fallen to a low of 216. Seven months later, he wrote Vieweg again, using identical phrases. He said he had been trying to exploit this theory for almost two years, with the help of some very good students, but distressingly small numbers of them.[4]

Were these students, admittedly smaller in number, nonetheless of higher caliber than those he had in his early years in Marburg? That might go far in explaining the great break. Let us attempt at least an impressionistic qualitative comparison of the two periods.[5] During the early Marburg period (1851–1858), he taught one man who would later be regarded as a master chemist, Peter Griess (1829–1888), and three others who could be fairly described as very good journeyman chemists, B. Wilhelm Gerland (ca. 1829–ca. 1905), the Englishman Frederick Guthrie (1833–1886), and the Irishman Maxwell Simpson (1815–1902). For two semesters in 1855–1856, the future industrialist Ludwig Mond (1839–1909) studied in Kolbe's classroom and laboratory, but he was only sixteen at the time and left no traces in publications or in Kolbe's correspondence. It is probable that his subsequent study with Bunsen at Heidelberg was of greater influence on him. Oddly, all three of these early German students of Kolbe would soon emigrate to England—Gerland probably in 1854, Griess in 1858, and

Mond in 1862—spending the rest of their lives working for various English chemical companies. After Marburg, Gerland and Guthrie studied with Frankland in Manchester, and Griess with Hofmann in London. Kolbe had inherited Gerland and Francis Wrightson (whom we met in chap. 6) from Bunsen's research group in 1851. Only a few of Kolbe's early Marburg students can be traced later than about 1858.

The foregoing applies to the period before 1858. During the period of his sudden efflorescence of activity, he had a very capable young man in his lab who later made a successful academic career at the Dresden Technische Hochschule, Rudolf Schmitt (1830–1898). Another worker, quite productive but probably in the "journeyman" category, was Eduard Lautemann, about whom little is known. He studied with Kolbe from 1857 to 1861, thereafter serving as assistant. He published his entire oeuvre of seventeen papers, some solo and many co-authored, during the period from 1859 to 1865, then traveled to India, began to study medicine, and vanished from sight. Adolf Claus (1840–1900), later a prominent structuralist, studied with Kolbe from 1858 and worked in the lab in 1859–1860 before transferring to Göttingen. He was a novice at the time, and there is no evidence that he did any significant research at Marburg.[6]

Bearing in mind that the second period is much shorter than the first, an unequivocal choice between the cast of characters before and after 1858 on the basis of their quality is difficult. However, it is probably fair to suggest that, on the whole, Kolbe had no better student material to work with in 1859–1860 than he did before this time, and as we have seen, he had fewer of them. Thus, we cannot explain the change by looking at the students.

Another possible explanation for the transformation is the influence of external events. I have already related the debilitating fevers and acute rheumatic attacks that plagued Kolbe during virtually all of 1857 and the first half of 1858. During this period as an invalid, Kolbe's frustration was intensified by the fact that it was in December 1856 that he had written a short prospectus of his carbonic acid theory and was then physically unable to substantiate it by experimental efforts. A mineral water "cure" at Wiesbaden in late spring 1858 made him vigorous and healthy again. His arrival back in Marburg on 16 June marks the precise point when he began to generate a prolific research program. On his return he wrote

> I have been back in Marburg for a week, and am fortunate to be able to tell you . . . that I am completely recovered, and feel healthier than I have in years. The Wiesbaden water really did me wonders, which I must all the more readily admit, since, to be honest, when I went to Wiesbaden I did not initially have much faith or belief in the therapeutic

> power of such an innocent appearing rivulet. . . . Giving lectures is [now] easy for me, and it seems to me that I have never lectured better than I do now. In short, I feel newly reborn, and for this I cannot be sufficiently grateful to Providence.[7]

Before 1857 he had been reasonably healthy, but he had had a number of other problems that tended to interfere with his scientific productivity: courtship, marriage, and founding a family; money troubles; rancorous collegial disputes; and efforts to make as rapid progress as possible on his voluminous textbook and on the *Handwörterbuch* for Vieweg.

But external events cannot provide anything approaching a full explanation. After 1858, as before, he was much occupied with his textbook, which still was far from complete, and with disputes with colleagues and acquaintances. He continued to be seriously underpaid, and ministerial support for his laboratory was so miserly that in the fall of 1860 he had to spend his own money to support it. His health, although improved, was still not good, and he continued to be afflicted by at least annual attacks of severe rheumatism, as well as regular influenzas and grippes. His mental health was also not good after 1858, and his wife suffered through two long and nearly fatal illnesses.[8]

Finally, there is strong presumptive evidence (detailed in the previous chapter) that publications as late as the middle of 1858 by Kekulé, Wurtz, and others influenced the formulation of the definitive version of his theory in that year. It is not unreasonable to conjecture that the medically enforced idleness in Wiesbaden of May and June 1858, complained of bitterly in a letter to Vieweg,[9] gave Kolbe the time to read and ponder the recent literature in a more relaxed fashion than would otherwise have been possible.

I have gone through this examination of candidate causes for the great break in order to support the thesis that the most obvious explanation for the change—the acquisition by 1858 of an extremely powerful theory that was absent before this time—stands virtually alone in importance. The difficulties with which Kolbe had to contend were as great during the years soon after 1858 as compared to before this time. But he had the principal prerequisite for productive research, a good new theory, and that made all the difference.

Polyfunctionality

Chapter 8 detailed Kolbe's route to his carbonic acid theory. There were several essential novelties in this theory, as compared to his earlier beliefs; he now fully accepted carbon tetravalence for most, but not all, organic compounds. He had also adopted a substitutionist viewpoint and given up all traces of the copula theory. The research of Wurtz and Debus in 1857 and 1858 had convinced him of the dibasicity of oxalic acid, hence the need to double his formula for it. Furthermore, by extension, he now accepted the generic category of polybasic acids.

Polybasic acids are examples of polyfunctional organic compounds. For years, Kolbe had resisted a direct theoretical confrontation with this kind of compound, and for good reason. His chemical instinct, developed to maturity in the electrochemical tradition, was to identify a single, central carbonaceous focus for a compound and to use that atom or radical as the theoretical centerpiece of the formula. Heteroatoms, especially oxygen, were grouped together as much as possible. Polyfunctionality was possible to formulate in this style, and Kolbe often did this before 1858, but only when all but one functional group remained in the background and hence were formulable as substituents within the radicals attached to the carbon group at the focal point. To relinquish this viewpoint would be to accept the structuralists' (Gerhardtian) thesis of the chemical equality of all carbon atoms in the molecule, thus to relinquish the last vestige of dualism. To be sure, by the time of Gerhardt's death Kolbe saw much of value in the Frenchman's system. Where the type theorists had gone seriously wrong, he thought, was in their subsequent development of multiple and mixed types. As it happens, multiple and mixed types were a result of the typists' struggle with polyfunctionality.

Wurtz in particular was generating phenomenal numbers of novel polyfunctional organic amines, alcohols, acids, and aldehydes in the years from 1855 to 1861. Chief among these were the two- and three-carbon organic acids and alcohols. Wurtz succeeded in oxidizing glycol to glycolic acid and then to oxalic acid. Since glycol had been prepared from ethylene, Wurtz argued that oxalic acid must have as much carbon as ethylene; moreover, glycol was seen to be as much the alcohol of glycolic acid as of oxalic acid, and glycolic acid was formulated as "diatomic" and dibasic (which is to say that both hydroxyl and carboxyl hydrogen atoms were acid in character). In modern terms,[10]

CH_2CH_2	CH_2OHCH_2OH	CH_2OHCO_2H	CO_2HCO_2H
ethylene	glycol	glycolic acid	oxalic acid

By analogous reactions, propylene was converted to propylene glycol and then to lactic acid, and analogous claims could be made for these as well:[11]

CH_3CHCH_2	$CH_3CHOHCH_2OH$	$CH_3CHOHCO_2H$
propylene	propylene glycol	lactic acid

Wurtz then found that phosphorus pentachloride could be used to replace both hydroxyl groups of lactic acid with chlorine atoms, to make "[chloro]lactyl chloride," and then by reaction with alcohol, "chlorolactic [ethyl] ester."[12] The existence of these compounds underlined for Wurtz the dibasic character of lactic acid and also its close relationship to propylene glycol, which underwent an analogous reaction with phosphorus pentachloride.

In the meantime, Heinrich Debus, a Bunsen protégé then at Queenwood College, had developed a means of oxidizing alcohol to produce many of these same polyfunctional two-carbon compounds, but also including glyoxal (the dialdehyde) and glyoxylic acid (the aldehyde–acid). Kekulé and his student R. Hoffmann demonstrated how to make monochloroacetic acid, from which could be derived glycolic acid and glycocoll (glycine or aminoacetic acid). Strecker showed how to oxidize glycocoll to glycolic acid, and alanine (aminopropionic acid) to lactic acid. There seemed to be clear genetic relationships among all these compounds, and all could be formulated using multiple and mixed types.[13]

Kolbe needed to respond to all of these novel reactions by providing interpretations consistent with his own ideas. Chapter 8 showed why Kolbe formulated glycol as a hydrated oxide,

$$2HO.\left.\begin{matrix}C_2H_3\\H\end{matrix}\right\}C_2O_2$$

rejecting Wurtz' claim that it is an alcohol. For Kolbe, the lone hydrogen in the brace meant that glycol cannot be oxidized to an acid since such oxidation requires a minimum of two hydrogen atoms attached to the carbonyl group. It could, however, be oxidized to a two-carbon homolog of glycerin, that is, replacing the lone hydrogen by a third oxygen function. In Kolbe's formulation, this would be $3HO.C_2H_3.C_2O_3$, a triple oxide hydrate. That this reaction had not yet been accomplished was not a problem for Kolbe.[14]

Wurtz had actually done what Kolbe considered impossible, in oxidizing ethylene glycol to glycolic acid, and propylene glycol to lactic acid. Kolbe responded that rearrangements must take place: in oxidizing glycol, one hydrogen of the methyl was replaced by a "hydrogen peroxide" radical (O_2H, i.e., hydroxyl) so that the other carbon now had room for full oxidation to the acid. Consequently, there must be a yet undiscovered alcohol whose oxidation yields glycolic acid without rearrangement, the "true" glycol or double alcohol, *isomeric* with Wurtz' compound and unrelated to ethylene. Milder oxidation of this hypothetical substance should also yield a new aldehyde isomeric with acetic acid. Finally, if Wurtz' compound, ethylene oxide hydrate, could be dehydrated, it should yield ordinary aldehyde. In short, Kolbe affirmed, ethylene glycol and propylene glycol are not alcohols at all and have no *substantive* genetic relationship to glycolic or lactic acids. The latter are derivatives of acetic and propionic acids, as Kekulé, Hoffmann, and Strecker had shown. To be sure, glycolic and lactic acids each possess a radical (O_2H) substituted for hydrogen, but the acids are monobasic, not dibasic as Wurtz had asserted.

Kolbe made all these claims in fascicle 8/9 of his textbook, written in the summer or fall of 1858 and published early in 1859. The driving force of his resurgent research program begun late in 1858 was the examination and substantiation of these ideas, which he regarded as a direct outgrowth of his carbonic acid theory. That this theory was generating interpretations and predictions different from those of the type theorists validated Kolbe's sense that he had developed a powerful theory different from and superior to the school he so heartily despised—to what would soon be known as structure theory.

As we have seen, there were indeed some distinctive aspects of Kolbe's approach. Kolbe took formulation far more seriously and literally than did the type theorists, believing that discerning the constitutions of molecules was a straightforward process of applying rigorous deduction to skillfully gathered hard evidence, and he was convinced that one could arrive at ultimate formulas in this fashion. He accepted carbon tetravalence ("tetratomicity") for most organic compounds, but he could not countenance carbon chain formation. Kolbe's proximate radicals (such as methyl and carboxyl in acetic acid, or methyl, hydroxymethylene, carbon dioxide, oxygen, and water in lactic acid) were combined with one another as discrete molecular units and presumably coulombically. In formulating any compound, the goal was to identify a single governing radical, which he named the fundamental radical that held as many other proximate radicals together as equaled its combining capacity.

His approach made it difficult for him to deal with polyfunctionality in organic compounds, especially when the functional groups in a single molecule were not all the same. Once he was compelled to confront polyfunctionality by the work of Wurtz, Debus, Kekulé, and others, his theory became much more powerful. But even thereafter, his discomfort with the phenomenon is revealed in many subtle ways. This discomfort was the ultimate source of his strong disagreement with

Wurtz over the nature of glycolic and lactic acids; rather than alcohol-acids, Kolbe was convinced that they were ordinary monofunctional acetic and propionic acids, merely with a substituent O_2H group. Kolbe even admitted publicly that it had been difficult for him to reach the conclusion that O_2H could replace H in organic compounds.[15]

Despite these differences, Wurtz' and Kolbe's chemical ideas were substantially similar. All of Kolbe's reasoning and all of his formulas could easily be translated into type notation, and vice versa. Their disagreements were usually unrelated to the distinctions that did exist between their theories. Wurtz could point to the genetic relationship between glycol and glycolic and oxalic acids to argue that glycol was a dialcohol corresponding to a reduced form of the two acids, but he then ran into trouble explaining—or rather, did not even try to resolve—the precise genetic relationship between ethylene and glycol, a point seized upon by Kolbe. Kolbe could easily account for the latter genetic relationship, but had to suppose a rearrangement and to posit undiscovered isomeric alcohols to explain the former. This set of problems was eventually solved, at least in principle, by Alexander Crum Brown's argument in 1864 that ethylene was not in fact CH_3CH, as virtually everyone had assumed by analogy to carbonic oxide, but rather CH_2CH_2.[16]

Predictions Unfulfilled: Hydroxyacids

In the fall of 1858, Kolbe put his student Carl Ulrich on the problem of hydrolyzing and reducing Wurtz' chlorolactyl chloride. Ulrich found a way to remove the two chlorine atoms in two clean stages, first by generating chloropropionic acid and then by reducing this to propionic acid by means of nascent hydrogen. The fact that the process occurred in two stages neatly underlined Kolbe's point that the two hydroxyl functions of lactic acid were not equivalent, hence lactic acid was not dibasic as Wurtz claimed. Moreover, Ulrich had transformed a lactic acid derivative into propionic acid, proving the close relationship Kolbe had been asserting.[17]

Simultaneously with Ulrich's paper, Kolbe published a summary of his views on the constitution of lactic acid. He rehearsed all of the arguments just published in his textbook; regarding the nonalcoholic character of glycols, he stressed that aldehyde had never been produced from glycol. Debus' glyoxal, interpreted by its creator as a double aldehyde, had been prepared from alcohol, not glycol, and in any case, there was no proof that it was an aldehyde.[18]

In a long article published simultaneously with Kolbe's and Ulrich's,

Wurtz developed his thoughts on reactions of polyatomic alcohols (derivatives of glycerin and glycol) on which he had done so much valuable work.[19] It was the first time Wurtz used Williamson's atomic weights (barred O's and C's indicating doubled conventional equivalents), and he thoroughly discussed the incipient theory of "molecular structure," as he termed it here. Wurtz attempted to build a case for the crucial role his research on polyatomic alcohols had played for the rise of this theory.

> I will force myself to be brief; for although I attribute a high value to theory, which must be the foundation and the end of all science, I believe that, above all, facts should be allowed to speak for themselves, and that in chemistry, theory consists only in the direct and judicious interpretation of that which experiment teaches us.

Among other things, "the theory" had suggested to Wurtz that there ought to be an intermediate member between glycerin and normal alcohol; this thought had led to his discovery of glycol. He reported that he was continuing to produce large numbers of glycol derivatives in his lab, as predicted by the theory.[20] The mere existence of glycols was unimportant; what mattered was that it fulfilled predictions, and transformed the "hitherto vague and unsupported hypothesis" of polyatomic radicals into substantiated fact.[21]

Wurtz noted that polyfunctionality means that type or radical formulas in general capture only partial views of molecules, but they are nonetheless of enormous value. "I realize that many people abuse them. But abuse does not condemn use."[22] He felt that one must not reject rational formulas, but that one must also not abuse them by putting too much trust in them. "It is wrong to present these things as the law and the prophets." Formulas cannot provide ultimate depictions of "the intimate constitutions of compounds," but rather should serve as guides to the prediction and interpretation of reactions.[23] Wurtz cited A. S. Couper and Kolbe as two men who had erred by believing their formulas too literally. Kolbe had often averred that multiple types are imaginary, hence useless, because (for instance) the double water type H_4O_2 as the basis for formulating sulfuric acid does not exist in nature. Wurtz responded that this objection "is not serious," for the advocates of such types, including himself, had always been careful to specify that the cause of cohesion of multiple types is a polyatomic radical. Williamson had formulated the hydroxyls of sulfuric acid as held together by a diatomic SO_2 radical, and Wurtz had formulated the three fatty acids of triglycerides as held together by the triatomic glyceryl radical.[24]

When Kolbe's and Ulrich's papers appeared in the spring of 1859,

Wurtz responded almost immediately with new compounds supporting new arguments. Chlorolactyl chloride was the starting material for both esterification of the acid group and acylation of the alcohol group, producing "lactobutyric ester" (butyryl ethyl lactate). Wurtz also prepared an ethyl ether/ethyl ester from the same starting material, using two moles of Williamson's sodium ethoxide. These reactions once more emphasized the difunctionality of lactic acid. However, Wurtz conceded the point deduced from Ulrich's reactions, namely, the non-equivalence of those two functions. Lactic acid is not dibasic in the same sense that oxalic acid is, Wurtz admitted. However, it is nonetheless *diatomic* (that is, it has two replaceable hydrogen atoms).[25] Here he was following Kekulé's ideas and language from a paper on glycolic acid published the preceding year.[26]

Wurtz took this occasion to stress once more the alcoholic character of the glycols and their close relationship to the hydroxyacids. Kolbe's name for ethyl alcohol was "ethyl oxide hydrate" and that for glycol "ethylene oxide hydrate." These two names show even more analogy than Wurtz' "alcohol" and "glycol." Kolbe had said he feared that the definition of "alcohol" would be stretched beyond all recognition or meaning by the inclusion of the glycols in this class. "Let him be reassured on this point," Wurtz wrote; polybasic acids such as oxalic acid have not destroyed the concept of acid! As for the fact that glycol had not yet been converted into an aldehyde, this was but a temporary situation. Debus' glyoxal was indeed the aldehyde analog of glycol, even though it had been prepared from alcohol and not glycol. In any case, no one was denying that methyl alcohol is an alcohol, even though no one had isolated a methyl aldehyde. Finally, several glycols had been converted to acids, analogous to ethyl alcohol being converted to acetic acid, which was Kolbe's own criterion for alcoholic character. As for Kolbe's argument that lactic acid is monobasic, "or rather monatomic," Wurtz thought that his reactions had shown this thesis to be untenable. Besides, where were all of Kolbe's hypothesized isomeric alcohols and aldehydes, the so-called true reduced analogs to the hydroxyacids? Wurtz believed that it was Kolbe, not he, who dwelt in the land of hypothesis.[27]

Kekulé was in complete accord with Wurtz on the matter of formulas and their interpretation, and also on the power of structure theory. As he wrote Lothar Meyer in 1860, "We and science quietly wend our way between the mischief of those who make a game of constitutional formulas, and the indolence of those who deny [rational] formulas, toward the star of a fundamental synthesis beckoning from afar."[28] As it happened, the first fascicle of Kekulé's textbook was being printed while Wurtz' papers just described were being published.

In this fascicle, Kekulé emphasized that rational formulas are derived solely from reactions, and that one must be allowed to write different formulas for the same compound, depending on what functionality was in question for a given reaction. He wrote, "It is clear that even for acetic acid—and all the more so for more complicated compounds—a completely comprehensive rational formula is not appropriate for ordinary use, even if one can be specified in the present state of the science." Rather, one uses whatever formula most clearly makes the point in question. Type notation is handy for many, even most situations, he felt. It clarifies, for example, the different chemical behavior of the "typical" hydrogen in alcohol, or the "typical" hydrogen in acetic acid, from the other hydrogens in the compound. However, type formulas, too, have their limitations.

> There are several cases where different hydrogen atoms should be equivalent according to the type theory, and are not. For instance, glycolic acid, as well as lactic acid, behave like monobasic acids, although they contain two typical hydrogen atoms. . . . One behaves just like the typical hydrogen of alcohol, the other just like the typical hydrogen of acetic acid. The different behavior of these two hydrogen atoms is apparently caused by the different positions they occupy with respect to the other atoms, particularly oxygen. One hydrogen atom lies in the neighborhood of two oxygen atoms, like that of acetic acid; the other lies in the neighborhood of one oxygen atom, like that of alcohol.[29]

Given the context of the rest of this fascicle, in which Kekulé laid out his founding version of structure theory, there can be little doubt what he had in mind here: lactic acid is hydroxyacetic acid, an alcohol–acid. He indicated here the via media between Wurtz and Kolbe, while also *implying* the first adequate fully resolved formula for the compound. The implication was made explicit in papers published by Kekulé, W. H. Perkin, and Alexander Crum Brown, all in 1861; Crum Brown made clear that he was only reading between the lines of Kekulé's 1859 quotation, just cited.[30]

But Kolbe never blinked. On the attack once more, he and his student Lautemann published another trio of articles on lactic acid in the February 1860 issue of the *Annalen*. Given the task by his mentor of converting lactic directly into propionic acid—in other words, foregoing the chloro intermediate—Lautemann found success with his fifth attempted reducing agent, hydrogen iodide. (This marked a significant methodological innovation in organic chemistry, for hydrogen iodide proved to be a very versatile reducing agent.) Concurrently, Kolbe discovered how to convert lactic acid to alanine (Strecker had accomplished the reverse). Both reactions tightened the analogies to the monobasic series, in this case propionic acid.[31]

In short, Kolbe thought that his proof of lactic acid as monobasic was irrefutable and that Wurtz' own work had only further confirmed this.[32] But just for good measure, he added more arguments to finish Wurtz off. Lactic acid forms no salts with two different metals and has no acid salts, nor does it have a diester. What Wurtz called a diester—his diethyl product using sodium ethoxide—is actually *oxyäthylpropionsaures Aethyloxyd*, a substituted monoester. As for Wurtz' butyryl ethyl lactate, Kolbe surmised that the compound was actually *Oxybutyroxylpropionsäureäther*, a hydroxy ketone monoester. Finally, Kolbe excoriated Wurtz for suggesting that both of their different formulas might apply equally to ethyl chloropropionate.

> I confess I do not have so broad a chemical conscience, and could never countenance such a doctrine, even if it had to do with more than simply a weak hypothesis. I believe that with these words Wurtz has passed judgment on his own hypothesis. . . . The symbolic expressions for our views on the proximate components of a compound and on their relative positions may of course change. But to assign a compound two different rational formulas *at the same time*, i.e., to maintain that it possesses sometimes one set of atomic groupings as proximate components, and at other times another set . . . is to maintain an impossible proposition.[33]

This gave Wurtz another opening. As far as their chemical consciences were concerned,

> Mine is less delicate concerning formulas. I envision them as expressing the mode of derivation, parental ties, and reactions of compounds, and in no way share the opinion of M. Kolbe, who endeavors to express the exact grouping of the atoms with the aid of his rational formulas. He pretends to know the nature and role of the groups in organic compounds. . . . I express merely parental ties. I express *certain* reactions, and everyone will agree that it is impossible to express *all* reactions by means of formulas of this kind.

Wurtz then carefully reiterated his position: lactic acid is indeed monobasic, which explains the absence of dimetal and acid salts. It is, however, diatomic, that is, it has alcohol character, a fact that Kolbe was trying to ignore. Wurtz had no intention of contesting Kolbe's key assertion that lactic acid is related to propionic acid; but it is just as clearly related to propylene glycol, for the latter oxidizes smoothly to lactic acid. Wurtz pointed out that the products of hydrolyzing butyryl ethyl lactate were consistent only with his, and not with Kolbe's, formulation of the compound. Finally, Wurtz presented a table directly comparing his and Kolbe's formulas for the same set of lactic acid derivatives and suggested that chemists choose between them. One might differ over issues of esthetics and informational content, but there is no question that Wurtz' were more compact and simpler.[34]

The following year Wurtz let fly another volley, in conjunction with his student Charles Friedel. They directly compared the two ethyl compounds of lactic acid, namely, the ethyl ester and the ethyl ether; the former was neutral, while the latter was fully as acidic as the parent acid. It would be hard to imagine a clearer demonstration of the replaceability of *both* "typical" hydrogen atoms and also their chemical nonequivalence.[35]

Kolbe let loose his own shot. "The efforts of some chemists," he wrote in a paper co-authored with Lautemann, "to demonstrate alcohols and aldehydes also for dibasic acids, e.g., to claim ethylene oxide hydrate as the alcohol and glyoxal as the aldehyde of the dibasic oxalic acid, are unscientific frivolities that deserve no notice here."[36] This was a strange outburst, both in the unjust violence of expression, as well as in its logic. Kolbe himself had repeatedly suggested that dibasic acids must have reduced forms—he had simply denied that glycol and glyoxal are the reduced forms of oxalic acid.[37]

Debus was moved to respond. It is possible, he wrote, to recast the Kolbe–Lautemann assertion into "a decent form." However,

> Before the judgment of Messrs. Kolbe and Lautemann can make the slightest claim for consideration, the concepts indicated by the words "aldehyde" and "alcohol" must be clarified. Then there must be derived from these concepts, or from a general principle, or from an a priori intuition, and not for example from any set of empirical observations, the impossibility that dibasic acids may correspond to aldehydes or alcohols. Messrs. Kolbe and Lautemann have not to my knowledge offered any demonstration of this sort, and therefore their verdict loses all foundation.

Debus also pointed out that Kolbe had predicted precisely what he was now claiming to be impossible. And since he had predicted such alcohols and aldehydes, what proof had he that glycol and glyoxal were not those compounds?[38]

Kolbe's search for these missing substances led him to speculate on a possible isomerism phenomenon in glycolic and lactic acids. Might it not be reasonable to think that they could exist in two modifications each? In particular, perhaps the conventional glycolic acid is hydroxymethyl formic acid and is produced from the oxidation of chloroacetic acid. In contrast, Debus' oxidized alcohol may not be identical to this compound but rather it may be the isomer methoxy formic acid, a monoester of carbonic acid. Similarly, perhaps conventional lactic acid is hydroxyethyl formic acid, while the known isomeric compound, lactic acid from meat, is ethoxy formic acid; or the other way around.[39] He privately speculated on the further possibility that oxalic acid may be analogous to glycolic acid, in the sense of being monobasic but diatomic. In this case, there should be two chemically distinct ethyl oxalates; a yet unknown isomer of oxalic acid should also exist that is truly dibasic and homologous with malonic and succinic acids.[40] Nothing concrete came of these ideas. Wurtz, too, had suggested the possibility of isomeric glycolic acids, but as early as 1858 Kekulé had asserted the identity of all candidate isomers.

Few of Kolbe's colleagues shared his sense of triumph over Wurtz in the matter of glycolic and lactic acids. Disagreeing with Wurtz over what must have seemed to most observers to be relatively subtle structural or even semantic distinctions, Kolbe generated many predictions, few of which were realized. None of his putative alcohols and aldehydes isomeric with Wurtz' and Debus' compounds, which he considered the true reduced analogs of the acids derived from glycols, were ever found, nor were the predicted isomers of glycolic, lactic, and oxalic acids or ethyl oxalates ever prepared. He himself, in conjunction with Guthrie, refuted his own prediction that one could dehydrate glycols to yield ordinary aldehydes.[41] He also conceded Wurtz' refutation of his interpretation of the constitution of butyryl ethyl lactate.[42] He ultimately adopted Wurtz' and Kekulé's view and language regarding the nature of lactic acid—that it was monobasic and diatomic—but he regarded this as his victory, not Wurtz'.[43]

In the end, his strong uncollegial language in a matter that was even under the most favorable interpretation contestable, and that many considered a losing cause, could only do Kolbe damage. This was one more repetition of the sort of unpleasant polemics that he had waged over the previous ten years against Gerhardt and Williamson. By March 1860, despite his newly productive research program, Kolbe felt isolated and under attack from most sides. The newer type theory continued to attract adherents, including such respected establishment figures as Will, Kopp, and Strecker, a fact that thoroughly mystified Kolbe. Liebig had been acting unfriendly toward Kolbe for years, Berzelius was long dead, and his own mentors Wöhler and Bunsen, although supportive, were neither interested nor active in theoretical matters. In later years, he often reminded the chemical community of this period in which, as he put it, he was considered a "crank."[44]

This unhappy situation was transformed in 1860 due to Kolbe's work with diacids and his predictions of secondary and tertiary alcohols.

Predictions Fulfilled: Diacids and Novel Alcohols

In his major theoretical article, "On the Natural Connection of Organic and Inorganic Compounds," Kolbe made predictions that were more successful. Having accepted the dibasic character of oxalic acid, he now proposed two "carbonic acid radicals" for other known diacids, such as succinic, malic, and tartaric acids. All three of these substances were known to have the same number of carbon atoms (four) and two carboxyl groups. Succinic acid has no other functional groups, whereas malic and tartaric acids have one and two additional atoms of oxygen (two and four equivalents), respectively. Kolbe suggested in his paper that malic and tartaric acids may have a similar relation to succinic acid as lactic and glyceric acids have to propionic acid, namely, that they may contain one and two (O_2H) groups, respectively, substituted for H. He named them *Oxy-* and *Dioxybernsteinsäure*.[45] When, late in 1859, Lautemann discovered that hydrogen iodide was capable of directly reducing the (O_2H) group of lactic acid to H, Kolbe reasoned that he could test his prediction by attempting to reduce malic and tartaric acids to succinic acid by using the same reagent. He assigned the task to his promising young assistant, Rudolf Schmitt.

The reduction occurred uneventfully, and Kolbe sent the paper to Liebig. Liebig rushed it into print, fitting it into the very next issue after the one that carried Kolbe's prediction.[46] Ironically, Liebig had recently published a different hypothesis regarding the constitutions of malic and tartaric acids.[47] He wrote Kolbe a very friendly letter:

> . . . the real purpose of this letter is to express to you the great satisfaction which your paper on the natural connection of organic and inorganic compounds gave me; the preparation of succinic from malic and tartaric acids is the triumph of your theory; I am only sorry that I recently gave a different interpretation of the constitution of these two acids, but I willingly recognize that yours is better.[48]

To say that Kolbe was pleased by this letter is more than a small understatement. For the past six years, Liebig had been cool and sometimes even hostile toward him, at first apparently because of the one-sided and polemical character of parts of his book and then because in an article in the *Handwörterbuch* Kolbe had ignored Liebig's analytical method for mercury. Kolbe had been mystified by Liebig's unfriendliness. At first he supposed Liebig's residence in the Bavarian capital and association with King Maximilian II had made him an arrogant courtier; he then suspected that enemies were whispering in Liebig's ear.[49] He replied to Liebig, delighted to see, as he put it, that his enemies had failed to sway Liebig's good opinion of him. A more public expression of approval, he hinted, would hasten the end to the influence of those who view the goal of chemistry as the "decoration of Gerhardt's schemata." To Vieweg he declared his intention of using Liebig's letter to pry more money out of the Kurhessian ministry.[50]

Kolbe was still earning the same miserable salary he had accepted in 1851.

Feeling like the proverbial cat that swallowed the canary, he immediately wrote his closest friend, Vieweg, enclosing Liebig's letter. He added that he had long been convinced that his ideas would eventually triumph, and he now predicted that in a few years no one would even mention "the completely unscientific manner of treating chemistry of Gerhardt and his consorts, which unbelievably even Strecker has adopted . . ." Liebig's letter could make this happen all the faster; he asked Vieweg to return it as soon as possible so that he could make appropriate use of it with his Ministerium.[51]

From this time onward, Liebig strongly supported Kolbe's research. Kolbe's discoveries, Liebig wrote him in December 1860, had the effect on him like that "of a trumpet on an old war horse," and he invited Kolbe to continue sending him his "gems" for publication. The following year Liebig averred to Kolbe that "the important thing is always that one starts down the right path, and all your work demonstrates that you are on the right path." There is much good work being done now in organic chemistry, Liebig continued, but also much playing around with formulas, and in most work one cannot discern the "scientific idea" that one must have as a goal, to clarify the physiological origins of various compounds.[52] Liebig meant this praise sincerely; to Vieweg, to Fehling, and to Volhard he wrote in much the same terms about Kolbe.[53] In February 1862 Liebig satisfied Kolbe's request for a public statement of approval, and in December of that year he proposed Kolbe as foreign member of the Bavarian Academy of Sciences.[54]

Liebig's approval appears to have been founded on two general areas of agreement. First, like Kolbe, Liebig took conventional equivalents to be the chemical atoms themselves, and so for both of them such Gerhardtian types as HOOH (two molecules of water in equivalents but one molecule of water in atomic weights) were imaginary and therefore absurd.[55] Second, Liebig felt that organic chemistry only made sense when it is pursued in conjunction with physiology. Since carbonic acid plays such a central role in physiology—especially for biosynthesis in plant physiology, from which most organic compounds dealt with in mid-nineteenth century organic laboratories were derived—Kolbe's carbonic acid theory made sense to Liebig in a way that the more abstract and schematic Gerhardtian theory did not.[56]

Other predictions by Kolbe concerned novel alcohols. Kolbe, like the structuralists, formulated ethyl alcohol as a carbon atom (which for Kolbe was the double atom C_2, his "Grundradikal") combined with two hydrogens, a methyl group, and what became known as a hydroxyl radical. In his major theoretical paper, Kolbe pointed out that if one or both of these two hydrogen atoms were replaced by one or two additional methyl radicals, two novel substances would be formed. They would retain the hydroxyl group, but they would not be oxidizable to an aldehyde or an acid since each of these oxidation reactions requires abstraction of *two* hydrogens from the same carbon. They would therefore fail the defining criterion for alcohols, but could be termed "pseudoalcohols."[57] These compounds are what modern chemists call isopropyl alcohol and tertiary butyl alcohol.

Of course, alcohols were familiar substances, long known to chemists chiefly as products of fermentation. The best known was ordinary (ethyl) alcohol, but it was also known that the so-called fusel oil, a higher boiling residue that remained after the redistillation of grain or potato alcohol, contained alcohol-like materials. The major constituent of fusel oil was found to be amyl alcohol ($C_5H_{11}OH$), but there was also a sizable amount of butyl alcohol (C_4H_9OH) and a much smaller amount of propyl alcohol (C_3H_7OH), along with a number of trace constituents. These three compounds were initially thought to be simple homologs of ethyl alcohol. The amyl and butyl alcohols were later found to possess branched-chain structures (for example, the latter is $[CH_3]_2CHCH_2OH$), and so they were eventually given the names *iso*-butyl and *iso*amyl alcohols, the prefix designed to distinguish them from the straight-chain (or "normal") primary alcohols such as ethyl alcohol, or propyl alcohol from fusel oil. In the same way that ethyl alcohol could be dehydrated to ethylene, so could propyl, butyl, and amyl alcohols be dehydrated to the homologous olefins propylene, butylene, and amylene. During the 1850s, Berthelot demonstrated how to convert these olefins back to the alcohols by aqueous distillation from dilute sulfuric acid solution.

In the summer of 1862, Wurtz' student Charles Friedel published a note in which he described the reduction of acetone to a three-carbon alcohol, using nascent hydrogen generated from sodium amalgam. He refused to identify his product with propyl alcohol from fusel oil, saying the matter needed study.[58] Four months later (12 November 1862) Kolbe learned of this paper by reading a German abstract. The next day he sent Emil Erlenmeyer, editor of the biweekly *Zeitschrift für Chemie*, a short article on Friedel's compound, asking that it appear in the next issue. "The subject interests me all the more," Kolbe wrote in his cover letter, "since I expect to find through Friedel's work a confirmation of my view (based on purely theoretical speculations) concerning the existence of such new alcohol-like compounds."[59] In this paper, Kolbe suggested that Friedel's new alcohol was the isomeric propyl alcohol he had predicted years earlier. The test would be to oxidize the compound; a secondary alcohol must yield acetone, whereas the known (primary) propyl alcohol would give propionic acid. He also suggested, presumably by comparing boiling points, that Friedel's product was the same as that propyl alcohol produced by Berthelot years earlier by hydrating propylene. But he said he did not want to forestall Friedel and so was leaving it for him to complete the investigation.[60]

A few months later Friedel reported the oxidation experiment and confirmed Kolbe's prediction. But he added a mild protest. What other product than acetone could have been expected when the reduction product of acetone was oxidized?! Obviously, no one could imagine this was common propyl alcohol. He had been perfectly well aware of what he had when he published his first paper; he only wanted to be able to present clean results devoid of conjecture and had been having troubles over impurities, so he had temporized. In rebuttal, Kolbe conceded he could not prove that his idea had formed the basis for Friedel's reaction, but thought it curious that Friedel had not made the discovery until Kolbe's interpretation had appeared in print. His conclusion was that the episode clearly demonstrated the fruitfulness of his own theory and the barrenness of type theory.[61]

Six months after Kolbe's rebuttal, in the summer of 1864, A. M. Butlerov identified a "tertiary pseudobutyl alcohol," trimethyl methyl (modern tertiary butyl) alcohol. He had obtained this new compound the previous year from the reaction of phosgene with methyl zinc, but had not immediately been able to specify its constitution. Butlerov noted that once more a prediction by Kolbe had been fulfilled. By this time, predictions of new isomers based on structure-theoretical precepts were rapidly proliferating. Butlerov argued, for example, that in addition to the known normal butyl alcohol and his new tertiary compound, exactly two more butyl alcohols should exist: a branched-chain primary (isobutyl) alcohol and a secondary butyl alcohol. (Here Butlerov was repeating a statement published a few months earlier by Kolbe.) As for the next higher homolog, no fewer than eight amyl alcohols should exist: four primary, three secondary, and one tertiary —or, by another manner of accounting, three associated with a straight-chain carbon skeleton, four with a branched-chain, and one containing a quaternary carbon atom. As far as lower homologs were concerned, structure theory appeared to predict a single methyl and a single ethyl alcohol. A similar analysis applied to the higher alcohols as well, and Butlerov was not slow to use the reaction that had given him *t*-butyl alcohol for the synthesis of new hexyl and octyl alcohols.[62]

The secondary butyl alcohol predicted by Kolbe and Butlerov was first prepared by V. H. de Luynes about the time of Butlerov's paper. This appears to have been a fortuitous event, and de Luynes did not attempt to determine the compound's constitution. Definitive identifications and structural assignments of all four isomers of butyl alcohol were first made in 1869 by Adolph Lieben, a student of Bunsen and Wurtz then working at the University of Turin.[63]

The formula for the butyl alcohol found in fusel oil by Wurtz was at first structurally indeterminate, but presumably the substance was assumed to be normal butyl alcohol.[64] Oxidation, however, did not yield normal butyric acid. In the meantime, Kolbe not only predicted the existence of an isobutyric acid (dimethyl acetic acid), but suggested no less than two different synthetic routes to it: isopropyl alcohol to isopropyl iodide to isopropyl cyanide, whose hydrolysis should yield isobutyric acid, or reduction of the hydroxyl group of acetonic acid (dimethylhydroxyacetic acid) using Lautemann's reducing reagent, hydrogen iodide.[65] Even while making corrections and exchanging proofs of the article containing these ideas with his editor Erlenmeyer, Erlenmeyer informed Kolbe of his current attempts to produce isobutyric acid by oxidizing the butyl alcohol from fusel oil. Erlenmeyer published this work shortly before V. V. Markovnikov independently published essentially the same reaction. Erlenmeyer and Markovnikov both concluded that their starting material contained a branched carbon chain, i.e., that it was isobutyl alcohol, because they found that Kolbe's suggested syntheses from isopropyl compounds yielded the same product.[66] In 1867 Frankland and Duppa removed any possible doubt about these structural assignments by ethylating and dimethylating ethyl acetate, yielding butyric and isobutyric acids, respectively.[67]

Of all these new alcohols, isoamyl proved perhaps the most interesting—and intractable. It had long been known (simply as amyl alcohol) as the major component of fusel oil, as had the associated olefin amylene. Wurtz published a series of articles in 1862–1864 on these compounds, and especially on rehydrated amylene, which he called amylene hydrate, an alcohol-like material differing in properties from the original natural alcohol.[68] Wurtz had the ill luck to have tackled a problem whose solution was beyond the capabilities of the science of his day. In fact, dehydration and rehydration of the natural alcohol had the effect of transferring the hydroxyl group two carbons down the chain, transforming a primary into a tertiary alcohol. The work was also hindered because fusel oils are complex mixtures from which it is difficult to isolate pure materials and because fusel oils from different sources often have quite different compositions (as Kolbe himself had discovered in his very first independent chemical research).

Erlenmeyer oxidized amyl alcohol from fusel oil; he obtained a valeric acid distinct from the normal variety. That this was isovaleric acid, i.e., a branched-chain structure, Erlenmeyer showed by producing the same compound by chain lengthening of isobutyl alcohol (converting to iodide, then to cyanide, and then hydrolyzing). Since isobutyl was by then known to be branched, the same had to be true for isovaleric acid.[69]

As for Wurtz' amylene hydrate, Erlenmeyer and Kolbe independently suggested that it was a secondary alcohol.[70] As early as December 1863, Kolbe was privately predicting that oxidation of amylene hydrate would yield diethyl ketone or, less likely, methyl propyl ketone. "Wurtz' most recent papers," he wrote Frankland, "are examples of how *not* to work, [they are] loose and sloppy." In February 1864 he claimed to have isolated what was "without question" the latter oxidation product and suggested that amylene hydrate was therefore methyl propyl carbinol (2-amyl alcohol); this was published in the *Annalen* later that year. However, the oxidation was not clean; from elemental analysis Kolbe concluded that he got only around fifty percent yield, the remainder being unreacted starting material. His formula assignment was based on an involved argument regarding boiling point regularities.[71] Wurtz' investigation of the same reaction, published in the meantime, was just as problematical, but not consistent with Kolbe's results; he found that the carbon chain was broken, with the chief products being acetic acid and acetone. Ever cautious, Wurtz refused to draw any conclusions regarding the constitution of amylene hydrate.[72] In fact, it was eventually established that amylene hydrate as prepared and identified by Wurtz does produce acetic acid and acetone upon oxidation. Straight-chain amyl alcohols are essentially absent in all fusel oils, and so Kolbe's published results are not easily explicable. He is vulnerable here to the suspicion of having found what he needed to find in order to verify his prediction.

At the same time that all of this was developing, Frankland and Baldwin F. Duppa published on the alkylation of oxalic ester, using zinc alkyls. A particularly interesting product was the diethyl compound, found to be isomeric with leucic acid (the latter was leucine, a six-carbon amino acid, converted by Strecker to a hydroxyacid by replacing its amino group by a hydroxyl group). Privately to Frankland and publicly in his paper on isomerism, Kolbe formulated it as diethyl glycolic acid. Although Kolbe did not realize it, this was consistent with Frankland's own formulation.[73]

The New Complexion of Organic Chemistry

A final brief case study from this period further illustrates the new atmosphere in the discipline and is one more curious instance of hot pursuit by simultaneous workers on a set of fruitful new structural investigations. A variety of actors were involved: Hugo Müller, a student of Wöhler and Liebig, assistant to Warren De la Rue and a close friend of Kekulé from his London years; Hans Hübner, a student of Wöhler and of Kekulé before becoming Privatdozent at Göttingen; Wurtz' protégés Lieben, Friedrich Beilstein, and Maxwell Simpson (the latter also a former Kolbe student); and Simpson's British countrymen Crum Brown, W. H. Perkin, and Duppa. The relevant reactions all involved the Kolbe–Frankland technique of replacing halogen with cyanide, then hydrolyzing the cyanide to carboxyl. The net effect of this sequence is to replace a halide with a carboxyl group, thus adding a carbon atom. By such means, the chlorine in Kekulé's monochloroacetic acid could be used to form a second carboxyl group and so transform acetic to malonic acid; an analogous route, apparently, could transform propionic to succinic acid—in short, monoacids into diacids. A closely related route is to start with ethylene bromide, which would give rise to succinic acid by double hydrolysis. Working with Kekulé in Ghent in 1862, Hübner published on cyanoacetyl compounds, but did not report hydrolysis to malonic acid. About the same time, Simpson published on the ethylene bromide to succinic acid route.[74]

From the summer of 1863 to January 1864, Kolbe described in letters to Frankland his work on this kind of malonic acid synthesis; he mentioned that on 17 January 1864 he had learned that his former student Crum Brown was working on the same problem. (Crum Brown's project in Kolbe's lab in 1862 had concerned diacids). On 12 February Kolbe sent Frankland and Erlenmeyer identical copies of a preliminary notice, with the request that the former translate and communicate it to the very next meeting of the Chemical Society, and the latter print it "*as soon as possible*."[75] Erlenmeyer dutifully inserted the note into the next issue of his *Zeitschrift für Chemie*, and Frankland had the paper read in English at the Chemical Society on 18 February.[76]

Two days later a few unnamed Marburgers visited Göttingen and informed Hübner of Kolbe's work. Hübner, who thought he had reserved the topic to himself by his earlier publication, was indignant and crestfallen over all the effort he had now spent in vain in trying to isolate pure malonic acid product. Beilstein, then a fellow Göttingen Privatdozent, was nearly as angry. He described the affair in a letter to Kekulé, with whom he had studied in Heidelberg before going to Wurtz, saying that Kolbe had "stolen" malonic acid from Hübner in the "unkindest possible manner." The situation was even worse than it may seem, Beilstein continued. In the fall of 1863 he had visited Marburg; Kolbe had asked him about Hübner's work and how far along it was, and Beilstein had reported to Kolbe fully and freely. A few short months later, Kolbe was now claiming the reaction as entirely new.[77]

Kekulé immediately reported the scandal to friend Müller in London, for he knew Müller was interested in the diacids. Müller responded by confessing to Kekulé that he had "committed the same sin as Kolbe, by preparing not only malonic acid, but also succinic acid." He had been working for some time on this reaction, he said. He remembered Hübner's work on similar approaches, but it had not given clean results and had not been continued; in any case, he knew of no publication by Hübner on cyanoacetic acid, which was the important point. He also happened to know that Perkin and Duppa were working on the same reaction, but so far without success. He happened to have been present at the meeting of the Chemical Society on 18 February when Kolbe's note was read on this very subject. He was disconcerted, but had the presence of mind to report extemporaneously on his own results; the Society then resolved to published Kolbe's and Müller's notes together.

> Ultimately this is a dirty trick, but I squandered the synthesis of succinic acid from ethylene [as Simpson's paper had already appeared]; the synthesis [of succinic] from propionic acid I intend to hang on to as tightly as possible. Kolbe may well be angry that I am contending with him about the further development of his own reaction, namely the introduction of CO_2 by the intermediate unit CN. In any case, I believe I can only do what my duty and obligation is. I am now working on preparing mono- and dibromosuccinic acid, possibly in order to produce the acids $C_5H_6O_6$ and $C_6H_6O_8$. I hope that ultimately I don't end up by encroaching on you in this direction.[78]

Frankland reported the events of 18 February to Kolbe. Kolbe responded by sending Frankland a letter to deliver to Müller proposing a collaborative approach. But in his cover letter to Frankland, Kolbe expressed concern that he might be too trusting. Was Müller a "Gentleman"? Perhaps Frankland should just keep the letter. "From what I hear, Müller is a good friend of Kekulé's, whom I do *not* consider a gentleman. I confess, this friendship between Müller and Kekulé makes me a bit doubtful. I place the matter entirely in your hands."[79] Frankland did deliver Kolbe's letter, and Müller agreed to Kolbe's proposal. Müller's article appeared together with Kolbe's in the April issue of the *Journal of the Chemical Society*, and they resolved to stay in touch about their research plans. They subsequently agreed that Kolbe would henceforth work on cyanoacetic and cyanomalonic acids, and Müller on dicyanoacetic and cyanosuccinic acids.[80]

Three months later Beilstein reported these events to Adolf Baeyer, whom he had befriended while both were working in Kekulé's Heidelberg lab in 1857. Not only had Kolbe acted unethically toward Hübner, Beilstein thought, but Simpson had also stolen a closely related reaction from Adolf Lieben. (Beilstein claimed to know this because he had been Simpson's neighbor in Wurtz' lab in 1858–1859, where Lieben also then worked.) Finally, Beilstein claimed that Müller had behaved even worse, for Beilstein knew (presumably through Hübner) that Hübner had told Müller directly about his work on this reaction as early as the fall of 1861.[81] Müller also failed to keep his word to Kolbe, publishing an article in the summer of 1864 on the reaction of trichloroacetic acid with potassium cyanide without letting Kolbe know. Kolbe was understandably wroth, and contact between the two ceased.[82]

It is diffcult to assess blame in such matters at this distance of time, nor for our purposes is it useful or necessary to do so. What seems clear, at least, is that the practice of organic chemistry had suddenly become far more ruthless than it had ever been, and traditional customs of courtesy and ethics were often slighted or even ignored. Müller provided an appropriate postscript to all of this in a letter to Kekulé: "Doing chemistry these days is an accursed business; you never know when you are going to be overtaken by someone else."[83] Kolbe and Erlenmeyer were not necessarily being paranoid when they speculated that some scientists used collegial visits to rival laboratories for espionage purposes.[84]

Beilstein's complaint was slightly different, but rooted in the same phenomenon:

> Everywhere we see indicated isomerisms and possibly existent compounds, which are becoming so numerous that even all the masses of material now known appears by comparison as but a drop. I was involuntarily reminded of the old saying, "Organic chemistry is the science of nonexistent bodies." Do these billions of compounds really exist? . . . I could never imagine that the dear Lord would want to make life so hard for chemists.[85]

Beilstein was fundamentally nontheoretical and was genuinely distressed at the sudden diversity. He criticized Butlerov's textbook privately for making too many predictions of novel structural isomers. "Our students are already frightened by the mass of material; imagine their faces when they have predicted for them the existence of God knows how many additional compounds. . . . I am firmly convinced that experiment will considerably diminish the numerous isomerisms."[86] Liebig's reaction was not very different; he wrote to Hofmann, "Through isomerisms, organic chemistry is gradually becoming enough to make one mad."[87] The explorers of Wöhler's primeval jungle were emerging with daunting numbers of new chemical species.

Whether the issue was subtle distinctions between predicted behavior of polyfunctional compounds (as in the matter of lactic acid derivatives), or a sudden proliferation of similar isomers, all of which had to be interpreted according to structure-theoretical precepts, and many of which were being worked on simultaneously (such as the new alcohols), or cases of multiple discoveries of new reactions with consequent hurt feelings and priority disputes (as in the diacids), the new pattern was clear. Chemistry, particularly organic chemistry, had seen a sudden acceleration, essentially an explosion, datable to around the year 1860. This acceleration produced noticeable changes in the way work was done and reported, and was felt by all who were active in the field.

Later Years in Marburg

All the research activity described in the foregoing sections had a significant and very positive effect on Kolbe's career. From an average of twelve students per semester in Kolbe's practica of 1859–1860, the number increased to an average of twenty-two during the years 1860–1865. Average attendance at his lectures increased in the same period from thirteen to nineteen, and chemistry majors increased from essentially none to about nine per semester. At the same time, the average enrollment at Marburg increased only about two percent.[88]

These changes came none too soon for Kolbe. On the last day of 1860, he wrote Vieweg

> I may boldly assert that over the last two years no German university laboratory, even including Göttingen, has produced so many truly good chemical papers as has mine; at the same time Göttingen, Heidelberg, and even Giessen (where for a decade almost nothing has been accomplished) are overflowing with students.

Given the political situation and the reputation of the despised Kurhessian prime minister Daniel Hassenpflug, no one wanted to come to Marburg, Kolbe thought. Foreign students, the real mark of success, were especially notable in their absence; only in rare semesters were there any at all.[89]

If Kolbe's great break as far as research is concerned occurred in 1858–1860, a similar break as far as student numbers are concerned came in summer semester 1862. Classroom enrollment and total numbers of Praktikanten were still about the same, but now he had five real chemistry majors as well as a number of advanced workers in the lab, including no fewer than six who already had Ph.Ds. Two of these six were his long-term prize students, his assistant Lautemann and his former assistant Schmitt. A third was a certain H. Scheuch, who published one article from the Marburg lab but is otherwise obscure. A fourth was Bunsen's former student Carl Graebe (1841–1927), who stayed just that one semester and never offcially matriculated; he returned to Heidelberg that fall as assistant to Bunsen. He subsequently spent four extraordinarily productive years as assistant to Baeyer; his teaching career included periods at Königsberg and Geneva, and he was also involved with the early history of the Hoechst dye firm.[90]

The fifth of Kolbe's Ph.D. workers that summer was Jacob Volhard (1834–1910), a former student of Will and Kopp at Giessen, who had served as assistant successively to Bunsen, Liebig, and Hofmann before his arrival in Marburg. Volhard's stints in Munich and London were relatively long ones (four and two years, respectively). He has been called Hofmann's oldest German student, but it appears that it was more Liebig's recommendation than Hofmann's that led him to Marburg. After working with Kolbe for three semesters, he returned to Munich, where he occupied the lower academic ranks for the next sixteen years. He accepted calls to Erlangen in 1879 and Halle in 1882. In his single publication from Kolbe's lab, Volhard insisted on using the reformed weights for carbon and oxygen—the first such publication from one of Kolbe's students. No doubt this was due to Hofmann's influence, from whom he had just come.[91]

The last of the six was the Scots chemist Alexander Crum Brown (1838–1922), who had studied with the English Liebigians William Gregory and Lyon Playfair at Edinburgh and London, but who had also been strongly (if indirectly) influenced by Frankland, Kekulé, and Wurtz. After leaving Kolbe, he spent a year with Bunsen before returning to Edinburgh, where he succeeded in making a fine academic career.[92] In addition to these advanced workers, we also know that Edmund Drechsel (1843–1897) was in Kolbe's lab in 1862 as well. Drechsel later worked with Volhard in Munich and with Kolbe and Carl Ludwig in Leipzig. He became a well-regarded physiological chemist and physiologist, teaching for many years at Leipzig and at Berne.[93]

It was just at this time that Kolbe began to attract a steady stream of Russian chemists. The first to come, in that busy summer of 1862, was Konstantin Zaitsev (or in German transliteration, Saytzeff); the following semester he was joined by his brother Aleksandr Mikhailovich Zaitsev (1841–1910), who subsequently became far more famous. After two semesters, Konstantin left the lab, but Aleksandr remained there until Kolbe left for Leipzig. In the spring of 1863 three new Russians arrived, and in the following year came three more; in Kolbe's last semester in Marburg, Nikolai Aleksandrovich Menshutkin (1842–1907) studied with him. Butlerov, then at Kazan where many of these men studied, must have advised several of them to travel to Marburg for foreign education.[94]

We can follow the rise in Kolbe's fortunes by following his increasingly proud reports to Vieweg and to Frankland. In winter semester 1862/63, there were eighteen Praktikanten, and in the following semester twenty. In summer semester 1864 no fewer than thirteen of the eighteen Praktikanten came from outside Kurhessen. In the fall of 1864 the overall number jumped to twenty-nine, which, counting a few unmatriculated advanced workers, exceeded the capacity of the lab. Moreover, ever larger numbers of foreign Germans—Prussians, Saxons, Bavarians, and so on—were enrolling. Even non-German foreigners such as the nine Russians named, three Englishmen, two Scotsman, two Swiss, and two Americans enrolled during Kolbe's last two Marburg years. One of the Americans was Charles W. Eliot (1834–1926), who attended Kolbe's practicum in winter semester 1864/65; four years after that he became the president of Harvard University.[95] One of the Englishmen was E. T. Chapman, who later published some excellent research in organic chemistry before his untimely death in 1872. Most of the foreigners at the entire university, Kolbe bragged in one letter, were his students, and the University's matriculation registry bears out the boast.[96]

By the early 1860s Kolbe had earned a remarkable international reputation. A knowledgeable anonymous reviewer writing in the *Westminster Review* in 1866 thought that "the Marburg laboratory has played a very considerable part in the chemical history of the last seven years," and he referred to Kolbe as "one of the few chemists who have succeeded in forming a school." Kolbe's pupils—he named Lautemann, Griess, Guthrie, and Ulrich—he thought were "distinguished for a certain kind of originality, and for great practical skill." Moreover, the writer regarded Kolbe as next to Liebig "the most successful chemical teacher in Germany."[97]

No detailed descriptions or university records of Kolbe's laboratory practicum have survived, but the general course was given in chapter 5. All surviving testimony indicates that Kolbe's teaching was conscientious, effective, and deeply appreciated by his students. We have cited comments regarding the Marburg years from Guthrie, Graebe, and Volhard, also in chapter 5. Kolbe hated the lab, referring to it as a "junk box" (*Rumpelkasten*). He suffered regularly from breathing the fumes—hydrogen cyanide was a common reagent—and considered the lack of ventilation to be positively dangerous, with good reason.[98]

In 1862 Kolbe was still working in essentially the same laboratory that he had inherited from Bunsen, which the latter had had constructed over twenty years earlier.[99] The lecture room was renovated in the summer of 1856, but the modest extent of the alterations is indicated by the fact that the work cost only 145 thalers. Moreover, the laboratory budget was wholly inadequate, a fact that Kolbe brought to the attention of the administration by ignoring it as regularly as possible. This did little to win him friends among his superiors and among colleagues in oversight committees. In the fall of 1860 the budget was increased from 700 to 800 thalers, but Kolbe was made personally responsible for any overages—a ruling that enraged him for weeks, so intensely that he could not work. Moreover, the budget was only designed to pay for equipment, instruments, and lecture experiments, not the consumption of chemicals by Praktikanten, who had to foot that bill themselves. He complained that his salary was barely half as much as would suffice to keep a family in proper style, particularly since his income from student fees was exceptionally low until around 1862.

But Kolbe's sudden success in research dramatically increased his local power and leverage; Liebig's public approval after 1860 undoubtedly also helped. In July 1861 Kolbe received his first raise in Marburg, from 600 to 700 thalers; eighteen months later it was raised again to 800. In early 1865 his salary increased once more, to 1000 thalers (but to add a note of perspective, Bunsen was earning 1200 thalers when he left in 1851). Finally, in May 1863 his often-reiterated proposal to renovate and expand his lab was finally accepted, and the work began immediately. Considering that no new spaces had to be constructed, the total cost of this renovation was relatively large, about 2000 thalers. The extent to which Kolbe's stock had risen can be seen by the fact that during that same summer, his laboratory budget was raised to 1000 thalers and a salary line for a second assistant was authorized. By the end of his Marburg years, Kolbe was making more than four times what he had been earning earlier from student fees, and his prestige was such that he could simply ignore all budgetary restrictions.

The laboratory space was completed astonishingly quickly, by the beginning of the next term (11 November 1863). The new lab was expanded to fill the entire west wing of the Deutsches Haus, including the space previously devoted to the lecture room, and could now be divided into sections for beginners, advanced workers, and large-scale general operations. The servant's residence in a small adjoining space was converted to a private lab for the institute director and was separated from the main lab by a glass partition. Smaller spaces were created to serve as a stockroom, a eudiometry room, a wardrobe, and a roofed open-air area for working with hazardous fumes. The east wing of the old building was converted to a large, light and airy lecture room, displacing the collections of the zoological institute. On the second floor of the east wing, spaces were prepared for a balance and equipment room, a general storeroom, and a darkroom for photographic and spectrographic work.

A significant expense must have been incurred by Kolbe's insistence that every two work spaces be provided with water taps (connected to a reservoir filled from the nearby Lahn River) and every station equipped with gas for Bunsen burners. This was a comparative novelty at that time and was made possible by the simultaneous laying of gaslight installation for the city of Marburg. The lab was also equipped with a hundred gas illumination burners, so for the first time, work could continue after sunset. The net effect of this renovation was to more than double the total actual laboratory space—all three sections together measured about eighty by twenty-five feet—and to increase capacity by half, from about twenty to around thirty Praktikanten. The only thing that Kolbe still desired was an on-site residence for himself.[100]

Kolbe was immoderately proud of his new lab. It was not just a good, it was an "elegantly outfitted laboratory," somewhat smaller but even better equipped than Bunsen's in Heidelberg and very similar to that in Göttingen. He reported to Vieweg that the gas burners, both at the bench and overhead, allowed him and his students to work about twice as fast as was previously possible. To his administration, he predicted a "new era" in the history of the chemical institute.[101]

This new era, at least under Kolbe's aegis, was of short duration. Less than two years later he exchanged Marburg for Leipzig.

10

The Theory of Chemical Structure and the Structure of Chemical Theory

We have now moved through more than half of our narrative, and this is an appropriate point to pause and consider in more detail four important themes that we have hitherto barely adumbrated: the development of synthetic methods in organic chemistry; philosophical questions connected with chemical theory; craft skills and tacit knowledge; and the problem of formula notation and interpretation.

Organic Synthesis

Histories of organic synthesis traditionally used to begin with Wöhler's preparation of artificial urea in 1828, which is supposed to have sent the enigmatic "vital force" into well-deserved oblivion. The situation was in fact extraordinarily complex, and a brief summary of this episode and its latter day interpretations is appropriate here.[1]

We now know that there were as many varieties of "vitalism" as there were of "materialism" and of "mechanism" in the eighteenth and early nineteenth centuries; that there is no simple—nor even any sophisticated—correlation between vitalistic and metaphysical, teleological, or theological habits of thought; that there were examples of the artificial preparation of "organic" materials from inorganic ones before 1828; that "total" and "direct" syntheses were only performed long after that date; that many chemists—including many "vitalists"—expressed confidence in the future possibility of unlimited organic syntheses during the first three decades of the nineteenth century; and that Wöhler's accomplishment in no way refuted vitalism at a stroke, nor could it have done so even in principle. The semantic problem alone is daunting, for it would require an army of interpreters to match the manifold senses of the word "synthesis" against the protean character of "vitalism," not to mention the necessity of a careful winnowing and analysis of historical events to judge which could be considered the critical tests. In conclusion, a full answer to the question "in what year and by what event was vitalism overthrown" is not possible and not worth the effort if it were.

All that said, with respect to the insignificance claim for Wöhler's urea, there is a danger in proving too much. Wöhler obviously thought that he had done something rather dramatic in his excited letter of 22 February 1828 to his mentor Berzelius, stressing that he could make urea without a kidney, or even a living creature. Berzelius' reply two weeks later is just as enthusiastic and just as focused on the issue of organic synthesis; Wöhler had produced a "jewel" for his "laurel-wreath" that would "immortalize" his name. Wöhler clearly noted in his letter that his accomplishment would fail to convert a committed skeptic, and his published article did not assert that vitalism had been refuted. Moreover, both Wöhler and Berzelius recognized that the reaction was very relevant to a separate issue, the emergent study of isomerism, for Wöhler had simultaneously discovered that urea and ammonium cyanate had identical compositions. However, these qualifications do not negate the fact that both men regarded the new reaction to be important for an evaluation of vitalist beliefs. Wöhler's pride in his feat was still in evidence thirty-five years later, when we find him urging his combative friend Liebig to respond to Berthelot's attempted usurpation of the entire field of organic synthesis, but modestly adding that Liebig should do this without explicit mention of the 1828 paper.[2]

The "myth" of Wöhler's overthrow of vitalism in 1828 (and myth it surely was) was not created, as has been argued, in the spate of celebratory articles published during the centennial year, nor even in Hofmann's 1882 obituary of Wöhler. It was created in the immediate aftermath of the event itself. As early as 1843, Hermann Kopp, writing as a historian, urged that this was the deed that destroyed vitalist belief, and ignored the reaction's relevance for isomerism; Kopp's portrayal became an important source for later writers.[3] The fact that most commentators on this "epochal" discovery during the first few years after the event emphasized its impact on the theory of isomerism may simply indicate an implicit recognition of the slippery character of the vitalist debate and simultaneous enthusiasm over a current live topic—not an assertion of its irrelevance for the demarcation between organic and inorganic chemistry.[4] Relevant it was; a refutation of vitalism it was not. However, by the early 1850s, if not before, the myth of a definitive refutation had become ensconced in the German textbook literature.[5]

Even Wöhler and his most partisan mythologists recognized that urea had not represented a total or direct synthesis from the elements themselves, that this was a very simple organic substance whose physiological significance could easily be marginalized, and that the event represented at best a first hesitant step toward a distant goal. For the next quarter century, chemists were ambivalent, at times predicting a glorious future of artificial organic compounds creating better living through chemistry, at other times racked with doubt about the possibility of extending Wöhler's work to ever more complex substances.[6]

Additional steps were eventually made, and here Wöhler's student Kolbe was in the forefront. We noted in chapter 3 that in 1845 Kolbe became the first to publish a total synthesis of an organic compound, namely, acetic acid from inorganic carbon and water, also using sulfur and chlorine. In this paper, Kolbe used the term *synthesis* perhaps for the first time in a chemical context and boldly predicted the artificial preparation of sugar, starch, and other organic products. A few years later, he and Frankland published what I have termed the first two important general synthetic manipulations (carboxylation through nitrile formation and the Kolbe electrolysis reaction).

From that time on, the synthetic repertoire of organic chemists quickly grew: Frankland's organometallic routes, Strecker's preparation of alanine and lactic acid from aldehyde, Hofmann's various eponymous amine reactions, Gerhardt's acid anhydrides, Williamson's ether synthesis, Kekulé's sulfurations, the Wurtz reaction and his glycol work, Wanklyn's preparation of propionic from carbonic acid, Berthelot's total syntheses of acetylene and other simple organics, the first commercial synthetic dyestuffs mauve and fuchsine, and so on. All of these examples are taken from the years 1847–1859; in the following decade the growing stream of organic-chemical novelties became a torrent. As late as 1850, Strecker thought it necessary to italicize his claim to have prepared artificial lactic acid; by 1858 Kolbe was confidently predicting that artificial indigo, alizarin, and quinine would soon make their appearance.[7] By the latter date—about the time of the formulation of structure theory—expressions of caution and doubt disappeared from organic textbooks. In sum, what undid chemical vitalism was not a single discovery, nor any small handful of them, but rather a gradually increasing sense of the grand possibilities of organic-chemical manipulations, closely connected with fast-growing empirical success.

Kolbe remained in the vanguard of synthetic organic chemistry during his burst of research in the years 1858–1865—in fact, he was arguably the leading personality. One of the most important of his contributions was a method for the preparation of salicylic acid from phenol, carbonic acid, and sodium metal, carried out in collaboration with his student Lautemann (this subject is explored in chap. 12). His role in the prediction and extended study of branched-chain secondary and tertiary alcohols and acids, and his participation in the further study of polyfunctional alcohols and acids, were discussed in chapter 9. Another instance began with Kolbe's published speculation that taurine, a component of a bovine bile acid discovered a generation earlier by Leopold Gmelin, was nothing other than aminoethylsulfonic acid. Since he regarded Magnus' isethionic acid (prepared from sulfuric acid and alcohol) as the analogous hydroxy compound, a reaction route was evident, and in 1862 he reported a synthesis of taurine from isethionic acid.[8] He also succeeded in reversing Strecker's synthesis of lactic acid, by converting the latter to alanine.[9] In both cases, he made important theoretical arguments, interrelating the constitutions of alanine and glycine with lactic, propionic, and acetic acids.

Jacob Volhard's project in Kolbe's lab in the eventful summer of 1862 was a study of sarcosine, long known as a hydrolysate of creatine (found in meat extract). Volhard proposed that sarcosine was N-methylglycine, and he proved the idea by preparing the compound from methylamine and Kekulé's monochloroacetic acid.[10] With another junior colleague, Rudolf Schmitt, Kolbe carried out a modification of his salicylic acid synthesis by reacting carbonic acid with metallic potassium in the presence of water and isolating formic acid from the reaction mixture.[11] In the same issue of the *Annalen* in which this paper appeared, Kolbe and Schmitt published a second article that signaled the isolation of a red dyestuff from the oxidation of commercial phenol by oxalic and sulfuric acids. The product, "rosolic acid," was subsequently shown to be identical to a compound first prepared by F. F. Runge many years earlier.[12]

These events occurred just at that heady time when a new science-based industry was burgeoning: the coal tar dye trade. There are indications that Kolbe was quite interested in principle in taking economic advantage of this development however he could, but for whatever reason, he failed to follow up on his and Schmitt's discovery. A few years later, several scientists and industrialists (including Heinrich Caro, James Wanklyn, Carl Graebe, Carl Schorlemmer, R. S. Dale, and J. F. Persoz) investigated the compound further, and this work led to the commercial dyes known as corallin and aurine. Between 1876 and 1880, the cousins Emil and Otto Fischer elucidated the complex structures of these and related dyes. They are triphenylmethane derivatives, just like fuchsine (magenta)—the second commercially successful color and the foundational product for several of the largest chemical companies today. Kolbe had failed to patent his method of preparing rosolic acid and his priority complaints were ignored. As for the Fischers' work, despite its structuralist basis Kolbe was "astonished" to find that it was "truly excellent." "It must only be translated into chemistry," he added.[13]

Such generosity of sentiment did not characterize Kolbe's reaction to Adolf Baeyer's structural and synthetic work on indigo. Kolbe himself had long been interested in indigo, both scientifically and for its potential pecuniary rewards;[14] he thought Baeyer's structural suggestions were unscientific absurdities. It must have bitterly rankled that Baeyer became the first to achieve a successful laboratory synthesis; mercifully, he did not live to see its lucrative commercial application.

After his transfer to Leipzig, Kolbe continued the sort of synthetic work that we have exemplified for the Marburg period; on the whole, however, his personal research activity declined. He did achieve fabulous financial success with one project, however—an improved version of his salicylic acid synthesis of 1859. This subject is covered in chapter 12.

Structure Theory and the Philosophy of Chemistry

During the eighteenth and much of the nineteenth centuries, the science of chemistry was more thoroughly imbued with the methodology and general culture of natural history than with that of the physical or mathematical sciences. Chemists were usually depicted—and depicted themselves—as naturalists exploring the earth's great garden of chemical species, in just the same way that the true naturalists were widening our horizons on biological species. This was the reason why systems of chemical classification were regarded as so central to the science, why Linnaeus' search for a *natural* botanical system was emulated by chemists, why genetic and parental relationships among chemical compounds were emphasized, and why biological metaphors were so often employed in chemistry.[15] It also partially explains the antitheoretical attitudes of many nineteenth-century chemists.

But this naturalist's scientific self-image was also mixed with some of the naturalist's scientific self-doubt. Many chemists felt the need for a general *theory* to guide their investigations, on the model of the exact sciences, but this goal proved to be elusive. Lavoisier, as much a physicist as a chemist, sought to create a general theory centered on oxygen and embracing physics and physiology as well as chemistry. Dalton's efforts to extend chemical atomism to physics were largely unsuccessful. His atomic theory became the closest thing to a generalized theory for the science, but most chemists innocently maintained the fallacy that this theory was coterminous with the law of definite proportions. Berzelius attempted to develop an electrochemical theory that would span the science, but he ran into increasing difficulties. Advocates of the radical and type theories of the 1830s and 1840s likewise tried to create general theories that could serve as guides to chemical investigations. Both theories came to be seen as insufficiently flexible and general until the advent of structure theory, which combined some of the best features of each.

As they began to search for a general theory analogous to that of the physicists, chemists also increasingly adopted the hypothetico-deductive methodology that had already begun to penetrate the exact sciences.[16] For a time, the inductivist commitments of most chemists kept hypothetico-deductivism suppressed, at least on the rhetorical level, but by mid-century it was breaking through into the science. As early as 1836 Dumas opined

> Theories have always been regarded as things quite different from truth; for this reason theories have long been accorded an importance proportional to the service that they render. . . . In chemistry, our theories are crutches; to show that they are good, they must be used to walk. . . . A theory established by twenty facts must explain thirty, and lead to the discovery of ten more; but nearly always it is modified or succumbs to ten new facts added to all that went before.[17]

Dumas may have acquired his taste for deductivism from Ampère, who was a particularly adept user of the method,[18] or from the physical chemist C. L. Berthollet. It appears that he communicated the same predilection to his protégé Wurtz.

In England, John Herschel's popular *Preliminary Discourse on the Study of Natural Philosophy* offered fervent praise of deductivism (despite the Baconian overlay); the book is known to have strongly influenced Michael Faraday. Both Herschel and Faraday were, of course, chemists as well as physicists. Herschel's book may also have been influential for Alexander Williamson; not only Williamson's hypothetico-deductivism, but also some of his language (the search for *verae causae* and for *experimenta crucis*), may well have derived from Herschel.[19]

As far as Germany is concerned, throughout the 1820s and 1830s Liebig was inclined toward hypothetico-deductivism, as was Berzelius, who was so influential for the direction of German chemical theory. Given the prevailing inductivist atmosphere in chemistry, however, both men were rhetorically cautious about their methodology and were also no doubt to a degree self-deluded. In 1862, after the method of hypothesis had become more generally accepted, Liebig composed a devastating critique of Baconian induction.[20] German physics was beginning to become perceptibly hypothetico-deductivist as early as the 1830s, with the work of such men as Gauss, Weber, and Neumann.[21] Kolbe, a student of the inductivists Wöhler and Bunsen, was an instinctive hypothetico-deductivist from the time of his earliest scientific work. His first extended paper contains repeated and explicit references to the method of hypothesis, at a time (1845) when this was uncommon in chemistry.[22] Liebig or Berzelius are the most probable sources for this predilection.

It is clear from all of this that the beginnings of a trend toward development of a strongly generalized and hypothetico-deductivist pattern of theorization modeled on the exact sciences could be discerned in chemistry at the time of Kolbe's entry into the field around 1840. These trends, however, had not yet entered the lifeblood of chemistry, certainly much less so than for physics and astronomy. The model of natural history was still dominant for chemists. In this methodological sense, it was the theory of structure that made the difference and created a new era for the field. Structure theory was a general theory that could be applied to inorganic as well as organic chemistry. In a stronger sense, its success also demonstrated the success of the underlying and more general *atomic* theory; after the 1860s, few chemists ever again seriously questioned the theory or equated it with the law of definite proportions.[23]

Moreover, structure theory was well suited for application of hypothetico-deductive method. Once its principles were understood and accepted, the theory generated hypotheses almost effortlessly, each of which could suggest one or more experimental tests. One of the qualities that made the theory so powerful for the method of hypothesis was the gradually acquired ability of organic chemists to synthesize new artificial compounds. As their synthetic repertoire rapidly enlarged during the middle decades of the century, chemists were increasingly able to ask and answer their scientific questions experimentally.

Perhaps the earliest legitimate example of this theoretical technique is Williamson's ether synthesis of 1850. Ether had first been prepared from alcohol in the sixteenth century, which illustrates the point that synthesis per se was nothing new to the nineteenth century. However, Williamson's innovation created a way to produce novel "tailored" ethers at will. In doing this, his intent was not the naturalist's ambition to add as many new chemical species as possible to the world's store of knowledge, but rather to provide a specific test, a crucial experiment, for two theories that could thereby be made to dictate two different experimental outcomes. Whether or not retrospective logical analysis can sustain Williamson's claim to have provided an irrefutable test, empirical history supersedes logical necessity, for as we have seen, his argument was soon universally regarded as compelling.

The polemic between Kolbe and Wurtz during the years 1857–1861 on the structure of lactic acid provides innumerable additional examples of this sort of approach. Kolbe and Wurtz not only explored new oxidations, reductions, and halogenations, they also created larger molecules by esterifications and ether syntheses, and each of these reactions represented an attempt to provide evidence for a particular point of view. Kolbe's predictions of the synthesis of new diacids and new secondary and tertiary alcohols, which soon led to their preparation, are particularly striking examples. Finally, it needs to be stressed that synthesis now could be placed beside analysis—indeed, by the 1860s it had already usurped the pride of place—in structural investigations. We have seen several examples of this trend in the previous section.

Craft Skills and Tacit Knowledge in Organic Chemistry

All of this suggests that a certain convergence of chemical toward physical hypothetico-deductive methodology took place around midcentury, attributable in substantial part to the emergence of the theory of structure. There is more to the methodological question than that, however. Physics and chemistry were distinguished by a host of different qualities, values, exemplars, habits of thought, details of practice, and so on, and the result of all of these particulars was that the two sciences maintained quite distinct cultures. Especially on the European continent, physics gradually became less concrete, more abstract, and more firmly based on an axiomatized mathematical foundation. The influence of technology was evident in increasingly complex instrumentation and apparatus, and the culture of precision measurements combined with rigorous error analysis imbued the practice of physics ever more strongly.[24]

None of these qualities characterized chemistry by and large, particularly not the organic field that became so dominant in the middle and later decades of the century. Today as a hundred years ago, most organic chemists have little need for higher mathematics. As far as precision measurements are concerned, they need to know how to weigh out stoichiometric quantities, measure densities, record melting and boiling points, and determine the correct atomic ratios in combustion analyses,[25] but these modest calculational and metrical demands cannot compare with those of the physicists. There was technology in the chemical laboratory, as in the physical laboratory, but it consisted mostly of simple materials in simple combinations and could not compare with the complex optical and electrical apparatus being used by the physicists.

What the chemists had, and needed in the fullest measure, were craft and observational skills. Chemistry has sometimes been compared to cookery, and the simile is apt in many respects. Substantial manual dexterity and technical know-how—for glass-blowing, devising and assembling apparatus, sealing connections, cutting rubber and cork, heating, cooling, pouring, mixing, grinding, and on and on—were imperative qualities. There was a standard repertoire of procedures, such as distillation, filtration, solvent extraction, titration, and recrystallization, that quickly became well-practiced habits for the novice. Precise observation was vital—color, viscosity, clarity, smell, taste, texture, crystal size and shape, and so on. An excellent memory was a virtual necessity to deal with the thousands of compounds regularly encountered.

There was much truth, therefore, in the perception that organic chemists were like naturalists exploring exotic ecosystems, only in the chemists' case, the object of study had often just been created for the first time by the investigator himself. When the discipline of physical chemistry was organized in the 1880s, the leading promoters advertised their field as the "chemistry of the future" and battled for social and institutional support with the well-entrenched organikers over the next two or three decades. Organic chemists resented having their lovely science depicted as a mere compounding of novelties, and their more laboratory-oriented values were even shared by a few of the physical chemists themselves. "Of course a man must specialize," W. D. Bancroft said. "He must be a chemist rather than a physicist." About the same time, he confessed privately, "I abominate exact measurements myself." But as a physical chemist, Bancroft was in a distinct minority.[26]

It must be emphasized that the organic chemist's laboratory repertoire was not a set of simile mechanical operations that all learned and performed equally. Cognitive activity, both conscious and unconscious, must accompany all craft operations. The ultimate quality and success of a reaction or of a recipe, even of a single operation in the process, is closely tied to what is happening in the chemist's or the cook's mind. Consequently, both the manual and the intellectual sides of the craft had to be mastered. This was a good reason for Kolbe's (and many others') emphasis on practical laboratory work as the centerpiece of chemical education. Precepts, ideas, and data could be learned in the classroom, but the craft skills and tacit knowledge of the professional could only be assimilated side by side with the master and his assistants.

There was a real *art* to be learned, and this is not simply a trope. Everyone with experience in the laboratory, including the greatest masters, knows of struggles with reactions that just will not go, products that emerge in gummy resins and refuse to give distinct crystals, yields that evaporate to a pittance, and product mixtures that resist every attempt at separation. The virtuosi, however, know how to overcome and circumvent problems and are often able somehow to coax success out of failure. The other side of the coin is the rapture of success, when the dazzling crystals of product suddenly and mysteriously blossom in the recrystallization flask. The ineffable "master's touch"—that of a Wöhler, a Bunsen, or a Kolbe—is the most difficult quality to teach and may well be impossible to impart.

There was an even deeper cognitive side to the chemist's craft. The epistemological technique of *transdiction*—inferring invisible submicroscopic details from macroscopic observables—was habitual with chemists long before physicists developed a similar art. Ever since chemical atomism arose in the early years of the nineteenth century, chemists were comfortable in routinely inferring the atomistic compositions of molecules from macroscopic gravimetric measurements. There was, of course, an epistemological debate, but most chemists refused to worry excessively about fine philosophical distinctions, and the radical phenomenalism that Gerhardt defended for a time never really became popular.[27] As the century progressed, transdiction became ever more elaborate, and inferences to composition were supplemented by inferences to molecular structure. When the chemist adds methyl iodide to an etherial solution of potassium ethoxide in a Williamson synthesis, for example, her mind's eye is focused on the activity at the molecular level, watching for signs of smooth assimilation of the two reactants, and hoping that she has sufficiently dried the solvent. Examples of such mental habits can be extracted from the host of reactions discussed in this book.

In general, reactions are assessed by analysis of the product—the payoff of an experiment—which also involves transdictive procedures and assumptions. The object of the game here is separating the components of a reaction mixture and purifying the substance on which interest is focused. Procedures for accomplishing these goals were developed, mostly empirically and many of them during the eighteenth century, and assessment criteria were well standardized. Every chemist knows, and knew in the period with which we are concerned, that constant sharp boiling and melting points are a primary indicator of purity and that admixture of impurities tends to lower and spread out these transition points. Every chemist knows, and knew, that attractive, well-formed, uniform crystals are another positive indicator. If such signs are not observed, then the chemist must try again—another distillation, sublimation, recrystallization, or repetition or modification of the reaction itself.

Elemental analysis is the final step, but only after product purity is attained. If a product matches the crystal appearance and sharp melting point of a known substance, and its percentage of carbon, hydrogen, and oxygen comes within the experimental error of that calculated for the relevant formula, then a match is achieved and the collegial community will probably be satisfied. If the purity indicators are good and the chemist has reason to expect a novel product, then the percentage composition of the observed product is again compared to the predicted formula, using accepted values for atomic weights. Such an identification was usually accompanied by at least a brief examination of chemical and physical properties and characteristic reactions, as well as preparation and characterization of simple derivatives such as salts of acidic or basic compounds, and so on.

Often the point of the organic chemist's investigation was achieved by establishing the very existence of a new compound—such as Williamson's asymmetric ethers. For other research, it was necessary to create and enumerate isomers corresponding to a single formula in order to test an idea. In still other cases, synthetic or analytical reactions needed to be explored to establish or refute conjectures, such as those concerning structural details of a molecule or genetic relationships among compounds. Synthesis was often a goal per se, especially of natural products. Finally, in some cases it was the particular properties of a new substance that provided the point of interest.

Although there certainly were distinctions in the details of laboratory practice in different universities and different countries, the generic chemical procedures, standards, and goals just outlined were quite uniform throughout the world during the period with which we are concerned. Thus, in principle, consensus was attainable regarding the general course of individual reactions and the existence of novel organic compounds. However, there were many potential points of challenge throughout the series of operations just described. Disputes over the "facts" were certainly not uncommon, and we have seen a number of them here, but what is striking is how *infrequently* experiments by members of the profession were successfully challenged. The story of Kolbe's chimerical salylic acid, which is related in chapter 12, is notable in its relative singularity.

I have gone through this material, at risk of trying the patience of chemically sophisticated readers, because I believe it is helpful for other readers to understand something of the mental equipment, operational repertoire, and epistemological habits of thought to properly appreciate and evaluate the historical action in this narrative. One theme of this book is pedagogy; chemistry students start out at the layman's level and have to acquire all of that equipment during their apprenticeship. We began in the first chapter with the guild model for the German university, and regarding chemistry as both a cognitive and a manual craft fits well in that model. It also needs to be stressed once more that, although chemistry had much in common with physics as a natural or exact science, it had a culture that was quite distinct in many ways, and acquisition of this culture was also part of the educational process.

Exploring Atomic Ecologies: Erlenmeyer and Kolbe

Ultimately, what made the theory of structure so powerful was an aspect that brought it closer to the physicists' scientific culture, namely, its ability to be axiomatized. In its simplest form, it could be reduced to a series of simple algorithmic or combinatorial rules, whereby possible molecular structures could be schematically derived by following straightforward valence assignments. The theory in this simple, powerful form is roughly the same as what Kekulé and Wurtz referred to in the late 1850s as the theory of "atomicity of the elements." It had roots in and similarities to both radical and type theories, but it was cleaner and more mechanical and was manipulable by simple algorithms rather than complex chemical intuition.

For example, most radical theorists regarded radicals as irreducible molecular moieties that were to be considered as wholes. A simple valence bond between two *atoms* was foreign, even antithetical, to the theory. Similarly, the type theory suggested more than mere genetic relationships among series of compounds: it asserted familial chemical similarities among members of the same type. This was one of the central claims of the theory that Kolbe (and others) began to contest in the 1840s and early 1850s. It was also what Wurtz was referring to when he wrote in 1861, "In a word, the idea of types is an artifice, a pure convention. In my opinion it must be subordinated to a more fundamental notion, that of the *atomicity of the elements*."[28]

Although Kekulé and Couper laid the foundation for the theory of atomicity of elements in 1857–1858, even advocates of this theory did not immediately eliminate old type/radical thought patterns from their minds and begin thinking in the simple structural combinatorics that would soon become second nature to organic chemists. There was a certain period of transition for the field, during which the language and thoughts of leading theorists were somewhere in between the two stages. By the time Hofmann was composing his *Introduction to Modern Chemistry* in 1865, this transition period was about over, and Hofmann could correctly refer there to the "revolution" that had taken place in the science.[29]

By the late 1860s, then, many if not most leading theoreticians had become full structuralists and thereafter thought largely in terms of those simple algorithmic patterns. However, this does not mean that organic chemistry had become a simple arithmetic problem, as Kolbe often charged. The schematic simplicity of the theory had to be compared and constantly checked and modified against a sometimes bewildering array of chemical peculiarities on the actual atomic–molecular level. The empirical contingency of atoms and their idiosyncratic chemical interrelationships contrasted starkly with the schematic simplicity of the theory. In effect, chemists had to emulate the naturalists once more, in exploring and determining what amounted to seemingly arbitrary (or at least then unmotivated) "ecological" relationships among the atoms in a molecule. In such a situation, it was vital that chemists maintain open attitudes. The theory was an indispensable guide in a general sense, but empirical flexibility was the most pragmatic course for questions of detail.

Nothing could have predicted, for instance, that "double bonds" would become the best way to view the relationship between adjacent carbon atoms in olefins, as opposed to maintaining free carbon affinities or two affinities satisfying each other, or even reducing the actual valence. Nothing could have predicted the eventually established circumstance that the four valences of carbon are all chemically equivalent. It was not at all clear why certain simple structures perfectly consistent with the theory, such as OCH_2 and $C(OH)_4$, resisted all attempts at realization. Kolbe asserted as early as 1850 that the carbonyl carbon of acetic acid was bound both to oxygen and to alcohol functions. In 1859, Kekulé made the same statement in structural terms and argued that this atomic arrangement defined organic acid behavior in the general case. But, again, exactly why this was true remained mysterious.

Chapter 9 examined the work of some of the leading organic chemists of the 1860s in exploring some of these contingent "chemical ecologies" and pondered the effects such activity had in reshaping the character of the science. This section and the next continue this line of investigation, concentrating on the thoughts and writings of two of the most avid converts to structural ideas: Emil Erlenmeyer and Aleksandr Mikhailovich Butlerov. Both men had interesting relationships with Hermann Kolbe.

A former apothecary and a student of Liebig and Will, Erlenmeyer first entered our story in chapter 7.[30] Closely associated with Kekulé as a fellow Privatdozent in Heidelberg in 1856–1858, Erlenmeyer was one of the most inventive of the early structural chemists. In 1859, Kekulé trusted Erlenmeyer so implicitly that he asked him to perform all copyediting and proofreading for the first sheets of his textbook, then being printed, and he did not even want the final version to be sent back to him before printing. Predictably, this procedure created friction between the two men.[31]

In 1858, Kekulé in collaboration with three other Heidelberg colleagues decided to found a new critical journal devoted principally to book reviews of the scientific literature and entitled *Kritische Zeitschrift für Chemie, Physik und Mathematik*. Kekulé's intent, as he wrote to Liebig, was to create a "dam" against the flood of "Schmierliteratur" then afflicting science.[32] However, neither Liebig nor any other first-rank chemist lent his name publicly to the enterprise.

Kekulé had departed for Ghent long before even the first volume of this journal was complete, and Erlenmeyer took over the chief editor's role. From this time on, the journal restricted its scope to chemistry alone and began to publish reprinted (and occasionally original) articles. Beginning with volume three in 1860, Erlenmeyer took sole command of editorial production. The dramatic increase in the pace of chemical research and in the number of chemists, Erlenmeyer noted in a preface to that volume, meant that researchers were now in constant danger of unintentionally trespassing on each other's work. All the more vital was it that an organ be established that provided for rapid publication of preliminary results and small projects. Consequently, the *Zeitschrift* was henceforth to appear twice monthly in numbers of two sheets (thirty-two pages) each.[33] What Erlenmeyer did not say publicly, but in private letters to colleagues in an attempt to boost circulation, was that he and others were also growing increasingly impatient with the slow publication schedule of Liebig's *Annalen*. His new model for the *Zeitschrift* was the *Comptes rendus* in Paris.[34]

Kolbe began to send Erlenmeyer offprints of his *Annalen* articles for reprinting. In 1862 they began a friendly scientific correspondence, and by the end of the next year this had developed into a real friendship, including regular visits. Kolbe perceived enough commonality in their views that he began to think of Erlenmeyer as a compatriot.[35] "I do not doubt," he wrote, "that my and your views will in time come into their own and win general approval. Without this certainty the isolation in which I now find myself with my views would have become unendurable to me."[36]

According to Erlenmeyer's later opinion, both he and Kekulé remained advocates of type theory (that is, they were not full structuralists) when Kekulé was in Heidelberg, even after he had enthusiastically adopted the point of view advocated in Kekulé's 1857–1858 papers. His conversion from this "one-sided" viewpoint to true structuralism came only late in 1861. The problem with the earlier viewpoint, Erlenmeyer said, was that its advocates always strove to identify a single determinative type, or atom defining the type, that could be used unambiguously to classify and characterize the molecule. He thought the correct viewpoint was to regard all pieces of a molecule, all of its radicals and atoms, to be ontologically equal, so that a molecule might be considered to be schematically dissectable in any possible fashion.[37]

Examination of volumes four and five (1861 and 1862) of Erlenmeyer's *Zeitschrift* supports this autobiographical contention. In a paper published in April 1861, Erlenmeyer stated that he considered Kolbe as much a typist as Williamson, Gerhardt, and Wurtz, and he argued, in a Gerhardtian vein, that a significant accomplishment of the type theory was to have shown that one cannot prove absolute molecular constitutions. However, the goal of the paper was to argue a point that Kolbe would have considered constitutional, namely, to demonstrate that all alcohols and acids are monobasic.[38] Erlenmeyer's thesis was pure Kolbean theory, and it would seem that at this time he would have described himself as a follower simultaneously of Gerhardt, Kekulé, and Kolbe, creating a useful syncretic viewpoint from these various influences.

However, early the following year Erlenmeyer began to publish a series of remarkable theoretical ideas that shows significant evolution from the previous year, ideas that are quite in the spirit of full structuralism. In a fervent defense of chemical theory in general and the newer theories in particular, Erlenmeyer provided one of the best early statements of the basic assumptions of the theory of atomicity of elements. The article contains the first suggestion (if not positive assertion) of the existence of carbon–carbon double bonds in olefins. He also proposed that the various affinity units (valences) of an atom have differing strengths, for on this assumption such otherwise anomalous compounds as carbonic oxide and nitrous gas could be assimilated into the theory of atomicity, since in such compounds not all affinities of some atoms are engaged with those of other atoms.[39] Soon thereafter, Erlenmeyer developed the idea, apparently derived from Wurtz and Kekulé, that n-valent atoms are accretions of n-monovalent subatoms, the subatoms differing in their absolute and relative affinities toward subatoms of various other atoms. This was a potentially powerful theory, but it proved to be empirically inaccessible and so was abandoned before the end of the decade.[40]

In April 1862 Erlenmeyer published another critical article, this time on the lactic acid controversy. He rhetorically asked "die Herren Typiker" (among whom he obviously no longer counted himself) why they believed their type formulas so literally when doing so generated many misleading predictions. The proper approach, he said, was to consider a chemical formula as a "positional diagram" (*Situationsplan*) for its constituent atoms, which specifies the "topographic positions" (*topographische Lage*) of all of its components. So as not to be misunderstood, he added,

> No one could imagine that with these formulas I intend to indicate the actual positions of the components. But I will remark that I very much intend thereby to express a topographic analogy of the derivative with the mother substance and to show that the mere reaction formulas of the typists no longer suffice, but rather we need to use *relative* constitutional formulas.[41]

Erlenmeyer's "relative constitutional formulas" look to modern eyes, and no doubt to many of his contemporaries as well, like structural formulas. Indeed, he became the earliest German advocate of the graphic formulas pioneered by Couper and Crum Brown, which proliferated in his articles especially after 1865.

At the end of 1863, Erlenmeyer published a defense of Kolbe's views. Kolbe was being misunderstood by nearly everyone, Erlenmeyer argued; he was a typist in the full sense and had lost credit for some of his discoveries only because of the unusual *form* of his theories, while the content was quite conventional. Indeed, Kolbe's types were in some respects superior to other versions, Erlenmeyer asserted; they were more consistent, more genetically and synthetically derived, and more exclusively focused on carbon. Kolbe's biggest real differences with the other typists were his insistence on the goal of establishing absolute constitutional formulas and his use of equivalents for carbon and oxygen. There was a certain irony in each of these points of conflict, Erlenmeyer continued. For one, the typists' "reaction" formulas, despite the conventional rhetoric of their users, do ordinarily specify constitutions in Kolbe's sense, and more often than not they correspond essentially with Kolbe's formulations. Moreover, Kolbe may well be right in viewing carbon as C_2 ($C = 6$) rather than

C (C = 12). (Here Erlenmeyer was making reference to his own subatomic speculations.)[42]

Kolbe was very grateful. After thanking Erlenmeyer, he added,

> By undertaking to prepare the way for a better understanding of my views on the composition of organic compounds, against which there is widespread prejudice, you are taking on a somewhat dangerous task, since a large number of chemists want to know *absolutely* nothing about them, and will resent you for demanding that they think seriously about chemical questions. . . . Even though our views differ in points of detail, I am pleased that they are nearly the same regarding carbon.[43]

Erlenmeyer was fast earning a reputation, and not altogether a favorable one, as a prolific but increasingly acerbic and incautious critic; even his formerly close relationship with Kekulé went on the rocks. It was partially for this reason that his journal was not flourishing. By the fall of 1864, there were only 150 subscribers (and more than half of them were Russians). He decided to give up his editorship and gave the journal over to three Göttinger Dozenten (Beilstein, Fittig, and Hübner).[44] In its place, he did what many of his theoretically inclined colleagues were doing and began writing a textbook. In 1868 Erlenmeyer's future was finally secured (he was then a forty-three year old Privatdozent) by a call to the Munich Technische Hochschule.

Erlenmeyer maintained a close relationship with Kolbe until the latter's increasingly strong attacks on structure theory finally drew his fire. In June 1871, he wrote Kolbe saying that he was composing a public lecture to be given in honor of King Ludwig's twenty-fifth birthday. Without giving away details, he wanted, among other things, to persuade Kolbe "that your formulas are nothing more than my formulas, that at most a difference between them arises when we have a different view of the constitution itself."[45]

The lecture was printed in the course of the summer, equipped with a ponderous 3500-word footnote dealing with Kolbe,[46] and Erlenmeyer sent a copy of it to his friend. Kolbe was one of the few who were still contesting structure theory, Erlenmeyer noted, but what Kolbe did not realize is that operationally, his formulas were nearly identical to those of the structuralists. He was a type theorist in the strict sense, the sort that he (Erlenmeyer) and Kekulé had been before both began to operate as full structuralists about 1861. According to Erlenmeyer, the only substantial difference between Kolbe and the structural chemists was that Kolbe denied the existence of not only direct connections between carbon atoms (chain theory) but also the very concept of valence bonds.

Erlenmeyer argued that Kolbe's position was not logical or consistent. Kolbe averred that his theory was based upon atomicity, that carbon atoms were (almost always) tetratomic, and that methyl, for example, was monatomic because one of the carbon atomicity units was unsatisfied. Kolbe had angrily rejected the conclusion that it was precisely that fourth valence of carbon that formed the point of attack or attachment for the methyl radical. It was the methyl radical *as a whole*, he said, that substituted for the hydrogen of water to form methyl alcohol, of methane to form ethane, of formic acid to form acetic acid, and so on. However, Erlenmeyer protested, it was Kolbe himself who had begun as early as 1850 to speak of the "point of attack" of chemical action, in this case referring to the carbonyl carbon atom of acetic acid.

Erlenmeyer also chided Kolbe for the one-sidedness that characterized all genuine type theories. Kolbe's search for the dominating "Grundradikal" or "Hauptkohlenstoffatom" resulted in an overemphasis on one radical or atom to the detriment of every other one in the molecule. It was downright silly to search for a single carbon atom in an alcohol that confers "family character" (Gattungscharakter) on the compound; it is rather functional group analysis (emphasis on hydroxyl groups, in this case) that has chemical significance. Kolbe had used various typographical techniques, including bolding, boxes, and size differentials, to distinguish among the carbon atoms in organic compounds. Were the alcohol of the eighteen-carbon stearic acid to be prepared, Erlenmeyer concluded, and were Kolbe to use sizing to distinguish all of the carbons, one would need a microscope to see the eighteenth.[47] Erlenmeyer summarized his position in the following way:

> Kolbe has remained a typist while most other chemists have abandoned the type viewpoint. He considers chemical compounds not according to their *entire* composition and constitution and *all* properties determined from these factors, but only as alcohols, as aldehydes, as carbonic acids, as sulfonic acids, as amines, etc. etc.[48]

That this conservatism had operational hazards Erlenmeyer was also able to show. Kolbe had predicted as a consequence of his point of view that the two chlorine atoms of the unknown compound 1,3-dichloropropane would be found to be chemically distinct, whereas chain theory (structure theory) would predict them to be chemically identical. Erlenmeyer succeeded in testing this prediction, at least for the bromine analog: he had the compound synthesized and reported that both bromine atoms hydrolyzed simultaneously to form the dialcohol.[49]

Erlenmeyer could hardly have devised a more stinging rebuke than to say that Kolbe was a one-sided type theorist and needed to think more seriously and more *chemically* about compounds and reactions, for this was precisely Kolbe's longstanding complaint against the type theorists and the structuralists. Kolbe's fury can only be imagined. In September he composed yet another general attack on structure theory for his new *Journal für praktische Chemie*. To Vieweg he commented, "In this piece I become personally unpleasant only once, namely against Erlenmeyer, a man who has not a trace of logic in his head. He challenged me and forced me to this. Liebig may not like it, but I cannot help that."[50] In that event, however, Kolbe decided to suppress his comment.[51] But he did not hesitate to disparage Erlenmeyer ever after in his private correspondence. "Once a pharmacist, always a pharmacist," he grumbled to Volhard.[52]

Butlerov, Kekulé, and Kolbe

The accession of Czar Alexander II to the Russian throne in 1855 soon led to a quickening interest in modern science, and his government embarked on an urgent program to encourage foreign education of eligible candidates, especially at German and Swiss universities.[53] One of the earliest of these scholars was Aleksandr Mikhailovich Butlerov,[54] who was already full professor at Kazan (although he had not yet done any significant chemical research) when he arrived in western Europe in 1857 on a government-sponsored *komandirovka* (educational tour). Butlerov was influenced by two visits with Kekulé in Heidelberg, but he spent his longest period (five months) in Wurtz' laboratory in Paris. This was just when Couper was there as well, formulating his ideas on what members of Wurtz' research group were already beginning to call the "structure" of chemical compounds. Butlerov participated in this program as well, presenting a lecture to the new Société Chimique (on 17 February 1858) that outlined a carbon type theory similar to that of Odling, Kekulé, Frankland, and Kolbe.[55] Unfortunately, this lecture was never published. By the end of May, Butlerov was back in Kazan, but the year that he had spent in Germany and France determined the course of his research for the rest of his life.

Three years later Butlerov was back in the West. In September 1861 he presented to the Speyer meeting of the Gesellschaft Deutscher Naturforscher und Ärzte a paper "On the Chemical Structure of Compounds" that Soviet historians have always considered the locus classicus of structure theory.[56] The real purpose of the paper was to try to move the chemical community from a halfway position, mixing older type and radical theories with the new theory of atomicity of elements, to the position of full structuralism. This was just the sort of transition to which Erlenmeyer later referred, and the timing matches, as well. It is therefore likely that Erlenmeyer's and Butlerov's conversions were connected, but it is not easy to determine who played the leading role for the other. Erlenmeyer wrote Butlerov eight months after the Speyer meeting to lodge a mild priority protest. "I think I can assume with certainty," he stated, "that the ideas [in Butlerov's Speyer paper] were, in large part at least, influenced by our conversations." These ideas included the constancy of tetratomicity of carbon, double bonds in unsaturated hydrocarbons, and the variable strength of the four valences of carbon, all of which, he asserted, he had been earlier to emphasize than Butlerov.[57]

It is in any case clear that Erlenmeyer and Butlerov agreed closely with each other from 1861 on. Erlenmeyer printed Butlerov's Speyer speech in full; it was otherwise unpublished, perhaps otherwise unpublishable due to its purely theoretical content. In part because Heidelberg was proving to be increasingly popular with Russian chemistry students, Erlenmeyer began to teach ever larger numbers of Russians; at times this national contingent comprised more than a third of his students. Similarly, his *Zeitschrift* was the most common publication outlet for Russian chemists in the early 1860s.[58]

In articles published in the *Zeitschrift* over the next few years, Butlerov instantiated his theoretical manifesto by means of innumerable examples from the science of organic chemistry, most of them having to do with the structural elucidation of isomers and many incorporating novel syntheses emanating from his laboratory.[59] In so doing, he established himself in the forefront of the fast-moving field of structural organic chemistry.

Butlerov was also concerned to integrate what seemed to be a fractured and fractious field, by attempting a substantive and semantic analysis of the language and theories of some of the leading chemists of the day. The leading personalities among the theoretically active chemists were Kekulé and Kolbe. Despite their strong apparent differences, Butlerov wrote, their "basic principles are nearly, or perhaps *entirely* identical," the conflicts arising only because of their incomplete or inconsistent application of those principles. Kolbe was inconsistent, Butlerov felt, because he sometimes assumed divalent or even trivalent carbon and because he used equivalents rather than atomic weights. Kekulé was inconsistent because of his assumption of "molecular compounds" for certain structures, which violate atomicity rules, and because he sometimes failed to propose a single unique structure for every compound.[60]

Semantically, Butlerov also tried to shed light rather than heat. If one really examines what each author *means* by his denotations, whether he uses the term "reaction formulas," "constitutions," "relative constitutions," or "topographic position" is of little consequence. Kekulé, Kolbe, Erlenmeyer, Wurtz, Heintz, and other advocates of the new theory, he claimed, were talking about essentially the same things and interpreting their formulas largely the same way. This was why Butlerov thought a relatively new and unused expression should be introduced, namely, *chemical structure*. Butlerov wrote

> In Kolbe, as in Kekulé, one sees the same tendency to judge the chemical connections of the individual atoms forming a complex molecule, or the points of attack of chemical affinity, which is saying the same thing. Kolbe defines constitution as the manner of this connection; I call this *chemical structure*. . . Be that as it may, most of Kekulé's as well as Kolbe's formulas are clearly based on the principle of chemical structure. I might add that ever since the importance and role of atomicity was recognized this principle has formed the real point of departure for almost all theoretical considerations.[61]

During the 1860s, Kolbe agreed with Butlerov's identification of "constitution" with "structure," for he commented in a letter to Frankland that the two terms had identical denotations.[62] After 1870, however, Kolbe began to draw increasingly sharp distinctions between them.

Butlerov had high respect for both Kolbe and Kekulé as leaders of the structuralist school. The influx of Russian chemists, mostly from Kazan, into Kolbe's lab during the years 1862–1867 can only have been due to Butlerov's recommendations to his students. Butlerov also knew of the importance of Kekulé's work for the rise of structure theory; he only resented Kekulé's attitude, especially his imperialistic manner of promoting his own work.

Kekulé had been reading all of this literature as it came out and was induced to add some clarifying comments. It makes little difference, he said, if one chooses to speak of "chemical structure," "constitutions," or "topographic positions"; the *ideas* were those on atomicity of elements that he had been the first to publish in 1857–1858.[63] He took strong exception to Erlenmeyer's and Butlerov's opinion that he had remained tainted by residues of type-theoretical thinking until 1861. Recent historical assessments suggest that Kekulé's claim to exclusive right to valence and structure theory is at the very least simplistic (as he later implicitly conceded), but that his avowal that he had been a full structuralist since 1857–1858 is consistent with what is known of his views in the late 1850s.[64]

It was during these years that Kekulé's relationship with Kolbe went from uneven to very bad. Despite his early close association with Kolbe's enemies such as Gerhardt and Williamson, Kekulé's view of Kolbe's work was not unremittingly negative. During the Kolbe–Wurtz polemic over lactic acid, Kekulé expressed to Erlenmeyer the wish to "negotiate" with Kolbe and to accustom his friend Wurtz to "better manners." In a letter to Lothar Meyer, he referred to Wöhler's and Kolbe's "*lovely*" syntheses (his emphasis), and a few years later publicly praised Kolbe's "sagacity" in successfully predicting the existence of secondary and tertiary alcohols.[65]

But Kolbe never had any use for Kekulé. He considered Kekulé a "foolish chatterer," a self-promoting and ostentatious pretender whose thin reputation would very soon collapse.[66] In 1861 and 1862 their work came uncomfortably close together, when both made tentative attempts to explain certain of the isomeric relationships of aromatic monoacids and unsaturated diacids. In one of his papers, Kekulé refused to comment on Kolbe's views on the first of these subjects, since the whole matter was still uncertain; moreover, he said, Kolbe's views changed so often that he had become a moving target, difficult to hit.[67] Kekulé had a second opportunity to upbraid Kolbe when he publicly identified two errors in one of Kolbe's papers. Kolbe conceded "a slip of the pen" for one of them, but tried to excuse the other by an unsatisfactory semantic ploy that did not satisfy the editor of the *Annalen*, Kolbe's good friend Hermann Kopp, who added an explanatory footnote.[68]

A third occasion caused Kolbe the most embarrassment. Shortly after arriving in Marburg, Kolbe had one of his students electrolyze succinic acid, expecting from theory to obtain ethylene. Kolbe was surprised to find his student reporting both eudiometric and gravimetric combustion analyses of the product as indicating methyl ether. He was never able to explain this result; moreover, he lost the data given to him by the student and even forgot the student's name. But he never forgot this curious result. In the fall of 1859, in the midst of that wondrous burst of activity that did so much to burnish his reputation, he appended a description of the experiment, including a statement only on the eudiometry and no analytical data, to a series of short notices published in the *Annalen*.[69]

In 1864 Kekulé repeated Kolbe's experiment and stated that Kolbe had misidentified the product: it was ethylene, after all. He then added insult to injury by expressing surprise that such a "skilled chemist" as Kolbe could confuse two such different compounds and pointing out that eudiometric combustion analysis is never sufficient to establish organic formulas.[70] Kekulé's reproof was printed in Erlenmeyer's journal, and Kolbe wrote Erlenmeyer soon thereafter:

> Kekulé is a cheeky insolent wretch to purport to teach me that eudiometric analysis alone does not suffice to determine the composition of a gas. But I will have a hard time answering him. . . . I can well imagine that Kekulé is very unhappy that his type theory no longer satisfies even him or gives him nourishment, and I can understand that such a petty man would loose his anger on someone who is traveling a better road with his chemical views. But regarding the succinic acid electrolysis he *may* be right.[71]

He *was* right—again—and Kolbe was forced to admit it publicly. Herr Kekulé "ought to have proceeded just a bit more modestly," but in the matter at hand he was "completely correct." Indeed, methyl ether as a product of succinic acid electrolysis had always seemed inconsistent with Kolbe's own theories. "By his demonstration of the error and by clearing the matter up Herr Kekulé has freed me from this embarrassment, and I am therefore doubly grateful to him."[72] Kolbe had done the right thing by graciously conceding to a man he despised. What is curious is how he had gotten himself in this position in the first place. He had published a highly anomalous observation that was eight years old, without data, without even remembering the name of the student who had performed the work, and without even attempting to repeat the relatively straightforward experiment. It was a careless and thoughtless act, and Kolbe was justly embarrassed by it. But years later, he would find repeated occasion to repay Kekulé for his insolence in pointing out Kolbe's mistake to all the world.

The Problem of Formulas and Their Interpretation

In a little-known passage, Albert Ladenburg, a student of Bunsen, Carius, Kekulé, and Wurtz, painted an intriguing picture of the Frankfurt Naturforscherversammlung of 1868. Butlerov had come to Germany for a third time, but Kekulé and several other Germans had felt aggrieved by Butlerov's critiques and were no longer very friendly. On a walk to a restaurant for lunch, Kekulé and Butlerov got into an angry argument, so when they arrived, some of the younger members of the group made sure to sit between the disputants. Later at one of the sessions, Kekulé and Erlenmeyer contested long and earnestly over questions of formulation. Ladenburg sat there silently, thinking that they were disputing unessential matters and had little disagreement in substance. He said that he was able to persuade both men of this in private conversation afterward.[73]

Such disputes were common in the 1850s and 1860s. Kekulé had personally contributed to some of the confusion, for in 1857–1858, he had formulated a substantially new theory (atomicity of the elements, or structure theory) from the socially safe haven of type theory, using type-theoretical language and formulas. This conventional (Gerhardtian) garb later led some colleagues, notably Erlenmeyer and Butlerov, to view Kekulé incorrectly as having come only halfway at first.[74]

In 1864 Kekulé attempted to straighten out what he regarded as misinterpretations of his formulas. Rational formulas are ordinarily written as reaction formulas, that is, incompletely resolved formulas designed to illustrate specific reactions, functions, or familial relationships. This is normally the most useful and appropriate way to write formulas, Kekulé argued, since excessive or unnecessary detail can degrade clarity of presentation. However, there are certain situations when one wishes to specify the bonding relationships of every atom in the molecule, and this is possible if the compound has been sufficiently investigated. One can write such a completely resolved formula using either type-theoretical or graphical notation (the latter being, for Kekulé, his curious sausage-shaped structures); graphical formulas are clearer but also more cumbersome. According to Kekulé, when one writes a completely resolved formula, whether in terms of types or sausages,

> compound radicals completely vanish. . . . Thus, one goes back to the elements themselves which compose the compound. But these are the very considerations that led to the view that carbon is a tetratomic element, and that carbon atoms have the property of bonding to themselves.[75]

It would appear that Kekulé was trying here not only to defend himself from Butlerov's critiques but also to explain Wurtz' position with regard to Kolbe, for Kolbe had scorned Wurtz' view that multiple rational formulas are possible for one and the same compound. Wurtz found himself making a similar defense against his Russian friend Butlerov. After Butlerov's "Erklärungsweisen" appeared, Wurtz published it in his *Répertoire de chimie pure* and sent him a letter complimenting him on the work:

> I find expressed there ideas which I share myself; among others, that the idea of atomicity of elements is based on Gerhardt's types and gives them their true sense. . . . But I would regret it if you were to renounce type notation before being able to replace it with a more advantageous

> one. Are you not struck by the simplicity of type formulas? . . . When bodies of complex composition are involved, these [structural] formulas are necessarily complicated, and in such cases indiscriminate use could hurt the clarity of the exposition. . . . Aside from these small reservations I am with you . . .[76]

Butlerov responded,

> I am very happy to know that you share my thoughts, and I must add at the same time that what I said about the signification of Gerhardt's types is of course your idea, an idea that you have long been expressing in your publications. I believe that those typical formulas that suffice for the majority of relations of bodies do nothing more than express their chemical structure, or at least the most salient part of that structure.[77]

Butlerov was writing a new textbook of organic chemistry just at this time, the first fascicle appearing in the course of the summer of 1864. Since (as Beilstein put it in a letter to Baeyer) the book was based on German theories, namely on a "Kolbe–Kekulé foundation," Butlerov sought and eventually found a publisher for a German edition.[78] Here he repeated what he had said in his 1863 article, acknowledging that Kekulé was the author of structure theory but averring that his partial allegiance to type theory had led him to a number of inconsistencies. It was these inconsistencies and gaps that he had tried to call attention to and remedy in 1861. Still, Kekulé "recognized the principle of chemical structure more completely and applied it more generally" than Kolbe, who nonetheless deserved major credit (in Butlerov's opinion) for developing the theory. However, it had been Erlenmeyer, Butlerov said, who had been the most consistent and definite in recent years.[79]

This was not Beilstein's view. He wrote Butlerov about this time,

> I also spoke during this vacation with Erlenmeyer about this point [atomicity of elements]. I told him that his papers often caused me a great deal of trouble, because each time I had to think myself into his theories, which cost a lot of time. The explanation for this is simple. These days there is no single orthodoxy or type theory, each chemist has his own beliefs & acts accordingly. Everyone is therefore used to thinking only in his own fashion & finds it difficult, i.e., is unaccustomed, to think in another fashion. For instance, Kolbe believes his formulas to be the simplest possible ones & once told Erlenmeyer that he has never been able to properly understand Gerhardt's theory.[80]

Erlenmeyer, for one, professed to have no trouble understanding other chemists' formulas. He wrote Butlerov,

> I think we are the only two chemists in the world who understand each other and everyone else. Besides us, I know of no one else who understands us, and none who understands everyone else, as we do. Most typists understand Kolbe not at all, Kolbe understands the typists not at all, Kekulé doesn't understand himself, Kolbe, or us. Wurtz is in the same position. Heintz understands Wislicenus and Wislicenus understands Heintz, but no one else understands anyone else. Heintz thinks he understands you, but he too misunderstands you. Who's left?[81]

Erlenmeyer's viewpoint seems to have been fairly common among his peers. Each thought he understood the various formula styles quite well, thank you; it was only the others who were guilty of misunderstanding. Beilstein's comment cited above complains not about the impossibility but merely the labor of formula translation; Ladenburg averred that he succeeded in serving as interpreter between Erlenmeyer and Kekulé in 1868 and that their disagreement had been insubstantial.

There always were, and still are, pitfalls in understanding caused solely by a lack of complete comprehension of the notation, formula, or language used. Despite Erlenmeyer's sarcasm, however, most of the misunderstandings in organic chemistry of the 1860s were relatively minor or short lived, such as those described between Kekulé, Erlenmeyer, Butlerov, and Wurtz. There is little basis for the thesis that the *primary* cause of chemical controversy at this time was due to competing formula styles. Priority for structure theory was a lively subject, to be sure, in part due to slippery semantics. But when the discourse concerned specific reactions and specific substances, competitors generally managed to understand each other pretty well. Moreover, by the end of the 1860s, graphical structural notation had largely supplanted the diverse type notations, and there was thereafter little cause for confusion among the majority of organic chemists.

The one significant exception to this generalization was Kolbe, who sometimes found it difficult to follow the details of structure-theoretical arguments and whose own idiosyncratic formulas were sometimes misinterpreted, especially during the 1850s when his notation kept shifting. This problem is explored in greater detail in the first section of chapter 13. But even here the difficulties can be exaggerated. There were some substantive differences between Kolbe and the structuralists. These distinctions were understood by both Kolbe and his rivals, and it was usually on these that they rightly focused. This was particularly the case after he finally adopted atomic weights in 1868.

Kolbe's birthplace and home from 1818 to 1826, the Elliehäuser parsonage. *Photograph taken in June 1992 by the author.*

Lutheran church in Stöckheim, Kolbe's home from 1826 to 1840. *Photograph taken in June 1992 by the author.*

Lutheran church in Lutterhausen, Kolbe's parents' home from 1840 to 1870. *Photograph taken in June 1992 by the author.*

Hermann Kolbe, ca. 1860, pen-and-ink drawing by F. Justi (1880), from a photograph.
Courtesy of the Erbengemeinschaft Justi, Bildarchiv Foto Marburg.

Deutsches Haus, Marburg—Kolbe's Institute, 1851–1865. *Photograph taken in June 1992 by the author.*

Chemical Institute of the University of Leipzig, ca. 1908. From the *Festschrift zur Feier des 500 jährigen Bestehens der Universität Leipzig* (Leipzig: Hirzel, 1909, page 72).

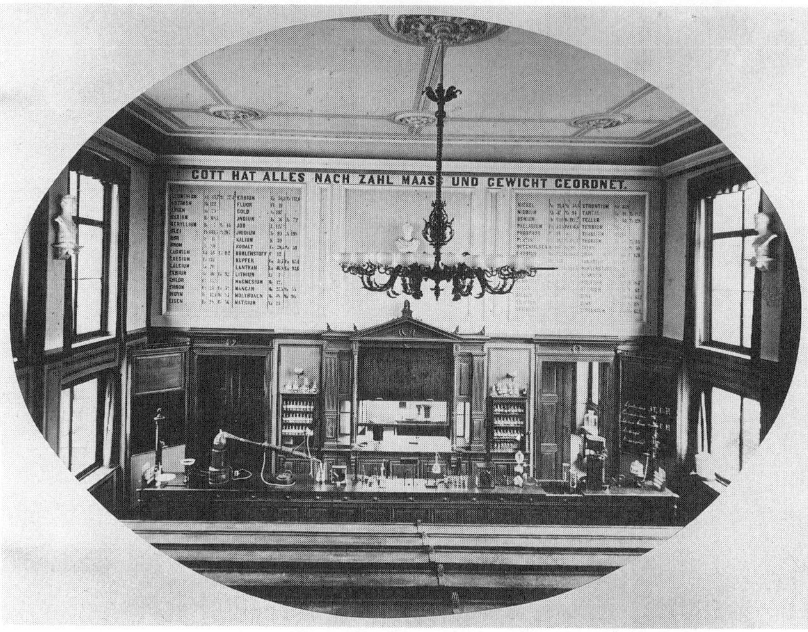

Kolbe's lecture theater, Leipzig, 1872. *Courtesy of the Deutsches Museum, Munich.*

Kolbe's teaching laboratory, Leipzig, 1872. *Courtesy of the Deutsches Museum, Munich.*

Villa Adolpha, Dresden, 1874: site of first production of salicylic acid. *Courtesy of the Deutsches Museum, Munich.*

Hermann Kolbe, ca. 1880. *Photograph from the author's collection.*

Justus Liebig. *Courtesy of The Edgar Fahs Smith Collection, Special Collections Department, Van Pelt-Dietrich Library Center, University of Pennsylvania.*

Friedrich Wöhler. *Courtesy of The Edgar Fahs Smith Collection, Special Collections Department, Van Pelt-Dietrich Library Center, University of Pennsylvania.*

Robert Bunsen. *Courtesy of The Edgar Fahs Smith Collection, Special Collections Department, Van Pelt-Dietrich Library Center, University of Pennsylvania.*

Edward Frankland. *Courtesy of The Edgar Fahs Smith Collection, Special Collections Department, Van Pelt-Dietrich Library Center, University of Pennsylvania.*

August Kekulé. *Courtesy of The Edgar Fahs Smith Collection, Special Collections Department, Van Pelt-Dietrich Library Center, University of Pennsylvania.*

August Wilhelm Hofmann. *Courtesy of The Edgar Fahs Smith Collection, Special Collections Department, Van Pelt-Dietrich Library Center, University of Pennsylvania.*

Adolphe Wurtz. *Courtesy of The Edgar Fahs Smith Collection, Special Collections Department, Van Pelt-Dietrich Library Center, University of Pennsylvania.*

Charles Gerhardt. *Courtesy of The Edgar Fahs Smith Collection, Special Collections Department, Van Pelt-Dietrich Library Center, University of Pennsylvania.*

11

Leipzig

The Kingdom of Saxony and Its University

At the time Kolbe was called to the University of Leipzig, Saxony was a prosperous kingdom of 2.4 million inhabitants. Of all the German states, it had by far the highest population density and had industrialized the earliest. Exactly half the population was engaged in industrial occupations of various kinds (compared to forty-one percent in the Rhineland, its nearest competitor, and seventeen percent in Prussia), especially in textiles, coal, iron, nonferrous mining, and heavy machinery.[1] Nowhere else in the German Confederation could be found such a concentration of industrialized cities as the likes of Leipzig, Dresden, Chemnitz, and Zwickau.

The country had come a long way during the previous two generations. At the beginning of the century, Saxony had been the last German state to turn against Napoleon, a policy that cost a good deal of its territory at the Congress of Vienna. As happened in many of the other German states during the Vormärz, the insurrections of 1830 ended a particularly reactionary period and resulted in a liberalized constitution the following year, along with emancipation of the serfs. The ensuing decades saw agricultural reforms and a gradual liberalization of the state, punctuated by the revolution of 1848–1849.

Economically, Saxony greatly benefited from entry in 1834 into the Prussian customs unions, tightly interconnected as she was with the surrounding much larger state of Prussia. Guided by the economic outlook of Friedrich List, the Saxon state built the first railroad line in Germany, between Dresden, Leipzig, and Magdeburg (1837–1840).[2] Thereafter, Saxony developed its railroad system more aggressively than any other German state, and this assisted the accelerating process of industrialization. King Johann (reigned 1854–1873), a Dante scholar and the most learned of all the nineteenth-century German princes, guided his country toward further liberalization and economic modernization, a process that continued to be encouraged by his successor, Albert. Johann's prime minister tilted increasingly toward Austria, and Saxony was its ally in the Austro-Prussian war. After Königgrätz, Prussia compelled Saxony to join the North German Confederation, but without engendering any significant anti-Prussian sentiment. The kingdom became a pliant member state of the German Empire in 1871.

A leading role in interior affairs was taken by Paul von Falkenstein (1801–1882), a member of an old aristocratic family.[3] As minister in charge of the Leipzig district during the late 1830s, it was Falkenstein who championed the first railroad. About the same time, he also succeeded in having the eminent legal scholar Wilhelm Albrecht called to the university, despite his sovereign's disapproval of this member of the "radical" Göttingen Seven group. For two decades following 1851, Falkenstein as Minister of Culture worked with remarkable effectiveness to raise the status of the University of Leipzig from a provincial to a nationally and even internationally respected institution.

Since its founding in the year 1409, the University of Leipzig had been one of the premier universities of Germany—the largest of them all during most of the seventeenth century, enrolling as many as sixteen percent of all German students.[4] However, during the Vormärz Leipzig was overtaken not only by the urban institutions at Berlin, Breslau, and Munich but also by Bonn as well. When Falkenstein became director of the Saxon ministry of culture ("Königliches Ministerium des Cultus und öffentlichen Unterrichts"), Leipzig University had reached its nadir, an enrollment of only 800 or about seven percent of German students (still far larger than the universities of most Kleinstaaten such as Kurhessen). Falkenstein's efforts produced an improvement in the university's fortunes that can only be described as spectacular. In 1865 there were a thousand students; six years later two thousand were enrolled, and by 1873 almost three thousand. During the mid- and late 1870s, Leipzig had half again as many students as its closest rival, Berlin, and one of every six German university students was enrolled there. Although the boom leveled off in the 1880s, the university maintained and even slowly increased its absolute numbers thereafter.

The two areas where Falkenstein's efforts bore the most dramatic fruit were in the percentage of foreign students (in both senses of the term "foreign," non-Saxon Germans and non-Germans), and in the number of science students. As far as the first category is concerned, for generations the university had maintained a ratio of about three-fourths Saxon and one-fourth non-Saxon German students. In contrast, during the 1870s well over half the Leipzig students came from outside the kingdom and one in seven came from outside the member states of the empire. Foreigners included especially Austrians, Russians, Swiss, Americans, and Britons; French students were notably absent.

As for science students, although in the early 1860s an average of about 50 per year matriculated, the number had jumped to over 200 by the early 1870s and to about 300 per year in the late 1870s. Regarding the intersection set of our two categories, the average number of non-Saxon science matriculations was only 18 per year in the early 1860s. This had mushroomed a decade later to 150 and by the late 1870s to over 200 per year. In this latter period, more than two-thirds of the science students came from outside Saxony.

Clearly the university had dramatically increased its attractiveness both inside and outside Germany. Part of this story has to do with the general expansion of German university enrollment. After decades of stagnation, the universities slowly began to expand in the 1860s, then exploded after the founding of the Reich. In the fifteen years after 1872, the empire's student population more than doubled.[5] However, Leipzig's boom began even before this national trend and exceeded it in magnitude.

Some of the attraction of Leipzig was undoubtedly due to local factors: the presence there of the Imperial Court of Justice; its reputation as a large, vital, and attractive city; and industrial and geographic factors as well. However, Leipzig could not compare with Berlin for political importance or with Munich for general ambience. A large part of this great success must therefore be ascribed to the work of the Saxon ministry of culture, with the cooperation of the sovereign and his legislature (the Ständesversammlung).

In the late 1850s, Falkenstein began self-consciously to dramatically expand the Leipzig science facilities. His first move was to locate a new observatory on a small hillock on the southeast edge of town, some distance from the main university district in the heart of the city. The hillock adjoined the so-called "Johannisthal," a city-owned garden district where townspeople could (and still do) rent small plots and summer houses. The university, one of the wealthiest in Germany, owned a good deal of rental property in and around Leipzig and was able to trade with the city for the real estate in this yet undeveloped district (there stood on the district's main Waisenhausstrasse only an institute for the deaf and the city orphanage).[6]

The observatory, outfitted by 1862, was only the first step in Falkenstein's elaborate and ambitious scheme; it formed the nucleus for what would be an extensive academic science and medical district. By locating the new campus in the suburbs, Falkenstein's architects could plan properly for optimal use of light and fresh air, and the buildings could be designed from the ground up with their dedicated end uses in mind. Moreover, the land was cheap and the streets were quiet, with none of the rumbling traffic that often interfered with precision measurements in the old laboratories on Universitätsstrasse.

Falkenstein's next opportunity came with the death of the professor of theoretical and pharmaceutical chemistry in the medical faculty, Otto B. Kühn (1800–1863). Kühn, a chemically inclined physician and the son of a professor of anatomy at Leipzig, had habilitated the same year (1825) as a fellow Saxon of about the same age named Otto Linné Erdmann (1804–1869). When in 1830 the professor of chemistry in the Medical Faculty, C. G. Eschenbach, retired, the faculty was not able to decide between the two young chemists. They compromised by awarding Eschenbach's lab and nominal successorship to Erdmann with the field of technical chemistry, now transferred to the Philosophical Faculty. Kühn got the position of general and theoretical chemistry in the Medical Faculty, initially without facilities. In 1842 Kühn inherited Erdmann's lab in the Pleissenburg on Universitätsstrasse, when Erdmann traded up for a new and superior facility in the Friedericianum, a couple hundred yards south of the main university buildings. Each of the two new ordentlicher professors was appointed at the paltry salary of 200 thalers.[7]

Neither of the Medical Faculty chemists, Eschenbach or Kühn, was ever able to achieve any real reputation beyond Leipzig during the eighty years of their combined careers. Falkenstein now had an opportunity to change this, all the more so since Kühn's death happened to coincide with a plan Falkenstein had developed to split physiology from anatomy by creating a new Ordinarius in the former field. To anticipate events related in the next section, he succeeded in calling Kolbe as Kühn's successor and Carl Ludwig as the new Ordinarius in physiology, both in the year 1865; each man was to get an elaborate new institute building in the Waisenhausstrasse. They were located right next door to each other, and both facilities were complete by 1869. Falkenstein did not even pause to catch his breath. The professor of mineralogy Ferdinand Zirkel had a new institute by 1872, likewise for the physicist Wilhelm Hankel by 1873, and for the anatomist Ernst L. Wagner by 1875.

Meanwhile Erdmann had died. At Kolbe's urging, the Philosophical Faculty proposed Rudolf Schmitt (Kolbe's former student, then at the Dresden Polytechnikum), Rudolf Fittig, or Hermann Wichelhaus for the position. Kolbe intentionally suggested only second-rank chemists, for he was convinced that a first-rate chemist would not accept a position that included Erdmann's old and poor laboratory. Someone (perhaps Carl Ludwig) convinced Falkenstein that it would be far better to hire a first-rank physical chemist, and Lothar Meyer and Carl Neumann came under consideration. Kolbe "protested energetically" against hiring such "superficial chatterers," especially since he understood that Kopp might be won for Leipzig. Falkenstein responded to Kolbe's complaint by calling Kopp, but Kopp declined; he merely wanted the call to improve his position in Heidelberg. Finally, Gustav Wiedemann was proposed (probably by Ludwig), and Falkenstein accepted the suggestion. The Faculty protested that Wiedemann was a physicist, not a chemist, but Falkenstein was adamant, and called him for the position—an Ordinarius for physical chemistry, the first such in Germany. Kolbe only asked that Wiedemann share the teaching of inorganic chemistry and that he conduct a chemical as well as a physical–chemical practicum, because by then Kolbe had far more students than he could handle. These conditions were granted.[8]

Wiedemann was in place by 1871 and soon took charge of his own new building—despite Kolbe's pessimistic assumption. Other institutes followed, especially in medical fields. Falkenstein actually stepped down in 1871 (or rather up, as he became Minister of the Royal House), but his plans were continued, albeit in a less grandiose manner, by his successor C. F. von Gerber. By the turn of the century, there were well over a dozen large scientific and medical institutes in the new university district near the Johannisthal. Shortly after Liebig's death in 1873, the Waisenhausstrasse was renamed Liebigstrasse, on Kolbe's urging.[9]

The vast expansion in the Leipzig science enrollments now can be seen at least partially as an understandable market response on the part of the German and non-German student clienteles. In the 1870s and 1880s, Leipzig could offer perhaps the largest and most modern scientific and medical facilities in the world. But how did Falkenstein come to command the large amounts of money required to build all of these institutes and pay the generous salaries necessary to acquire these academic stars?

First, Falkenstein was highly respected and extraordinarily persuasive in Saxon governing circles. He always believed that economic and bureaucratic modernization and continued industrialization offered political prestige, prosperity, and social stability for Saxony in the long term, and he regarded university reform as an element of this larger goal. Second, he happened to have an enlightened sovereign and a relatively cooperative Ständesversammlung, who normally provided whatever he could make a persuasive case for. Saxony was sufficiently prosperous to bear such financial costs. A third more contextual factor was the academic entrepreneurial fever that caught hold in many of the German states in the 1860s and 1870s. As Joseph Ben-David and others have emphasized, the competitiveness of the decentralized German academic marketplace goes some good distance in explaining the phenomenal rise of the German universities in the nineteenth century.[10]

Finally, it has been mentioned that Leipzig was already one of the wealthiest universities in Germany, with an income during the 1860s of over 100,000 thalers per year. The state had always had to supplement this income, of course, but when Falkenstein's expansion plan got going, the state subsidy increased dramatically, until by 1888 it was over a million marks per year (the currency conversion upon the founding of the Empire established 3 marks to the thaler). Nonetheless, the "academic-mercantilist" case could be (and was) made that this was an investment that yielded immediate dividends, for it was estimated that every non-Saxon student spent about 300 thalers per year attending university, and the non-Saxon contingent mushroomed as a direct result of the investment program. One observer thought that this meant that the home boys were educated essentially for free, subsidized (in effect) by foreigners. Moreover, the indirect and long-term benefits from heightened academic prestige were by no means negligible.[11]

Kolbe's Call to Leipzig and Its Context

In an influential monograph published in 1976, Peter Borscheid argued that the reason the academic study of chemistry in Baden took off after 1850 was that the government was concerned to stabilize the country after the economic and agricultural disasters of the 1840s that had led to revolution in 1848–1849; it was thought that Liebig's interrelated prescriptions for reform of academic and agricultural chemistry had potential to do just that, by raising agricultural fertility and productivity. A similar socioeconomic model for the promotion of scientific medicine in Baden has been convincingly articulated by Arleen Tuchman, although she appears to have weakened Borscheid's case inadvertently by commenting that the reform plans for the Heidelberg science institutes began as early as 1844, that is, before the crises began.[12] In any event, there is no question that after 1850, not just Baden but a number of German states began to move aggressively to upgrade academic chemistry.

The case for state economic interest as helping to justify this academic reform in Baden (especially at its university in Heidelberg) is strong, as long as this single element is not considered as a full explanation. Above and beyond any particular state agenda was the obvious fact that, for whatever reason, chemistry enrollments in most German universities had begun an upward climb around 1840 and showed no sign of leveling off. The retirement of Leopold Gmelin in 1851 gave the Baden authorities the chance to make a real change. Determined to win a "recognized celebrity," they went all out after Liebig, prepared to make any concession necessary, including the promise of a large new laboratory. Liebig, however, played Heidelberg against Munich for several months, until the latter won out. The choice for Heidelberg then came down to Hofmann or Bunsen; Bunsen got the call, arriving from Breslau in 1852.[13] As for Liebig's arrangement at Munich, he was asked to teach one lecture course per semester with no laboratory instruction and was given a 5000 thaler salary.

The Prussian Ministry of Culture had already caught the fever, for they had attracted Bunsen from Marburg to Breslau the previous year by building for him the largest chemistry institute up to that time, at the high cost of 34,000 thalers. (Bunsen's new laboratory in Heidelberg, however, was to cost 44,000 thalers by the time it was completed in 1854.) Liebig had publicly lambasted Prussian academic chemistry in 1840 in an article that raised a storm in the Berlin Ministry and in all the Prussian universities. The deal that temporarily brought Bunsen to Breslau reflected the altered political climate in Prussia regarding the study of chemistry.[14]

Indeed, as in Baden, the Prussian authorities carried out a heroic overhaul of their chemistry facilities during the 1850s and 1860s. In addition to the Breslau upgrade, a new lab at Königsberg cost 16,000 thalers in 1857, Greifswald had a new institute three years later at a cost of no less than 70,000 thalers, and a new laboratory was built at Halle in 1863 for 35,000 thalers.[15] The flagship of the Prussian educational system, the University of Berlin, had a number of local factors inhibiting reform,[16] but the nearly simultaneous deaths of the two Ordinarien for chemistry, Eilhard Mitscherlich and Heinrich Rose in 1863–1864, provided the necessary opening. Coincidentally, the retirement in 1863 of C. G. Bischof at Bonn gave the Prussian Kultusministerium yet another open post to deal with.

It was in fact the Bonn position that first came open.[17] The Prussian authorities had tired of waiting for Bischof to retire or die, and they began discussions with Hofmann about a possible call as early as the summer of 1861. Besides having become by this time the most famous chemist of his generation, Hofmann had taken a leave of absence from Bonn nearly twenty years earlier to accept his professorship in London and so was the obvious choice. The call to Bonn apparently came in January 1862, but negotiations dragged on well over a year until Hofmann finally assented in March 1863. The deal was princely: essential features included a salary of 2000 thalers (at the time of his retirement, Mitscherlich was earning only 800), a generous budget, and a new laboratory far larger and more expensive (projected to cost 100,000 thalers) than any hitherto contemplated. Accompanied by his Bonn architect, Hofmann toured Germany for a month in the summer of 1863, gathering design ideas from all the modern chemical laboratories.[18]

The situation became more complicated when, in January of that year, Mitscherlich retired and Gustav Magnus began promoting Hofmann's candidacy at Berlin. After Bunsen was offered the position and declined, Hofmann got this second call about November 1863; it included everything in the Bonn contract, plus an even better salary of 2500 thalers and a larger and more expensive lab. Hofmann was assured that the second choice candidate for Bonn, whoever it might be, would receive exactly the same deal as he had been offered even if he were to go to Berlin, so that choosing this alternative would not have the effect of damaging the future of chemistry at Bonn. Still, Hofmann felt bound by his promise, and by 19 December he had "definitively" decided to go to Bonn. Further negotiations with Berlin over the Christmas holidays, however, led to an arrangement whereby Hofmann would not be forced to choose between these positions until both new institutes were built, or at least definitively designed.[19]

Apparently the death of Heinrich Rose in January 1864 shifted the negotiating ground yet again, for Hofmann now was offered what was in effect the successorship of both Berlin chemists. This made a significant difference, since it was not the custom in German universities to have two Ordinarien in the same field, and it was thought that such split professorships create difficult or even impossible situations. This conviction was only reinforced by the circumstances that Rose and Mitscherlich never got along and that their feud had stymied reform of chemistry in Berlin for many years.

Now that Hofmann was being offered the sole chair of chemistry, it seemed inevitable to all as early as February 1864 that he would end up in Berlin, and by June it was regarded in the chemists' gossip network to have been officially settled.[20] They were right, of course. Remarkably, though, Hofmann's leverage was so great that he was able to exact the explicit condition that the Bonn position remain open and available to him until the end of 1866, should he find the Berliner life not to his liking.[21] The Berlin lab was begun in May 1865, simultaneous with Hofmann's first semester of teaching there; it took four years to complete, accommodated 70 Praktikanten, and cost the amazing sum of 200,000 thalers, not including the cost of the land, furnishings, and equipment.[22]

Meanwhile, Hofmann did not even definitively give up his London post, taking a three-year leave of absence instead. In April 1865, Kolbe expressed frustration to Vieweg that Hofmann now was holding on to no less than three professorships (four, if both the Rose and Mitscherlich chairs were counted).[23] On 30 August 1865, Kolbe and Hofmann both were visiting Bonn, where Hofmann told his friend that he was undecided whether or not he would stay in Berlin. In the following summer, there was still speculation in Germany that he would take the Bonn position after all. Kolbe described the Bonn lab, still in construction, as "a chemical palace, exceeding all my expectations," though he later thought that it was "somewhat too luxuriously and spaciously appointed." As late as March 1868, some English colleagues were still convinced that Hofmann would return to London.[24]

I have gone into the details of the Hofmann calls partly because Kolbe was a highly interested spectator to these events but also because this episode represents a critical transition in the history of German chemistry, indeed in the history of German academic science. The new Hofmann institutes at Bonn and Berlin were on a hitherto unseen scale, and they provided the models for future institutes, not only in chemistry but also in other fields of science and medicine, not least at Leipzig. A recent scholar has rightly referred to the Hofmann laboratories as well as Kolbe's at Leipzig, which was roughly contemporaneous, as inaugurating the "second generation" of German chemistry institutes.[25]

Kolbe, of course, had long been unhappy in Marburg and was desperate to get a call to either Berlin or Bonn. Once Hofmann was out of the picture, and with Liebig, Wöhler, and Bunsen now permanently situated, a number of names kept appearing as possible candidates for Bonn: Adolf Strecker, Georg Staedeler, Carl Löwig, Heinrich Limpricht, and Hermann von Fehling. However, Kolbe and Kekulé were always the leading candidates. From the start, Kolbe made it clear to everyone that his strong preference would be Bonn. "Es ist Alles verkehrt in Preussen," Kolbe complained, meaning of course Berlin. He felt that there were too many "crossing interests" there, a real "intellectual swamp"; it wore people out at a young age; in contrast, at Bonn one could found a true chemical research school.[26]

Kolbe thought that the Bonn succession entirely depended on whom Hofmann recommended, once Hofmann had definitively declined and the post was no longer encumbered. Frankland reported to Kolbe that Hofmann had enthusiastically endorsed Kolbe's candidacy, but Hofmann told Kekulé's friend Hugo Müller that he had impartially proposed both Kolbe and Kekulé, refusing to state a preference between them. He subsequently told Kolbe directly that he had endorsed Kolbe for Bonn.[27] Kolbe could scarcely believe it when he heard that Kekulé was under consideration. He wrote to Vieweg, "His choice would be in my opinion the grossest blunder, but for that reason is all the more probable under the present regime, which always does the most wrong-headed things." In any case, Kolbe thought that (Gerhardtian) type theorists in general had the advantage in professorial calls at that time; indeed, all the candidates under consideration were "typists" except Kolbe.[28]

In the spring of 1867 the call finally came, and it was to Kolbe. Ironically, Kekulé's candidacy appears to have been derailed by one of his rare responses to Kolbe's insults: the passage had come to the attention of an "influential personage" involved in the deliberations, who concluded that Kekulé was no gentleman.[29] Although by this time Kolbe was well ensconced in Leipzig, with a large new laboratory about to be constructed, he very nearly accepted the call. The sticking point was the insistence on the part of the Prussian ministry that Kolbe share a "portion" of the new lab with the physical chemist Hans Landolt, who was being promoted to Ordinarius. There would thus be another split professorship, which Kolbe regarded as unacceptable, and for this reason alone he declined the offer.[30] "Man kann zu Zwei in einem Bette schlafen, aber nicht zusammen ein Institut benutzen," Kolbe observed sourly to Hofmann, or else he would have gladly accepted.[31]

The call to Bonn then went to Kekulé in June 1867, who quickly and successfully negotiated the conditions. In contrast to Kolbe, Kekulé had no trouble coming to agreement with Landolt over how the lab and the fees were to be divided (Kekulé took two-thirds), for he and Landolt were old friends. In the event, this shared arrangement was not of long duration. The lab was completed in 1868—at a final cost of 144,000 thalers exclusive of furnishings—and two years later Landolt accepted a call to Aachen. Kekulé made sure that Landolt's successor was an ausserordentlicher Professor, subservient to him.[32]

Let us now return to Leipzig, where the Kühn succession had been hanging fire all during the Bonn and Berlin negotiations. Asked (according to tradition) for their recommendations, the Medical Faculty declined to advocate an Ordinarius successor to Kühn, largely because Erdmann was already professor of chemistry and they thought it inappropriate to repeat the mistake of 1830 and divide the professorship. Instead, they suggested a "provisional arrangement" (presumably looking toward Erdmann's retirement or death) whereby Kühn's Assistent, the Privatdozent Heinrich Hirzel, was to become ausserordentlicher Professor for pharmaceutical chemistry and take over direction of Kühn's laboratory, and Wilhelm Knop, then ausserordentlicher Professor for agricultural chemistry in the Philosophical Faculty, was to be further supported in salary, facilities, and lab budget, and be asked to teach organic chemistry as well.[33]

The professor of anatomy Ernst Wagner wrote a dissenting opinion, suggesting that Knop be called as Kühn's successor and that Hirzel be a second Ordinarius in pharmaceutical chemistry. He would have considered proposing a non-Saxon for the first position, were one to be had (he wrote), except that the Berlin episode (the Mitscherlich–Rose enmity and Hofmann's reluctance before Rose's death to accept a split chair) suggested that going for national candidates would never succeed under a split professorship.[34]

Hirzel (1828–1908) was born and educated in Zurich and had been working with Kühn for fifteen years. He had developed a modest reputation in Leipzig, especially for his teaching and popular science writing. Knop (1817–1891), a Ph.D. under Wöhler and an Assistent successively under Wöhler and Erdmann, taught agricultural chemistry for several years at the public trade school in Leipzig. In 1853 he habilitated in the Philosophical Faculty, then three years later became Director of the recently established agricultural experiment station in Möckern, just outside Leipzig. In 1861 he was promoted to ausserordentlicher Professor. In the late 1850s and 1860s, the value of experiment stations began to be appreciated throughout Germany, and agricultural chemistry penetrated the university system. Knop was particularly valued in Saxony for his important contributions to the discipline, and he was thought (not only in collegial circles but also in the Dresden Kultusministerium) to be essentially irreplaceable.[35]

Falkenstein was unhappy with the faculty report. Although Hirzel had been assistant to Kühn, he was in the Philosophical Faculty. Upon his request for promotion the previous year, that faculty (including Erdmann) had judged him to be inadequately qualified for an Extraordinarius, much less—as Falkenstein argued—for what would amount to be a successor to an Ordinarius. For a field of the importance of chemistry, moreover a subject whose enrollments continue to increase every semester, a certifiably superior academic candidate must be found, and Falkenstein asked the faculty once more for names of such candidates, who surely cannot be rare. The appointment could be either in the Medical or in the Philosophical Faculty, depending on the candidate's background. Regarding Knop, Falkenstein agreed fully with the faculty's plea for further support and promised to provide it for this excellent researcher. He thought, however, that Knop had more than his hands full without asking him to add organic chemistry to his teaching load. Finally, regarding Erdmann, no slight was intended in going for a second Ordinarius in chemistry; considering the enormous significance of the field, he argued, two full professors are by no means too many.[36]

Four documents were submitted in reply to Falkenstein's second directive: an official faculty response, a second dissenting report by Wagner, a long brief by Erdmann (who, being in the Philosophical Faculty, was not officially involved), and a separate report by the medical faculty concerning the unsatisfactory character of Kühn's old lab on the Universitätsstrasse.[37]

In their official response, a substantial thirty-four-page memorandum, the faculty had little to add or to change from their previous position. They did, however, expatiate at length on their reasons for opposing a "doubled" or "split" professorship. Since the deaths of Mitscherlich in Berlin and Kühn in Leipzig, such an arrangement was found nowhere in Germany, and for good reason. Lines of authority were confused, fee rights were uncertain, and an unhealthy competition ensued for students, lab space, budget dollars, salary, and so on. Consequently, resources were splintered and wasted. Recent history demonstrated, they wrote, that it was next to impossible to attract an eminent chemist to a shared post. Certainly no good chemist would ever want to take over Kühn's lab, which was dark, poorly ventilated, continually disturbed by the noise and shaking of heavy traffic in the Universitätsstrasse, and had poor and outmoded furnishings and equipment. Were a substantial name to be attracted to Leipzig, he would undoubtedly demand a new lab building, which would cost 40,000–50,000 thalers or even more. Finally, the faculty argued, why look outside Saxony when such fine local talents as Hirzel and Knop are available?

In his response, Erdmann related the unusual circumstances behind his and Kühn's joint call in 1830, averring that the results had been "unhappy," "inexpedient," and "crippling." He then rehearsed all of the same arguments that the Medical Faculty had used to oppose a second Ordinarius. The custom of one Ordinarius per discipline is all the more compelling for a field of the growing size and importance of chemistry, Erdmann stressed. He also took the opportunity to complain about his lab. A model of excellence when it opened two decades ago, it was now far behind the rapidly advancing state of the art.

(Erdmann related a recent conversation he had had with Bunsen, during which the latter asked him in surprise, "You mean you don't get everything you ask for?")[38]

Erdmann did explore one possible compromise position: to retain him as sole professor of chemistry, but restrict his teaching to inorganic chemistry and appoint a second Ordinarius for organic chemistry. To be sure, organic is not in principle so very distinct, "but in recent times this field has been pursued so preferentially by younger scientists, and their discipline is characterized by such rapid changes, that it seems entirely justifiable to give it a specialized representation by a younger talent"—just as Wöhler and Bunsen had done in Göttingen and Heidelberg, he added.[39]

Falkenstein finally lost patience, and his irritation showed.[40] His third directive represented a firm rejection of the faculty's proposal and a reiteration of his desire to replace Kühn with a second Ordinarius, indeed with the most eminent chemist anywhere to be found. No one could possibly object to a second Ordinarius for a field with the demand and the importance of chemistry; Leipzig already had high enrollments in the field, and they would no doubt continue to grow. The faculty's opinion notwithstanding, a provisional arrangement with local Extraordinarien would have the effect of limiting and diminishing the field. Nor could one Ordinarius be limited to inorganic chemistry, the other to organic chemistry, for that would violate the cherished principle of Lehrfreiheit.

Falkenstein's response indicates that he was trying to loose the faculty's hold on the unstated assumption that money and resources were highly circumscribed (a mistaken notion, as we have seen, that was still made during the negotiations over the Erdmann succession five years later). Leipzig was a large and wealthy university, Falkenstein argued, and we should be thinking in expansive terms. He was untroubled by the faculty's frightened warning about the need for a new institute were the search to be national; if a new institute is needed, we will provide it, he affirmed. He once again asked for a proposed set of national candidates, if necessary coordinated with the Philosophical Faculty, "and [the ministry] does not want to neglect drawing attention among others to the name of Professor Kolbe in Marburg, who has been recommended to us by a number of knowledgeable men."[41]

Since Kolbe was a nonphysician, following receipt of this directive the Medical Faculty asked a committee from the Philosophical Faculty to deliberate with them. In May 1865 they fully capitulated to Falkenstein's requests, reaching a quick and unanimous decision to accept Kolbe's nomination and to make the appointment in the Philosophical

Faculty. The decision was communicated to the Ministerium in memos of 31 May and 13 June from the Medical and Philosophical Faculties, respectively. On the former date, Erdmann wrote Kolbe to tell him of the call.[42]

Kolbe knew as early as 7 February from a letter from his old Marburg acquaintance Carl Ludwig (whom Falkenstein had just called to Leipzig) that his name was under consideration for a call to Leipzig, along with Kekulé and Strecker.[43] Kolbe was inclined to accept such a call, given the proper conditions; he strongly preferred Bonn, but that post was still encumbered to Hofmann.[44] He investigated whether a Bonn call might actually come (the Prussian minister Olshausen later told Kolbe that he wanted to call him, but that his hands were still tied by the Hofmann commitment), while at the same time using Ludwig to help him formulate conditions of acceptance.[45]

Those conditions were to include full equality with Erdmann as second Ordinarius for chemistry; a new laboratory, comparable to the Berlin and Bonn institutes, to be built within two years; a 2500-thaler start-up budget for new equipment and supplies; an annual lab budget of 1000 thalers; three assistants; and a salary matching Bonn's offer to Hofmann of 2000 thalers. The salary was by any measure generous; just in March—presumably in direct response to the early feelers from Dresden—the Kurhessian ministry had given Kolbe a raise from 800 to 1000 thalers.[46] As early as 16 June, he was indicating privately to friends that he intended to accept the Leipzig offer, with or without a new lab, but he continued negotiating for nearly two more months before formally accepting. He traveled to Leipzig on 26 September, staying at the Hotel de Prusse while renting winter lodgings for his family in Königsplatz 14, near Kühn's old lab in the Universitätsstrasse, which he would temporarily inherit. A week later he returned to Marburg to arrange his move. On the first of October, he declared to Vieweg, "Today I am a Saxon!" and on the fourteenth he arrived in Leipzig with his wife and three children. On the sixth of November, he had a highly gratifying audience with King Johann that lasted a quarter hour. The next day he gave the first lecture of his Leipzig career.[47]

Establishing the Leipzig Laboratory

As mentioned, until Kolbe's new lab was built, he was forced to work in Kühn's unsatisfactory facility. One condition he made was to ask for an immediate expansion of this lab, into the ground floor space of the adjoining building (a private house acquired by the university). This addition was finished with "fabulous speed," in time for the Praktikum to begin in early November. But Kolbe was shocked at the conditions in Kühn's lab: it contained not a single retort, condenser, filter stand, or piece of useable glassware. He was amazed that Kühn could even pretend to teach laboratory procedures there; he had to spend twice his start-up budget just to get to the point where he could accept students. Still, he felt happy to be in Leipzig, "newly invigorated" by the change, impressed by the excellent collegial spirit among the professors, and delighted to be free of the "miserable relationships" in Marburg. In the spring, Kolbe moved his family into an apartment directly adjacent to the lab, at Universitätsstrasse 20.[48]

Problems soon arose. The Saxon Kultusministerium that had been so generous with its thalers now began to pinch groschen, complaining about the projected cost of the new lab and the salaries of assistants. It delayed the start of construction from the promised date of Easter 1866, first to that summer, then upon the outbreak of war to the summer of 1867. Meanwhile, Kolbe was gratified by the demand for his talents: seventy Praktikanten per semester were registering and more than eighty auditors for his lectures, which stretched even the expanded lab to its limits and beyond. This was more than Erdmann and Knop were attracting in both of their laboratories combined. Even the war made no difference to his enrollments.[49]

The funding for the new institute was finally formally approved by the Ständesversammlung in January 1867, and construction in the Waisenhausstrasse began in late August 1867, with a budget of 80,000 thalers. The large plot (220 by 250 Fuss[50]) was acquired by the university from the city by means of a real estate trade. In March and April 1867, Kolbe toured other laboratories in Berlin, Greifswald, Marburg, Heidelberg, Munich, Stuttgart, Bonn, and Zurich in the company of his architect, Zocher; in March 1868 he spent three weeks in London on the same assignment. Remarkably, his favorite lab was Staedeler's, in Zurich. The plans were approved in July 1867, and construction began the following month. After the long delay, rapid progress was made, so that by the time of his second architectural tour, the structure was already roofed. The residence was ready for occupancy by October 1868, and the official dedication of the lab took place on 16 November. The final cost was about 85,000 thalers, plus 15,000 for furnishings and equipment.[51] This amount was considerably less than what the Prussians had spent on either the Bonn or Berlin institutes, even though the Leipzig lab could accommodate as many students as both put together.

Kolbe was now master of a wonderful edifice, by far the largest chemistry institute in Germany and probably the world.[52] Above the large, well-lit basement (housing the furnace, coal bins, laundry, baths, and pantry) was the main floor, then a second story above which were attic storage rooms. The two main floors held 44 rooms in all—altogether 51,000 square Fuss (4660 square meters) of working and living space—with high fifteen-foot ceilings. Kolbe made certain that the building was designed for maximum light and ventilation in all rooms. It was equipped with coal-fired steam heat, running water, gas for illumination and burners, and all of the most modern conveniences. Fully a seventh of this space was taken up by an opulent director's residence along the front wing of the building, open to the light and air on three sides, and consisting of fourteen large rooms plus a cellar and attic, with a small garden outside. A hundred feet of second-floor corridor separated Kolbe's airy private study in his residence from his well-equipped private laboratory.

The building was laid out roughly in a letter "E" shape, a configuration that was much admired and copied, particularly in the United States, for its rational management of natural light. The "E" was arrayed with its bottom line along the east–west Waisenhausstrasse; four main laboratory rooms for the Praktikanten were laid out along the main north–south axis, two on each main floor, with the beginners below and the advanced students on somewhat larger benches above. In addition to an impressive variety of special purpose laboratory spaces for the students, the building also provided residences for the superintendent and three assistants, and private labs for the latter as well as for the director. Two auditoriums, seating 60 and 160 students, respectively, were also located on the ground floor. The smaller one was essentially for the Privatdozenten, while the larger one—boasting a nineteen-foot ceiling and a hundred illumination burners for evening lectures–was for the director.

Projecting upward from his average Praktikum enrollments of 70 per semester during 1865–1867, Kolbe planned initially for a capacity of 100. Falkenstein, however, urged Kolbe to think more expansively, to plan for a maximum of 130. Kolbe was doubtful, knowing this was twice the size of any institute yet planned or built and fearing later recriminations regarding unused space, but he swallowed hard and accepted Falkenstein's proposal.[53] To Kolbe's shock and amazement, even this huge capacity was soon oversubscribed. It took only a year after the new institute was opened for the number to hit 100, and by summer semester 1872 the capacity of 130 was reached.[54]

The following semester no fewer than 170 students attempted to register for the Praktikum. Kolbe at first wanted simply to turn 40 of them away, but a means was devised to fit all of them into the existing space. Since there was no way he could directly oversee the work of so many, Kolbe arranged that the 40 extra students would pay their honoraria directly to the assistants supervising them—of whom he now had five—and not to him. As Kolbe admitted to Liebig, this was all part of his plan to prove to the Kultusministerium under the new leadership of Gerber how important he was and what sort of student demand he could command.[55] In the summer of 1873, he had no fewer than 210 attempted registrations, of whom he accepted 170. For one more semester, winter semestor 1873/74, he accepted 170, then, his point having been driven home, he ruled that thereafter only the first 130 were to be accepted.[56] His enrollments in lecture have not been documented; only once have I found a mention in correspondence, namely, that in winter semester 1881/82 he had 270 auditors (in 1873 Kolbe's auditorium had been expanded to seat 250).[57]

Kolbe had certainly come up in the world, and in a hurry. As late as 1860 in Marburg, Kolbe was attracting merely a dozen or so Praktikanten per semester into his laboratory and a similar number into his auditorium. Six years later he had increased these numbers sixfold, or looking at a longer period, the demand jumped fifteenfold in the course of thirteen years. During the 1870s, Kolbe's institute was by far the largest of all seventeen institutes, seminars, and Sektionen at the University of Leipzig.[58] Kolbe was successful in using these numbers as leverage, for in 1866 his budget was doubled to 2000 thalers, then in 1872 doubled again to 4000. He did not worry about increasing his salary, for with all the students at his disposal, his honoraria were substantial, over 5000 thalers per year. From his teaching alone, Kolbe was earning close to 8000 thalers per year in 1873, six or seven times what he had been making a dozen years earlier.[59]

What factors operated to produce this fabulous success story? No doubt Kolbe himself ascribed it principally to his success as a teacher and scholar, and to some degree he was correct. However, other considerations were also important. All the universities of the German Empire were booming in the early 1870s, and none more so than Leipzig. After 1868 Kolbe had the largest and most modern institute in Germany to serve as a draw. Even so, other institute directors at major universities were having similar experiences. Wislicenus at Würzburg reported 102 Praktikanten for summer semester 1874, including 28 professionalizing chemists; Volhard had 151 for winter 1877/78.[60]

Moreover, even in the era of one Ordinarius per institute, Kolbe had a degree of monopoly on the teaching of chemistry in Leipzig that was rare in Germany. For example, Wöhler, Liebig, and Bunsen at Göttingen, Munich, and Heidelberg, respectively, had created decentralized institutes where much basic chemical instruction—and virtually all of it in the burgeoning field of organic chemistry—was tendered by Extraordinarien and Privatdozenten. Erdmann's successor, hired in

1871, was Gustav Wiedemann, who was much more a physicist than a chemist and could attract little chemistry clientele—despite Kolbe's efforts to have him share the load. Knop was still around; in 1870 he was promoted to Ordinarius, but was simultaneously absorbed within the new Landwirtschaftliches Institut to teach agricultural chemistry only. Hirzel had given up academia altogether and had henceforth applied himself to entrepreneurial activities. Therefore, not only did the professionalizing chemists, whose numbers were exploding, find themselves in Kolbe's chemistry classes but also every student of pharmacy and medicine and many students in other applied fields such as veterinary medicine, forestry, economics, metallurgy, and agriculture. Finally, Kolbe was one of the few eminent and active organic chemists around who were accepting personal students, his major rivals being Kekulé in Bonn and Hofmann in Berlin, and this at a time when organic chemistry was booming.

Kolbe also knew how to promote his subject. In the introduction to his book *Das chemische Laboratorium der Universität Leipzig*, he aggressively defended the pedagogical utility of the study of chemistry, not only for science students, but also for philosophers, philologists, and law students. Chemistry is often criticized by laymen as a crude and empirical craft, whereas, Kolbe proudly affirmed, it is a true science and an essential liberal art, an integral part of natural philosophy with well-developed theoretical structures; the time cannot be distant when all will be expected to study chemistry to become truly educated.[61] This is especially true for theology students, he thought. Directly reflecting the views of his father, the broadminded pastor Carl Kolbe (whose death had occurred just two years previously), Kolbe wrote

> It is becoming ever more apparent these days that there must be a change in the academic training of our theologians, whose orthodoxy has alienated the public and is the principal culprit in the much-lamented indifference of the masses toward the institutions of the church as well of those of the state, whose intolerant rigidity and narrowmindedness repels even the educated classes. The young theologian requires a broader education than he has customarily received, he must above all become acquainted with the natural world in which he lives, and study the Book of Nature, this other divine revelation, as well as the Book of Books.[62]

Kolbe concluded his brief by arguing for the long-term economic significance of chemistry. Saxony was highly respected for its industrial might, hence its wealth, especially in the machine and chemical industries. Regarding the latter, whence came this fortunate situation? His answer: from the Saxon academic laboratories where chemists are properly educated—during the last generation especially by Erdmann.

It was because Saxony and its sister states had wisely invested in chemical education that German chemical industry had become stronger than the English or French industries, where comparable investments had not been made. Moreover, he continued, it is prudent for a state to invest heavily in academic laboratories, for they serve as models for private industrial labs, which increase the prosperity of a state.[63]

It must be noted that Kolbe's contention was neither obvious nor irrefutable. He admitted as much by conceding that German academic chemistry had often been treated in a stepmotherly way in comparison to the more traditional fields. In sharp contrast to other fields, even other laboratory disciplines, chemistry students were required to pay for their own materials, a burden that added anywhere from 5 to 100 thalers or more per semester onto their costs. In addition, the Prussian and some other German universities demanded a pro rata contribution from each student to defray the laboratory's general budget. This special "tax" was "not only unfair and unjust, but also impolitic from a national-economic viewpoint," as Kolbe attempted to argue. Rather than voicing alarm at the "epidemic" of interest in the study of chemistry with its attendant costs, university administrators should greet this development with joy for the future national prosperity that it portends.[64]

The importance of the science-versus-technology question here broached by Kolbe warrants discussion. To Edward Frankland, sitting in the capital of the unenlightened British Empire, Kolbe's arguments were welcome but also not self-evident. He wrote to Kolbe:

> I wish you could in some way demonstrate this, which, a priori ought certainly to be the case. Just the opposite opinion unfortunately prevails here & greatly impedes the progress of experimental science in this country. When I urge upon Politicians here the disgraceful position of science in this country as compared with Germany, they reply contemptuously "What has science done for German commerce & manufactures? To whom are due the invention of the two greatest modern chemical manufactures—Paraffin oil & Aniline colours? Are they not due to Englishmen? Why, if all this science does her any good, does Germany still dread the competition of our manufacturers, and impose enormous duties on our goods and why, notwithstanding these high protective duties, and the greater cost of labour and coal in this country, is Germany still our best customer for chemicals and steel? Why is Germany obliged to apply to England to get her capital supplied with Water and Gas?" These are awkward questions, & when I say "There is a want of enterprise in Germany (to which the wretched condition of the streets of Berlin & the overcrowding bear testimony)" the reply of course is "Yes! and this want of enterprise is caused by too much attention to, & dependence upon, abstract science."[65]

As it happened, of course, this debate was occuring right at the watershed period in which German chemical industry was first beginning to be strongly influenced by academic science, and the latter was climbing toward world leadership. Frankland's unnamed politicians were right that it had been an Englishman, William Henry Perkin, who first developed a commercially successful coal tar dye, in 1856–1858. What they failed to note was that Perkin had been a student of Hofmann at the Royal College of Chemistry, that Hofmann represented a partial transplanting of German (Liebigian) academic science into England, and that Hofmann and several other German expatriates had since returned to Germany.

Many entrepreneurs correctly perceived that Perkin's discovery promised to transform a plentiful but worthless waste product of gas and coke production into a valuable feedstock material, and so a large number of dye companies were formed in Britain, France, and Germany during the late 1850s and the early 1860s. The pace of innovation in these early years was furious, but experimentation was not guided by well-understood or well-articulated theory. That began to change in 1863, with the introduction by Edward Nicholson's firm of systematically alkylated aromatic dyes appropriately named "Hofmann's violets." After Kekulé's benzene theory was proposed, the influence of theory became much stronger. The first dramatic payoff of structure theory for the chemical industry came in 1868–1869 with the synthetic production of the important natural dye alizarin, in which Baeyer, Graebe, Carl Liebermann, Heinrich Caro, and Perkin all played significant roles. At this time, alizarin was second only to indigo as a dye, and its source, madder, was the major crop in Provence and other parts of the world. With the introduction of the chemists' far cheaper and purer alizarin, the madder farmers were ruined. From this time on, a clear trend toward the amalgamation of theory and practice was much in evidence, especially in Germany. The English, however, were slow to see the connection. In another generation, the relationship was evident to all, but not yet in 1872.[66]

I have said nothing yet about the details of Kolbe's classes and practica in Leipzig. Unfortunately, documents that could illuminate such details do not seem to have survived. The semester enrollments previously discussed suggest that Kolbe instructed about 1600 Praktikanten during his nineteen years in Leipzig. Records of promotions to the doctorate do survive and indicate that he served as advisor to 68 Doktoranden.[67] Among the best known of his practicum students in Leipzig were Ernst von Meyer, Henry Armstrong, A. M. Zaitsev, V. V. Markovnikov, Constantin Fahlberg, Hermann Ost, Ernst Beckmann, Rudolf Leuckart, and Theodor Curtius.

Every winter semester Kolbe lectured on inorganic chemistry six days a week at nine A.M., every summer semester on organic chemistry four days a week at eight. His beginners' Praktikum was four or five days per week, two to three hours per day; advanced students worked from nine until one and two until five o'clock, six days a week.[68] From three assistants in 1865, Kolbe worked his way up to four in 1870, then five by late 1872. Advanced Praktikanten numbered a dozen or so per semester early in Kolbe's Leipzig period, and more like thirty to fifty during the 1870s.[69] During this latter period, Kolbe had three of his assistants carry out the primary supervision of the beginners on the ground floor, while a fourth helped him with the advanced workers upstairs; the fifth assistant was dedicated to the lecture experiments. But Kolbe kept abreast of the progress of every Praktikant, no matter how crowded his laboratory became nor how much a novice the student was.[70]

Indeed, Kolbe regarded such personal attention as a fundamental pedagogical principle. It led, he averred, to a certain "patriarchal relationship" between him and his students, which developed over the long term into a permanent esprit de corps that reinforced the positive qualities he was seeking to instill. His strict hierarchical bureaucratic organization thus had a certain pedagogical justification, and moreover it seems to have operated well, without the authoritarian tone that many expected.

Kolbe explicitly stated that his pedagogical philosophy and teaching methods as developed in Marburg continued without essential change in Leipzig.[71] Those methods have been described in chapter 5, with firsthand reminiscences by Leipzig students, Markovnikov in 1866–1867 and Armstrong in 1867–1870. One more witness may be introduced here, that of his student and eventual son-in-law Ernst von Meyer (1847–1916).

Meyer, descended from an aristocratic Kurhessian family, was a second cousin of Kolbe's wife and knew her slightly from Marburg. After arriving in Leipzig and attending Kolbe's first lecture for winter semester 1866/67, he introduced himself, was received in a formal fashion (typical of northern Germans, commented Meyer), and was invited to visit. He soon became intimate with the entire family. His reminiscences of the laboratory routine dovetail with those of Markovnikov and Armstrong, as well as with Kolbe's descriptions. Meyer wrote

> A better school than Kolbe's I could not wish for. . . . With the help of able assistants (Finkelstein and Drechsel) he was able to devote himself to everyone, even the beginners; he would not tolerate any of the many

> reactions we observed to remain unclear to us. At that time there were still no printed short introductions to qualitative analysis, as are ordinarily in use today. Our handwritten observations were examined and reviewed by an assistant, also by Kolbe himself. This kind of instruction instilled in the beginner a firm foundation for further development.

Meyer stayed in Leipzig two years, then attended Heidelberg before serving as an officer in the Franco-Prussian war. On his return to Leipzig, he resumed his studies, earning a doctorate in 1872, whereupon he became one of Kolbe's assistants. In 1875 he became engaged to Kolbe's eighteen-year-old daughter Johanna, and they married the following year.[72]

Acquiring a Journal

During the 1840s and 1850s, Liebig's monthly *Annalen der Chemie und Pharmacie* dominated the field of academic chemistry in Germany. (Poggendorff accepted principally physical and physical–chemical papers for his *Annalen der Physik und Chemie*, and Erdmann's *Journal für praktische Chemie* was oriented toward technology and consisted mostly of reprinted articles.) When Liebig went to Munich, he gave over the effective editorship to Hermann Kopp (though Liebig's and Wöhler's names remained on the title page), and Kopp took those duties with him when he transferred to Heidelberg. Kopp was very conscientious, but his authors grew increasingly restive at his dilatory publication schedule. This was all the more galling to active researchers in the fast-moving and increasingly cutthroat field of organic chemistry, where notices could be published by French rivals in the biweekly *Comptes rendus* or by English rivals in the weekly *Chemical Gazette* in a fraction of the time. When Erlenmeyer took over Kekulé's *Zeitschrift für Chemie* in 1860, he reconstituted it as a fast publication outlet for original short papers, preliminary communications, foreign notices, and republished pieces, loosely patterned after the *Comptes rendus*.[73]

Erlenmeyer's *Zeitschrift* might well have given the Liebig–Kopp *Annalen* stiff competition were it not for Erlenmeyer's unwise editorial practices. For one thing, he used the journal indiscriminately as an outlet for his own profuse and often incautious theoretical speculations. As Beilstein later put it to Butlerov, Erlenmeyer's "occasional good thoughts were simply drowned in the great amount of sauce that he poured over everything. I finally just stopped reading his long essays . . ."[74]

More seriously as far as the commercial success of the endeavor was concerned, Erlenmeyer increasingly began to write critical editorial notes, some of them lengthy and some quite sarcastic, and he occasionally used exclamation points and question marks inserted in parentheses into his authors' printed manuscripts. He did this even with his friends and theoretical comrades-in-arms; his hitherto close personal and professional relationship with Kekulé was nearly destroyed by these practices.[75] Ultimately they destroyed his journal.[76]

It should be noted that Erlenmeyer's critical style of editorship conformed to a well-established German tradition, in science as in other fields. As far as chemistry is concerned, Berzelius' *Jahresberichte* (a foreign product in origin but having its widest promulgation and greatest influence in Germany) and Liebig's *Annalen* of the 1830s might be mentioned as examples of periodicals having a principal goal of regular critical evaluation of the literature. Moreover, both Liebig and Berzelius could be quite sharp in their judgments. But both became worn out and worn down by the strain.[77]

In March 1864 Erlenmeyer told Kolbe, who was at that time a real friend, that he had nearly decided to quit his editorship. He complained that an editor's position was inherently delicate and awkward (*misslich*) and really should be performed only by a real authority in the field. Kolbe agreed, only adding that the editor must be an *elder* authority. The forty-five-year-old chemist wrote,

> Frankly, I would think that even I am too young to edit a *critical* journal. For years I have toyed with the idea of writing short annual reports on chemistry, more specifically of a critical character, for which those of Berzelius would serve as a model. But first I want to seek a firm foundation for myself. Perhaps in ten years, if I live to see it, I may seriously consider it.[78]

Kolbe also agreed that Erlenmeyer's critiques had hurt him and his journal and commented that the same reproach had often been made to him. He added that he had resolved to deal henceforth less with polemics, critiques, and theories and more with facts and observations, and he suggested the same course for Erlenmeyer.

Erlenmeyer did not want to give up without a fight. With Butlerov he explored the idea of transforming the *Zeitschrift* into a Russian journal—many of the subscribers were Russian, anyway—but Butlerov wanted none of that.[79] With Kolbe he explored the possibility of making the journal a weekly, analogous to the *Chemical Gazette* (now renamed, under William Crookes' editorship, the *Chemical News*). Kolbe got Vieweg interested in this idea, and he even volunteered to be head editor for an interim period. However, it appears that Erlenmeyer's requested honorarium was more than Vieweg wanted to pay, at least not before he could see what sales would be.[80] In any event, the journal was taken over, beginning in January 1865, by three younger associates of Wöhler in Göttingen: Ausserordentlicher Professor Friedrich Beilstein, Privatdozent Rudolf Fittig, and Assistent Hans Hübner, who ultimately became Wöhler's successor. Hübner became the chief editor and made a success of the second series of the *Zeitschrift*, more than tripling subscriptions in six years.

Kolbe was unhappy with the change. He advised Erlenmeyer to have nothing at all to do with what promised to be a "watery and colorless" journal. He was certain in 1864 that the editors would fail in a year or two, at which point Erlenmeyer could pick the enterprise up again; in the meantime they could all expect to see a good deal of "Göttinger Mist." He also advised Erlenmeyer to have nothing to do with Erdmann's *Journal für praktische Chemie*. This journal still had a respectable readership, which was mystifying to Kolbe, who thought the editing was extremely poor. "I hear that Erdmann has almost nothing to do with it himself, but rather leaves the editing to one of his assistants."[81]

In addition to the *Annalen*, the *Zeitschrift*, and the *Journal*, a fourth chemical periodical began in 1867—the *Berichte* of the new Deutsche Chemische Gesellschaft (DCG). The DCG was founded in Berlin as a German analog of the Chemical Society or the Société Chimique, and its *Berichte* was at first merely a small and essentially local "proceedings of sessions" of the society. However, the deficiencies of the *Annalen* under Kopp's editorship as well as the explosive growth in the science of chemistry led to an explosive growth in subsequent annual volumes of the *Berichte*. By the early 1870s, it was beginning to serve as the periodical of choice among German chemists for rapid publication of short communications. The resulting competition forced the *Zeitschrift* out of business after December 1871.

Meanwhile, the general dissatisfaction with the *Annalen* was coming to a head. In the late 1860s, Kekulé conceived the notion of founding a companion journal for the *Berichte*, allowing the latter to specialize in short preliminary notes while the new periodical would publish the detailed versions of the same research at a later time—much like the relationship of the *Comptes rendus* to the *Annales de chimie*. Before Kekulé made his idea public, Kopp decided in March 1871 to retire as editor of the *Annalen*. Liebig asked his (and Kolbe's) former student Jacob Volhard if he would take on the day-to-day editorial work; Volhard accepted on the condition that Erlenmeyer be joint editor. This created a logistically favorable arrangement since all three men were working in Munich. Kekulé decided this was the propitious moment to let colleagues know of his plan for a new journal. The reaction, however, was largely negative, since many had a sentimental attachment to the *Annalen* and it was thought that a new journal would kill Liebig's historic progeny. Consequently, Kekulé's project failed.[82] One of the most energetic protesters was Kolbe, who in the meantime had acquired a new journal himself, Erdmann's *Journal für praktische Chemie*.

Kolbe promised Erdmann on the latter's deathbed that he would complete the editing of the current volume of the journal. This placed Kolbe in an enviable position since he was the logical person for the publisher (Barth) to ask to take over in a permanent capacity. Moreover, Kolbe was aware that his friends Heinrich Vieweg and Franz Varrentrapp of Vieweg Verlag (Eduard Vieweg had suffered a stroke and was then terminally ill) were interested in principle in publishing a chemical periodical. For three weeks in October 1869, Kolbe negotiated between Barth and Vieweg for the best possible conditions. Regarding a possible new Vieweg journal, Kolbe promised the partners that he would employ "respectful criticism with as little polemics as possible." He also wanted a managing editor who would do essentially all the daily tasks (as Volhard and Erlenmeyer did for the *Annalen*). He thought it likely that he could get Falkenstein to agree to Rudolf Schmitt as Erdmann's successor, who would be the perfect man for the role. He figured around 600 thalers per year for all editing honoraria sounded about right.[83]

Kolbe well knew the advantages of a personal journal, having in mind the models of Berzelius' *Jahresberichte* as well as Liebig's *Annalen* in its early years. We have seen that he had been thinking of a personal journal at least for several years before 1869. "What particularly attracts me about editing a journal," he wrote Varrentrapp,

> . . . is the value it has for a large chemical laboratory to have a journal at its disposal at all times, and not to be dependent on the mercy or mood of another when it is a question of rapid publication and defense and advocacy of a viewpoint. . . . You know that it matters to me to have the Journal für pr. Chemie at my disposal, and that it would be very unfortunate were it to end up in the hands of another director of a large chemical institute, such as Hofmann.[84]

Negotiations with both Vieweg and Barth nearly collapsed in November, and Kolbe temporarily gave up all hopes; but in January 1870 he reached agreement with Barth to take over the *Journal*.[85] He was to retain the editorship until his death almost fifteen years later, and he certainly made the most of his permanent platform; but he did not, as he promised Vieweg, avoid polemics.

12

Aromatic Chemistry

Present-day organic chemists divide their millions of known compounds into two broad categories, aliphatics ("fatty" substances) and aromatics (compounds based on the benzene family, some of which have notable aromas). Aromatic substances are important both because they are the larger class and because they have more general application to chemical industry, being essential for most important dyes and pharmaceuticals. Up until Kolbe's entry into the science circa 1840, most organic chemical research had been devoted to what later became known as aliphatic compounds, and the few aromatics known were not thought to be taxonomically distinct. This began to change in the 1840s. Hofmann's first major research (1842) was a pathbreaking investigation of compounds in coal tar, many of which are aromatic.[1] In the years that followed, aromatics were aggressively pursued in many laboratories, notably by Hofmann, Laurent, and Gerhardt. It was noticed that these compounds appeared to have a minimum carbon content (twelve equivalents or six atoms), that they had far less hydrogen relative to carbon than most other organic substances, and that they had a distinctive set of chemical characteristics.

In 1855 Hofmann introduced the term *aromatic* to signal these distinctions, and the term quickly spread.[2] By this time, the isomeric relationships of aromatic compounds were becoming interesting, and investigations into the constitutions of organic molecules were becoming widespread and successful. But the constitutional question regarding aromatic compounds proved intractable: the base radical for aromatics consisted of $C_{12}H_5$ in equivalents or C_6H_5 in atoms, but how were these atoms configured? Kekulé simply side-stepped the issue in his structure theory of 1857–1858, devoting himself almost entirely to aliphatic compounds. Only A. S. Couper and J. Loschmidt proposed possible structures for benzene during this time period, but these were only speculations for which neither man offered any empirical justification. Their proposals were not influential.[3]

Early Work on Salicylic and Salylic Acids

Salicylic acid (2-hydroxybenzoic acid), a constituent of a number of plant extracts with noted febrifugal properties, had been studied since the late 1820s, as had the related substance phenol (carbolic acid or hydroxybenzene). Benzoic acid (benzene carboxylic acid) had long been known. For some reason, Kolbe became interested in these compounds shortly after his arrival in Marburg.

Since salicylic acid was known to decarboxylate easily (mild heating converted the substance to phenol, with evolution of carbon dioxide), Kolbe speculated that salicylic acid might be simply an ester of phenol and carbonic acid. Accordingly, it should have been possible to synthesize the acid by esterifying the two components (e.g., reacting sodium phenolate with phosgene, then hydrolyzing the remaining chlorine). Repeated experiments along this line consistently failed, but without shaking Kolbe's assumption.[4]

Kolbe put his first Ph.D. student, Wilhelm Gerland, to work on closely related problems. Gerland succeeded in making a second aminobenzoic acid, "benzaminic acid" (the first such was anthranilic acid). Since anthranilic acid could be oxidized smoothly to salicylic acid, Kolbe suggested publicly that these two compounds had analogous structures, that is, the former must be an amide if the latter (as he still assumed) was an ester. This would explain the isomerism of anthranilic with benzaminic acid, the latter being the true aminobenzoic acid. But Kolbe confessed continued failure in proving this notion by making salicylic acid through esterification.[5] Gerland then found that oxidation of benzaminic produced an isomer of salicylic acid, which he named "Oxybenzoësäure." He suggested that it was this new isomer, not salicylic acid, that was the carbonic acid ester. This seems to have been Gerland's proposal and not Kolbe's.[6]

By 1859, after a hiatus caused partially by his ill health, Kolbe had put two new students to work on related problems. One was Rudolf Schmitt, not yet Ph.D. but already an assistant, who attempted to expand the isomerism of anthranilic/benzaminic acids to analogous aromatic sulfoacids. The other was Eduard Lautemann, then an advanced Praktikant, who was assigned the study of a number of salicylic acid homologs.[7] This work was going on simultaneously with Kolbe's attempts, with Lautemann's and Carl Ulrich's help, to show that glycolic and lactic acids were monobasic hydroxy derivatives of acetic and propionic acids, respectively, to counter Wurtz' contentions that they were dibasic, or rather (as Wurtz later argued) diatomic and monobasic.

I have suggested in chapter 9 that Kolbe's discomfort with the idea of polybasic organic acids originated in his theoretical commitments to radical theories and the earliest version of the type theory. Kolbe himself stated that his effort to show that salicylic acid was an ester was motivated not so much by an otherwise unexplained isomerism but by his disagreement with those who viewed the compound as dibasic.[8] Piria's demonstration in 1855 of double salts of salicylic acid appeared to provide an irrefutable demonstration of its dibasic character. However, late in 1859 Kolbe and Lautemann finally succeeded in synthesizing sodium salicylate from phenol by simultaneously introducing gaseous carbon dioxide and finely divided sodium metal into hot phenol.[9]

Kolbe initially thought that this synthesis demonstrated his ester hypothesis.[10] However, by April 1860 he abandoned that idea because he realized that his conjectural structure could not account for the known reduced forms of salicylic acid: salicylaldehyde, and salicyl alcohol. Nonetheless, the new synthesis provided proof (he thought) that salicylic acid was indeed monobasic. The substance was, in fact, entirely analogous to glycolic and lactic acids, in that it consisted of benzoic acid in which one hydrogen atom of the phenyl radical was substituted by the radical HO_2. That the hydrogen of this "hydrogen peroxide" radical could be replaced by a metal atom in the double salts of salicylic acid did not mean that the acid was dibasic. He went so far as to assert that *only* monoacids, such as salicylic acid, have corresponding aldehydes and alcohols and that any assertions to the contrary were "unscientific frivolities" that did not even deserve to be discussed. This remarkable statement contradicted Kolbe's own published views, and he soon had to retract it.[11]

Kolbe's new view of the constitution of salicylic acid seemed to have solved one problem, but it created another: now Kolbe was designating both salicylic and its isomer oxybenzoic acid by the same rational formula and so the cause of the isomerism was mysterious. Fortunately, a recent discovery by Lautemann suggested a resolution. Reduction products of salicylic acid included not only salicylaldehyde but also what appeared to be an isomer of benzoic acid, for which Lautemann and Kolbe suggested the name "salylic acid." Thus, benzoic acid could be converted to benzaminic and then to oxybenzoic acid, whereas salylic acid was related genetically to both anthranilic and salicylic acids. There were therefore two independent isomeric series. All of this could easily be explained if there were two isomeric parent aromatic hydrocarbons having the equivalent formula $C_{12}H_6$, and Kolbe suggested precisely that. The parent radical of the salylic acid series Kolbe suggested should be called "phenyl," which is the radical in phenol; the progenitor of the benzoic acid series could then be called "benzyl," the radical of benzoic acid. Such a hypothesis could also explain the known isomerism of benzyl alcohol versus cresol.[12]

Kolbe admitted that there were problems with this suggestion. No second isomer of benzene was actually known (decarboxylation of salylic acid, for instance, apparently yielded ordinary benzene), and there was no second phenol. Moreover, there should be a distinct salyl aldehyde and salyl alcohol, a sulfosalylic acid, and so on. Nor was there any way to know in what the isomerism of the parent hydrocarbon may consist. However, Kolbe was willing to speculate: benzyl may be

$$\left.\begin{matrix} C_{10}H_3 \\ H_2 \end{matrix}\right\} C_2$$

while phenyl could be $(C_{10}H_5)'''C_2$. One radical thus has a structure in which one or two additional functional groups may substitute for two hydrogens of methylene, while the other contains a carbon (C_2) that is bound only to a single triatomic group; he symbolized these isomeric radicals as $b(C_{12}H_5)$ and $p(C_{12}H_5)$. Kolbe offered no empirical justification for these structural details, nor is one evident to the historian.

Kekulé's former roommate in London, Hugo Müller, verified the existence of salylic acid soon thereafter, or so he thought at the time, and wrote Kekulé about it.[13] Kekulé then published an article that followed up on Kolbe's research, expanding Kolbe's suggestion of two isomeric series to include such compounds as the two known chlorobenzoic acids. He stated that the existence of an isomer of benzoic, i.e., salylic acid, was no longer in doubt, and he thought that there may also be two isomers of benzene.

> However, I consider it inappropriate even now to enter into a theoretical discussion, since the necessary factual foundation for this is missing. Even Kolbe, who has to date penetrated the furthest toward true knowledge of the actual arrangements of atoms, was able to discover no other difference in the constitutions of these substances than that, in addition

> to the elements that otherwise compose organic compounds, the first contains a soft "b," the other a hard "p."

Kekulé had evidently been dismayed by Kolbe's public charge (presumably directed at Wurtz, but arguably also at Kekulé) of "unscientific frivolities," for he followed this little witticism with a more direct barb:

> But what further deters me from discussing this question is the circumstance that I probably could not do that without criticizing Kolbe's theoretical views (which hitherto from one paper to the next have constantly changed), and also his manner of dealing with other chemists in his publications—a critique I wish to avoid, if possible entirely, or at least as long as possible.[14]

Since 1857 Kekulé had been working with benzoic and salicylic acids himself, and he had hinted at a "denser" and "next-simplest" arrangement of the carbon skeleton of benzene in his structure theory paper of 1858. But he was not yet ready to enter into an explicit theoretical discussion, presumably for the reason stated—the still scanty empirical record. It is probable that he had not yet formed any satisfactory hypothesis regarding the isomerisms.[15]

Kolbe's conjecture that his "phenyl" (in contrast to his "benzyl") radical had a carbon with no hydrogen atoms suggested to him that one might be able to hydrogenate this carbon, producing compounds with more hydrogen than aromatic compounds but less than any aliphatic compounds. He had had good results with sodium amalgam as a reducing agent, and so in December 1860 he tried it on phenol, salylic acid, and salicylic acid, and other "phenyl" compounds. In February 1861, he reported excitedly to Liebig that he thought he had succeeded in his goal and that the discovery promised to yield years worth of exciting research. Reduction of salylic acid yielded much aldehyde, but also a certain percentage of a weakly smelling liquid that appeared to be an acid with a larger percentage of hydrogen than the starting material.[16] Liebig was enthralled; this was an "extremely fortunate idea, which opens up a treasure-trove of new discoveries."[17]

Kolbe published a short preliminary communication in the *Annalen* on this subject and tried to follow it up, but he was unable to isolate enough of the compound to prove his idea of an intermediate link between aromatic and aliphatic compounds. As he wrote to Liebig, there were severe experimental difficulties, including a multiplicity of small amounts of products that were hard to separate. In the meantime, he and his students were more successful (or so they initially thought) at further substantiating his claim of two isomeric series of aromatic compounds. For example, Schmitt applied the idea to benzenesulfonic acids, and Lautemann prepared a variety of new iodo and hydroxy derivatives of salicylic acid and of phenol, all depicted as "phenyl" or "p" compounds.

Unfortunately, both lines of research soon reached dead ends. The chief difficulty was with salylic acid. Kolbe had thought he had more than sufficient evidence to identify this substance as an isomer of benzoic acid. The two solids were different in appearance and their crystal forms were distinct; salylic was only about a third as soluble in water as benzoic, it was more volatile, and it melted at 119°C rather than 121°C.[18] The existence of the acid was accepted by nearly everyone, including De La Rue, Müller, and Kekulé (although Kolbe later revealed that Lautemann himself, the actual discoverer, had always expressed doubt regarding the real existence of salylic acid).[19] But as early as the spring of 1861, Cannizzaro argued convincingly that decarboxylation of salylic and benzoic acids both produce a single substance, benzene, and that there is no other isomer of benzene. Later that year, Rudolf Fittig, an assistant of Limpricht and Wöhler at Göttingen, discovered that even tiny amounts of impurities in benzoic acid could dramatically alter its properties, such as crystal structure and solubility; he thought that salylic was simply impure benzoic acid. Two years later, Friedrich Beilstein established that whereas nitrodracylic was a true isomer of nitrobenzoic acid, "dracylic" and benzoic acids were identical. The following year he definitively confirmed Fittig's observations on "salylic acid."[20]

For some time, Kolbe attempted to defend the existence of salylic acid and its putative parent hydrocarbon. Forgetting the reasoning that had led him to abandon the ester hypothesis for the constitution of salicylic acid, he argued in 1863 that there must be four different isomers possessing the composition of salicylic acid. There are two parent hydrocarbons, hence there must be two different ester compounds (salicylic being one), as well as two true (hydr)oxybenzoic acids. After he had failed to crack the intractable aromatic reduction problem, Kolbe gave it to a student, Max Herrmann. In 1864 Herrmann published a paper claiming to have succeeded in adding four atoms of hydrogen to benzoic acid. The next year he thought he had accomplished an analogous transformation with hippuric acid, but an "unfortunate accident" prevented him from precisely characterizing the novel reduction product.[21]

But just about the time this last paper was published, Kolbe became convinced by Beilstein's work: salylic acid did not exist and benzoic acid had no known isomer. With the fall of salylic acid and the increasing evidence against a second isomer of benzene, Kolbe's isomerism hypothesis appeared to be in trouble. When in 1865 Kolbe published a monograph reprinting his papers and those of his students since 1859, he omitted five aromatic papers, including the one announcing the synthesis of salicylic acid and the preliminary note that proclaimed the existence of intermediate links between aromatic and aliphatic compounds. When salylic acid was mentioned in another reprinted paper, Kolbe attached editorial notes denying its real existence and explaining the source of his error.[22]

"Kekulé Always Rides a Fiery Steed"

Although by 1862 Kolbe was encountering difficulties in substantiating his early aromatic conjectures and had partly repudiated them by 1865, he continued to work (and to assign his students projects) on aromatic compounds, and his laboratory produced many novel substances in the early 1860s. In the spring of 1863, two of his students in Marburg simultaneously and independently discovered the first case of triple isomerism in the aromatic series. Konstantin Zaitsev (in German transliteration, C. Saytseff), one of the first of numerous Russian students, never made a name for himself in chemistry, but he was the brother of the much more famous Aleksandr Mikhailovich Zaitsev, who arrived at Kolbe's lab a semester later. Early in 1863, Konstantin prepared from anisic acid (modern *p*-methoxybenzoic acid) a new hydroxybenzoic acid that was demonstrably distinct both from salicylic acid and also from Gerland's "Oxybenzoësäure." He named it "Paraoxybenzoësäure," the prefix simply indicating that it was an isomer.[23]

About the same time at a neighboring lab bench, Georg Fischer (who is even more obscure than K. Zaitsev) obtained the same compound starting from toluene through nitrodracylic (modern *p*-nitrobenzoic) and a novel para-aminobenzoic acid. The fact that there were now *three* hydroxybenzoic acids suggested to Fischer that, in addition to benzoic and salylic acids, there must be a yet undiscovered third isomer, "Parabenzoësäure."[24] Zaitsev's and Fischer's articles were published back to back in the *Annalen*. In that same issue, Beilstein argued that the aminobenzoic acids were now three in number. Shortly thereafter, H. Hlasiwetz and L. Barth revealed the existence of a third dihydroxybenzene, christened "resorcinol," and N. Sokolov concluded that three nitrobenzoic acids could be sharply distinguished. Finally, in a communication dated August 1864, Beilstein announced the discovery of a third distinct chlorobenzoic acid and concluded that there must also be three bromobenzoic and three iodobenzoic acids.[25]

What all this means is that within the space of about three years, the isomer problem in the aromatic series had suddenly and radically changed. In 1861, it seemed highly probable that there were two of each species, in other words, two isomers of benzene itself, two isomers of each monoderivative such as benzoic acid, and two of each diderivative such as salicylic acid. By the end of 1864, it was becoming clear that there was only one benzene and only one of each monoderivative of benzene, but that there were in general three of each diderivative such as hydroxy benzoic or nitrobenzoic acid. Kolbe's conjectures were no longer viable, and Fischer noted plaintively that the "very important question" of explaining aromatic isomerism was not presently answerable. In extenuation of his mistaken identification of salylic acid, Kolbe likewise commented in a note written in March 1865 that he had been led astray by his urgent desire to explain the *genuine* generic isomer problem for aromatic compounds.[26] But it would appear that Kolbe had given up on the problem as hopeless.

It was none other than Kolbe's nemesis Kekulé who first succeeded in finding a satisfactory solution to this problem, and this was precisely as Kolbe was writing the remark just cited. Kekulé hinted at double bonds in benzene in 1858, and he later implied that he had already formulated his benzene theory at that time. This claim, however, was probably disingenuous because, as we have seen, he accepted the existence of salylic acid and of two isomeric benzenes in 1861. But the new discoveries of 1861–1864 dramatically altered the requirements for a successful theory. If we can believe his famous dream anecdote (and circumstances suggest that we can), he had the idea for his benzene theory around early 1862, but this was still too early in the chronology to found a successful empirically based theory. By the end of 1864, it was time.

Early in 1865 Kekulé published a paper (in French, since he was then working in Francophone Belgium) that argued that a cyclohexatriene formula for benzene could probably account for all the known isomeric relationships in the aromatic series. The "Kekulé formula" would allow only one benzene and one benzoic acid, but would appear to predict three isomers (today referred to as ortho, meta, and para isomers) of every disubstituted benzene. Kekulé expressed himself cautiously in this article and spent most of his time discussing how the theory could explain isomer relationships of side-chains rather than positional (ortho/meta/para) isomers. But he was more definite and more confident on the crucial question of positional isomers in a long German-language article published in Liebig's *Annalen* in February 1866. A definitive treatment was published in the sixth fascicle of his textbook, which appeared that summer.[27]

Kolbe read Kekulé's *Annalen* paper with interest (he probably had not seen the earlier French papers, since he rarely read French journals). "What do you think about Kekulé's new philosophy in the February issue of the *Annalen*?" he asked Erlenmeyer. "He always rides a fiery steed."[28] We do not possess Erlenmeyer's reply to this question, but there can be little doubt that it was favorable, for an article appeared in the very next issue by Erlenmeyer praising Kekulé's theory as the "most probable" representation of aromatic compounds (although Erlenmeyer also made a number of suggestions modifying and extending the theory). Kolbe must have been astounded at the overwhelmingly favorable response to Kekulé's theory. Not only did students and friends of Kekulé, such as Erlenmeyer, Williamson, Müller, Beilstein, Baeyer, Ladenburg, L. Meyer, Dewar, Hübner, and Körner, become advocates of the theory but also many who had no direct personal relationship to him, such as Wilbrand, Watts, Naquet, Fittig, and V. Meyer. In fact, three former members of Kolbe's own research group became strong advocates of the new theory, namely, Graebe, Crum Brown, and Claus, and Kolbe's good friend and theoretical comrade Frankland also joined the crowd.[29]

Some historians of chemistry have been misled by the lengthy but often minor disagreements over certain structural details into thinking that Kekulé's theory was poorly accepted until close to the end of the century. In fact, the proposers of principal "rivals" to Kekulé's structure thought of themselves essentially as advocates of the theory who were offering only reinterpretations of it. Ladenburg, the author of the "prism" formula, always stressed the similarities of his and Kekulé's formulas. In 1874 he conceded that Kekulé's formula was in most respects superior to his own; in 1876 he stated that virtually all chemists had accepted Kekulé's theory, at least to a certain degree, and that the cyclohexatriene formula was "at least as appropriate, if not more so, than the prism." Even Kolbe conceded in 1874 that Kekulé's theory was "accepted by the great majority of chemists."[30]

Kolbe's concession was by no means a surrender. Despite his repudiation of salylic acid in 1865, he did not immediately abandon his belief that there are isomeric "phenyl" and "benzyl" radicals corresponding to the formula $C_{12}H_5$, reaffirming this idea two years later.[31] Appalled by the growing popularity of Kekulé's "chaining" theory of structure and deeply concerned that Kekulé's cyclohexatriene interpretation of benzene was already then being viewed by many as the successful capstone of that theory, Kolbe was desperate to produce a coup in the aromatic realm. His interest in aromatics was of even longer standing than Kekulé's, and he shared Kekulé's identification of the central desideratum, namely, a general theoretical explanation for isomerism. He found it particularly galling that Kekulé, who had been a thorn in his side on many occasions and whom he considered an upstart and a "superficial chatterer," had now become his chief rival. Even many of his own advanced students and junior associates during the years 1866–1870, such as Ernst Carstanjen, Carl Graebe, Hermann Ost, and Henry Armstrong, disagreed with his views on aromatics, and some of them preferred Kekulé's.

It was in response to the arguments and urgings of Markovnikov in 1866–1867 and possibly also Graebe in the summer of 1868 that Kolbe finally accepted the larger atomic weights for carbon, oxygen, and sulfur, which by that time had been adopted by virtually all German chemists. The first occasion when Kolbe employed the new weights was in his dedicatory lecture for his new institute on the Waisenhausstrasse, on 16 November 1868. The topic of this printed academic dissertation was his conception of the constitution of hydrocarbons; its centerpiece was a novel theory of aromatic compounds. Both were offered explicitly as alternatives to Kekulé's ideas. Indeed, Kolbe's second sentence averred that his ideas were "essentially different from the views of the chemists of so-called modern chemistry, and especially from those based on Kekulé's chain theory." He continued,

> I do not intend here to subject the latter to a thorough criticism; I merely state clearly and succinctly, in order at once to remove any doubt concerning my estimation of its significance, that I consider the chain theory, just like the earlier type theory, to be an ephemera among recent chemical hypotheses, and its principles to be untenable and erroneous, without denying that it is ingeniously devised, and, as experience confirms, is well suited to blind younger chemists.[32]

The next few pages of Kolbe's lecture recounted his views on aliphatic hydrocarbons, in their current state of evolution. Kolbe's readers (as well as the readers of this book) had seen this material before.

Trimethine–Trimethane

What they had not seen was Kolbe's new benzene theory. Just as three diatomic methylene radicals, he wrote, might substitute for three hydrogen atoms in each of two methyl radicals:

$$\left.\begin{matrix}(CH_2)''\\(CH_2)''\\(CH_2)''\\H_2\end{matrix}\right\}C_2$$

so might benzene be considered to be constituted from three trivalent methine (CH) radicals, which substitute for three hydrogen atoms in each of three methyl radicals:

$$\left.\begin{matrix}(CH)''' \\ (CH)''' \\ (CH)''' \\ H_3\end{matrix}\right\} C_3$$

Kolbe considered this to be the true constitution of benzene: a trimethine–trimethane.[33]

In structural terms, the first of the compounds just described would be considered to be a bicyclic cyclopentane in which two methines are symmetrically bridged by three methylene groups. The second would be a tricyclic cyclohexane in which each of three methines forms three bonds to three other methines. Thus, Kolbe's proposed benzene formula (interpreted structurally) was a species of the genus cyclohexamethine. Remarkably, nearly all nineteenth-century benzene theories, and all even modestly successful ones—Kekulé's, Ladenburg's, Dewar's, and so on—were cyclohexamethine theories.[34] More particularly, Kolbe's structural benzene formula was equivalent to Adolf Claus' "diagonal" structure, proposed two years earlier (cyclohexamethine with each carbon atom bonded not only to its immediate neighbors but also to a partner directly across the ring). Claus' structure did not prove to be popular, presumably because it appeared to predict only two isomers of every disubstituted benzene.[35]

Claus had been one of Kolbe's Marburg students (1858–1861), but he had then gone to Göttingen for his Ph.D. (1862), had become a structuralist, and in 1868 was ausserordentlicher Professor in Freiburg/Breisgau. Claus was much influenced by Kolbe and was one of the few who championed his and Frankland's priority claims to valence theory. Nonetheless, he was never intimidated by his former teacher. In 1871 he published a note in Kolbe's journal in which he claimed priority for Kolbe's recent formulation of diglycolic acid. In doing so, he also asserted in a strong and unambiguous way that Kolbe's formulas were essentially identical to those of the structuralists.[36] Given these circumstances, it is curious that Claus never claimed priority for Kolbe's benzene theory.[37] As for Kolbe, he admitted in response to Claus that he had been unaware of the existence of Claus' 1866 monograph in which both the diglycolic acid and the benzene formulas appeared; having received a copy from its author, he commented that Claus' benzene formula is "less improbable" than Kekulé's.[38]

Since Kolbe disagreed with all structuralist interpretations of his constitutional notions, it is not surprising that he did not recognize Claus' structural formula as identical with his own. Indeed, there are some real distinctions between the two. Kolbe refused to speak or even think of bonds between atoms (only between radicals), and he asserted a chemical distinction for every notational difference in his formula. Consequently, he distinguished chemically between the three methine hydrogens (those within the parentheses) and the three methyl hydrogens (those at the bottom of his formula), and so he predicted isomeric relationships that were quite different from Kekulé's. Not only should there be two isomeric benzoic acids, two phenols, two toluenes, and two of every other monosubstituted benzene, but there should be no fewer than four of each disubstituted benzene when the two substituting atoms or groups are not identical—as in salicylic acid, for instance.

Kolbe thought that this view of benzene was far more beautiful, natural, and intrinsically probable than Kekulé's cyclohexatriene. The latter had been ingeniously "invented" (*erfunden*), but certainly not "discovered" (*gefunden*) in nature. It was highly artificial, he felt, exhibiting no analogies with inorganic chemistry or even with anything else in organic chemistry. No one had shown any reason to believe that there were both single and double carbon–carbon bonds in the molecule. There was no reason to believe that mere positional distinctions (ortho, meta, and para isomers) could produce the dramatic differences in chemical properties that were often observed. The formula was even misnamed: "It is called a benzene 'ring,' even though it is painted as a hexagon."[39]

Kolbe was sensitive to the question of the empirical adequacy of his own theory. In its favor, he cited the synthesis of benzene from three molecules of acetylene and of mesitylene from three molecules of acetone. He also pointed out that three of the six atoms of hydrogen seem to be more readily substitutable by other atoms or groups.[40] All of these arguments were also used by Kekulé and his allies—none could be said to clearly favor Kolbe's over Kekulé's theory. Even worse for Kolbe, known isomeric relationships strongly favored his opponent. For example, no second isomer of a monosubstituted benzene, nor an instance of four isomers of a disubstituted benzene, was yet known. Kolbe forthrightly recognized, even flaunted, these lacunae in his theory:

> In contrast to the chain theory, and to the way that one effortlessly finds explanations for everything by using the so-called bonding of atoms alone, I am not now able to explain *everything* with the above hypothesis. But it seems to me to be not a weakness, rather to a certain degree

> an advantage of a hypothesis, when it is capable of further expansion and so leaves something left over to explain. . . . Every attempt at explanation and every hypothesis which promises too much in this direction awakens initial mistrust. What matters is not that one explains everything, but *how* one interprets.[41]

Kolbe could have hoped for a more enthusiastic reception of his theory. He sent Hofmann and Liebig offprints and asked them for their reactions. His chief purpose, he told Hofmann, was to attack the structuralists and especially the benzene ring, "the show-horse of Kekulé's chain theory." He admitted to Liebig that he was thereby entering a "wasps' nest," but could no longer sit in silence. Apparently neither Hofmann nor Liebig satisfied Kolbe with an endorsement of his views.[42] The first published comment to appear was from Carl Glaser, perhaps Kekulé's most valued assistant in Ghent and Bonn during his critical "benzene years" of 1864–1869, who professed astonishment to find that Kolbe still relied on the Berzelian dicta that all organic compounds must be formulated to bear analogies to inorganic substances and be viewed as substitution products of the latter. Kolbe replied with deep scorn and contempt. He welcomed honest criticism from fledgling chemists, he said,

> But when a young man with only a couple of chemical papers to his name undertakes to criticize an older experienced chemist who was publishing chemical papers when that young man was scarcely born . . . that testifies to an overabundance of confidence and a deficiency of caution and modesty.[43]

It is no exaggeration to say that virtually no one accepted Kolbe's theory, not even most of his own students and junior colleagues. As Armstrong, always a staunch defender of Kolbe, recollected sixty years after the events, Kolbe "held most peculiar views as to [benzene's] structure, which we [students] often disputed with him." A contemporary letter from Armstrong to his father supports this account.[44] Few textbook and review authors of the day even mentioned the theory, and when they did, it was usually to demonstrate its empirical inadequacy. Kekulé used Kolbe's words against him when he wrote sarcastically (but not for print) that the theory must have been very good indeed to leave so much left over to explain.[45]

Nonetheless, Kolbe never relinquished his theory. In fact, armed with what seemed to him to be the definitive interpretation, he was renewed in his conviction that there must exist, after all, a second undiscovered isomer of benzoic acid and of phenol and a fourth undiscovered isomer of salicylic acid. Mindful that a single such discovery would weaken if not destroy Kekulé's brainchild, he spared no effort to produce one. This proved to be a hard row to hoe. He returned once more to salylic acid, his discredited isomer of benzoic acid, and to salicylic acid. In fiddling with these reactions, he encountered a new synthetic approach to the latter and thought for a time that he had promising results on the isomer issue.[46]

In 1873, he hit upon a crucial experiment to definitively confirm or refute Beilstein's argument for the nonexistence of salylic acid. Beilstein had shown that when salylic acid was distilled, pure benzoic acid was found in the receiver and a small residue of impurities remained in the distilling flask, from which he concluded that salylic was impure benzoic acid. Kolbe thought it possible that there had been an intramolecular rearrangement during the distillation from salylic to benzoic acid. Here was the possibility of a definitive test: recombine the distillate with the residue. If, as Kolbe thought, a rearrangement had occurred, then the recombined material should retain the properties of benzoic acid. If Beilstein were right that it was the impurities that produce the apparently distinct properties of "salylic acid," then those properties should reappear in the recombined material. Kolbe performed the experiment; the properties of salylic acid were regenerated, and even Kolbe regarded this as "unambiguous proof" that Beilstein had been right all along. Kolbe deserves credit for his honesty in forthrightly publishing this result—even if he did delay the publication for two years.[47]

Kolbe had lost a battle, but not the war; even if salylic acid was a phantom, that did not mean that benzoic acid had no isomer. He had put his nephew, namesake, godson, and student (and eventual assistant and biographer) Hermann Ost on the problem, and Ost was coming up with interesting results.[48] But Kolbe suffered another personal setback in 1880. Having persuaded Heinrich Vieweg to hire Ost as editor of the organic portion of the sixth (posthumous) edition of E. F. Gorup-Besanez's popular textbook, Kolbe was dismayed to find that Ost discussed only Kekulé's benzene theory and failed to treat his own! When Kolbe asked him to include a discussion of the latter, Ost flatly refused, on the grounds that it would damage the reception of the book. Kolbe described these events in a letter to Ost's mother, his sister Bertha; the letter was written in a furious rage and is nearly illegible. He did not mind when students disagreed with him, in fact he encouraged it, but he thought it only common courtesy for Hermann to at least mention his uncle's theory. He told Bertha that he was "disinheriting" Hermann, chemically speaking. Fortunately, a second letter written on the same day reveals that shortly after the first was posted, Hermann came and apologized, promising to include the material that Kolbe had requested.[49] Still, this must have been a bitter pill to swallow.

To the end of his life, Kolbe kept searching for the missing isomers. He never found them. However, the quest led him to one of his greatest scientific (and certainly his most lucrative technological) discoveries.

The Salicylic Acid Craze

Kolbe proposed his benzene theory on the very day his new laboratory opened, and for the first few years thereafter it was all he could do to keep up with the crush of students. The number of papers emanating from his institute mushroomed, but virtually all of the research was performed and much of it was also designed by his students and junior colleagues. Finally in mid-September 1873, refreshed by a long vacation on the Baltic Sea, he set to work with great enthusiasm and optimism to crack Kekulé's theory and prove his own. He opined to Volhard that

> . . . Kekulé's benzene theory (which is indeed very beautifully contrived and invented, but not discovered) with everything that follows from it, will sooner or later be disproven. In ten years it will be as little discussed as Gerhardt's type theory has been ignored these many years. This winter I intend with my students to carry out a major study of the derivatives of salicylic acid . . . I do *in fact* believe that [salylic] is not identical with benzoic acid, but rather is easily converted to the latter. Establishing this isomerism would overthrow Kekulé's benzene theory. Perhaps I shall also succeed in forming an isophenol.[50]

Fourteen years earlier, Kolbe and Lautemann had tried and failed to synthesize salicylic acid by reacting phenol with carbon dioxide and sodium hydroxide; they had then turned with success to the much more expensive sodium metal itself. Within two weeks of starting his new investigation, Kolbe discovered the conditions under which the sodium hydroxide route would work. An equimolar solution of phenol and sodium hydroxide was dried under heat to a fine powder (sodium phenolate) and protected from moisture. The powder was then placed in an iron retort on an oil bath, and a stream of dry carbon dioxide was introduced. The retort was heated over the course of twelve hours, starting at 100°C and gradually increasing to 250°. During the process, half of the original phenol distilled off, so that the stoichiometric yield was fifty percent; in practice, it was about forty percent, still an efficient process. The residue was then dissolved in water, acidified with mineral acid, and extracted and recrystallized to obtain the free salicylic acid.

By this new process, salicylic acid could be obtained far more quickly and easily, and in far larger batches, than by the earlier synthesis. Moreover, whereas salicylic acid prepared from willow bark or wintergreen oil—or by the 1859 synthesis—cost around 50 thalers per pound, the new process could produce pure product at around a thaler per pound. This cost saving opened up scientific as well as entrepreneurial vistas. For example, one barrier to a thorough exploration of the nature of salylic acid had always been the high cost of the starting material from which it is made, salicylic acid. With the new process on line, Kolbe could make almost any quantity desired and could mount a definitive assault on the compound. A second point of scientific interest was Kolbe's mystifying discovery that simply substituting caustic potash for soda results in the formation of paraoxybenzoic instead of salicylic acid.

A third point of interest was potentially the most important of all. Kolbe found that when the phenol that distills off during the reaction is reused with fresh caustic soda and carbon dioxide, the yield of product was much reduced, and even further reduced the third time around. He came to the ready conclusion, as he wrote Varrentrapp, that he was here dealing with a hitherto unknown isomer of phenol that is unreactive with carbon dioxide. Viewing commercial phenol as a mixture of two isomers would also explain why nitrating the substance yields two isomeric nitrophenols and why the yield of picric acid (trinitrophenol) is always so poor.[51] Certainly Kolbe knew very well that other circumstances, such as the constant sharp melting point of phenol, spoke against his conjecture. As hard as he tried—for the discovery would destroy Kekulé's theory at a stroke—he was unable to prepare two distinct phenols. He eventually gave up on this idea, after publishing only a preliminary note on the subject.[52] He was never able to unravel the mysteries of his reaction; even today the mechanism is not fully understood. As for his definitive assault on salylic acid, despite his disappointed private assessment in 1873 that "proof" of its *non*existence had been established, he continued to work on the putative compound for the rest of his life, but was never able to demonstrate its real existence.[53]

But even if the scientific content of his discovery did not have the explosive impact Kolbe had hoped, the technological interest did not escape him, even from the very beginning. He thought first of the possibilities of using his salicylic acid to make inexpensive artificial oil of wintergreen of hitherto unattainably high purity, and he also considered possible applications to dyes.

Within weeks of developing the new reaction, Kolbe entered into partnership with Dr. Friedrich von Heyden (1837–1926), a student of Rudolf Schmitt, who told Kolbe that he himself had neither the time nor the resources for the endeavor. Heyden set up an industrial lab in the carriage house of a family estate on the outskirts of Dresden. After several months of intense labor, he began to produce and sell product in the spring of 1874, at which time a decision was made to build a proper factory in Radebeul, seven miles northwest of Dresden. Heyden provided most of the development work and all of the capital, around 15,000 thalers in all. Production at Radebeul began early in 1875. The two entrepreneurs took out patents in eighteen German and ten additional foreign states, issued licenses to other producers, and began to build further capacity. By 1878 the factory employed twenty-seven workers and six managers and was producing 50,000 pounds annually.[54]

The Radebeul firm expanded and diversified repeatedly in the last quarter of the century, growing rapidly even in the midst of the depression of the late 1870s and 1880s. In 1899, then the largest chemical factory in Saxony, it was transformed into a joint stock company. The next year the company had five million marks of stock capital and 780 employees, and it was producing no fewer than fifty-six products. The company advertised at the Paris Exposition of 1900 that their salicylic acid had enjoyed "a true victory march through the entire civilized world." Be that as it may, the Radebeul factory had become the preeminent site for the Saxon chemical industry, and it made both Heyden and Kolbe a great deal of money. At the time of German reunification in 1989, the company was a state-owned pharmaceutical works.[55]

It is possible to come up with at least a rough estimate of how lucrative the undertaking was for Kolbe. He mentioned to Frankland his expectation that the product could be obtained for about 1 thaler per pound, and Kolbe and Heyden established their initial retail price at 5 thalers per pound.[56] Combining an estimate for Kolbe's share of per-pound profit with the known capacity figures, it would appear that from around 1878, Kolbe may have earned on the order of 10,000 thalers per year from salicylic acid manufacture. Such an amount would double the income from his already lucrative salary and student fees; it alone was several times his *total* income in Marburg. This sort of calculation can only be extremely approximate, but what it can do is make clear that in the last decade of his life Kolbe became a wealthy man.

This calculation does not include income from licensing the process. A few entrepreneurs who wanted to produce salicylic acid without paying fees to Kolbe and Heyden attempted to have the patent declared invalid because, as they claimed, it was based on an existing process, namely, the 1859 Kolbe–Lautemann synthesis. After Kolbe successfully argued in Prussian court that the 1873 reaction was distinct, the same problem arose in England and in Belgium. Kolbe enlisted Frankland's and Armstrong's aid for the English litigation.[57] Wurtz supported Kolbe in Belgium, even though Wurtz had been directly approached by the patent infringer. Kolbe was surprised and appreciative, but not so grateful as to cease his polemical attacks on Wurtz in his journal.[58]

Kolbe's reaction remained limited by the fifty percent stoichiometry until Rudolf Schmitt found how to avoid distilling off unreacted phenol (by lowering the temperature and increasing the reaction time). The yield doubled, and the cost of production nearly halved, which dramatically increased income from the process. Schmitt's paper was dated five months after Kolbe's death. Today the synthesis is known to organic chemists as the "Kolbe–Schmitt reaction," and it has proven to be a flexible and important synthetic route to aromatic carboxylic acids.[59]

What fueled the increasingly strong demand for salicylic acid was aggressive market-oriented research and promotion by Kolbe. He made for himself several useful products: a tooth powder containing salicylic acid and its methyl ester (oil of wintergreen); a mouthwash with the same ingredients that he proclaimed to be an instant cure for bad breath; a foot powder to prevent sweating and odor, which he thought should be mandatory for all soldiers; a bath salt formulation that he found extremely refreshing; and a tonic. In his reports in the *Journal für praktische Chemie*, he was careful to specify from whom salicylic acid could be purchased inexpensively. He found by experiments on himself and eight volunteers among his assistants and advanced Praktikanten that the human body could tolerate daily salicylic acid consumption of at least 1 to 1½ grams. In 1877 he began to drink as much as one liter of "salicylated water" per day, amounting to ingesting close to a gram of salicylic acid per day. He testified that this habit cured his digestive upset, blisters on his mouth and tongue, and his kidney stones and that it gave him a feeling of general vigor and well-being. This was not hucksterism; he believed it. "With this treatment I hope to grow very old," he wrote Heinrich Vieweg.[60] Curiously, considering his regular bouts with severe rheumatism, he never mentioned any anti-inflammatory or analgesic effect of the compound.

All of these uses were relatively minor compared to the two largest potential markets for salicylic acid: as an antiseptic and a food preservative. The formula for salicylic acid exhibits a family relationship both to the strong disinfectant phenol (a.k.a. Dr. Lister's carbolic acid) and to benzoic acid, which had already had various applications in pharmaceuticals and food technology. This gave Kolbe the idea to test salicylic acid for similar properties. Examples of pure chemist-entrepreneurs such as Liebig, who had made large profits from his extract of beef, and Hofmann, who had a number of lucrative dye patents, must also have been in his mind.

Asking his assistant Ernst von Meyer to collaborate with him, Kolbe began these investigations in March 1874. They found that very small amounts of the substance would halt, retard, or prevent fermentations or spoilage in a variety of materials, including amygdalin, urine, milk, butter, beer, wine, eggs (treated by soaking in salicylated water), bread, fruits, and meats. Kolbe's optimism soared, for the need for food preservation has been constant throughout history, and salicylic acid appeared to be far more innocuous and bland and just as efficacious as such obviously toxic and unpalatable additives as phenol, boric acid, and formaldehyde—all of which found application in nineteenth-century food technology.

Kolbe's initial optimism was tempered by some of his subsequent findings. He persuaded a sea captain to take salicylated water on a voyage to the tropics, but the crew found that this water went bad just as quickly as untreated water (Kolbe concluded that the salicylic acid became bound to the wood of the casks). Similarly, despite his promising initial experiments and much subsequent effort, by 1882 Kolbe reluctantly gave up on developing a commercially attractive process to preserve fresh meat using salicylic or carbonic acid. The preservatives did retard putrefaction remarkably well, but after only a few days the meat acquired an unpleasant aroma and taste (not from spoilage, Kolbe carefully noted, but problematical nonetheless) that survived cooking.[61] He had better luck with bread, butter, beer, and wine, and for these products salicylic acid successfully entered the food industry. Even so, some governments were unconvinced of the safety of the additive; for example, although approved in the Imperial German food law, it was refused entry into Bavarian foods. Hermann Ost regarded this as a foolish prejudice and predicted in 1885 that reason would soon prevail.[62] Salicylic acid as a food preservative was most successful in Germany; in most other countries, however, benzoic acid was preferred for its gentler action on the stomach.

Application as an antiseptic carried a different set of standards and problems. Lister's innovations of the 1860s were more rapidly adopted in Germany than in Britain; Carl Thiersch, professor of surgery at Leipzig, was one of the first prominent surgeons after Lister himself to adopt antiseptic techniques using phenol. In the late 1860s, shortly after Thiersch's arrival, the Surgical Institute was built just a few steps from Kolbe's institute, and Kolbe and Thiersch maintained cordial relations. In April 1874, Thiersch began a series of clinical trials with salicylic acid, at Kolbe's behest. Bandages soaked in a salicylate solution appeared to maintain antisepsis as well as phenol, with far less irritation to the skin or the nose, and Thiersch became an enthusiastic proponent of the new application.[63] The Leipzig professor of gynecology, C. S. F. Credé, adopted the new antiseptic for his clinic as well and reported excellent results. Kolbe publicly recommended salicylated water as a disinfecting agent for hospitals and sickrooms (its disadvantages compared to phenol were lower solubility as well as higher price, and so it found no market in this application). Thiersch and others began to explore possible pharmacological application as an internal bactericide. Kolbe also contacted Max von Pettenkofer in Munich about the same time, but without result.[64]

On 31 July 1875, Kolbe's eighteen-year-old daughter (and Ernst von Meyer's fiancée) Johanna became seriously ill with diphtheria. Since the family doctor (Thiersch) was not immediately available, Kolbe began treating her himself. He had her take 0.3 grams of salicylic acid every one and a half hours, gave her salicylated water with which to gargle, and sprayed it topically on her throat. When Thiersch arrived, he let Kolbe's cure proceed, and Johanna quickly recovered.[65] This experience convinced Kolbe that salicylic acid would become an invaluable internal medicine against a number of epidemic diseases: in addition to diphtheria, it was tested against cholera, typhus, and a number of animal diseases. However, the results were negative, and Kolbe's hopes were not fulfilled. Moreover, after a few years of enthusiasm, most Listerian surgeons decided that antisepsis using salicylic acid was not as reliable as that using phenol, and by the early 1880s, most were returning to Lister's original material.[66]

In the short term, salicylic acid proved to be an important antiseptic, food preservative, and antipyretic and analgesic. Even more important for the long term, it became an indispensable intermediate in the manufacture of a wide variety of perfumes, flavors, pharmaceuticals, and dyes. It was a principal element in the rise of the German fine chemical industry during the last quarter of the century. Thirteen years after Kolbe's death, a Bayer company chemist named Felix Hoffmann acetylated Kolbe's compound, and the most widely used (and least expensive) pharmaceutical in history was born: aspirin.[67]

13

Life and Work in Leipzig

Kolbe's "Merry Celestials"

The phrase in the section head ("hellgeborene heitere Joviskinder") derives from Otto Erdmann and designates in an avuncular and humorous way—but not favorably—the modernist followers of Kekulé in the chemical community.[1] Like Kolbe, Erdmann and many other older chemists were bewildered by the explosion of new ideas and distrusted their power to specify details of molecular architecture. It is a bitter irony that many of the best chemists who passed through Kolbe's lab were (or later became) structural organic chemists: in a semichronological list, we can name Griess, Claus, Crum Brown, Volhard, Graebe, Zaitsev, Menshutkin, Markovnikov, Armstrong, Meyer, Ost, Curtius, and Beckmann. Even Edward Frankland, one of Kolbe's oldest and best friends and whom he always considered in some sense a protégé, was distinctly structuralist by the time Kolbe arrived in Leipzig. All of these men had the highest regard for Kolbe, and he reciprocated that feeling. Several of them tried to persuade Kolbe that many structuralist theories were not very different from his ideas and that he ought to pay more attention to them. Some also remonstrated with him about his retention of the older equivalent weights long after nearly all of his colleagues had switched to atomic weights.

A discussion of Kolbe's views on chemical constitution and a comparison with structuralist ideas can be found in chapters 8 and 9. Kolbe's essential peculiarities were his absolute denial of direct carbon–carbon bonds—the basis of what he called Kekulé's "chain theory"—and his conviction that there was a strict hierarchy of constituent radicals in every compound, with only one being the molecule's "fundamental radical" (*Grundradikal* or *Stammradikal*). We have seen how Kolbe was able during his later Marburg years to more than hold his own in his encounters with the structuralists on a variety of fronts. However, as structural chemistry blossomed and flourished during the late 1860s, Kolbe began to lose pace with the field—as we have already seen in the last chapter as regards aromatic chemistry.

One example of this trend involves the question whether the four valences of carbon might be chemically distinct. Many structuralists, including Butlerov, Erlenmeyer, Crum Brown, and even Kekulé, thought for a few years that this was probable, for there were several cases of apparent isomerisms that could not otherwise be explained. The most notorious such case was the isomerism of "methyl gas" (CH_3–CH_3) and "ethyl hydride" (CH_3CH_2–H), the first produced by electrolysis and the second by reduction of ethyl compounds, where the two indicated bonds were presumed to engage carbon valences of different chemical value. It was Frankland who discovered and defended this isomerism most vigorously, for the two compounds, although very similar themselves, seemed to result in quite distinct chloro derivatives.

In 1864 Carl Schorlemmer, an expatriate German working at Owens College Manchester, demonstrated the probable identity of methyl gas and ethyl hydride and argued that they formed an identical series of chloro derivatives. Schorlemmer's paper made a great impression, and little was said thereafter by structural chemists about differences between carbon valences. Frankland himself was largely convinced.[2] For two years he had been writing formulas using the reformed weights, and he was then on the verge of adopting Crum Brown's graphical formulas and becoming a structural chemist in the full sense. He wrote to Kolbe, asking him what he thought about Schorlemmer's work.

Kolbe was profoundly *un*convinced. He would only grant that Schorlemmer had demonstrated that both hydrocarbons yield the same ethyl chloride, still maintaining that "ethyl hydride" is a derivative of methane:

$$\left.\begin{matrix} C_2H_3 \\ H \\ H \\ H \end{matrix}\right\} C_2$$

whereas "methyl" consists of two methyl radicals

$$\left.\begin{matrix} H \\ H \\ H \end{matrix}\right\} C_2, \left.\begin{matrix} H \\ H \\ H \end{matrix}\right\} C_2$$

This was was a departure for Kolbe, for it is the earliest indication that he had finally accepted the dimeric character of hydrocarbon "radicals" (we have seen that he conceded the importance of Wurtz' evidence for this thesis as early as 1855). To maintain his distinction between the hydrocarbons while simultaneously conceding the identity of their chlorinated derivative, he suggested to Frankland that there was a molecular rearrangement of "methyl" to the "ethyl hydride" constitution during the course of the reaction.[3]

To put it simply, from the time he first accepted the tetravalence of carbon (in 1858) to the end of his life, Kolbe assumed that the four radicals around a carbon atom were held with different degrees of affinity. This was also true for other atoms, he thought; the oxygens of sulfuric acid, the hydrogens of ammonia, and so on, were all chemically distinguishable.[4] After Schorlemmer's work, there was little evidence for this idea that most chemists found compelling. Kolbe, however, kept coming back to a single argument again and again; the fact, for instance, that methane can be monochlorinated proved to him that one of the hydrogens of methane is held less tightly than the others. Similarly, the existence of monoderivatives of benzene demonstrated conclusively, he thought, that the six hydrogens are chemically distinguishable, hence Kekulé's theory cannot be right.[5] The empirical failure of the assumption of chemically distinguishable valences, namely, the fact that it predicts numerous isomers none of which were ever found, was excused by Kolbe again and again by one of two gambits: molecular rearrangements or insufficient empirical experience. The structuralists' explanation for monoderivativization, that one hydrogen is *randomly* selected and the process stops after the first substitution, meant nothing to Kolbe.

These issues were brought to a head for Kolbe in 1866 and 1867 as a result of work on the base-catalyzed self-condensation of ethyl acetate performed independently by Anton Geuther at Jena and by Frankland and Duppa in London. Frankland's work arose out of his successful alkylations of esters using either zinc alkyls or sodium plus alkyl iodides—the final payoff from twelve years of trying to "ascend the homologous series of organic bodies" (see chap. 8). The diethyl substitution product in oxalic ester acquired the name "isoleucic acid,"

$Et_2C(OH)CO_2H$, because it was an isomer of the hydroxyacid derived from leucine; reduction of the hydroxyl group led to a branched-chain isomer of the straight-chain caproic acid, $C_5H_{11}CO_2H$. Substituting one and two ethyl groups into acetic ester yielded butyric and isocaproic acids directly.[6] In a remarkable paper published in 1866, Frankland and Duppa announced the synthesis by a similar route of acetoacetic ester, as well as of ethyl- and diethylacetoacetic acid, $(CH_3CO)Et_2CCO_2H$. They found that these compounds decarboxylate in base to produce acetone and ethyl- and diethylacetone, respectively.[7]

It was during the course of this research program (in 1863–1866) that Frankland came over to structural ideas and to Crum Brown's graphical formulas. The details of his intellectual odyssey are not known, but one may presume that he found the modern ideas and formulations simpler and heuristically more valuable than the older ones. Frankland is a good model to compare with Kolbe because both men started from a similar set of views. No one understood Kolbe and his ideas better than Frankland, and no one was better situated than he to help Kolbe to follow along the same path that he had trod. He sent Kolbe an offprint of the culminating paper of this series.

Kolbe studied the paper with care and responded privately at length.[8] That Kolbe had truly invested sincere effort in understanding the paper is evident, not only from the fact that he was able to translate all of Frankland's formulas accurately into his own notational style but also because he suggested a reaction mechanism for acetoacetic ester synthesis that Frankland immediately recognized as superior to his own published conjecture.[9] But it is just as clear that Kolbe failed to understand the structure-theoretical principles upon which Frankland was reasoning. He objected to many of Frankland's condensed formulas because they seemed to posit groups of carbon atoms whose total combining capacity appeared to be short of the number required by tetravalence (for example, a group of two carbon atoms [C = 12] with a valence of six rather than eight, or four carbons with a valence of ten rather than sixteen). The structuralist interpretation of these formulas required subtracting from the sum of all the valences the number necessary to form carbon–carbon bonds, as Kekulé had discussed seven years earlier in the pages of his *Lehrbuch der organischen Chemie*. Kekulé's treatment was surely the clearest and most influential early exposition of the principles of structure theory, but Kolbe never read it, or at least not until many years had passed.[10]

Kolbe also disagreed with Frankland's formulation of the decarboxylated compounds as derivatives of acetone. Rather, Kolbe averred, they were derivatives of methane, for example,

$$C_2\begin{cases} H \\ C_4H_3O_2 \\ C_4H_5 \\ C_4H_5 \end{cases}$$

One could conceive of an ethylated or diethylated acetone, but that is a different compound, he argued, because it is derived from a different fundamental radical (*Stammradikal*). Kolbe's systematic and elementary way of attempting gently to disabuse Frankland of his putative errors indicates that Kolbe had simply failed to follow Frankland's and Duppa's structure-theoretical exposition. He was also strongly opposed to the use of graphical formulas:

> Frankly, I believe that all of these graphic representations are inappropriate for the times and are even dangerous, because they leave too much scope for the imagination, as for example happened with Kekulé: his imagination bolted with his understanding long ago. It is impossible, and will ever remain so, to arrive at a notion of the spatial arrangement of atoms. We must therefore also take care to avoid drawing a picture [of that putative arrangement] for ourselves, just as the Bible warns us from making a visual depiction of the Godhead.[11]

Kolbe laid all this out (without the metaphors) in a publication a year later. He wrote, "I expect to hear the objection which is repeatedly made to me orally," that his formulas are identical with Frankland's and Duppa's. "To be sure, they may appear so on superficial examination," but he and Frankland had chosen different fundamental radicals, and so it was clear that they were referring to isomeric and not identical compounds—just like, for example, ethyl acetate versus methyl propionate.[12] For Kolbe, choice of the base radical appears to have been crucial because of the presumed differences in carbon valences, but he did not make this assumption explicit.

Kolbe's friends and old students commiserated with each other. Frankland wrote to Crum Brown about this correspondence: "I am just now endeavoring to get Kolbe to express certain of his fundamental formulae graphically. We should then understand each other better." Crum Brown replied,

> I quite agree with what you say of Kolbe. I worked with him for a summer session at Marburg. I was always able to explain any theoretical views of mine to him by first translating them (sometimes with a good deal of trouble) into his language, but I am quite sure he would fail to recognize his own ideas if translated into our language. For instance as to his HO...[C_2O_2]O he certainly regards the HO as water, but he recog-

nizes the fact that there must be as many HO's at the beginning of the formula as there are O's outside the bracket at the end of it, & he points out (for the first time I believe) that [C_2O_2] or CO is a group capable of combining with two equivalents & that in acetic acid it is combined with one equivalent (or atom) of C_2H_3 methyl & one equivalent (or as we sh[d] say 1/2 atom) of oxygen. To this he adds HO & as far as I can see the only difference between his view & ours is that he does not *explain* why. We say it is because the 1/2 atom of Oxygen in the HO is the other half of that which is combined with the CO. His obstinate opposition to O = 16 prevents his accepting such an explanation.[13]

Kolbe's article forced Frankland and Duppa to respond publicly. They praised Kolbe's proposed mechanism, agreeing that it was more probable than their suggestion, but they fulfilled Kolbe's prediction by arguing in great detail how the formulas they employed for the new compounds, both condensed and graphical, were entirely equivalent to his, providing that "Kolbe does not invest in his symbols some meaning that we cannot understand." They noted that the dividing up of a constitutional formula into radicals was a purely arbitrary operation, a matter of expository convenience and not ontology. Consequently, they were uncertain of the exact significance that Kolbe attached to his apparently important word *Stammradikal.* But even if there were a real difference in the compounds due to a difference in the "fundamental radical," they argued, this is certainly not analogous to the ethyl acetate versus methyl propionate case: the latter compounds have differences in the bonding order of the atoms in the molecule, whereas the Frankland–Duppa and the Kolbe formulas indicate the same atoms combined in apparently the same way.[14]

Not only Crum Brown, but also Graebe, Volhard, and other senior workers who passed through Kolbe's Marburg laboratory had to do the same sort of routine formula translations as Frankland did in order to converse (and argue) with the master. Referring to the same summer semester as Crum Brown had (1862), Graebe later wrote,

> The radical theory as he developed it was at that time an excellent point of departure from which to understand structure theory, which was just then developing. It was only necessary to rewrite his formulas (at that time still being written in equivalents) into the new atomic weights, in order easily to understand the structural formulas.[15]

Volhard, too, did this sort of routine formula translation and became the first from Kolbe's lab actually to publish in atomic weights.[16] The same was true, of course, in Leipzig. Armstrong arrived there from Frankland's laboratory in 1867, and several Butlerov students passed in and out during the early Leipzig years, especially A. M. Zaitsev and V. V. Markovnikov; all had absorbed structural ideas from their first mentors. Kolbe stated repeatedly, and all evidence from his students substantiates the claim, that he encouraged his students to follow their own ideas, to disagree and argue with him. A surprising number of publications by his students in Leipzig contain views with which he was known to disagree. At times this went so far that Kolbe attached long footnotes to student papers outlining his disagreement with them.[17]

Markovnikov's memoirs contain the most revealing anecdotes along these lines. Soon after his arrival in Leipzig in the fall of 1866, he wrote Butlerov (for whom he had served as assistant and lecturer in Kazan) that Kolbe was far from a martinet, that he positively enjoyed arguing with students, and that he (Markovnikov) had already succeeded in locking horns with him more than once. Moreover, Kolbe actually used Markovnikov to help him interpret modern formulas.[18] As it happened, Markovnikov's research project was related to the Frankland–Duppa material, for Markovnikov was attempting to show that Frankland and Duppa's product from the dimethylation of oxalic ester was identical to Staedeler's "acetonic acid" and that both had the structure of what could be more rationally named hydroxyisobutyric acid. This project was successful.[19]

Markovnikov related many years later that as he was writing up these results and preparing to leave Leipzig for Kazan (this must have been in the summer or early fall of 1867), he had one more tussle with his mentor. Kolbe had claimed that one of his formulas was wrong, but Markovnikov was emboldened by one of Kolbe's assistants who privately sided with him. He was invited into Kolbe's private office to discuss the matter.

> The heart of the dispute was the old oxygen theory. "You don't understand me because you are not used to my formulas," said Kolbe; "I will express your thoughts in your own formulas." "Aha," I thought, "now, Herr Professor, you are mine." . . . He began to write, stopped halfway through the formula, thought a minute, then set the pencil down. "Ja, Sie haben Recht." Then he completed the formula and said once more, "Yes, yes, this is true; you are right," and somewhat confusedly began to explain something. I quickly retired, to spare the self-esteem of an honored teacher. A year later, I received from him a pamphlet on another of our disputed questions. In it he developed his theoretical ideas in detail, but now he wrote the weight of oxygen in the new way.[20]

It would be nice to know more details. What was the disputed point, and what was the formula? There is no particular warrant for the accuracy of this story, but the timing seems to fit. It was a little more than a year after Markovnikov's departure that Kolbe sent colleagues offprints of his dissertation on hydrocarbons, which was the first time he published in the new atomic weights. On the other hand, Markovnikov was only one of several who were putting pressure on Kolbe at the time; Frankland's article in rebuttal to Kolbe's public response may have had more influence on him, at least regarding the atomic weight issue if not the more general formula question.[21]

The actual point of conversion for Kolbe from the older to the newer weights appears to have been the summer of 1868, just when Carl Graebe was beginning to pass the various stages in the habilitation procedure. Graebe had studied with Kolbe for a semester in Marburg, but had been most strongly influenced by the structuralist school, especially by Baeyer; his Habilitationsschrift was entirely based on the Kekulé benzene theory. The phrase "*hellgeborene heitere Joviskinder*" (quoted at the beginning of this chapter) was directed sarcastically to Graebe, but its author (Erdmann) and Kolbe both enthusiastically approved the habilitation. The fact that Graebe chose Kolbe's lab for habilitation, certainly with full knowledge of Kolbe's contrary theories, indicates once more that Kolbe was trusted implicitly (by those who knew him) to respect opposing points of view.[22]

In the event, Graebe was Privatdozent in Leipzig for only one semester, winter 1869/70. One of his auditors was Kolbe's student Ernst von Meyer, who was then in his fourth semester of study. In his memoirs, Meyer stated that he derived great profit from Graebe's instruction, absorbing and learning to appreciate the new structural chemistry. From then on, he added, he could read with full understanding articles in both the older and the newer styles and had no difficulty in translating one into the other. Although he agreed with Kolbe's criticism of the sloppy and conjectural character of much modern chemistry, Meyer respected and valued structure theory, a point that he stressed in correspondence and that emerges clearly from his historical writing.[23]

The Leipzig Research Group

As we did for the Marburg period, let us begin our analysis with some numerical measures of research productivity using the standard unit of productivity, the published paper. This time we take data across Kolbe's entire career, as shown in table 3.[24]

Some patterns are apparent in this table. In the early Marburg period—before acquisition of the "carbonic acid theory"—Kolbe's research productivity was low, as was that of his students. The high point of Kolbe's career came during the 1860s, when he was publishing

Table 3. Kolbe Research Group Publications

Years	*Kolbe Solo*	*Kolbe Papers per Year*	*Kolbe with Co-author*	*Student Papers*	*Total Papers per Year*
1841–1850:	9	1	2	—	1
1851–1858:	4	0.5	1	10	2
1859–1865:	22	3	9	37	10
1865–1869:	15	3	2	38	11
1870–1874:	34	7	2	82	24
1875–1879:	25	5	2	58	17
1880–1884:	47	9	2	62	22
Totals:	156		20	287	

around three good papers per year and his group authored about ten per year. This overall level of productivity roughly doubled after the great surge in enrollment in the early 1870s, so that he began to publish half a dozen or more solo papers per year and his group about twenty per year. However, this late period was not as productive as it might first appear. For one thing, he had far more students with which to work—five to ten times as many as in Marburg—so that in this light a doubling of total yield appears modest. Moreover, the great majority of his own publications after 1875 were either short notes with no experimental results or polemical critiques.

How does this overall productivity compare to his contemporaries and near contemporaries? His numbers were modest compared to those of Liebig or Wöhler a half generation earlier, to Hofmann in his generation, or to Baeyer a half generation later, each of whom had several hundred personal and collaborative papers; Baeyer's research group at Munich alone is said to have published over 1600 papers.[25] However, Kekulé himself, certainly the most theoretically important chemist during the third quarter of the century, published only 131 papers, all but 18 of these during the years 1850–1873.[26] As has often been rightly remarked, numbers of publications do not necessarily correlate to quality or significance.

So let us now attempt to characterize the Leipzig school in terms of subject matter and its significance. Two areas that we have already explored are the synthesis of natural products (chap. 10) and experimental investigations connected with theoretical issues involving romatic compounds (chap. 12), both of which were also represented in Marburg. Another continuation from the Marburg period was the prediction and production of novel aliphatic isomers, especially secondary and tertiary alcohols and acids. One important field largely new to Leipzig and forming the topic of dozens of his students' dissertation projects was organosulfur compounds. Included in this group were studies on aliphatic and aromatic sulfides, sulfates, sulfonic acids and esters, sulfinic acids and esters, and other organic sulfonyl and sulfoxyl compounds. In the process, scores of new substances were discovered, including the first members of entirely new classes of compounds, such as the dialkyl sulfones, dialkyl sulfoxides, and nitroalkyls. One of the leading ideas motivating this research was Kolbe's conviction, contrary to Kekulé, that the valence of an element was not constant but rather could vary. Indeed, the novel organosulfur compounds named include substances containing divalent, tetravalent, and hexavalent sulfur. The last field that deserves mention is a long series of investigations after 1873 on the preservative and antiseptic properties of salicylic acid, as discussed in the preceding chapter.

In terms of quality and significance, nearly all of this large body of research was quite competent and most of it was scientifically important. However, little of the work done after around 1870 opened up new theoretically significant opportunities. The most exciting topics elsewhere in Germany—positional isomerism in the aromatic field, synthetic methods, structure determinations, and after 1876, stereochemistry—were all outgrowths of the classical structure theory that Kolbe so abhorred. Simply stated, in the 1840s and 1850s, Kolbe had been a principal founder of the investigation of "constitutions" of organic compounds; in the 1860s, he was well able to keep pace with the leaders of the field and to make substantial contributions; but after 1870, the contemporary significance of his work declined dramatically.

Besides research, the other significant activity of an academic institute is education, and so we must now turn to Kolbe's students and junior colleagues in Leipzig. Unfortunately, class lists have not survived, and so a complete analysis by name or even by precise statistics is not possible. From unsystematic indications of the course of overall enrollments, we can presume that Kolbe must have taught something like 1500–2000 Praktikum students during his nineteen years in Leipzig. Many of these were not chemistry majors and had no further contact with the science after their one brief exposure; even some of the chemistry and pharmacy majors doubtless had virtually no impact on the institute. How many people, then, can be considered members of Kolbe's "school"? The widest definition would include all of the following groups: (1) all who published at least one paper from the lab, regardless if they took a degree; (2) all who received a Ph.D. under Kolbe (for which records do exist), regardless if they published; (3) all

Table 4. Kolbe Research Group Size

	Marburg (1851–1865)	*Leipzig (1865–1884)*
Total Praktikanten	~240	~1600
Total research group	28	137
Average number of Praktikanten	~16	~125
Average number of advanced Praktikanten	~4	~40
Total recorded Ph.D.s	18	68

assistants, Privatdozenten, and postdoctoral guests, regardless if they published; and (4) all who were mentioned as having assisted in any of the 369 papers in table 3 that date from the Leipzig years. The Leipzig research group in this widest sense consists of 137 identifiable people, just half of whom (68) completed Ph.D. dissertations under Kolbe and all but a dozen of whom appeared as paper authors. These numbers suggest that an average of something like 30 of them were present at any given time, so that his group of advanced Praktikanten may have averaged around 40. Collating these numbers with those from Marburg gives the summary shown in table 4.

Whereas Kolbe's Marburg group was, until his last few years there, almost exclusively composed of Kurhessian students, the group in Leipzig was far more cosmopolitan. About one-fourth of the group was foreign, and of the Germans only about one-third came from Saxony. In his entire career, Kolbe could boast of having taught twenty-one Russian students, twenty Brits, ten Americans, seven Swiss, three Austrians, and a smattering from five other countries (but no French, Italian, or Spanish students). Of these, five Russian and seven British students took their Ph.D. degrees with Kolbe.[27]

Indeed, it was with his foreign students that Kolbe had many of his greatest educational successes. Edward Frankland (as sort of an unofficial Kolbean) and Henry Armstrong (1848–1937) were two of the most influential science educators of their day in England, and they were considered the deans of late nineteenth- and early twentieth-century British chemistry, respectively. Zaitsev (1841–1910), Menshutkin (1842–1907), and Markovnikov (1838–1904), all world-class chemists, made a great impact in Kazan, St. Petersburg, and Moscow during and after the life of their teacher Butlerov. After receiving his Leipzig Ph.D., the Russian Constantin Fahlberg (1850–1910) bounced around several positions before spending a year as an assistant to Ira Remsen at Johns Hopkins University in Baltimore, where he (systematically) discovered saccharin and (serendipitously) its sweetening properties. In 1886, he founded a factory near Magdeburg to manufacture the substance, with great success and profit.

As for Americans, we have noted that future Harvard president Charles Eliot (1834–1926) spent a semester with Kolbe in Marburg. H. P. Armsby (1853–1921), who became a noted agricultural chemist at the University of Wisconsin and at Pennsylvania State College, was a Leipzig Praktikant in 1875–1876, as was the New York private analyst Gideon Moore (1842–1895). One last American Praktikant worth mentioning is Sidney A. Norton (1835–1918), who for twenty years was head (and only member) of the chemistry department of Ohio State University.[28] Perhaps the most curious foreign career path was traveled by the Bavarian Oscar Loew (1844–1941), who had stints at the City College of New York, the United States Geological Survey, the Department of Agriculture, and Tokyo University before returning to his homeland at the age of seventy as honorary professor at Berlin. He was the last significant Kolbe student to die.

The highest prestige that a German professor could wish for a student was that he become an ordentlicher professor himself at another German university. Here Kolbe had little success. Only three of his students ever gained a university Ordinarius: Ernst Beckmann (1853–1923) at Erlangen, Leipzig, and Berlin; Theodor Curtius (1857–1928) at Kiel, Bonn, and Heidelberg; and Ernst Schmidt (1845–1921) at Marburg. Schmidt, the only undistinguished chemist of the three, became the first to achieve this rank—ironically, at Kolbe's former university and in the very year of Kolbe's death (this was as Zwenger's successor as director of the pharmaceutical institute).[29] This poor record of spawning new full professors contrasts with that of Liebig a half generation earlier, and with Baeyer a half generation later, each of whom taught close to thirty future German university Ordinarien.[30]

This last enumeration counts only those who actually received a Ph.D. under Kolbe, not his entire research group. Moreover, there were other educational institutions besides universities, other ranks besides the Ordinarius, and other countries besides Germany. Of the total group of 137 students, 31 (twenty-three percent) pursued academic careers. Among the non-Ph.D. students in Leipzig were such future academics as Carl Graebe, Edmund Drechsel, Hermann Credner, and Gustav Hüfner, as well as many of the foreigners mentioned earlier. Among those who never made it to Ordinarius but nevertheless built reputations in chemistry were Conrad Laar (1853–1929), who coined the term *tautomerism*; Rudolf Leuckart (1854–1889), who developed an eponymous synthetic reaction that yields complex aliphatic amines; and Friedrich Fittica (1850–1912) at Marburg.

The last group of academics comprises those who worked at the trade schools and technische Hochschulen, which were gradually raising their status during the latter part of the century. Kolbe's Marburg student Rudolf Schmitt had an excellent career at the Dresden Polytechnic, and when ill health forced him to retire, his successor was Ernst von Meyer (1847–1916). Hermann Ost (1852–1931) was at the Hanover Technische Hochschule for nearly forty years, and in all, about a dozen of the 137 Leipzig Kolbeans spent major portions of their careers at technical schools.

It has been noted that only twenty-three percent of Kolbe's Leipzig group—or to put the matter more starkly, only two percent of his Praktikanten—had a subsequent academic career of any sort. The majority of Kolbe's Praktikanten were instead future physicians, pharmacists, schoolteachers, businessmen, civil servants, and so on, no doubt even including a few law and theology students.[31] The future academics were in the minority even among the future professional chemists, for in the last third of the century, a university education was becoming common, even expected, preparation for a career in the chemical industry. Unfortunately, biographical sources for industrial employees are poor in comparison to reference works for academics, and so it is difficult to identify Kolbe's budding industrialists even from an accurate list of names.

However, some identifications, at least, are clear. Kolbe's Marburg student Wilhelm Kalle founded what would become an extremely successful dye firm two years after obtaining his Ph.D., in 1863. Ludwig Mond, the later ammonia–soda magnate, also studied in Marburg. Griess, Graebe, and Gerland were all Marburg students who worked in the chemical industry. In Leipzig, there was a larger (and increasing) percentage of technical students. The case of Fahlberg has already been mentioned, but this example is only representative. Some two dozen of the Leipzig students can be shown to have entered industry, but the true number is certainly very much higher. A rough estimation can be made that perhaps twenty-five percent (around 400) of his Leipzig Praktikanten became industrial chemists, contrasting with perhaps nine percent (about 20–25) of his Marburg students.[32]

Meyer, Ost, and other junior colleagues contributed far more to the liveliness and success of Kolbe's institute than has been appreciated. Both Meyer and Ost were assistants and Privatdozenten in the institute from the early 1870s until Kolbe's death in 1884, Meyer being promoted to Extraordinarius in 1878. They had capable colleagues in fellow long-term assistants Anton Weddige (1843–ca. 1904) and Ernst Carstanjen (1836–1884), both of whom also became ausserordentlicher professors at Leipzig, but neither of whom progressed further along the academic ladder.

Although all doctoral degrees in chemistry were officially granted with the Kolbe imprimatur, it appears that Ost and especially Meyer began increasingly to take over day-to-day direction of the Doktoranden, especially after Meyer's promotion to Extraordinarius. Judging by acknowledgments in doctoral dissertations and by Meyer's later statements, it seems that the majority of the doctoral students after 1877 got most of their advice, and even many of their original topics, from the junior members of the institute. Meyer mentioned that his Ph.D. topic, selected in 1871, was of his own devising. However, the trend was certainly stronger after the mid-1870s, a period that corresponds to a serious decline in Kolbe's health. The fact that the two most eminent Kolbe students, Ernst Beckmann and Theodor Curtius, were in this late group, speaks to the quality of Meyer's mentoring.[33]

During the first Leipzig decade, of course, Kolbe's role was stronger and more direct. There are many indications (from Kolbe's acknowledgments of specific students' assistance in some of his papers, from students' acknowledgments to Kolbe in their papers, and from retrospective accounts) that Kolbe often used his students as his "hands" to pursue his own concerns; examples include much of the organosulfur research and his search for aromatic isomers. Indeed, Kolbe's Leipzig period might be viewed as exhibiting all the prerequisties for an ideal research school. Taking Jack Morrell's well-known criteria for such an entity,[34] we can affirm that Kolbe was a man of eminence and personal charm if not charisma; that his students formed a cohesive group with excellent esprit; that they were allowed to publish under their own names, and after 1870, in Kolbe's own proprietary journal; that Kolbe had a distinctive theoretical research program to be elaborated, with a set of dependable and predictable techniques; and finally, that he had plenty of manpower, more than adequate physical facilities, and generous institutional and financial support. Why, then, was the Leipzig school not more successful than it was?

In the course of a compelling endorsement of research schools as a unit of historical analysis, Gerald Geison notes some inherent difficulties with this kind of approach, to which additional ones may be added.[35] For instance, it is by no means clear that research groups are always the cohesive and distinctive units they are sometimes assumed to be. Indications of transmission of ideas from teacher to student may be inferred, but dependable statistical data can rarely if ever be constructed. Exactly which themes were suggested by Kolbe, which by his assistants, and which by the students themselves? A formal acknowledgment by a student to the group leader does not really tell us very much. Even for those cases where we can determine that the project started as a definite assignment, scientific research is such that few projects lead in a straight line from conception to conclusion, and the twists and turns along the way are often the real points of interest; who was doing the twisting at each point?

If such are typical of the historian's analytical problems for the Doktoranden, the difficulties are even worse for guest Ph.D. workers, assistants, Privatdozenten, and other junior colleagues. The German university system was fluid; junior colleagues frequently traveled from university to university, and many had not been Kolbe's students originally. Nor was it just the junior colleagues, for the Doktoranden themselves were also often a peripatetic lot. Moreover, organic chemistry, at least, had developed a national scientific culture that was highly similar at the various German universities. Enough has been said in this chapter (and in chaps. 5, 11, and 12) to demonstrate that structural chemistry was alive and well in Kolbe's institute, even while Kolbe was turning apoplectic over it. Themes of students' projects were frquently irrelevant (or even contrary) to Kolbe's pet concerns, and this is true even in the early Leipzig period. As extreme cases of independence, some notable disagreements between Kolbe and his students were published under the auspices of the Leipzig lab.[36]

Kolbe's group research was notably moribund until he acquired a powerful theory, largely equivalent to structure theory, which he then exploited with energy and mastery. He transferred to Leipzig in the middle of the period of his greatest productivity. The sudden and dramatic change in institutional setting, resources, and numbers of students—nicely paralleling the maturation of the field of organic chemistry in Germany—made remarkably little difference to his success rate and overall standing in the field, or even to the productivity of his group research if measured by a proper (intensive) yardstick. Productivity and standing only began to decline when Kolbe began to focus exclusively on those points of difference between his and the structuralists' theories.

In this context, it must be reiterated that Kolbe's theoretical approach was so distinctive as to be characteristic of only himself, for he did not convince even his students of the advantages of his approach to the study of chemical constitutions. Kolbe's ideas were not powerful in comparison to those of his competitors at other universities. To speak more precisely and with only a touch of hyperbole, the ideas were powerful only to the extent that they happened to coincide with structuralist notions. The projects that were predicated on Kolbe's conviction of a difference between carbon valences, on a denial of chain formation, on his trimethine–trimethane benzene theory, or on the uniqueness of the *Stammradikale*, all proved scientifically sterile.

Here, finally, is the principal reason why the Leipzig research school was not more scientifically productive and influential. For Kolbe, at least, the focus should not be on distinctiveness as an advantage—especially not as an essential prerequisite for a successful school as some have viewed it—for we have seen that in this case his distinctiveness could only hurt him in the long run. Rather, attention needs to be directed to the power and empirical stature of the ideas driving the research. Kolbe's increasing concern—virtually an obsession—with precisely those details of his carbonic acid theory that proved most sterile ensured that his own research would be moribund. One consequence was that direction of the laboratory in more fruitful areas was left exclusively to junior associates, who were less able to lead and inspire. Another was that as Kolbe's stature in the field fell, his word meant less in recommendations for his students. The net result was that Kolbe's Leipzig school was gradually transformed from an exciting and productive example of group research into a combination of a quixotic personal research program of the director and an undistinguished mass-research group of a rather conventional character. What made his decline even more precipitous was the holy war he waged against his adversaries during the last sixteen years of his life.

The Crusade

Kolbe switched to the new atomic weights in 1868, at the same point that he decided to take a stand against the theory of chains, valence bonds, and benzene rings. Coincidentally, at the beginning of 1870 editorship of the *Journal für praktische Chemie* fell into his lap, which provided a bully pulpit for educating and haranguing the chemical world. He quickly penned a manifesto to open his first volume, then found repeated occasion during his first two years as editor to compare his views with those of the structuralists.

He now understood, correctly, that structure theory posited a sort of chemical "democracy," in which every atom is in principle as important as every other. Kolbe's own model was that of an army: a methyl group, for example, is like a "commando" unit consisting of a corporal (carbon) and three privates (hydrogen); in propane there are two more carbon atoms, but these are of higher rank than corporal and hence are chemically more central. The following year he used another metaphor, that of an autocratic state, which is effective precisely because it is hierarchical, in contrast to a democracy.[37]

At first, these discussions were carried on without evident rancor on either side. At the 1867 Naturforscherversammlung in Frankfurt, Kekulé treated Kolbe with extreme cordiality and Kolbe resolved privately to reciprocate in the future. By 1874, this was more difficult; at a chance meeting at a resort in Interlaken, Kolbe brushed off Baeyer's friendliness.[38] The transition appears to have been precipitated by national and personal events that occurred in 1870–1871.

Kolbe was unwell most of the summer of 1871, suffering from dizziness and nausea. Finally, he traveled to Marienbad for a five-week cure, which did him much good. Upon his return, he wrote Varrentrapp,

> I used the involuntary leisure in Marienbad to give my chemical heart and conscience some relief and to expose the great flaws and weaknesses of the new chemical fashions. Much real mischief in this line is being done by both older and especially younger chemists, and since no one else is opening his mouth to stand up against this swindle, I have considered it my duty once more to burn all ten fingers by portraying this modern child in its true flaws.[39]

In this new essay, "Fashions of Modern Chemistry," Kolbe indicted structure theory for being at once too empirical and too speculative. The structuralists were overly schematic, hence guilty of "crass empiricism," in immediately turning to pencil-and-paper manipulations once they had an empirical formula. They failed to investigate the hierarchy of radicals forming a molecule, merely drawing pretty pictures that purport to explain all chemical relationships of the compound. At the same time, he thought, they were overly speculative in that they presumed to have the ability to specify spatial arrangements of the constituent atoms. Once he gave his amanuensis the empirical formulas for three novel compounds for which their discoverer had just assigned structural formulas. The man reported back in a half hour with several more candidate structures, some of which looked more probable to Kolbe. Kolbe concluded that structure theory is a dangerous toy, especially for inexperienced chemists, and that structures are often assigned "in one's sleep," with little or no empirical warrant.[40]

Privately, Frankland "entirely dissent[ed]" from this judgment:

> [I]t seems to me that your experiment with your amanuensis resulted in a great triumph for these formulae; since, without any previous knowledge of the subject, he at once found the only possible constitutional formula for one of the compounds whilst his formulae for the two others would be at once modified by a chemist as unnecessarily complex. Surely the more simple, & free from possible misconstruction, such formulae can

> be, the better. And I think they are generally used by chemists, not as means of investigation but as expressions of the writer's ideas of the constitution of the bodies he is describing. But even as instruments of investigation they are not altogether useless . . .[41]

Volhard also objected privately; he joined Frankland in expressing regret over the personal character of many of Kolbe's remarks. Kolbe's response was straightforward:

> Scientific matters must not be taken personally. I cannot help myself, I *must* criticize and contest the chemical ideas that I consider false and worthless, just as I tolerate others' opinions, and I am happy to see my views contested, when the controversy is pursued in a gentlemanly fashion. Science always benefits from that. . . . You say that my critiques will only succeed in gradually alienating all the chemists of Germany. That may be the case, if not for all, at least for many, i.e., for some time; but I ask you, did Liebig ever hesitate to express his convictions, in critiques and otherwise, for fear of thereby alienating many?

As he later wrote Varrentrapp, he would rather be considered sharp-tongued than cowardly.[42] He now fulfilled a longstanding desire to follow the examples of Berzelius and Liebig by starting a series of annual critical retrospective essays, published each December in his journal. These gave him additional opportunities for spicy polemical harangues.

In Kolbe's third and fourth retrospects, those for 1873 and 1874, he went after his favorite example of structuralist excess: Kekulé's benzene theory. He absolved Kekulé himself of much of the blame,[43] for he was convinced that Kekulé regarded the theory merely as an intriguing and useful hypothesis. However, most chemists by this time viewed the hexagon as "infallible dogma," as "the Pope is for Catholics." They were true fanatics, Kolbe wrote, and viewed him (Kolbe) as a rank beginner, of weak understanding. They were right, he thought, for he could not understand arguments built "in the air" or "on loose shifting sand." The end of their sand castles was not distant. Kolbe continued,

> The modern chemist, who knows exactly what a chemical compound looks like in its middle and its end, how the six carbon atoms of benzene are symmetrically linked together in a plane, who then further purports . . . to have a clear conception of the spatial arrangement of the atoms, of their ortho, meta, and para positions, who determines the positions of all of the atoms in the compound, has long since abandoned the solid ground of exact science; the scientist has become a metaphysician. Rhetoric which is bereft both of content and of value but which sounds profound has begun to displace solid research and sober judgment.[44]

Kolbe later developed this image by defining true physicists as "ortho" physicists, versus "paraphysicists" who pursued fanciful notions such as kinetic theory and metaphysicists who have no use whatever for experimental confirmation. In his own view, he was an "orthochemist" and Kekulé was a "parachemist."[45]

Kolbe found his ideal model for the "metachemist" pursuing "transcendental chemistry" when the then-unknown J. H. van't Hoff unveiled his theory of the asymmetric carbon atom, the first step toward chemistry considered in three dimensions—soon to be called stereochemistry. Van't Hoff, who had studied with Kekulé and Wurtz, first found employment at the Utrecht Veterinary College; in 1878 he was appointed at the University of Amsterdam. A sketch of the theory was published in Dutch in 1874, and a longer French version appeared the next year. The first enthusiastic advocate of van't Hoff's theory was Johannes Wislicenus, professor at Würzburg, who was a mid-career structural organic chemist with a fine reputation; he had even published some thoughts on three-dimensional (physical) isomerism himself. Wislicenus asked his student F. Herrmann to prepare a German translation of van't Hoff's "Chemistry in Space," wrote an enthusiastic preface, and sold the work to Vieweg Verlag.[46] Kolbe found out about this translation almost immediately because the Vieweg company had a long-established policy of automatically sending proof sheets of all their organic-chemical publications to Kolbe.[47]

Kolbe was not a happy man at this time. Suffering repeated bouts of ill health himself, he had seen in the past few years the deaths of many of his closest friends and relatives: Otto Erdmann and Eduard Vieweg in 1869, F. J. Otto in 1870, his own father also in 1870, and Liebig in 1873. By far the hardest blow was the death from cancer of his beloved wife on 26 December 1876. He heard about Franz Varrentrapp's death on 3 March 1877, while he was still severely depressed about his wife. Kolbe was close to sixty himself, exhausted and in poor health. To Heinrich Vieweg's business manager Herr Lücke Kolbe wrote, "The older one gets, the more frequently one looks around himself in the circle of his friends and close relatives, watching Death carry out his sad and terrible office, until one's own turn comes around."[48] After losing out to Baeyer as Liebig's successor in Munich (1875), Kolbe knew that he was in Leipzig for the duration.

It was in such a mood, no doubt feeling that he no longer had anything to lose or anyone to please, nor any time to waste, that Kolbe sat down to compose under the heading "Sign of the Times" a thundering reproof against van't Hoff and Wislicenus, a devastating critique that would once and for all extirpate this "cancer" of structuralism (for that is what he considered and named it). Using another pathological metaphor, he claimed that modern chemistry was nothing less than a revisiting of *Naturphilosophie*, the "plague of the century" as Liebig called it, promoted by pseudoscientists who wish to smuggle their wild notions into the science, like introducing a fashionably dressed prostitute "into good society where she does not belong."

> Whoever thinks this worry seems exaggerated should read, if he is capable of it, the recent phantasmagorically frivolous puffery . . . on "The Arrangement of Atoms in Space." . . . A Dr. J. H. van't Hoff, of the Veterinary School of Utrecht, finds, it seems, no taste for exact chemical research. He has considered it more convenient to mount Pegasus (apparently on loan from the Veterinary School) and to proclaim in his "La chimie dans l'espace" how, during his bold flight to the top of the chemical Parnassus, the atoms appeared to him to be arranged in cosmic space. The prosaic chemical world had no taste for these hallucinations, so Dr. F. Herrmann, assistant at the Heidelberg Agricultural Institute, undertook a German edition to give the work a wider audience. . . . It is typical of these uncritical and anti-critical times that two virtually unknown chemists, one of them at a veterinary school and the other at an agricultural institute, pursue and attempt to answer the deepest problems of chemistry which probably will never be resolved (especially the question of the *spatial* arrangement of atoms), and moreover with an assurance and an impudence which literally astounds the true scientist.

These notions would have been quietly buried, Kolbe noted, had not a chemist of reputation, Wislicenus, taken van't Hoff under his wing. Wislicenus had thereby placed himself in imminent danger of squandering that reputation, of no longer being considered a true scientist but a "spiritist of the first water."[49] To his friend the publisher, Kolbe was unrepentant, for he wrote Heinrich Vieweg that his critique could only increase sales. "I don't understand Wislicenus," he added. "Sometimes I fear that he may not be of sound mind."[50]

Wislicenus was understandably upset. A fundamentally kind and broadminded man, he wrote Kolbe a long, emotional letter, trying hard not to show open anger.

> You cannot possibly have studied van't Hoff's essay . . . [for] how else could you have reproached me (by logic I do not understand) for a tendency toward spiritualism, or held against the young van't Hoff his position at a veterinary school, or against the translator Herrmann, who was my assistant and solely due to pressing external circumstances accepted a position at the agricultural institute in Heidelberg! I have never doubted that it is a holy zeal for the truth that guides your critical pen; but on the other hand I regret that you do not seem to concede any possibility of your own fallibility, which everyone must grant. . . . I

> know that I can err, but I also know that I have no cause to allow myself to be struck from the ranks of exact scientists, for I as well as you have the will to serve the truth . . .[51]

Van't Hoff responded publicly, and very effectively, in the pages of the *Berichte*:

> A theory that so far is contradicted by no single fact can only be further examined experimentally. Thus when someone, even so fine a chemist as Kolbe, avers that a chemist who is not yet well known and who is employed at a veterinary school should not bother himself with theories . . . I can only say that such behavior fortunately is not a sign of the times, but rather must be regarded as a contribution to understanding a single individual.[52]

Kolbe's critique of van't Hoff was his most famous diatribe, certainly his most humorous, and one of his most vicious. It was at this time that many began to wonder if Kolbe had become mentally ill.[53] Whether literally ill or not, there is no question that he was sick at heart at what he saw happening to his beloved science. To be sure, he was not alone in thinking that structural chemists often went overboard. The following October he had a conversation with his old mentor Wöhler, during which Wöhler commented that "what is published these days as chemistry, is not chemistry at all." However, when Kolbe pleaded with him to allow his words to be quoted directly, Wöhler quickly and strongly demurred, saying that he hardly even read the literature any more. Kolbe responded that he didn't either, and had just as little understanding of "modern" chemistry. Similarly, Kolbe urged his true Doktorvater Bunsen, who was likewise sympathetic with Kolbe's position, to stand with him against the structuralists, but Bunsen also firmly declined, even (like Wöhler) to have his name mentioned. With Berzelius and Liebig dead and Wöhler and Bunsen unwilling even to be named, Kolbe felt very much alone.[54] Ernst von Meyer, Kolbe's loyal assistant and loving son-in-law, became co-editor of the *Journal für praktische Chemie* at the beginning of 1879 and, as he later related, often tried to exert a moderating influence—only occasionally with success.[55]

Wislicenus' and van't Hoff's complaints may have had some influence, at least in redirecting Kolbe's fire toward more prominent chemists. Kolbe decided, after all, that the only way to destroy the weed was to get to its roots, so from this time on he went right after Kekulé, as well as Kekulé's most famous student, Baeyer. Seven months after the van't Hoff polemic, Kolbe published a "Confidential Letter to Professor Kolbe," purportedly written by a structuralist named "Dr. R.," but in fact written by himself as a parody of structure theory.[56] Kekulé, whose recent rectoral address at Bonn (18 October 1877) was well roasted in the piece and who could easily divine the real author, wrote an "open letter" in rebuttal and asked Kolbe to publish it. To twist the sword in the wound as best he could, Kekulé pretended ignorance of the identity of "Dr. R." and added that, in contrast to Dr. R., he should be identified as the author of the rebuttal, for "I have always been of the opinion that anyone who respects himself must also have the courage to accept responsibility for his actions and words." He then said that he did not doubt that Kolbe would print the piece, knowing Kolbe's sense of fair play.[57]

Kekulé had trapped Kolbe, but Kolbe was resourceful and squirmed away. As it happens, he had just composed a "Critique of Kekulé's Rectoral Address," and so he published this together with Kekulé's rebuttal in the same issue.[58] Kolbe now really let loose. Kekulé's speech was poorly constructed, he said, almost illiterate, the obvious product of a former *Realschüler*;[59] Kolbe cited a number of what he thought were egregious solecisms. More substantively, Kekulé's chemistry was not only colored by "crude Haekelism" but was also filled with "wild phantasies without any real basis." His latest hypothesis of intramolecular atomic vibrations illustrated to Kolbe "to what monstrosities an intellectually gifted man can let himself be carried, who has not learned early to order his thoughts, to think logically, and to rein his imagination." He twice ridiculed Kekulé's "chemical dreams" and concluded by offering a "dream" of his own. The carbon atoms of benzene, you see, are constrained by three bonds each, so they must move in the *fourth* dimension! "Das ist *meine Theorie*," Kolbe proclaimed triumphantly, but confessed that he had not the courage to develop this idea any further, so he would leave it to the "*most* modern chemists" to do so.[60] Graebe was dismayed by this article. He wrote Rudolf Schmitt, "I always regret that such a significant scientist, who is personally so amiable, puts himself in such a false light with articles like this. Those who don't know him imagine him to be an unpleasant person."[61]

A few months after this episode, Kolbe found occasion to heap ridicule on Baeyer in a similar fashion. Baeyer had given an address on chemical synthesis in honor of King Ludwig's thirty-second birthday, in which he portrayed for a lay audience some of the leading ideas of recent chemistry and physics. To make the concepts accessible, he eschewed scientific terminology and epistemologically cautious circumlocutions, speaking, for example, of valence bonds as analogous to "fishhooks," atomic "glue," and so on. In his published critique, Kolbe made much of Baeyer following his teacher's example—the speech was just as baroque, illogical, and dreamlike.[62] He wrote Wöhler, "I'm no spiritist, but I would not have been surprised if Liebig's ghost had seized him by the collar after this speech and thrown him from the podium as an unworthy successor." He told Frankland to read it for amusement—"much wilder than Kekulé's rectoral address . . . I think my critique will somewhat ameliorate his obscurity; Kekulé has also already become tamer." He thought Baeyer might be suffering from softening of the brain or megalomania.[63]

Volhard, like Graebe, was appalled by Kolbe's attacks and wrote his former mentor: "I ask you please for all the world no more critiques like that of Kekulé's rectoral address! I cannot agree with this critique in any way." He pointed out that an annual address is compulsory for the university's rector and that it must be directed to a lay audience; consequently, "es ist nicht fair zu kritisieren" as if it were a chemical treatise.

> And as for the form, I ask you: What do you care about Kekulé's style, or his classical education? Consideration for K's scientific accomplishments, indeed mere collegial respect ought to have stopped you from treating this man as if you were a teacher looking over a schoolboy's assignments and correcting his mistakes. . . . I beseech you, no more such intemperate critiques! More tolerance and respect for scientists who have made and are still making their contributions!

Volhard was a good friend who had enormous respect for Kolbe; indeed, he regarded Kolbe as one of the greatest chemists of the century, and Kolbe knew it.[64] Kolbe also liked and respected Volhard. Only such a man could direct such words to Kolbe.

Even so, they put Kolbe in a rage, answering Volhard with visible effort to control his pen. It particularly galled him that Volhard had accused him of being *unfair*, when in fact he was placing himself bravely and alone in the line of fire for the sake of his beloved science, while Volhard and everyone else were sitting comfortably on the sidelines. This time there were no horsey metaphors, but sexual, militaristic, religious, and political ones. Kolbe wrote,

> Your letter troubled me, for I see from it that you now number yourself among the (in a word) weaklings who are not troubled when our chemical social democrats, Kekulé and Baeyer, slap the face of our science and soil it, but who break out in a sweat when a pure hand is raised to put a stop to it. . . . I cannot sit quietly and see these two make our science a footstool of their vanity and misuse it to satisfy their arrogance. Have you too really come so far in these feeble materialistic times that you can no longer be inspired by higher goals, by ideals? . . . Believe me, I enjoy criticizing; can you not imagine, since all others out of convenience or

> cowardice are silent, that someone feels the calling and the obligation to science to stand up for it publicly? . . . I close by assuring you that wherever our science is violated you will *always* find me, as long as I have the power, as the *first* of her defenders.[65]

Kolbe was increasingly frustrated. He continued to write to friends, seeking comrades for his crusade, but found at best silent support, more commonly demurs and even rebukes. These made no difference to his actions, for he became if possible even more personal. One of his favorite hobby horses in the last few years of his life was linguistic critiques. Kolbe was an excellent writer; however, he harbored the common delusions (consistent with his conservatism in other areas) that there was a happy time, long before the present corrupt age, when all educated people wrote "classic" German, and that philology was properly a prescriptive rather than a descriptive enterprise. In some cases his remarks were well founded, even if petty and unnecessary; in other cases he was simply closed-minded and parochial.[66] He often exclaimed in print, sarcastically remarking on his victims' lapses, "*This* is how a professor at a German university writes!" He also campaigned against Realschulen, for he was convinced that only exposure to classical neohumanism in the traditional Gymnasium could truly educate.

Kolbe's old friend and comrade-in-arms Frankland provided the final straw. The first sign of real trouble was when Kolbe heard that Frankland was sending his son Percy to study with Wislicenus, rather than with him or Bunsen. Kolbe could hardly believe it. Wislicenus "has long since ceased to be an exact scientist, but rather is a Naturphilosoph, a metaphysician"; a boring chemist, even if a fine man.[67] Then Kolbe found out from reading proofs from Vieweg that, in a historical introduction to Roscoe and Schorlemmer's major treatise of organic chemistry, they had uncritically accepted Kekulé's view of the last forty years. This was highly worrisome to Kolbe, for Roscoe and Schorlemmer had both been Bunsen students, and the inorganic portion of their textbook had already proven to be influential. He fired off a letter to Roscoe. "Kekulé has *deliberately* falsified history, in order to place himself and the French chemists he is so fond of in the foreground." It is Frankland and he, Kolbe continued, who deserve credit for valence theory. Moreover, Kekulé is uneducated, cannot think clearly, lies, steals, and seeks to derogate and slander great men such as Berzelius. His *Lehrbuch* is the "worst textbook that we have." Roscoe sent Kolbe's letter to Schorlemmer, who was simply amused, and thought Kolbe to be "mad as a march-hare." His accusation that Kekulé had stolen carbon tetravalence from him and Frankland was "best proof that he is mad!"[68]

Having failed to receive satisfaction from Roscoe, Kolbe now set to work to right the historical record, also with the intent of defending Frankland. He wrote a long essay on "My Participation in the Development of Theoretical Chemistry" and published it piecemeal in his journal.[69] Here he portrayed the whole of modern chemistry as a highly injurious product of the unscientific and sterile French chemistry of Dumas, Laurent, and Gerhardt. Kekulé, Baeyer, Wislicenus, Fischer, and others had substituted flights of fancy and the painting of pretty pictures for the exact scientific principles that Berzelius had introduced into chemistry and whose further development had been due to such men as Liebig, Wöhler, Bunsen, Frankland, and himself. Kolbe felt that his seminal work had been systematically ignored by the structuralists—which is not far from the truth. He did not hesitate to accuse Kekulé of both the largest and the smallest transgressions: he repeatedly and explicitly proclaimed that Kekulé's conduct could only be viewed as intentional usurpation of the theories of others, then added insult to injury by filling page after page with stylistic criticisms. He even cruelly drew attention to Kekulé's premature aging, which he maliciously ascribed to his guilt for self-consciously leading German chemistry down the structuralist cul de sac.[70]

As for structural formulas, he characterized them no fewer than three times as "grob sinnlich" (coarsely sensual) and materialistic, moreover, as a symbol of an unachievable goal.

> The sober prudent scientist will tell [Kekulé] that *the object for which he and the majority of modern chemists strive is a chimera, that we will never succeed in gaining a conception of the arrangement of the atoms in the molecule, and that chemists should set for themselves a more modest goal: the investigation of chemical constitution in the sense of Berzelius*. The goal for which Kekulé strives, and which he considers accessible, is actually even more inaccessible for us than the moon, for we can see the moon and determine its form, but atoms we cannot see, and their form is perceivable with none of our senses.[71]

These words were echoed (and partly contradicted) in the foreword to Kolbe's abridged organic chemistry textbook, written in February 1883. Structural formulas are mechanical and coarsely sensual, a symptom of the modern "crassly materialistic treatment of scientific matters; the latter ought to be conceived by the mind and not mechanically."[72]

Having defended Frankland's priority, Kolbe now tried to get him on board. He asked Frankland for permission to title the brochure version of his essay "Frankland's and My Participation in the Development of Theoretical Chemistry." Frankland had no objection, as long as Kolbe would state clearly in the preface that he alone was responsible for the contents. Kolbe was taken aback at this request and withdrew his suggestion. "Why are you so fearful? You want me to pull your chestnuts out of the fire; I do not hesitate to do that, having no fear of the fire of truth. . . . We should have stood up against the insolent behavior of these people years ago. But perhaps even now it is not too late."[73] Now it was Frankland's turn to feel a bit insulted. He replied that he was perfectly capable of taking care of his own chestnuts; indeed, *he* had been the first, four years earlier, to assert his and Kolbe's right to valence theory; but by "mixing up Dumas and Wurtz with the protest," he felt (no doubt rightly) that this would weaken the case that needed to be reiterated.[74]

Here the matter rested for two years. Finally, Kolbe dusted off the chestnuts once more and wrote his friend. "If only other chemists, and above all I am thinking of you, would not always leave me alone to pull the chestnuts from the fire, but would instead come to my aid in energetically fighting Baeyer and the thoughtless hollow schematism of structural chemistry." We could "finish Kekulé off" once and for all, Kolbe promised, and show structuralism for the humbug it is. Frankland was incredulous.

> Your letter astonished me not a little for it had never entered my head to imagine that you could for one moment think of me as in any degree an antistruktur Chemiker. Still less that you should think of me as one likely gegen die Strukturchemie "*energisch zu kämpfen*" or as one sharing your opinion that die Strukturchemie ein Humbug ist! Turn over the leaves of my "Lecture Notes for Chemical Students" and you see Strukturchemie in its extremest development on almost every page! . . . That there should be no more such mistake in future, I here record my Glaubensbekänntniss: *Chemistry owes its progress from empiricism to exact science entirely* (so far as theoretical conceptions are concerned) *to Strukturchemie. Without Strukturchemie there is no science of chemistry.* And allow me to add,—two of the first Strukturchemiker to whom this progress is due were Berzelius & Kolbe![75]

And now it was Kolbe's turn to be astonished. Who had seduced his friend (or, rather, *former* friend), "*you*, who were once upon a time, with me, a champion of *positive* exact chemistry," into becoming a "spiritist"? Did not Frankland remember those golden days of great scientific discovery in Marburg? And now Frankland had not only turned tail, but had insulted Kolbe by branding him with the despised epithet "structuralist." "I now see that we do not understand each other."[76] This was the end of their correspondence.

Neither Kekulé nor Baeyer ever really responded to Kolbe's insults. Kekulé wrote up a long response, and sent it to his friend Volhard, then editor of the *Annalen*. Volhard strongly urged him to withdraw it.

> Don't you see that you will only thereby legitimize his attack, which can only properly be condemned by maintaining silence. You may be sure that no one will welcome your defense more than Kolbe himself, for it would finally break this terrifying silence, and give him the opportunity for a salty reply. And, as before, you will be sure to draw the short end of the stick, for you are no match for Kolbe when it comes to coarseness and ruthlessness.

"That's what I call friendship," Kekulé gratefully replied, and took Volhard's advice.[77] As regards Baeyer, Volhard again offered advice and consolation, this time to Baeyer's wife.

> My old friend Kolbe is behaving truly irresponsibly. A pity on the man; since he began to devote himself to insults he has produced nothing more of value. Moreover, one is in good company when one is upbraided by him, so one may always console oneself this way. Everyone who has achieved some reputation in science should take K. as a cautionary example: he believes it sincerely, considers it his duty to behave this way; he does not see what immoderate overestimation of himself is involved, although he is otherwise a very clever and intelligent man.[78]

As Volhard had predicted in 1876, Kolbe had now succeeded in alienating himself from most prominent German chemists, including several hitherto good friends (we will see how he and Hofmann parted ways in the next chapter). Volhard was still left, but not for long. On 20 July 1884, they had a conversation during which apparently they could agree on nothing; Kolbe's letter the following day (one of his last surviving letters) complained that Volhard, too, had now been "badly infected by Kekulé's and Baeyer's dogmas."[79]

Kolbe had other shocks as well. Even after his devastating denunciation of van't Hoff's small book, Heinrich Vieweg accepted a larger work by the same author, *Ansichten über die organische Chemie*, just a year later. ("The *greatest nonsense* I have ever read," fumed Kolbe to Vieweg. "The author is definitely out of his mind, ready for the madhouse.") Then Vieweg selected Wislicenus to edit—and structuralize!—Strecker's organic chemistry, which had been Kolbe's favorite text. Kolbe was appalled at this "disfigurement." He feared that Wislicenus might be called to Halle as Heintz' successor, thus making him a close neighbor (in fact, *Volhard* was called to Halle, but Wislicenus became *Kolbe's* successor three years later!). The *Hand-*

wörterbuch der reinen und angewandten Chemie, for years Kolbe's brainchild, had long since become a structuralist reference work. Kolbe made Vieweg add a clause to his contract for his own *Kurzes Lehrbuch der Chemie*, stipulating that in the event of his death, Vieweg would not choose a structuralist editor.[80] He now knew that he would not live to see the revolution that for the last quarter century he had been predicting as imminent. "We live now in a time of barbarization [*Verwilderung*] in chemistry, like never before," Kolbe mused sadly. "Soon the crash will come, and then the loudmouths will vanish from the scene."[81]

The last meeting between Kekulé and Kolbe was bitter, although Kolbe enjoyed every moment. On 4 April 1884, Kekulé and his wife arrived at a resort (in Bodighera, on the Italian Riviera) where Kolbe also happened to be vacationing. "You should have seen his pale, frightened face the next morning . . . when he first saw me," Kolbe wrote his nephew. "He is an *old man*, stooped over; I would never have recognized him had I not known it was him." The Kekulés had intended to stay a week, but left after three days: "I smoked him out," Kolbe related gleefully.[82] Kekulé was then fifty-four, Kolbe sixty-five. We do not know Kekulé's reaction when he heard of Kolbe's death seven months later.

It is now time to take a step back from the fray and directly ask an important question that we have been skirting. Was Kolbe simply a sour old man whose crusade was motivated by an irrational obsession or monomania (as Armstrong later called it)—in short, was hc acting as a poor scientist? As much as this places one under vehement suspicion of Whiggery—for Kolbe was so unfortunate as to contest much of what constitutes organic chemical theory today—it is difficult to avoid giving an affirmative response to this question. In defending oneself from Whig opprobrium, one might note that as regards *contemporary* opinion after 1870, Kolbe lost nearly all of his specific factual arguments and convinced absolutely no one of the truth of his modified radical theory of organic constitutions, even his own students and closest friends. Moreover, the previous discussion has shown that he himself recognized the completeness of his defeat by the time of his death.

That said, it must be noted that while most of Kolbe's theoretical affirmations after 1870 were unsuccessful, by no means did all of his criticisms of his opponents miss the mark. We have seen that Kolbe had substantial silent support for many of these criticisms among such men as Liebig, Erdmann, Varrentrapp, Wöhler, Bunsen, Beilstein, Erlenmeyer, Frankland, and even Volhard and Lother Meyer on occasion. I have found no Kolbe correspondent who denied that there was a great deal of inferior, sloppy, and excessively conjectural work being published, and they all agreed that the amount of such unsatisfactory research was increasing. (Volhard believed that this had nothing to do with structure theory but rather simply with the great expansion of the field; it was no longer a self-selected elite group, as it was a generation earlier, but a mass of average workers creating mass-produced science.[83]) In fact, Kolbe was justified in his accusation that structural formulas were sometimes bandied about thoughtlessly and superficially, with little regard to empirical evidence and reasoned justification.[84]

Of course, the same charge could be laid at Kolbe's door, to the extent that he may be considered a structural chemist himself. Was he? Contemporary opinion on this point was unanimously in the affirmative, not only in the view of structuralist opponents such as Kekulé, Wislicenus, Lothar Meyer, and so on but also in the opinion of those who knew him best: his students, former students, friends, colleagues, and family members, including Frankland, Claus, Crum Brown, Volhard, Ernst von Meyer, and Ost.

But despite this unanimity, the assertion is not strictly true. To be sure, Kolbe's views and those of the structuralists coincided on many points, and formula translation was always possible between the two systems. However, there were also real differences, as I have been at pains to argue in this and the preceding five chapters. In the course of his career, Kolbe made many predictions of the possible existence of new compounds and new isomers; some of these predictions were fulfilled, some were not. Some of his predictions were fully equivalent to those of the structuralists, while some highlighted the distinctions between the two systems. It is remarkable that every time Kolbe attempted a crucial test, that is, of a prediction of the latter type, the result disappointed him.

Let us review some examples, most of which we have seen before. As early as 1857 and 1858, Kolbe was predicting the existence of a variety of isomers of alcohols, glycols, acids, and aldehydes, including isomers of oxalic, glycolic, and lactic acids. He thought that Wurtz' glycol ought to be dehydratable to aldehyde, but that it could never be oxidized to an acid. In his 1868 treatise, he predicted two isomeric propylenes and no fewer than fifteen isomeric pentanes.[85] He tried to find an isomalonic acid and a second carbon oxysulfide, and he thought that a chemical distinction between the two chlorine atoms in 1,3-dichloropropylene ought to be demonstrable. He looked for reduction products of benzene, an isomer of benzene, and isomers of all monosubstituted benzenes, and he thought there should be four isomers of each disubstituted benzene. He was able to find none of these compounds or reactions.

Kolbe was continually predicting isomers in excess of those envisioned by classical structure theory because of his hierarchical view of organic molecules (each carbon atom in a molecule, considered as the *Stammradikal*, should give rise to a distinct series of isomers) and because he was convinced that the four valences of carbon were intrinsically distinct. Chemists of the 1860s and 1870s were already being deluged by a flood of new isomers, and most greeted the apparent proof of equivalence of carbon valences with considerable relief. But Kolbe simply refused to accept the evidence provided by the nonexistence of isomers. This was true even though many positive refutations of his predictions emanated from his own laboratory (for some examples, see note 36 in this chapter). Logically, his position was impeccable: the missing isomers are simply too labile to be isolated, transforming themselves into known isomers before they can be characterized; or we simply have failed to find the right reagents or conditions to produce them. But the accumulating empirical evidence became more and more difficult to ignore or brush aside. It is little wonder that he was unable to convince his colleagues in the field, and even his own students, of the advantages of his system over that of the structuralists.

14

Pride and Prejudice

Chauvinism[1]

Kekulé, Hofmann, and Kolbe, the three premier German chemists in the generation after Liebig, form interesting contrasts, in their personal lives as in their science. Kekulé was cosmopolitan and patrician in style, and much inclined toward internationalism. After Giessen, he enjoyed four and a half years' worth of three successive foreign *Wanderjahre*; then a brief period in Heidelberg was but a prelude to nine years as a professor in French-speaking Belgium. By the time he was called to Bonn, he had spent thirteen of the previous sixteen years abroad; he could speak English and French almost without accent and fluent Italian as well. He was also principal organizer of the first international chemical congress. Hofmann, for his part, spent twenty happy and productive years in England. Like Kekulé a suave sophisticate, Hofmann's oral and written English was so masterly that he did not hesitate to correct the language of his English students. Armstrong's thumbnail sketches are apt:

> Kekulé was a born aristocrat in manner. An intellectual of a high order, many-sided in his interests, he was too critical and cynical to be a leader of men in the way that Hofmann was, though even superior to him as an orator; he attracted through his clear-cut talent, his gift of precise speech and his great command of knowledge. . . . Kolbe was equally simple [as Frankland], never a man of the world, a good lecturer and a far better writer but not an orator: the best chemist of them all. Hofmann and Kekulé were cosmopolitans; . . . Kolbe—just the dear old German,

academic pedagogue of the highest class: there is no other way of describing him.[2]

Indeed, Kolbe's was a very different character. With the single exception of Jacob Berzelius, whom he considered an honorary countryman, all of Kolbe's models were German, above all, the heroes of the classical period of the rise of German chemistry: Wöhler, Liebig, and Bunsen. Linguistically as well, Kolbe forms a contrast: although he learned a reasonable amount of English in the eighteen months he spent in London, he soon forgot most of it, at least as far as oral communication is concerned.[3] There is no evidence he ever mastered or even seriously studied any other foreign language. Apparently he could read French, although certainly he avoided doing so as much as possible. As for foreign travel, aside from his one postdoctoral stint and a brief laboratory tour to England, a fishing vacation in Norway with Eduard Vieweg, and his semiannual "cures" taken often in Swiss resorts, he did not leave the German Confederation or Empire. He particularly avoided the Catholic countries of Austria and France.

Kolbe's first recorded derogation of the French dates from the period in 1848 just after the February revolution in Paris and the "March days" in Germany, but his language became much sharper when it appeared that the reformers might really carry the day. His concern and anger can be discerned in the first fascicle of his textbook, published in 1854. To Eduard Vieweg, he confessed his desire to continue Berzelius' critical tradition against the "extravagances" of foreigners, now that the heroic Swede was no longer alive.

But Berzelius was not Kolbe's only model for ferocious critiques; he also followed the pattern established by his other great hero, Liebig. Liebig's views of foreign chemistry are best exemplified by examining his relationship with his greatest rival, J. B. Dumas—as we have done at the beginning of chapter 4. In their worst period, the late 1830s and the 1840s, Liebig continually accused Dumas of the vilest motives and actions. Dumas and his friends returned the sentiments. J. B. Boussingault wrote Dumas, "I am never so good a Frenchman as when I am along the banks of the Rhine, it is truly shameful that an evil hole like Giessen is a focal point of science . . ."[4] Despite the vehemence of these opinions, Liebig and Dumas eventually reestablished their friendship. However, this reconciliation was still in the future when Kolbe imbibed his extremely negative views of Dumas from Liebig, whose diatribes were often openly published in the scientific literature. Berzelius and Wöhler, two other major influences on Kolbe, also had opinions of Dumas and other French chemists which were not much more positive than Liebig's.

Kolbe's prejudices against foreigners, especially the French, were not necessarily tied to conservative political sentiments. We have seen in chapter 3 that Kolbe's general political orientation during his thirties was quite typical of his class and time period, namely, center to center-right liberalism. He had nothing but contempt for the reactionary Kurhessian regime, vaguely distrusted Prussia but despised Austria, feared republicans, extreme democrats, and socialists, and hoped for German unification, presumably under Prussian leadership but with constitutional guarantees. He looked with deep suspicion on Bismarck's and King Wilhelm's struggles of the 1860s with the Prussian *Landtag*. When in the spring of 1866 war with Austria threatened, Kolbe (with most fellow Germans) feared a catastrophe, for it was by no means clear that the Prussian army was sufficient to the task, and the Austrian yoke promised to be infinitely more onerous than that of Prussia. "Lieber Bismarckisch (so schlimm das auch ist)," commented Kolbe to Frankland about the alternative outcomes of the approaching war, "als österreichisch-jesuitisch!" Moreover, Saxony was sandwiched ominously between Prussia and Austria, and everyone expected the battle zone to be close to Leipzig.[5]

In the event, the decisive battle occurred at Sadowa (Königgrätz), two hundred miles southeast of Leipzig, and was handily won by the Prussian army. Kolbe's sentiments, again like those of most of his countrymen, were profoundly altered by this military success and by the prospect of a unified German nation. "Say what you like against Bismarck," Kolbe wrote Frankland, "one cannot deny that he is a decisive, quietly reflective man, *the premier statesman of Europe*."

> The situation is perhaps the following. Had Austria won the upper hand and destroyed Prussia, Germany would be lost and we would have Austrian conditions: lies, Jesuitism, concordat, systematic corruption, general moral disintegration, destruction of material prosperity, abolition of free scientific research, etc. With the battle of Königgrätz a new star rose over Germany; from this day Germany is a unified nation. Further, our political, material, moral and scientific development will receive a new impetus.

In short, Kolbe was convinced that "Prussia's victory signifies freedom and free development in every direction."[6]

Chemistry: A French or German Science?

Kolbe's long-simmering hatreds burst into the public domain at the time of the Franco-Prussian War. The decline of his influence in theoretical chemistry, along with his general isolation in the collegial community, must have increased Kolbe's ill temper, and after January 1870 he had his own journal to express his unexpurgated opinions. The war, along with the uproar over Wurtz' opening of his recently published history of chemistry, that "chemistry is a French science," provided the occasion for his outbursts.[7] In a polemical article "On the State of Chemistry in France" published simultaneously with the French declaration of war (and obviously modeled on Liebig's similarly titled essays on Prussian and Austrian chemistry), Kolbe lambasted the French for their dissolute ways and their feeble scientific establishment. There is *no* French university, he declared, that can compare with *any* German university for chemical education.[8]

As the war proceeded, Kolbe was even further radicalized. He was delighted by the Prussian victories at Sedan and Metz, but impatiently abided the long siege of Paris; he did not understand why Moltke held off on the bombardment for so long.[9] To Varrentrapp he wrote,

> The French are truly a nation of half children, half madmen. I have had deep hatred and contempt for the French, but I had never considered them so uncivilized, barbarous and base as we now see them to be. I believe France is now in a rapid decline, and will never recover. . . . The whole nation puts no value at all on honor, only on gloire.[10]

The sharpest contrast in this respect could be drawn between the French and the Germans, Kolbe thought, as he wrote to Frankland,

> The Germans, who seek their gloire in the arts of peace, and go to war only as a last resort, would never sacrifice their sons to the whim of anyone, even if a narrow-minded, fanatical, bellicose German emperor should one day accede to the throne. In our country the only kind of war that will be popular and possible is one that defends the fatherland.

Frankland ought therefore to have no fear of future German aggression. Furthermore, Kolbe bristled at Frankland's sentiments in favor of a republic, for the example of the United States illustrates that a republic is no more than "a playground for swindlers and adventurers, on which the insolent mediocrity bring their influence to bear, a language in whose dictionary the word 'gentleman' does not appear. . . . My dear friend, for heaven's sake no republic." We, like you (Kolbe concluded), would rather have a king than an emperor, and not one from Prussia; "aber die Nothwendigkeit hat eiserne Arme," and he and his compatriots were delighted with their new situation.[11]

When the French Academy of Sciences neglected to remove from the wrapper of their *Comptes rendus* mention of the Alsatian cities of Strasbourg and Mulhouse, and Metz in Lorraine, after their transfer to Germany, Kolbe was enraged.[12] He wrote Liebig,

> My contempt for the whole contemporary French chemical world is beginning more and more to turn into pity. Even the Parisian Academy appears to have no idea how ridiculous it appears to the scholarly world by this miserable bickering, for which Herr Pasteur constituted the ferment. Forgive my expectoration. The behavior of this lost and lying nation sometimes makes me a little passionate.[13]

But Liebig's was a sympathetic ear. The French, Liebig complained, were displaying "insane arrogance," demonstrating that they were a "dissolute race"; the "megalomania of this unfortunate nation is certainly capable of anything."[14] "How terrible it must be for this vain and arrogant nation to have achieved *not a single* advantage in battle."[15] Bismarck's adroit behind-the-scenes manipulations maneuvering both countries toward crisis had been essentially invisible to the German public, and the war propaganda was skillful. Even Kekulé was induced to denounce the "nation of scoundrels" they were fighting.[16]

Emotions began to cool, at least on the German side, after peace was concluded, but Kolbe kept up the heat, continuing his Francophobic polemics for more than two years. Having been elected, along with Liebig, Wöhler, and Bunsen, a charter honorary member of the German Chemical Society, Kolbe resigned in 1871 out of anger that the Society had not defended his critique of Wurtz' dictum when that critique had met public foreign opposition. Meanwhile Kekulé, together with Volhard and Erlenmeyer, successfully persuaded the Society to become less provincial. Among other reforms suggested by this group, after 1872 the Society only named foreigners as Honorary Members. But to Kolbe the Society had already been far too internationally oriented.[17]

Hofmann, who very much wished to soothe the raw feelings between the two countries, picked up the cue at this point, proposing Auguste Cahours as the first Frenchman to receive such an honorary membership after the war ended. This was the last straw for Kolbe, who protested loudly, both publicly and privately (but without effect, partly because he had now resigned). In his journal he asserted that there were "dozens" of more deserving Germans. "What a disgrace," he wrote Varrentrapp, "again with Cahours; what is the purpose of this international coquetting with France? Hofmann unfortunately lost the fatherland in England."[18]

Kolbe's tone became even harsher in his final years, when he became truly irrationally preoccupied with his various crusades. Ironically, the French were far less oriented toward structure theory than the Germans; Kolbe noticed this fact with alarm, for to him it indicated a surprising source of French strength that was dangerous for the future health of the German chemical community. "I know full well," he wrote Volhard,

> . . . that if Prussia continues to ruin chemistry . . . the time will soon return when, as in the second decade of this century, German chemists will go to Paris to educate themselves in chemistry. As at that time, when everyone in Germany was crazy about the Naturphilosophie of Hegel and Schelling, this swindle made no headway in France, and for that very reason France was far superior to us in science, so today, with the single exception of Wurtz, French chemists keep away from the naturphilosophische swindle of the modern structural and bonding chemistry, and therefore they will gain a head start on us once more.[19]

The irony was, as Kolbe well knew and loved to point out, that this same unscientific structural chemistry was a direct product of French chemistry—namely, an outgrowth of the type theories of Dumas, Laurent, Gerhardt, and Wurtz. Kolbe thought this was where Kekulé had gone wrong; he had followed not only the bankrupt theories of the French but also their larcenous behavior. The more highly Kekulé's textbook was valued, the more Kolbe railed against the "tendentious forgeries" committed by its author.[20]

Despite Kolbe's quirkiness, he saw a number of points quite clearly. Kekulé was indeed an internationalist at heart, and he had been decisively influenced by the French chemists Dumas, Laurent, Gerhardt, and Wurtz. He and other (predominantly German and German-influenced) chemists—such as Erlenmeyer, Crum Brown, Frankland, Ladenburg, Butlerov, Baeyer, Fischer, Victor Meyer, Graebe, and Wislicenus—had developed structural chemistry from that essentially French background. Kolbe was also correct in viewing Kekulé and Wurtz as flawed historians, for the latter did have hidden agendas in mind and neglected the very real contributions of those they disagreed with—especially Kolbe, Frankland, and Couper. Finally, Kolbe was right to see Wurtz as one of the few prominent representatives of structural chemistry in France.

Indeed, Wurtz' isolation in France was sort of a mirror image of Kolbe's in Germany, placing the contretemps over his chauvinist historical comment in even sharper relief. Read with attention to the thematic orientation of the entire work and placed in context with Wurtz' other interpretive, historical, and polemical writings of the 1860s, the apparently gratuitous chauvinism of his opening motto is subject to a different, or at least additional, interpretation. Wurtz had accepted essential parts of the Gerhardtian reform in 1853; by 1858 he was a full and enthusiastic convert. But continued opposition among his colleagues led him, rather isolated in France, to initiate a concerted campaign for the new chemistry, including structural ideas. He started a new journal (*Répertoire de chimie pure*) and a new society (Société Chimique de Paris); became a leader, along with Kekulé, of the Karlsruhe Congress organizers; wrote a heavily subtexted éloge for Gerhardt and Laurent; presented invited historical lectures to the Société Chimique, the Collège de France, and the Chemical Society of London; wrote a textbook; and finally, published a full, formal history prefacing a multivolume dictionary. All were designed to propagate the new chemistry in a country still dominated by older ideas. None were notably successful.[21]

I want to suggest, in short, that Wurtz' "chemistry is a French science" has a thematic load that was heavier than mere chauvinism. It was not so much Lavoisier and the first chemical revolution that Wurtz wanted to promote, but rather Lavoisier's countrymen Laurent and Gerhardt (not to mention Wurtz himself, aided by foreign Francophiles such as Williamson and Kekulé)—these being the authors of the still incompletely consummated *second* revolution. The work was directed inward rather than outward, its intended audience Wurtz' fellow Frenchmen. What better way to persuade them to join the new movement than to appeal to their patriotism by arguing for the *continued* dominance of French chemistry in the international arena? If I am right, we have here an example of nationalism put to rhetorical purposes, but for a cognitive goal—and not for mere chauvinist puffery. But it was difficult for foreigners to get past that first fearsome line.

Kekulé practiced the same technique. His 1859 history of chemical theory, prefacing his textbook, had a number of significant omissions. As was the case with Wurtz, these were partly due to selfish priority interests, and chauvinism may have also played a role; but there was also a rational didactic or rhetorical intent promoted by the distortions. He had a new theory to push, and he needed to tell the history behind it in such a way as to make the theory appear rational, even inevitable. The work of Kolbe and Frankland in particular failed to fit into the neat story Kekulé wanted to tell. This historical–didactic technique was, of course, very old and well attested.[22] It had been practiced with particular skill by Lavoisier himself. Although such a procedure may be devious and covert (or perhaps self-deluding), chauvinism was only at best a secondary motive.

The historical work of Hermann Kopp, a close friend to Kolbe, Hofmann, and Liebig, forms a sharp contrast to Kekulé's and Wurtz' partisan histories. Despite having been commissioned to write a history of chemistry *in Germany*, moreover just at the time of the Franco-

Prussian War and in the immediate aftermath of Wurtz' apparent chauvinism, Kopp's *Entwickelung der Chemie in der neueren Zeit* was aggressively and explicitly international in orientation. The case of Kopp is sufficient to show that chauvinist currents were by no means all-pervading, even during the most jingoistic of times.[23]

The optimistic interpretation to which such considerations lead—that chauvinism in science is perhaps less damaging than has hitherto been thought—can be further supported by looking again at some of the protagonists in our story. Liebig, for instance, exhibited prominent elements of Francophilia as well as Francophobia, and not only because his first rigorous scientific education took place in Paris. His biographers have emphasized his international outlook, which was often in evidence.[24] As the war with France progressed, Liebig expressed compassion and concern for his French colleagues, some of whom were good friends. In September 1870, Liebig told Wöhler that he had just written his brother-in-law, the army physician Karl Thiersch, then with the Prussians in Versailles,

> . . . that he might seek out Regnault and offer him his help. I wonder how our friends in Paris, Dumas, Peligot, Boussingault, etc. are doing? If only it were possible to do something for them, but they will not be allowed out of Paris. The lovely city, what suffering she faces![25]

Through Thiersch, Liebig succeeded in getting a letter to Deville in Paris from his wife, a refugee in Geneva. He sent 500 francs to C. L. Barreswil's wife in Boulogne, under the presumption that she needed it; he considered the same charity for Madame Deville.[26]

In the first meeting of the Bavarian Academy of Sciences after cessation of hostilities, Liebig delivered a speech assessing the causes of Prussian victory and French defeat. He suggested that German superiority was an indirect but very real consequence of wise governmental policies that, inter alia, gave sufficient support to academic research, which led in the long term to efficacious *scientific* rather than mere rote applications; but he made a particular point of praising the glories of French science. Liebig, like Kekulé, had begun his career as a Francophile, showing nothing but contempt for his previous German teachers; he always revered his French mentors Arago, Dulong, Thenard, and above all Gay-Lussac. He subsequently formed an exceedingly close relationship with J. T. Pelouze and others, spoke and wrote French fluently, and until his death kept in close contact with the leading figures of the Parisian establishment. In 1845 he wrote Wöhler, "Indeed, Frenchmen have something exceptionally appealing and amiable that is generally missing from the Germans."[27] As we have seen, he successfully reconciled with Dumas. Even Gerhardt, whom he had accused publicly of being an assassin and a highwayman, eventually managed to elicit kind and generous comments from his former teacher and became fully reconciled before his death in 1856.

Liebig concluded his speech by saying

> A warm sympathy for all that is noble and great and an unselfish hospitality are among the finest traits of the French character; these features will be rekindled and reactivated on the neutral ground of science, on which the best minds of the two nations must meet in their endeavors toward the high goal common to both; thus will the ineradicable feeling of brotherhood gradually contribute in the field of science to soothe the bitterness that the deeply wounded French national pride feels toward Germany, as a result of the war which they forced upon us.[28]

Partisan emotion was clearly showing through here, but we must grant that Liebig's heart was in the right place and at a difficult time for German as well as French hearts. Liebig's good side often eventually won out over his outbursts of chauvinism, selfishness, and temper.

It may be noted parenthetically that Liebig's relations with English chemists were also very close. Despite disparaging comments on English dilettantism and their lack of attention to pure science, as well as a public attack on the idol of English experimentalism, Francis Bacon, Liebig's high regard for English chemists and his continuous collegial contact with them has prompted one prominent English Liebig scholar to refer to Liebig quite justly as "very much an honorary Englishman."[29]

In conclusion, there is no evidence that Liebig was prey to the sort of *pathological* national prejudice that might have chronically interfered with his appreciation of foreigners' work and thus with his pursuit of science. None of this is to deny a certain hot-headed and instinctual chauvinism at the heart of Liebig's character, but the judgment of one historian that "Liebig was the undisputed champion of this growing and squalid German nationalism in scientific affairs"[30] is quite unjust.

Many would want to award such championship honors to Hermann Kolbe, and in truth it would be hard to find a better candidate. And yet, close examination of Kolbe's career reveals an interesting irony. No one had more contempt for the French or their theories in the late 1840s and early 1850s than Kolbe. However, the striking new reactions and brilliant arguments by Gerhardt, Williamson, Wurtz, and Frankland during the early 1850s that convinced most of Kolbe's German colleagues to accept the French–English theories were by no means lost on Kolbe either. By 1857, he had developed a theory of his own that was strikingly similar to the Williamson–Gerhardt newer type theory, namely, that all common organic compounds could be regarded as substitution products of carbonic acid. He retained this theory almost without modification for the rest of his life.

Colleagues, friends, and rivals all pointed out, from the late 1850s until Kolbe's death, both publicly and privately, that Kolbe had become a de facto convert to Gerhardt's system. Kolbe denied it with all the energy at his command. Despite some substantive distinctions between what Kolbe called his own "real types" and the purely "formal types" of Gerhardt's theory, however, the similarities were striking, both to Kolbe's contemporaries and to modern observers. In 1868, two years before the war broke out, Kolbe even converted to modern atomic weight formulas, the last highly visible difference between him and the structuralists–and a step that most French chemists did not take for another quarter century.

To put the matter a bit simplistically, Kolbe's pathological chauvinism had failed to prevent him from understanding and being persuaded by the hated French ideas; it had only operated to prevent him from *believing* he had adopted them. Using his faux types during his most productive years in the 1860s, Kolbe practiced substantively and very successfully the same sort of theoretical chemistry being pursued simultaneously by the structuralists. In short, to the extent that he was an exceptionally good scientist—and there is little doubt that he was—he was also an internationalist in spite of himself. It would be wrong to suggest that Kolbe's bigotry did not damage the quality of his science, for I believe it is clear that it did, especially after 1870. But what is striking is that a man of such violent and ineradicable prejudices against the very direction that we have come to know as modern chemistry was able essentially to become a modern chemist in spite of himself.

I would not want to push my argument too far, for there are well known instances in the history of science when national feelings have seriously damaged the free interplay of scientific ideas. But the present case demonstrates that the Germans accepted the French–English chemical reforms of the 1850s astonishingly rapidly. In fact, it is a striking irony that these essentially French reforms were pursued much more aggressively and enthusiastically in Germany than in France; by the 1860s, structure theory had become a quintessentially German field, while Wurtz felt his to be a voice in the French wilderness. Thus, the prevalence of nationalist fervor provides much less predictive guidance in explaining the growth, development, and differential national reception of some scientific theories than one might have expected.

The German Jew as Chemist

Besides xenophobic chauvinism, Kolbe's other great prejudice was antisemitism. Of course, he was not the only one in his culture to suffer from this malady, and so some background will be helpful in establishing context.

The autobiography of the great German organic chemist Richard Willstätter (1873–1942) provides a sobering reminder of how hard it was for a talented German Jewish scholar to build a career in the first third of the twentieth century.[31] It is insufficiently appreciated how different the academic world looked a generation earlier. Albert Ladenburg (1842–1911) received perhaps the earliest call for an unconverted Jew to be a full professor of chemistry in a German university—to Kiel, in 1872. Yet there is not a single mention of antisemitism or even prejudice throughout his autobiography; he was astonished to receive the call, he wrote—but only because he was still young and very little known.[32] In fact, our views of the position of Jewish scientists in Germany before the rise of the antisemitic movement have been excessively influenced by our knowledge of the horrors that came after. Moreover, there were some unique aspects about chemistry in particular that for a time held hope for Jewish academic careers, aspects that were not present to the same extent in other fields—not even in other fields of science.

Traditionally barred from owning land and from the civil service—hence also from academia—German Jews concentrated their efforts in many of the professions that became stereotypically Jewish, especially business, banking, law, medicine, journalism, and the book trade. Liberated from the ghetto after the Napoleonic wars, full "emancipation"—strict civil and legal equality with gentiles—was granted only upon establishment of the North German Confederation in 1869, which was extended to the entire German Empire two years later.[33] Before this time, most Jews who felt the lure of chemistry naturally gravitated to the chemical industry, such as Heinrich Caro, Franz Oppenheim, Paul Mendelssohn-Bartholdy, and Ludwig Mond. The fact that chemistry was the earliest of the pure sciences to have broad applications to industry clearly served as an incentive for the study of that science at universities. In the great expansion of chemistry enrollments around mid-century, Jews were represented in far greater percentage than their proportion of the population.

In view of the circumstances that there were virtually no Jewish students before 1800 and that before 1869 true professorial calls were essentially impossible for Jews, there were surprisingly many Jews in the periphery of German academia during the middle decades of the century. The more effective exclusion of Jews from other fields meant a concentration of Jewish students in the sciences as well as in medicine and law, a trend that was reinforced by a certain congruence between German Jewish culture and the scientific Weltanschauung. This nineteenth-century German phenomenon represented the first significant Jewish scientific cohort in history. In the field of chemistry, the cohort group was very large indeed. Many of these Jews were simply learning the trade preparatory to entering industry, but many also had been smitten by love for pure science and were willing to tolerate a dead-end career. One recent compilation of "important" German Jewish scientists and mathematicians of the second half of the century suggests that nearly two-thirds were chemists—an amazing proportion.[34]

Even casual prosopography reveals commonalities in Jewish academic career paths in the years before emancipation.[35] Perpetual residence among the ranks of the Privatdozenten, occasional teaching posts at trade academies or technische Hochschulen, or scientific writing and editing provided some of the means of maintaining a presence in academia despite being barred from entering the mainstream. Conversion by baptism was always an option and provided a route into that mainstream for many; however, silent prejudice against converts who obviously still retained their "Jewishness" proved a powerful barrier, both before and after legal restrictions were removed.

There were, to be sure, some exceptions to the exclusion of Jews from academia. A partial loosening of the Prussian legal barriers soon after Friedrich Wilhelm IV acceded to power in 1840 made entry at least possible. Gustav Magnus, a student of Berzelius and a prolific, wide-ranging researcher, became ausserordentlicher Professor at Berlin in 1833 and attained the Ordinarius in 1845. K. F. Rammelsberg ascended from Privatdozent to Extraordinarius in 1846. F. L. Sonnenschein, a third Jewish Dozent at Berlin, had the best chemical laboratory at the university during the 1850s, but did not become ausserordentlicher professor until 1869. Any or all three of these men may have undergone baptism.[36] It is said that the first call to an Ordinarius position for an unconverted Jew was the Göttingen mathematician M. A. Stern, in 1859.[37]

Jews in all walks of life, no matter how well assimilated, had to fight against widely held negative stereotypes in bourgeois society; even those Germans without active prejudices were influenced by these caricatures. Partially emancipated early in the century, many German Jews did achieve status and wealth by the middle decades, and many were apparently fully assimilated into society. However, the negative stereotypes, deepened by such novelists as Gustav Freitag and Wilhelm Raabe, slowed the movement toward granting of complete civil equality. Active antisemitism such as that of Richard Wagner was unusual during the economically prosperous years of the 1850s and 1860s, but the increase in fervor of German ethnic nationalism and aspirations toward a unified state, as well as the new perception of a connection between Jews and radical or anarchic politics, allied with the older perception of even assimilated Jews as "foreigners," laid the groundwork (even among liberals) for the later emergence of an organized antisemitic movement.

Bismarck included Jewish emancipation as an element of the liberal side of his agenda, and full emancipation followed the founding of the Reich. Prosperous Jews played a full role in the overheated economic expansion of the *Gründerjahre* in the early 1870s and were correspondingly popular scapegoats in the inevitable crash and depression, which plagued the German economy for over two decades after 1873. These circumstances help explain why it was precisely this period that saw the emergence of an organized and virulent antisemitic movement in Germany.[38]

Emancipation produced an influx of Jews into the German professoriate, at first very modest in scope and concentrated largely in the sciences and in medicine. Antisemitism, usually of the silent variety, still provided a rather effective barrier for the most prestigious universities, even for baptized Jews, but less so for the more peripheral and especially eastern universities such as Berlin, Breslau, Königsberg, Rostock, and Kiel. Baden, with the most progressive political tradition of all the German states, also provided relatively greater opportunities for Jews at its University of Heidelberg. Albert Ladenburg, as mentioned, was one of the first Jewish chemists to receive a respectable call (although Kiel was then the newest of the German universities). The first to break into the "majors" was probably Victor Meyer, who was called to Wöhler's old chair at Göttingen in 1885 and then to be Bunsen's successor at Heidelberg in 1889. Kekulé's protégé Otto Wallach, an openly professing Jew, was named Extraordinarius at Bonn in 1876; he succeeded Meyer at Göttingen. By 1874, four percent of German Ordinarien were of Jewish descent, the large majority of these Jews having been baptized. Fifteen years later the percentage had risen to more than six percent. By 1909, twenty-one percent of all German Privatdozenten, although only seven percent of Ordinarien, were Jewish by conviction or immediate descent.[39]

Berlin was the one major university that saw Jews enter its professorial ranks soon after 1869. Among those in the chemical community, in addition to Magnus, Rammelsberg, and Sonnenschein, were the pharmacologist Oscar Liebreich and the organic chemist Alphons Oppenheim; Carl Liebermann was successor to Adolf Baeyer (whom the Nazis were to consider a Jew) at the Berlin Technische Hochschule. Together with the steady stream of Privatdozenten that passed through the university and the large number of chemical industrialists and entrepreneurs, Berlin had by far the largest chemical community in the Reich, and a good many of them were Jewish. By 1874, a fourth of the teaching force (including Privatdozenten) and six percent of the Ordinarien at Berlin were Jewish.[40]

As mentioned, antisemitism in Germany grew during the depression years after 1873 and explosively after 1878; however much the ground had been prepared for this movement, it was perceived as novel by proponents and opponents alike. In his justly celebrated study of the German nonscience professoriate during the Wilhelmian and Weimar periods, Fritz Ringer has emphasized the instinctively conservative, socially snobbish, and antimodernist tendencies among the normally "liberal" academic community, a set of attitudes he compares to that of the mandarin elite of China.[41] The sudden lurch of German students into full-throated nationalism, political illiberalism, and antisemitism is more puzzling than that of their elders because the predominant tone of student culture had been, since the days of the early nineteenth-century Burschenschaft, notably liberal. As recently as the opening years of the depression, many had openly flirted with socialism and Marxism. But from 1879 on, there was no mistaking the grass-roots popularity of academic antisemitism, from both sides of the lectern.

The Collision of Kolbe and Hofmann

Kolbe and Hofmann viewed this new movement in very different ways. After his return from England, Hofmann energetically pursued his own agenda in Berlin, which after the Austro-Prussian war became the leading center of political and military power, not only in Germany but in Europe as a whole. With his scientific brilliance, his rock-solid health, and his remarkable natural eloquence, poise, and diplomacy, Hofmann inevitably became the universally acknowledged dean of Berliner, and indeed of German, chemists. Hofmann was benevolent and broad-minded, but also savvy and ambitious. In the years after 1866, all sorts of pan-German associations began to spring up in Berlin, including the Deutsche Chemische Gesellschaft. Hofmann became the first president of the DCG and served as either president or vice-president the last twenty-five years of his life. In his own estimation, as well as in everyone else's, Hofmann *was* the German Chemical Society. From modest beginnings, within a handful of years, the DCG became a large and extremely successful organization, its *Berichte* surpassing Liebig's venerable *Annalen* as the leading European chemical journal.

Kolbe was not alone in having mixed feelings regarding Prussia after the founding of the Reich. Ardent in his support of Bismarck and Kaiser Wilhelm and immoderately proud of his powerful and newly unified country, he was at the same time suspicious of *domestic* Prussian imperialism. Hofmann's decision to name the new society "German" rather than "of Berlin" revealed to him an ugly element of hubris and overreaching ambition, as we shall see later. Hofmann's relative neglect of his old friend (probably a result of Kolbe's increasingly intemperate attacks) did not help; Kolbe's visits to Berlin were not reciprocated, and their correspondence became spotty and stiff. Privately and orally Hofmann began to refer to Kolbe's periodical as the "Journal für *polizeiliche* Chemie."[42] It was perhaps partly to quiet his obstreperous friend that Hofmann added Kolbe's name to those of the elder spokesmen Liebig, Wöhler, and Bunsen as charter Honorary Members of the society.

Matters came to a head after Wurtz' nationalist dictum was published. However much Wurtz tried subsequently to exonerate himself by claiming only to have indicated the *birthplace* of chemistry, the chauvinist implications drew fire, especially from across the Rhine. In 1870, Kolbe wrote a blistering and highly insulting critique of contemporary French chemistry, and his former student Volhard published an interesting historical essay whose premise was that Lavoisier was more a physicist than a chemist. Thereupon four prominent Russian chemists published a declaration in the *St. Petersburger Zeitung* castigating Kolbe and Volhard for further inflaming nationalist sentiments during wartime.[43]

With Liebig's assistance, Volhard wrote a temperate response to the Russians, which Kolbe published in his journal. There the matter might have rested, had a second protest by the entire Russian Chemical Society not been translated and printed in the *Berichte* as part of a correspondent's report on chemical news from Russia.[44] Kolbe's fury—against the Russians, against the DCG, against Hofmann, and against Hofmann's "henchmen"—knew no limits. His reaction to this incident revealed a burning hatred for Berlin Jewry. Kolbe wrote Baeyer, then president of the Deutsche Chemische Gesellschaft,

> I won't say anything about the fact that the [German] Chemical Society uttered not a word of disapproval or rejection of the shameless insolent statement by Wurtz that "chemistry is a French science," or against his book that begins with these words, because it is well known that the international tendencies within the pale of the Society are too great. But that . . . this inappropriate communication [by the Russian Chemical

> Society] was ultimately accepted even into the reports of the Berlin Chemical Society testifies truly to a great disrespect to its Honorary Members, which understandably makes it impossible for me to continue any further relationship with it. . . . Moreover, I regret the fact, as an indication of the ever apparent lack of self-confidence and self-respect among us Germans, that the Berlin Chemical Society has given its approval to the Petersburg declaration by accepting it in silence, and has thereby further expressed the view that it finds nothing worthy of reproach in Wurtz' book.[45]

Baeyer replied with a tactful letter, pointing out that the Society per se never takes responsibility for authors' views and commenting that most members disagreed with the Russian statement.[46]

But Kolbe could not be mollified. The Society had insulted not only him but also Liebig, since he understood that the Society was considering a plan either to purchase Liebig's *Annalen* outright or to found a new companion journal for the *Berichte* that would be devoted to longer articles and would therefore compete directly with the *Annalen*. Baeyer replied that such an objection sounded odd coming from the editor of another journal that already competed with the *Annalen*.[47] Kolbe shot back a letter claiming (quite disingenuously, as surviving correspondence documents[48]) that he had only accepted the editorship of *Journal für praktische Chemie* under the prior express approval of Liebig. Moreover,

> If the Chemical Society has really founded a new journal, it will soon no longer be Hofmann and Wichelhaus who decide the direction and orientation of the journal, but the masses, the rabble of young chemists preparing themselves in the Chemical Society. The Chemical Society is after all already known as a hotbed of Jewry in chemistry.[49]

It was just at this time that Hofmann returned to Berlin from London, where he had been visiting his seriously ill wife. Brought up to date by Baeyer, he wrote his friend, feigning astonishment that Kolbe had taken offense:

> You expressed your views on French chemists, this was not to the taste of the Russians, they expressed themselves in the Chemical Society in St. Petersburg, and our correspondent, who reports regularly on the meetings of the society, mentions this announcement as well. How could this constitute an insult to our valued honorary members? I must confess I have a different view. To me it seems only fair that, after one has expressed himself on a question concerning which (it cannot be denied) very different views are possible, others be permitted to express their opinions as well.[50]

Kolbe could not be swayed. "Don't tell me that I am judging the matter falsely," he retorted with anger. "I am no child, forming my judgment without mature reflection."[51] He demanded that the DCG publish an expression of regret over the incident; Baeyer and the Executive Committee refused. Kolbe then resigned. The incident was discussed in the Society's annual report, printed in the *Berichte*, with the final comment by President Baeyer that the editor (Wichelhaus) had acted fully within the Society's bylaws in publishing the Russian declaration. Kolbe described the affair in a letter to Liebig, saying that he felt he had gotten the better of "die Berliner Herren," that they had never expected he would actually resign.[52] To Volhard he wrote,

> Let us stick closely together, that the (this between us) fraudulent spirit of the Berlin Chemical Society and the Jewry that flourishes there not take root in Germany. The editor of a journal has a good weapon, and should indeed use it for defense.[53]

He did use this weapon, with all the skill of a dedicated polemicist, especially in his annual retrospectives. The DCG came in for repeated criticism: for having the arrogance and imperialist ambition to call itself "German," for its internationalist and social–democratic tendencies, and for electing Cahours and then (even worse!) Wurtz himself as honorary members. The Berliners, he suggested, were trying to grab power and centralize the discipline in their city; but this was a mistake that the overly centralized French had committed, to their sorrow. He even lambasted the Society for its mode of organization and the style of editing the *Berichte*.[54] In their annual meetings, Hofmann and other officers of the Society defended themselves in the pages of the *Berichte* against these attacks.[55] To various correspondents, Kolbe complained of Hofmann's towering ambition and vanity. He would openly challenge Hofmann, he suggested, were it not for his old friendship; moreover, Hofmann habitually used any of a number of lackeys to do his dirty work, Kolbe thought.[56]

Kolbe was by no means alone in his dismay over the events in Berlin. He succeeded in eliciting sympathetic comments against the DCG from Bunsen, Liebig, Kekulé, Lothar Meyer, Franz Varrentrapp, and Volhard, although neither Bunsen nor Liebig would allow his name to be used openly.[57] Friedrich Beilstein needed no prompting; he despised Hofmann, as well as his obsequious "personal footman," Wichelhaus, and detailed his outrage at their conduct in letters to Erlenmeyer.[58] But Kolbe pushed his position too strongly, and some of his language, especially in published pieces, exceeded the bounds of decency. Lother Meyer wrote Baeyer,

> You can imagine that I really had to laugh over Kolbe's silly exit from the society of cultured chemists. But the matter also has its serious side. K. now writes so much strange noise in the style of infallibility that I begin to fear for his health. I have given his new piece against Virchow to a number of impartial people, and every one used the expression, "This man is certainly crazy." Indeed, I am beginning to believe that K. is suffering softening of the brain, whose symptom is of course megalomania. A pity on the man, who deserves something better.[59]

In his correspondence with good friends such as Volhard, Varrentrapp, and Kopp, Kolbe's language was even cruder. He was disgusted, he said, over the actions of the "Berlin Chemical Jew-Society"; moreover, the "Jew-boys" seemed to have Hofmann under their control as much as the other way around. But they would not succeed; once Baeyer leaves, Kolbe predicted, Hofmann would be the only chemist of note left in Berlin, and the Society would collapse of its own weight.[60] "It's sad," he wrote Varrentrapp, "that he was so long in England and forgot the customs [Sitte] of German scholars. . . . Hofmann unfortunately lost the fatherland in England."[61] In fact, he said he feared for Hofmann, for the Jews might well want to discard him after using him for their own purposes. He told Hofmann this directly in a letter of February 1873.[62] Hofmann never answered this letter. On Kolbe's next visit to Berlin four months later, Hofmann managed to avoid him.[63] Their relationship was essentially over.

Hofmann Versus the Antisemites

All this occurred just before antisemitism began to flourish as a distinct movement in Germany. Its outbreak in Berlin in 1879 caught Hofmann in its wake, for in the following year he was elected rector of the university.

The birth of modern German antisemitism in that year began with the agitation of journalist Wilhelm Marr and preacher Adolf Stöcker, but what made the movement especially powerful was the academic respectability given it by the famous historian Heinrich von Treitschke, one of the most popular and respected professors at the University of Berlin. While distancing himself from rabid and racist agitators such as Marr, Treitschke considered the Jews "an element of national disintegration" and called on them to assimilate fully to their fatherland as the price of emancipation.[64]

The liberals were not silent in the face of these developments. A "declaration of notables" condemning the attacks was issued by 73 prominent Berliners, among them 17 of Treitschke's faculty colleagues—including fellow historian Theodor Mommsen, biologist Rudolf Virchow (both founding members of the Progressive Party), and Rector August Wilhem Hofmann. But the movement caught fire among the students. In Leipzig in October 1880, Bernhard Förster began to circulate an academic petition with radical demands for restrictions of the rights of Jews. Within eighteen months, 255,000 German students, about one-fourth of all matriculants, signed.[65]

At the same time in Berlin, an eloquent and popular law student named Erich von Schramm began to organize a "German Students' Union" (Verein deutscher Studenten), whose main purpose was to be antisemitic agitation. Fortunately, he needed official approval for registration of the society, either by the university or by the police, so that Hofmann had the power to prevent the group from forming at the university. Throughout the 1880–1881 academic year, he rejected three successive drafts of the proposed bylaws, disguising his fundamental opposition in legal technicalities and in the neutral argument that "political" organizations, of whatever orientation, are inappropriate in the academy and would lead only to discord among the students. Most of Hofmann's colleagues supported this reasoning. Finally Schramm gave up and registered the Union with the police. However, after a final revision of statutes in 1881, the university officials were forced to admit the Union as a registered student society.

The German Student's Union, born in Berlin, spread quickly to other universities, even before its official certification. In Leipzig, Halle, Göttingen, Kiel, and Greifswald, many hundreds of students joined.[66] Hofmann was sorely troubled by these events; it is said that early in 1881 he contemplated resigning his office.[67] For its part, the Union never forgave Hofmann for his interference and boycotted every Hofmannfest in Berlin, even his funeral in 1892.[68]

No doubt Hofmann was delighted that the antisemites stayed away from the various appreciations given him in his last years. Despite the hard feelings and jealousies engendered by Hofmann's crowning himself "king" of chemistry and creating a sort of chemical court around himself, few chemists have been so well and so vocally feted during their lifetimes. His scientific productivity was little diminished even into his old age, and he was honored by most scientific societies around the world. Even his death was fortunate in a sense, for his health continued to be nearly perfect until a sudden stroke or heart attack killed him at the age of seventy-four. The flowers sent from around the world filled his home to overflowing. The funeral was attended by 1500 mourners.[69]

Kolbe's mature years were very different in tone. His friends began to slowly disassociate themselves from the "thunderer of Waisenhausstrasse," as Lothar Meyer called him.[70] Jacob Volhard was caught in the middle. Kolbe regarded his former student as a good friend and loyal comrade in arms, with whom he could be completely open and trusting—after all, Volhard had shared the Russians' censure for his nationalistic piece on Lavoisier in 1870.[71] But Kolbe had forgotten, or perhaps was unaware, that Volhard was the son of an extremely liberal and idealistic attorney, that he had an internationalist outlook similar to that of Hofmann, and that both he and his parents had long been intimate friends with the Hofmanns and with the Kekulés (as well as with the Liebigs) from their hometown relationships in Darmstadt.[72] By the mid-1870s, Volhard was also remonstrating with Kolbe and sometimes not altogether gently, despite his feelings of respect and friendship.

We have seen that Kekulé became Kolbe's favorite target for vilification. Kolbe even claimed that Volhard himself was suffering from Kekulé's machinations, since Kekulé had engineered a call for his student Theodor Zincke to Marburg—like Kekulé, an uncultured, incapable man—and Kolbe maintained that Volhard would otherwise have gotten the call. His bitterness was no doubt heightened by the fact that this had been his chair until 1865. Kekulé, he concluded with characteristic coarseness, "has succeeded in transforming the chemical chair at Marburg University into a night chair [*Nachtstuhl*, or chamber pot] for a large number of years."[73] In the fall of 1882, Kolbe blasted Baeyer with the same viciousness as he had Kekulé.[74] In a letter to Frau Baeyer, Volhard expressed sympathy:

> My old friend Kolbe is behaving truly irresponsibly. A pity on the man; since he began to devote himself to insults he has produced nothing more of value. . . . Once one has delivered oneself up to vanity, one does not realize to what degree of madness this will lead him.[75]

Obituaries provide some estimation of contemporary respect for the departed, especially if one knows how to read between the lines. Volhard wrote a long, highly laudatory biography of Hofmann, one of three book-length treatments of his life and career that were published in the first decade after his death (a fourth appeared in 1918, simultaneously Hofmann's hundredth birthday and the fiftieth anniversary of the founding of the Deutsche Chemische Gesellschaft).[76] In his description of Hofmann's battles with the antisemitic students, Volhard left no doubt regarding his own hatred for the "disgusting Jew-baiting" of the students, and his admiration for Hofmann's persistent resistance.[77]

As for Kolbe, Volhard made an appreciative speech at the unveiling of a plaque in his honor, but it was never published. Hofmann composed a somewhat stiff and very short obituary for the *Berichte*, departing notably from his customary exuberant style.[78] It is well known that Hofmann was an enthusiastic and energetic obituarist in his old age, filling three substantial volumes of collected "Memories of Departed Friends."[79] Kolbe was not among these friends; Kolbe's hated French rivals Dumas and Wurtz, along with three Jews (Gustav Magnus, Alphons Oppenheim, and Paul Mendelssohn-Bartholdy), were. In fact, there may have been an unstated message in the fact that he chose to memorialize both Wurtz, author of the line about French chemistry that caused so much pain, and Oppenheim, Wurtz' Jewish student who translated the work into German.[80] The Wurtz biography, published in German, is by far the longest and best biography ever written of this major French chemist.

In general, Kolbe has fared very poorly regarding obituaries and biographies. The authors of the two most authoritative obituaries were at once relatives and former students of Kolbe (his son-in-law Ernst von Meyer and his nephew Hermann Ost). Significantly, a primary objective of both of these obituaries was to resurrect Kolbe's tarnished reputation and to try to place his attacks in the best possible light. With the presumed exception of the present effort, Kolbe's life and career have not been properly studied—a fact that is surely in some measure a consequence of his conduct, especially after 1870.

A principal focus of this chapter has been intolerance, antisemitism in particular. If that prejudice is a disease from which one suffers (as many, even antisemites themselves, have described it),[81] then Kolbe can be said to have contracted a serious, even pathological, case. To what extent was he typical of his peers in the academic chemical community? Any direct answer must be carefully qualified. It is obvious that most Christians in German academia, even those who worked closely and happily with Jews, were influenced by the negative stereotypes that prevailed in Bismarckian society;[82] however, I have found little evidence for widespread and active prejudice among the peer community described here.[83] Hofmann stands out as a particular example of almost aggressive philosemitism, but many others in the community—Bunsen, Wöhler, Kekulé, and Volhard, for example—were invariably kind and fair-minded regarding their Jewish students and colleagues.

Many of the great names in German chemistry, including those just cited, had numerous Jewish students, assistants, and associated Privatdozenten. Bunsen recommended Victor Meyer as his successor at Heidelberg, and Kekulé did everything he could to advance the career of Wallach. It is certain that many recipients of letters from Kolbe containing antisemitic slurs did not welcome them—namely, correspondents such as Volhard, Kopp, Hofmann, and Baeyer. Even noted antisemites could act kindly toward individual Jews: Oppenheim was one of Treitschke's few friends during his student days, and Treitschke loved him dearly his entire life.[84] Even Kolbe could have kind words of respect for Ladenburg and Victor Meyer.[85]

Were scientists in general any different from humanist scholars during this period? An unequivocal answer is not possible. Scientists are certainly full members of their wider collegial community as well as the general culture of the society of their day. As Ringer has shown, the Wilhelmian academic community, imbued with the neohumanist ethos of *Kultur*, had a number of characteristics that appear to us as arrogant, narrow-minded, and antimodernist. Most academic scientists during our somewhat earlier period shared many of these values, especially a sense of the ineffable qualities and inestimable importance of the neohumanist Gymnasium as a means of forming a sensitive, cultured, and broadly educated mind. Both Hofmann and Kolbe fought tenaciously—Hofmann with his accustomed tact and Kolbe with his usual abrasiveness—against the movement to weaken Gymnasium education as preparatory to university. "To have read Homer," waxed Hofmann lyrical,

> . . . quickens one's life. The face of a gray-headed public servant, upon which the pencil of time has engraved the unmistakable traces of official monotony, will lighten, if perchance the full-sounding hexameter of the Iliad strikes unexpectedly upon his ear. It is as though his youth suddenly flickered up again within him. What the Bible is to the common people, such is in many respects Homer to the educated.[86]

Clearly, in many respects scientists shared the humanists' value system. Nor do I mean to set up chemists as paragons of virtue and tolerance during this period of increasing illiberalism. Kolbe is a particularly virulent counterexample on the side of the chemists, and the historian Mommsen's staunch progressivism was a counterweight to colleague Treitschke's bigotry.

And yet, there is a certain danger in underestimating the cultural differences between academic scientists and humanists considered as groups, in the last century as in this. William Coleman has written vaguely, but I think accurately, about "a veritable culture of science" which took root in early nineteenth-century Germany, a thesis I have attempted to address more fully in the first chapter of this book.[87] This scientific culture gathered force through the century, especially as the sciences and (until late in the century) chemistry above all looked increasingly important for promoting German industrialization and political–military power. David Rowe has also argued that Ringer's conclusions from his study of the nonscience German professoriate cannot unproblematically be extended to the science community.[88] This suggestion is reinforced by the fact that however hard it may have been for Jews to succeed in science, it was virtually impossible in fields such as literature or languages, where the first Jewish Ordinarius was not appointed until the Weimar period.

It must be noted, however, that real strains had begun to develop in Bismarckian scientific culture, and the Kolbe case illustrates some of these. In a sensitive and revealing monograph on the astrophysicist Karl Friedrich Zöllner, a colleague of Kolbe's at the University of Leipzig, Christoph Meinel has explored "the fragility of the culture of science in Imperial Germany, and its hidden antinomies."[89] Zöllner, like Kolbe, was a vehement (and antisemitic) critic of modernity, who attacked Hofmann and other prominent liberal scientists of his day. (Hofmann himself thought the two were in league, though apparently this was not the case.)[90] The situation thus had become sufficiently complex to frustrate any simple generalizations.

The price of prejudice, in every discipline, must not be underestimated. Even the perpetrators of prejudice were victims: one object lesson from this study, a "cautionary example" to use Volhard's phrase, is the self-inflicted demolition of Kolbe's erstwhile brilliant career. Much more destructive than the actions of a single man, though difficult to gauge, were the institutional barriers that existed until the Empire was founded, which acted to exclude a talented community of intellectuals from German universities. Institutional and legal barriers were, of course, re-erected after Hitler's accession to power in 1933; Alan Beyerchen has masterfully depicted the consequent impact on the German physics community.[91] A final factor that is most difficult to document or study in detail is silent prejudice, since it may not even be revealed in private letters. There is no question that qualified Jews in every field continued to be discriminated against even after full emancipation. Nor was antisemitism confined to Germany; the difficulty of Charles Gerhardt to achieve career success in France was surely at least in part a function of his Jewish heritage. It is also true that antisemitism was a serious and in many cases insuperable barrier to Jewish advancement in the elite universities of the United States during this same period.

Historians sometimes forget that, however much we try to depict the apparent inevitability of later historical events from earlier contexts, factors, and trends, history is actually a contingent process, manufactured by freely acting and often unpredictable human beings.

The apparent progress that Jews made in academia during the 1870s and even in the 1880s was not *foreordained* to be reversed. When Friedrich III ascended the throne early in 1888, he represented the great hope of German liberals. One of his first acts was to force the resignation of the reactionary interior minister R. V. von Puttkamer. Consequently, there was real reason to believe that the elements of liberal democracy were finally in the ascendant, in the academy as elsewhere. Tragically, Friedrich was already dying when he became emperor, and he ruled only ninety-nine days; he was succeeded by Wilhelm II, a very different man from his father.

Five months after Friedrich's death, Hofmann published his collected biographies, with its liberal subtext, and dedicated the volumes to Friedrich's widow, Victoria. Victoria was the eldest daughter of Queen Victoria and had studied informally with Hofmann in her youth. Upon Hofmann's death four years later, Victoria wrote a long letter of condolence to Hofmann's widow and directed that busts of two scientists be placed flanking a statue of her to be erected in the Tiergarten: Helmholtz and Hofmann.[92] About Wilhelm, the new emperor, she is said to have commented, "Don't for a moment imagine that my son does anything from any motive but vanity."[93] The German Reich then had begun its trajectory toward war and holocaust.

Last Years

When Kolbe arrived in Leipzig in 1865, he brought with him his wife Charlotte, nine-year-old Carl, eight-year-old Johanna, and five-year-old Maria. (He also brought several members of his Praktikanten "family" from Marburg, including Drechsel, Finkelstein, Glutz, Wischin, Zaitsev, and Ziegler.) A third daughter, Elisabeth, was born in January 1868. The family lived in an apartment next to the old laboratory on Universitätsstrasse for three years, then in October 1868 moved into the spacious and elegant residence in the new institute on Waisenhausstrasse. At about this time, an unmarried sister of Kolbe's, "Tante Rutsch," moved in with the family permanently.

Kolbe's first decade in Leipzig was a time of great satisfaction. His research group was extremely productive, he had literally crowds of students (both auditors and Praktikanten), and after January 1870 he even had a journal at his disposal. He was regarded rightly as one of the preeminent chemists in Germany, and plenty of distinctions came his way. The Russian government granted him the Stanislaus Medal for his work with Russian students, and the Universities of Kazan and Moscow gave him honorary degrees. He was given the Davy Medal of the Royal Society, Wöhler successfully nominated him Correspondent of the Göttingen Societät der Wissenschaften, and dearest to his heart, Liebig proposed him as Knight of the Bavarian Maximilians-Orden. That same month (December 1872) he was appointed a Saxon Geheimrat (privy councilor), the highest honorific a German state could confer on a university professor. He was also made Honorary Member of the new Deutsche Chemische Gesellschaft at its creation, along with Liebig, Wöhler, and Bunsen. When the Prussian Akademie der Wissenschaften made him Corresponding Member in 1875, he declined and sent the certificate back—partially because the document read *Heinrich* Kolbe.[94]

Then as now, the hard currency of reputation is the demand for services. The University of Bonn called Kolbe in 1867, but he declined, leaving this prize to Kekulé. It was known that Liebig always desired that Kolbe should be his successor at Munich. When Liebig died in April 1873, his junior colleague ausserordentlicher Professor Volhard wrote to Kolbe, delicately inquiring about his possible conditions for a possible call. Kolbe made it clear (or at least tried to) that he would be inclined to accept a call were the conditions sufficiently generous, and outlined his current emoluments and facilities. To Kolbe's disappointment, these expensive hints were enough to put the Bavarian authorities off the scent. Nonetheless, Kolbe was able to use the feeler at home to gain a raise in his laboratory budget (to 4500 thalers) and an expansion of his auditorium.[95] (The Munich call actually went to Kekulé, who declined and used it for leverage in Bonn;[96] finally, in 1875 Baeyer was called as Liebig's successor and he accepted.) The Liebig succession understandably excited the chemical rumor mills. In the middle of the negotiations, Adolf Lieben wrote Erlenmeyer, then at the Munich Technische Hochschule,

> What's happening with Liebig's succession? You no doubt have heard that the mere thought of Kolbe (for an actual call is said not to have taken place) sufficed to make him privy councilor. I am glad of it, with not the slightest touch of envy. As long as the earth lasts, perhaps no man lives who is more appropriate, who is better equipped by nature, to be a privy councilor than Kolbe. God, who ordered all things by measure and number and weight, once created Hermann Kolbe, not in his own image, but rather after the type of the Geheimrat, and said that it was good![97]

It is because of Volhard's inquiry that we know that Kolbe was earning close to 8000 thalers in salary, honoraria, and fees, several times what he had been making in Marburg.[98] Certainly his relatives were impressed. In a letter to his sister, he emphasized the various financial obligations that were connected with his position, arguing that the income was not as magnificent as it might appear: 200 thalers annually for entertaining, 300 each year for a medically necessary spa "cure," 100 for school, 100 for piano lessons, 100 each for a gardener and a maid, 240 for utility bills, 600 for life insurance, 100 for taxes, and 100 for medical expenses, not to mention food and clothing for seven people.[99] This enumeration alone suggests quite comfortable bourgeois domesticity, especially in light of the fact that his residence was gratis. Considering that he was not able (or in any case failed) to pay off the last of his personal debt to the Vieweg family until 1877,[100] it is clear that he had learned how to spend money. This is especially true after 1874, when he was able to earn many thousands of thalers additional income from salicylic acid manufacture, as we saw in chapter 12.

All of this does not include income from writing and publishing, for which figures are not available, but which could not have been negligible. In addition to his journal, Kolbe had a number of writing projects. In the spring of 1868, he published a brochure describing his new (then not yet completed) laboratory, and four years later he edited a large volume containing reprints of all the articles he and his students had published since 1865. Both works were based on existing published models.[101] In the meantime, Kolbe's detailed textbook of chemistry, begun about 1848(!), was still not finished. He continued to work away on the third volume until Heinrich Vieweg finally persuaded him in 1871 to give it over to others, at which time he lost all control over the project.

About the time the textbook was finally completed (1878), Kolbe put his son-in-law to work on a second edition. That this work was more an obligation than a free choice for Meyer is indicated by the fact that, upon Kolbe's death, Meyer put a halt to the project, at a time when only two parts of the work had appeared. This was not just an updating but an extensive rewrite of the first edition, and Meyer had full authority for its contents. Throughout this work, Meyer used linear structural formulas (e.g., leucine as $CH(NH_2)[CH_2CH(CH_3)_2]COOH$) and propounded an aromatic theory that was neither Kolbe's nor Kekulé's. He did, however, use ortho/meta/para nomenclature, assert the chemical equivalence of all benzene hydrogen atoms, deny the existence of salylic acid, and affirm the need for three isomers for every diderivative of benzene. All of these were Kolbean heresies, which illustrates once more the extent to which committed members of Kolbe's school often proceeded in a very non-Kolbean fashion.[102]

Meanwhile, a new project began to take up Kolbe's time and occupied him off and on the rest of his life: a *short* textbook of chemistry.

Begun in 1872, the first (inorganic) volume was finished by 1877 and the second (organic) volume was out by 1883. This work proved far more popular than his long textbook of organic chemistry. A second edition and an English translation had begun to appear before his death in 1884.[103]

The last source of Kolbe's income that needs to be mentioned is a category that was gradually becoming more significant in the lives of scientists: connections with industry. Evidence from his correspondence suggests that from the late 1850s on, Kolbe was looking for possible applications from his scientific work. His investigation of salicylic acid in 1873 was at least partially motivated by his desire to find a route to indigo synthesis, and he always paid attention to novel compounds with potential as dyes. None of these ideas bore fruit until the salicylic acid work created an ideal entrepreneurial opportunity. Finally, it appears that in 1869–1870 he was financially involved in an American company that was attempting to apply Liebig's formula for extract of beef to Texas cattle (the company failed), and shortly thereafter he did consulting and quality control work for a similar concern in Montevideo.[104]

In many respects, Kolbe's personal life hit apogee around 1875. His wife had never enjoyed robust health but was then doing acceptably, as might be judged by the fact that the Kolbes, who lived very quietly between 1857 and 1869, had rediscovered an active social life. The two oldest children were teenagers in the early 1870s, and their parents gave them occasional large parties and balls.[105] The oldest child, Carl, graduated from the Leipzig Gymnasium in March 1875, then absolved his one-year military service before studying chemistry with his father. He made a successful career in the chemical industry.[106] The next oldest, Johanna, was courted by Ernst von Meyer, and they were married in March 1876. Meyer eventually succeeded Rudolf Schmitt at the Dresden Technische Hochschule. The two youngest daughters, Maria and Elisabeth, were married within two months of each other in the fall of 1887.[107] Unfortunately, Charlotte Kolbe barely lived to see Johanna married. She suffered from a variety of ailments, especially a severe lung infection in late 1875. Two months later, not yet recovered, she was diagnosed with breast cancer. A successful operation by Carl Thiersch failed to cure the disease; after a period of apparent recovery she declined rapidly in the fall of the year before succumbing on 26 December 1876.

About the same time, Kolbe's own health began to decline. He had always suffered from periodic severe colds and influenzas and at least annual bouts of painful rheumatism, sometimes lasting two weeks or more. He was overweight, suffered from atherosclerosis, and ate unhealthily (by modern standards, at least).[108] The death of his wife led to a month of serious mental and physical illness, which gradually abated, but he never quite recovered from the grief.[109] As we have seen, several close friends had died in the early 1870s, and his war with the structuralists was not going well.

Then in early May 1879, a nearly fatal poisoning episode (breathing phosphorus pentachloride) precipitated severe bronchitis that developed into chronic asthma and emphysema. Kolbe had long become accustomed to semiannual "cures" at such resorts as Nauheim, Marienbad, Sassnitz on the Baltic, and Gersau or Brunnen on the Lake of Lucerne; now his holidays became even more frequent and extended. During the last six years of Kolbe's life, he constantly shuttled between Leipzig and these resorts. He was able to do a substantial amount of literary work (especially on the *Kurzes Lehrbuch* and his various polemical articles), but little actual laboratory supervision or research. After a bad winter of 1883–1884, Kolbe's health became better the following summer than it had been in years. On 25 November 1884, Meyer wrote to tell Ostwald that Kolbe was doing exceedingly well so far that winter. A hurriedly written postscript reported a massive heart attack suffered that evening by "my excellent father-in-law."[110] The thunderer of Waisenhausstrasse was dead.

Issues and Reflections

Henry Armstrong was a partisan observer, but there is much truth in his advocacy of his mentor:

> Kolbe is chiefly known to-day [in 1930, but the point is just as valid two generations later] by his attacks on Kekulé and van't Hoff: the fundamental importance of his work is overlooked. Some day when—as Sir Thomas Browne might urge us to do—we omit the "improperations and terms of scurrility" which he launched at Kekulé, he may come to be regarded as the parent of the modern system of resolved structural formulae: we here have to thank him for having made Frankland, whose senior he was, what be became.[1]

No chemist was a more skillful or fruitful worker in the middle decades of the last century than Kolbe, who was in the very forefront among those investigating molecular "constitutions" and devising some of the earliest synthetic methods in organic chemistry. Moreover, it was precisely these directions—structure and synthesis—that proved to be the key to the great burst of scientific and technological innovation described in our story. Even though it was the true "structuralists" such as Kekulé, Butlerov, Crum Brown, Frankland, and Erlenmeyer who formulated and developed the theory in its essentially modern form, Kolbe's pioneering work needs to be recognized as such.

In pedagogy, too, Kolbe played a leading role. It is difficult to disentangle Kolbe's contributions from those of his predecessors, especially Liebig and Wöhler, but it is a truism that the German system of science education which all of these men helped to develop was gradually exported (with admixture of important indigenous elements) to Britain, France, the United States, Russia, and other countries. This borrowing became quite self-conscious beginning in the 1860s, when German superiority began clearly to be sensed abroad; government commissions in Britain, Russia, and France set off alarms and produced reforms. Liebig was particularly influential in Britain, and Wöhler was the most prolific chemist breeder for the United States. However, characteristic Kolbean elements can be discerned especially in Russia, where Kolbe was the chemist of choice for *komandirovky* in the 1860s and 1870s,[2] and in England, through the mediating influence of the Kolbeans Frankland and Armstrong. Armstrong was the leading English science educator at the end of the last and the beginning of this century and did much to spread enthusiasm for what he called the "heuristic" system. By this term, he meant that students should not be passive receptacles for knowledge by listening to lectures and memorizing facts (what he called the older "didactic" or "chalk and talk" method) but rather they should *create* their own knowledge by personal discovery and with their own hands, through a combination of intensive expert-guided and self-guided experiments. Much about Armstrong's teaching philosophy appears to come straight from the Kolbe laboratory in Leipzig.[3]

Kolbe's mentors, colleagues, and students well appreciated the high significance of his work. The three heroic chemists of the "classical" generation—Liebig, Wöhler, and Bunsen—all agreed that Kolbe was the very best among the younger generation, regarding him above even Hofmann and Kekulé. (Liebig favored Hofmann during the 1840s and 1850s, but then their relationship cooled somewhat, after which Liebig transferred his primary loyalty to Kolbe.) Evidence for Wöhler's regard is his attempt in 1862 to persuade Kolbe to edit a new edition of his *Grundriss der Chemie*.[4] From the beginning of their relationship until Kolbe's death, Bunsen's loyalty never wavered. Moreover, all three chemists strongly (though silently) sympathized with Kolbe's attacks on structural formulas and structural chemistry. In addition, Kolbe students such as Ernst von Meyer, Ost, Schmitt, Armstrong, Volhard, Graebe, Crum Brown, Curtius, and Beckmann prized their relationships with their mentor and knew well the important historical role that Kolbe had played.

* * *

In retrospect, however, the most salient aspect of Kolbe's career is the thoroughness of its destruction. As we saw in the last chapter, the beginning of the end was Kolbe's public exit in 1871 from the Deutsche

Chemische Gesellschaft and also, as Lothar Meyer put it, from the "Gesellschaft gebildeter Chemiker,"[5] over a really minor matter that Kolbe managed to whip into a storm. Two years later, upon the death of Liebig, a committee comprised of the most prominent German chemists was set up to plan commemorative events. Kolbe was pointedly excluded, even though he was apparently at that time the leading candidate to succeed Liebig; this exclusion seems to have been due to the action of the chair, Hofmann.[6] Upon Wöhler's seventy-fifth birthday (which was also his fiftieth anniversary of teaching), an elaborate appreciation was given him, but again Kolbe was snubbed, even though Kolbe was Wöhler's oldest student of any prominence and all were aware of Wöhler's regard for his former student.[7] Even many of Kolbe's hitherto closest friends, such as Hofmann, Volhard, and Frankland, became alienated during the last years of Kolbe's life. After 1873, Liebig and Eduard Vieweg were no longer around, and those who still supported him, especially Wöhler and Bunsen, refused to do so publicly. In his last decade, Kolbe was isolated from the collegial community in an extraordinarily complete fashion.

Dramatic decline in the influence of a hitherto major figure in a field is a familiar story in the history of science, and such decline often takes place, as with Kolbe, in middle age. In many cases, this occurs through such obvious phenomena as psychological burn-out, physical exhaustion, or the complacency bred by a successful career—Kekulé, for example, suffered from all of these and published little personal research after the age of forty-four. What is unusual about Kolbe's story is the degree to which he remained engaged with both the theoretical and experimental sides of the science until the age of sixty, and the degree to which empirical evidence, including even many of his own experiments, failed to support his scientific beliefs. These factors transform an otherwise prosaic case into an interesting anomaly.

Some causal elements of Kolbe's tragic dénouement can be suggested from social and psychological factors. Kolbe's simple rural upbringing contrasted with the upper-bourgeois urban origins of such men as Hofmann, Kekulé, and Baeyer; in this sense he found a natural counterpart in Frankland, who had a similarly modest background. As Armstrong commented, he was plain spoken and straightforward, not at all eloquent or cosmopolitan. He must have bitterly resented Kekulé's dismissive and elitist responses in some of their polemics, for, whether conscious or not, that sort of treatment succeeded in accentuating the social distance between them. Kolbe's rejoinder was to highlight Kekulé's putative linguistic solecisms, a tactic that was intended to demonstrate a lack of *Bildung*. As time passed, Kolbe became ever more proud, imperious, and distant, and he lost the mental and emotional elasticity to adjust to change. Moreover, his sincere religious faith, increasingly conservative politics, and open prejudices rankled with the generally centrist, materialistic, and agnostic peer community. In Germany there was nothing like the clear center–periphery dichotomy that was (and is) so prominent in French and British social geography, but Kolbe's career trajectory as a sort of satellite orbiting the dominant state of Prussia is remarkable, and we find in Kolbe a decided aversion to things Prussian and Berliner.

Psychologically, there appear to have been pathological elements, especially after 1870. Kolbe was nothing if not conservative in his theoretical preferences,[8] and he began to view novel developments in chemistry as just another aspect of modernism. Somehow he began to associate structural formulas with sensualism and materialism, possibly even with irreligion. His whole life was devoted to the science of organic chemistry, and he saw that science almost in the personification of a pure virgin being seduced and destroyed by meretricious villains, by liberals, social democrats, traitors, atheists, Catholics, and Jews. In the 1850s and 1860s, he suffered periods of paranoia and severe depression, and after 1870 he appears to have had delusions of grandeur. He became a caustic critic of modernity tout court, a personality type that was by no means unique in Bismarckian society—as we saw in the preceding chapter.

* * *

What perspective on our subject is provided by simple considerations of generational change? It is remarkable that Kolbe's birth year (1818) distinguishes almost without exception nonstructuralists from structuralists in Germany: virtually all of the former were born before this date, the latter all after.[9] This means that nearly all of our modernist protagonists were in their twenties and thirties during the crucial decade of the 1850s. It is also interesting to note the high degree of antipathy to theory in general expressed by the leading chemists of the older generation. Wöhler and Bunsen both were non- or even antitheoretical their whole lives, and Liebig and Dumas ostentatiously rejected theory around 1840; Williamson and Wurtz also retreated from the theoretical realm. After about 1860, Liebig and Wöhler ceased to make even the most perfunctory attempts to remain current with the literature. They were viscerally repelled by the new system of formulas and had no understanding for them. In contrast, by the mid-1860s nearly all theoretically active (that is, younger) German chemists had become structuralists. In this sense, we can say that Kolbe was both typical and atypical of his generation: typical for being on the nonstructuralist side of the cusp, even if barely, and thus finding most of his contemporaries and older colleagues agreeing with him in rejecting structural ideas, but atypical of the same peer group in being profoundly interested in theory, and virtually unique in being a nonstructural organic theorist of *any* age in Germany in the years after 1865. To put the matter more simply, everyone in his older peer group other than Kolbe rejected structure theory simply by rejecting *all* organic-chemical theory.

A similar ambiguity is encountered when one examines the issue of *style* of theorization. Chapter 10 explored the suggestion that a self-conscious hypothetico-deductive theoretical style began to displace Baconian inductivism in Europe in the 1840s and 1850s and asserted that Kolbe was one of the leaders of this trend, at least in the chemical community. However, a concurrent trend was also in evidence, namely, an increasing tendency toward explicit use of conventionalist theory, and this Kolbe tenaciously resisted. Proceeding from a standpoint of naive realism, he objected that the structuralists' radicals were "idealized entities invented for convenience," whereas his radicals were "facts." The carbonic acid theory was indeed a kind of type theory, he averred, but his types were "real," while those of his opponents were merely "formal," "conventional," and "ideal."[10] This predilection for manipulating conventional entities, he averred, had led modern chemists into making statements about matters that were simply beyond human ken.

> The goal for which Kekulé strives, and which he considers accessible, is actually even more inaccessible for us than the moon, for we can see the moon and determine its form, but atoms we cannot see, and their form is perceivable with none of our senses.[11]

The structuralists agreed that atomic arrangements are invisible to the physical eye, but responded that we must use our *mental* eye. "Such prophetic vision," Kolbe acidly commented, "has not been given to me."[12]

On the rhetorical level, this sort of language is clear enough, but Kolbe got into problems of self-consistency in attempting to instantiate his points. One way in principle of distinguishing real from conventional entities might be to demonstrate genetic interrelationships for the former versus merely schematic interrelationships for the latter. Despite the emphasis in his fundamental paper of 1860 on genetic chemical relationships (usually substitution reactions) as the basis of his theory, many of his type assignments were not founded upon direct interconversions of one compound to another, and Kolbe recognized and conceded that fact.[13] A related tactic might have been to operationalize his theoretical entities, demonstrating their more direct relationship to macroscopic experiments and concrete data. Kolbe certainly believed that he could make this sort of case, but in fact his radicals (oxatyl, carbonyl, hydrogen peroxide, methine, ethyl, and so on) were no more isolable than those of his opponents. Kolbe conceded this point, as well—ironically, in the course of an attempt to indict structuralism in a different sense. Structural formulas are mechanical and coarsely sensual, he wrote, a symptom of the modern "crassly materialistic treatment of scientific matters; the latter ought to be conceived by the mind and not mechanically."[14] In sum, Kolbe tried but failed to make ontological and epistemological distinctions between his and his opponents' theories; what remained was nothing more than an objection to conventionalist rhetorical *style*.

Kolbe had little sympathy or understanding for physics or for the field of physical chemistry that was beginning to mature in his last years, partially because these disciplines also used conventional theory and sought to investigate the (apparently) empirically distant microworld. It was appropriate that it was Lothar Meyer, the physical chemist and protégé of the early conventionalist and hypothetico-deductivist Franz Neumann, who remonstrated most effectively with Kolbe on exactly these points:

> Why do you, you who have contributed so much to the correct understanding of the chemical constitution of compounds, wish to erect an a priori barrier for the research of yourself and others at a spot which we have not yet reached, and therefore cannot know whether a navigable channel or an impenetrable wall of ice stands before us? You say we will never attain knowledge concerning the spatial arrangement of atoms. I ask, why not? When the microscope was perfected we learned to see things of which we had no inkling, and also these days we measure things that are far too small to be seen even with the most powerful microscope. Why do you want to lame the wings of research with a categorical "To this point and no further!"? Did such a warning tablet succeed in sparing the so-called vital force from analytical dismemberment? . . . If only we are careful to beware of the (unfortunately quite frequent) confusion of our hypotheses and theories with the absolute truth, which we have perhaps never yet found, then speculations surely hurt very little, even the very boldest ones.[15]

Frankland had similar words for his friend:

> I cannot coincide in your condemnation of the use of constitutional formulae by young chemists. It seems to me that chemists who make no use of their imagination (Odling & Brodie for instance) do but very little orig-

> inal work. The hard dry facts of the science are not likely to excite the enthusiasm of anyone. . . . The real progress of science depends upon the accession of new facts, and theories are only of use as incentives to discovery, and, looking back at chemistry during the past 25 years, it really does not seem to make much difference, as regards progress, whether the theories, which act the part of stimulants, be true or false.[16]

Both Frankland and Meyer understood that they were speaking to a passionate theorist and that the real point in dispute was conventionalism, whereas others, even to this day, have mistaken Kolbe's position by labeling it radically empiricist. He was, in fact, strongly inclined toward hypothetico-deductivism and philosophical realism; the former fit in well with the second half of the nineteenth century, but the latter did not.

Here we may have one means of understanding his (and many of his older compatriots') revulsion for structure theory. Benzene theory during the last third of the century can stand as a case in point. To a remarkable degree, Kekulé's cyclohexatriene structure was distrusted ontologically, while at the same time used in a confident heuristic fashion by most theoretically active organic chemists. The apparent contradiction is resolved by viewing most structuralists as (conscious or unconscious) conventionalists.[17] More generally, structure theory did not make much sense from the standpoint of well-developed global theories such as electromagnetism or Newtonian mechanics, which may have increased the attraction of the conventionalist standpoint for structuralists. Most of them understood that structures proposed in the literature were usually partial and provisional, especially for complicated molecules. Such provisional structures were gradually refined and improved, often by their original proposer. This gradualist approach continued to characterize synthetic organic chemistry well into the twentieth century, indeed to a large degree even today. It contrasted, however, with the older style of proposing and then "proving" a particular point of view, and the new style stuck in the craw of Kolbe—as also of Liebig, Wöhler, and many other older chemists.

We can summarize all of this by affirming that issues of generational change offer us some helpful perspectives on the enigma of Hermann Kolbe. Significantly, however, Kolbe's actual patterns of theoretical manipulation—aside from questions of methodological style, rhetoric, and language—were essentially the same as those of his opponents. His theoretical entities were not more easily operationalized to the macroworld of the laboratory nor were they more thoroughly founded on genetic chemical relationships. For all of his emphasis on definitive "proofs," not one of his peers exhibited more twists and turns along the theoretical pathways than he, nor more of an emphasis on hypothetico-deductivism, nor more heuristic use of chemical formulas, even if those formulas were different from those of the structuralists.

Kolbe also exhibited far more flexibility during the course of his early and mid-career than has often been supposed. He kept track of the literature and heeded evidentiary arguments, and during the 1840s and 1850s, he continually modified his positions in accordance with the advancing state of the art. These events have been detailed in chapters 3, 6, and 8. The position he had arrived at by 1860 was largely equivalent to that of the newer type theorists. Kolbe vehemently denied the arguments of both friends and rivals that he had joined rather than defeated his enemies, but the critics were essentially right. Using his "carbonic acid theory," Kolbe was able to make fundamental contributions to organic chemistry throughout the 1860s.

Nevertheless, the vestiges of electrochemical dualism retained by Kolbe resulted in some real differences between his and the structuralists' ideas: Kolbe avoided the idea of anisotropic bonds between atoms and affirmed strictly hierarchical molecular constitutions. These differences led to predictions and retrodictions that in certain cases were empirically distinguishable from those of the structure theorists. As we have seen in chapters 9, 12, and 13, in every such distinguishable case Kolbe failed to produce empirical confirmation of his own theory or refutation of the structuralists', and the cumulative anomalies in his position became ever more serious through the 1860s and 1870s. The remarkably one-sided character of these results helps to explain why structure theory became increasingly popular and powerful[18] and why Kolbe succeeded in persuading not a single prominent scientist of the advantages of his variant, not even his own students. The real mystery of the matter is why Kolbe himself was not converted to structuralism and why he retained his theory of 1860 until the end of his life. Ultimately, the tentative social and psychological rationalizations suggested above must suffice.

* * *

The previous two sentences are undergirded by the assumption that evidence matters to scientists and, if sufficiently compelling, will prompt conversions even from a comfortable and well-loved theoretical position to one that is novel and initially distasteful. A generation ago such an assumption would have been regarded as unsurprising, but as historians and other people know, times change. A central and now well-accepted element of post-positivist philosophy is the claim that theories are in principle seriously underdetermined by empirical evidence.

This thesis has been expanded by advocates of the "strong program in the sociology of knowledge" to the position that scientific knowledge, like any and all other forms of human belief and doctrine, is entirely socially constructed, fully independent of empirical investigations.

In coining the quoted phrase and defining the program, David Bloor was careful to specify that causes other than social might in particular situations be determinative. This appears to allow entry to the influence of empirical evidence. In practice, however, the social constructivists tend to proceed as if no such influence were ever important. In his quest for Bloor's injunctions toward "symmetry" and "impartiality," Bruno Latour was moved to the position that "nothing extraordinary and nothing 'scientific'" ever happens in laboratories, and so we must "abolish the distinction between science and fiction." Similarly, Harry Collins believes that "the natural world has a small or nonexistent role in the construction of scientific knowledge." As Barry Barnes and Steven Shapin put it, "The intense concern of earlier generations with the special status of science and its allegedly distinctive characteristics has begun to ebb away."[19]

The strong programmers have been met by strong criticism, including charges of purveying "voodoo epistemology" and the "pseudoscience of science."[20] The social constructivists have also been known to use terms of opprobrium, referring to the "paralysis" of philosophical minds and the "bleatings of recusant epistemologists and methodologists."[21] Difficulties with the positions of some sociologists of scientific knowledge include a tendency to overshoot in their correctives, a proclivity toward the construction of straw men, and their generally unconvincing responses to a variety of reflexivity problems. With regards to the first of these issues, no one, I think, would deny that the overall historiographic changes in the profession over the last generation have been highly salutary, providing a more contextual, more realistic, and more fully historicist vision of the past. Nearly all historians of science advocate the underdetermination thesis and regard social causation as a central element of historical explanation. But admitting underdetermination does not mean that the resulting gap must necessarily be filled *exclusively* by social causation, and even gross underdetermination does not mean *zero* determination.

Moreover, the assertion of a distinctive character to science need not be taken in the way that some positivist–Whig historians once viewed it, as a sort of purely cerebral and bloodless pursuit of ontological truth. Such a caricature is often the model attacked by strong programmers, but in fact modern historians and philosophers have long been constructing much more sophisticated and carefully nuanced conceptions of the nature of the scientific enterprise. The apparent general conformability of the world with our socio-culturally based investigations of it does indeed suggest a certain epistemological asymmetry. I believe that there is in fact a degree of asymmetry here. A central paradox then requires resolution, namely, how those thoroughly socially grounded investigations can nonetheless exhibit certain special characteristics—success, power, predictive heuristics, technological applications, and ultimately what appears to be gradual ontological progress—that we have come to associate with science and that are dismissed by social constructivists. This problem is still an open and difficult one.

But the position of the constructivists also contains paradoxes. For instance, a thoroughly symmetrized approach conceals an implicit scientism more radical than any they are attacking, for many of the strong programmers confidently assert that *all* knowledge, including social relationships and social epistemology, is as transparent to human reason and empirical effort as the scientists' far more delimited territory. But ironically and paradoxically, in order to be consistent, they must at the same time relinquish all claims to have succeeded in erecting anything other than a socially determined castle of cards. Such circularities and reflexivities have often been discussed, both by advocates and by opponents of the sociology of knowledge movement.[22] Many, including the present author, believe that the constructivists' attempts to show that such circularities are not vicious have not been successful.

In this book (especially in chap. 6), I have portrayed the "quiet revolution" that played out in Germany in the 1850s as a major change in the science of chemistry that was consummated surprisingly quickly and easily, hence quietly. There was so little controversy, I suggest, mostly because the force of evidence drove conversions efficiently, often in the face of social and career interests. This was even true for Kolbe himself. No one had more contempt for the French or their theories in the late 1840s and early 1850s than Kolbe. However, the striking new reactions and brilliant arguments by Williamson, Gerhardt, Wurtz, and Frankland during the early 1850s that convinced most of Kolbe's German colleagues to accept the French–English theories were by no means lost on Kolbe either. We have seen that by 1857 he had developed a theory of his own that was strikingly similar to the Williamson–Gerhardt newer type theory. Kolbe's pathological prejudices had failed to prevent him from understanding and being persuaded by the hated French ideas; they had only operated to prevent him from *believing* he had adopted them. He never became a card-carrying structuralist, for which one needed to relinquish all vestiges of electrochemical dualism; but he was a de facto fellow traveler for a great part of the journey. Later, his attempts to stir controversy and to win support for his variant position were quite unsuccessful.

In according a determinative role to the impact of new empirical evidence and reasoning as explaining conversions to this revolution and thus urging, by extension to the general case, a stronger role for purely cognitive factors in the process of scientific change than is granted by many social historians of science, I do not advocate a return to the bad old days of untrammeled internalism. Indeed, sociologists of science, including no doubt the social constructivists, could—and, I hope, will—have a great deal of fun with this case as well, despite its having been particularly suited for my thematic purposes. (If scientific theories are strongly underdetermined by empirical evidence, all the more so must be historical interpretations.)

* * *

One case of particular interest in this regard is France; by focusing in this study on the favorable reception of the Gerhardt–Laurent reforms in Germany, I have said little about why they made such slow gains at home. The differential reception in the two countries must have been due to social, cultural, and institutional factors, since the French had essentially equal access to the same papers of the early 1850s that so quickly converted their German neighbors. Indeed, as I have emphasized, they had even more direct access to the personal influence of the leaders of the reform, Laurent and Gerhardt, and subsequently Wurtz. Even Williamson, the key figure in the transformation of the evidence from circumstantial to decisive, was a Francophile. Why were the French so oblivious to this movement?

A full answer to this important question has not yet emerged.[23] Clearly, the facts that Laurent and Gerhardt were social misfits (provincials, ardent republicans, materialists, and possibly both of Jewish extraction) and that they almost defiantly refused to play the conventional social games, had the result that their true merits were much slower to be recognized than might otherwise have been the case. There might even have been an element of psychological backlash, in which the Laurent–Gerhardt reforms were regarded, in the time of political reaction after their deaths, as "tainted" or even dangerous—for look what a tragic fate the protagonists shared! Moreover, after the reforms caught on in Germany—which, as I have argued, happened even before the death of Gerhardt—they could be viewed, perversely, as German, hence foreign and suspect. Nor can it be mere coincidence that the first French chemist to base his textbook on the work of Laurent and Gerhardt was also the republican and scientifically peripheral

Alfred Naquet. In addition, it appears that the positivistic tone of French science, noticeable since before the beginning of the nineteenth century and continuing for more than a hundred years, worked to the disadvantage of the obviously theory-based movement described here. Why French science had this particular complexion is, again, incompletely understood, but it must be due primarily to social and cultural forces.

Two final non-cognitive factors deserve mention. The decentralization that was so characteristic of Germany in the decades before 1871 has often been depicted as driving vigorous and ultimately beneficial rivalry between the various states. Healthy competition characterized German academic science and university administration, as we have seen at many points in our story and as has been noted by other authors.[24] It was less prominent in France[25] or in Britain. This kind of competition operated on many levels—local, regional, national, and international, as well as in matters of prestige, prosperity, institutional structures, military and technical superiority, and so on. German as well as foreign observers saw an important cause of Prussian dominance over Austria and France in the superiority of German science, propelled by a competitive research ethos. Whether this kind of case can be sustained or not, it would appear that the research ideal had declined in France during the decades that it had increased so powerfully in Germany, and after 1871 the French began aggressively to address the problem.

Finally, it behooves the author of a biography, at the risk of belaboring the obvious, to stress the contingency of history and to note that individuals can notably succeed or fail to exert powerful influence in their peer communities. For whatever reason, after the death of Gerhardt in 1856 France lacked a critical mass of reform-minded chemists who could make the case for the new ideas.[26] Wurtz was professionally lonely, just as Kolbe was isolated on the other side of the debate and on the other side of the Rhine. More generally, the rise of the German research ideal so prized and emulated was associated with only a handful of leading actors.

* * *

I am far from alone in urging a flexible approach as regards the outworn dichotomy between cognitive and social history of science.[27] Flexibility, pluralism, and an eclectic and empirical approach are important elements for success in science. I have argued in chapters 7 and 10 that these attributes were precisely the necessary ones to accomodate the rise of structure theory and that they were the ones that Kolbe notably lacked after 1870. The same attributes are also beneficial for the historian. Social and cultural forces are powerful, pervasive, and efficacious, and so is the strength of ideas and evidence as pursued by conscientious scientists on the "agonistic field" in which one marshals resources. In the valuable perspectives provided by recent sociological studies, the power and vitality of scientific ideas and logic themselves, the constant regulating appeal to the empirical world, and the contingent influence of individual actions should not be underestimated.

Abbreviations for Frequently Cited Sources

Annalen	*Annalen der Chemie und Pharmacie*
Ann. chim.	*Annales de chimie et de physique*
August-Kekulé-Sammlung	Institut für Organische Chemie, Technische Hochschule, Darmstadt
Baeyer Collection	Adolf Baeyer Collection, privately held
Berichte	*Berichte der Deutschen Chemischen Gesellschaft*
Brock, *LHB*	W. H. Brock, ed., *Justus von Liebig und August Wilhelm Hofmann in ihren Briefen (1841–1873)* (Weinheim: Verlag Chemie, 1984)
BSC	*Bulletin de la Société Chimique*
Carrière, *BLB*	Justus Carrière, ed., *Berzelius und Liebig: Ihre Briefe von 1831–1845*, 2d ed. (Munich: Lehmann, 1898)
Chemiker-Briefe	Letters of Kolbe to A. W. Hofmann, Sammlung Chemiker-Briefe 38, Archiv der Akademie der Wissenschaften zu Berlin Brandenburg
C.r.	*Comptes rendus hebdomadaires des séances de l'Académie des Sciences, Paris*
DSB	*Dictionary of Scientific Biography*, 18 vols. (New York: Scribners', 1970–1990)
Frankland Archive	Edward Frankland microfilmed papers, Open University, Milton Keynes (references to this collection are followed by a microfilm frame index number)

Dingler Nachlass	Erlenmeyer letters, Hugo-Dingler-Stiftung, Hofbibliothek, Aschaffenburg
GUA	Universitätsarchiv, Göttingen
Handwörterbuch	*Handwörterbuch der reinen und angewandten Chemie*, 9 vols. (Braunschweig: Vieweg, 1837–1864)
Hofmann, *LWB*	A. W. Hofmann, ed., *Aus Justus Liebig's und Friedrich Wöhler's Briefwechsel in den Jahren 1829–1873*, 2 vols. (Braunschweig: Vieweg, 1888)
HSA	Hessisches Staatsarchiv, Marburg
JB	J. J. Berzelius, *Jahresberichte über die Fortschritte der physischen Wissenschaften*
JCS	*Journal of the Chemical Society*
JpC	*Journal für praktische Chemie*
Lehrbuch	Kolbe, *Ausführliches Lehrbuch der organischen Chemie*, 3 vols. (Braunschweig: Vieweg, 1854–1878)
Liebigiana IIB	Letters of Kolbe to J. Liebig, Liebigiana IIB, Handschriftenabteilung, Bayerische Staatsbibliothek, Munich
Lockemann, HK	Georg Lockemann, "Hermann Kolbe," in G. Bugge, ed., *Das Buch der grossen Chemiker*, 2 vols. (Berlin: Verlag Chemie, 1930), **2**, 124–135
Meyer, HK	Ernst von Meyer, "Zur Erinnerung an Hermann Kolbe," *Journal für praktische Chemie*, **138** (1885), 417–466
Ost, HK	Hermann Ost, "Hermann Kolbe: Ein Lebensbild," *Westermann illustrierte deutsche Monatshefte*, **30** (1885), 118–133
PTRS	*Philosophical Transactions of the Royal Society of London*
Roscoe Collection	H. E. Roscoe Collection, Royal Society of Chemistry, London
SSDM	Sondersammlungen, Deutsches Museum, Munich
UAL	Universitätsarchiv, Leipzig
UBM	Universitätsbibliothek, Marburg
VA	Kolbe file, 311K, Vieweg Verlag Archiv, Wiesbaden
Wallach, *BWB*	O. Wallach, ed., *Briefwechsel zwischen J. Berzelius und F. Wöhler*, 2 vols. (Leipzig: Engelmann, 1901)

Wöhler Nachlass	Letters of Kolbe to F. Wöhler, Cod. Ms. F. Wöhler, 74, Niedersächsische Staats- und Universitätsbibliothek, Göttingen
ZfC	*Zeitschrift für Chemie und Pharmacie*, 1860–1864; continued *Kritische Zeitschrift für Chemie, Physik und Mathematik* (1858–1859), and continued by *Zeitschrift für Chemie* (1865–1871)

Notes

Notes to Introduction

1. This is especially true for German chemistry. See Peter Borscheid, *Naturwissenschaft, Staat und Industrie in Baden (1848–1914)* (Stuttgart: Klett, 1976); Otto Krätz, "Der Chemiker in den Gründerjahren," in E. Schmauderer, ed., *Der Chemiker im Wandel der Zeiten* (Weinheim: Verlag Chemie, 1973), pp. 259–284; and Jeffrey Johnson, "Academic Chemistry in Imperial Germany," *Isis*, **76** (1985), 500–524.

2. The raw numbers are given in Krätz, ibid., pp. 269–270, which I have converted to doubling periods.

3. These two words had differing connotations in the 1850s and 1860s, meanings that will be explored in the chapters to come. For now, it suffices to note that both indicated the disposition of the atoms and groups of atoms within chemical molecules.

4. Clearly, a market economy model of academia, of the sort used by Joseph Ben-David and Avraham Zloczower, is useful in helping to explain the rapid expansion of organic chemistry in Germany after 1850. As I argue in this book, this social dynamic was only one factor, albeit an important one. For a concise description, analysis, and critique of the Zloczower thesis, see Steven Turner, Edward Kerwin, and David Woolwine, "Careers and Creativity in Nineteenth-Century Physiology: Zloczower Redux," *Isis*, **75** (1984), 523–529.

5. Laurent, *Méthode de chimie* (Paris: Mallet & Bachelier, 1854), p. 28; Kekulé, *Lehrbuch der organischen Chemie*, **1** (Erlangen: Enke, 1859), 58; L. Meyer, ed. notes in S. Cannizzaro, *Abriss eines Lehrganges der theoretischen Chemie* (Leipzig: Engelmann, 1891), pp. 53–58.

6. This applies both to my own treatment in *Chemical Atomism in the Nineteenth Century* (Columbus: Ohio State University Press, 1984), chaps.

7–10, and to John Brooke, "Avogadro's Hypothesis and its Fate," *History of Science*, **19** (1981), 235–273 (on p. 257), although both of us drew attention to the extended and multipartite character of the "Karlsruhe revolution."

7. Jeffrey Johnson, "Academic Chemistry in Imperial Germany."

8. H. Kiesewetter, *Industrielle Revolution in Deutschland, 1815–1914* (Frankfurt: Suhrkamp, 1989).

9. Liebig's career in Giessen has been much studied, although the regional context still needs to be explored further. There are a few fine studies of chemistry in other non-Prussian locales, e.g., Christoph Meinel, *Die Chemie an der Universität Marburg seit Beginn des 19. Jahrhunderts* (Marburg: Elwert, 1978), and Peter Borscheid, *Naturwissenschaft, Staat und Industrie in Baden.* For Prussia, see the writings of R. S. Turner, especially his masterly essay, "Justus Liebig versus Prussian Chemistry: Reflections on Early Institute-Building in Germany," *Historical Studies in the Physical Sciences*, **13** (1982), 129–162.

10. Robert C. Tucker, *Stalin as Revolutionary* (New York: Norton, 1973); idem, "A Stalin Biographer's Memoir," in S. H. Baron and C. Pletsch, eds., *Introspection in Biography: The Biographer's Quest for Self-Awareness* (Analytic Press, 1985), pp. 249–271 (on p. 270).

Notes to Chapter 1

1. Justus Liebig, "Der Zustand der Chemie in Preussen," *Annalen*, **34** (1840), 97–136 (on p. 100).

2. For background regarding eighteenth-century chemistry, see Karl Hufbauer, *The Formation of the German Chemical Community (1720–1795)* (Berkeley: Univ. of California Press, 1982); E. Schmauderer, ed., *Der Chemiker im Wandel der Zeiten* (Weinheim: Verlag Chemie, 1973); and Christoph Meinel, "*Artibus Academicis Inserenda*: Chemistry's Place in Eighteenth and Early Nineteenth Century Universities," *History of Universities*, **7** (1988), 89–115.

3. The following discussion of the German universities in the eighteenth and nineteenth centuries is based on a body of recent high-quality English-language research on this subject. See especially R. Steven Turner, "The Growth of Professorial Research in Prussia, 1818–1848—Causes and Context," *Historical Studies in the Physical Sciences*, **3** (1971), 137–182; idem, "University Reformers and Professorial Scholarship in Germany, 1760–1806," in Lawrence Stone, ed., *The University in Society*, 2 vols. (Princeton: Princeton Univ. Press, 1974), **2**, 495–531; Charles McClelland, *State, Society, and University in Germany, 1700–1914* (New York: Cambridge Univ. Press, 1980); and Turner, "The Prussian Professoriate and the Research Imperative," in H. N. Jahnke and M. Otte, eds., *Epistemological and Social Problems of the Sciences in the Early Nineteenth Century* (Dordrecht: Reidel, 1981), pp. 109–121. A recent and excellent German compilation focusing on the professoriate is Klaus Schwabe, ed., *Deutsche Hochschullehrer als Elite 1815–1945* (Boppard: Boldt, 1988).

4. A. G. Kästner, *Briefe aus sechs Jahrzehnten, 1745–1800* (Berlin, 1912),

cited in J. L. Heilbron, "Experimental Natural Philosophy," in G. Rousseau and R. Porter, eds., *The Ferment of Knowledge* (New York: Cambridge Univ. Press, 1980), pp. 378–379.

5. For example, William Coleman, "Prussian Pedagogy: Purkyně at Breslau, 1823–1839," in Coleman and F. L. Holmes, eds., *The Investigative Enterprise: Experimental Physiology in Nineteenth-Century Medicine* (Berkeley: Univ. of California Press, 1988), pp. 15–64.

6. Ibid., pp. 45–53; Kathryn Olesko, "On Institutes, Investigations, and Scientific Training," ibid., pp. 295–332 (on pp. 297–299 and 309–310).

7. McClelland, *State, Society, and University*, pp. 174–180.

8. Arleen Tuchman, "From the Lecture to the Laboratory: The Institutionalization of Scientific Medicine at the University of Heidelberg," in Coleman and Holmes, *Investigative Enterprise*, pp. 65–99 (on pp. 85–86 and 91–92); Coleman, "Prussian Pedagogy," ibid., p. 40.

9. The best general works on Berzelius in major languages are J. E. Jorpes, *Jac. Berzelius: His Life and Work* (Stockholm: Almqvist & Wiksell, 1966), and Berzelius, tr. O. Larsell, *Autobiographical Notes* (Baltimore: Williams & Wilkins, 1934). Other works on aspects of Berzelius relevant to the material I treat here are H. G. Söderbaum, "Berzelius und Hwasser, ein Blatt aus der Geschichte der schwedischen Naturforschung," in Julius Ruska, ed., *Studien zur Geschichte der Chemie* (Berlin: Springer, 1927), pp. 176–186; Evan Melhado, *Jacob Berzelius: The Emergence of His Chemical System* (Stockholm: Almqvist & Wiksell, 1981); Anders Lundgren, *Berzelius och den kemiska atomteorin* (Uppsala: Almqvist & Wiksell, 1979); Vladislav Kruta, "Berzelius' Interest in Physiology," *Lychnos*, **1973–1974**, 256–262; and T. Frängsmyr and E. Melhado, eds., *Enlightenment Science in a Romantic Age: Berzelius and His Science in International Context* (New York: Cambridge Univ. Press, 1992).

10. Söderbaum, "Berzelius und Hwasser"; Kruta, "Physiology"; and Berzelius, *Autobiographical Notes*, pp. 61–62, 96, 123–128, and 180–181n.

11. Melhado, *Berzelius*; Lundgren, *Berzelius*; Frängsmyr and Melhado, *Berzelius*; and A. J. Rocke, *Chemical Atomism in the Nineteenth Century: From Dalton to Cannizzaro* (Columbus: Ohio State Univ. Press, 1984), pp. 66–79 and 153–190.

12. On Wöhler, see Robin Keen, "The Life and Work of Friedrich Wöhler," Ph.D. dissertation, Univ. College London, 1976; idem, "Friedrich Wöhler," *DSB*, **14**, 474–479; A. W. Hofmann, "Zur Erinnerung an Friedrich Wöhler," in *Erinnerung an vorangegangene Freunde*, 3 vols. (Braunschweig: Vieweg, 1888), **2**, 1–205; and Johannes Valentin, *Friedrich Wöhler* (Stuttgart: Wissenschaftliche Verlagsgesellschaft, 1949).

13. Keen, "Life and Work," p. 40.

14. Wallach, *BWB*, **1**, 105, 137, 157, 191–192, 235–236, etc.; see also Wöhler's similar comments to Liebig, in Hofmann, *LWB*, **2**, 150–151.

15. Keen, "Life and Work," p. 115. This was when Wöhler visited Liebig in Giessen for two weeks in November 1831 and again for two weeks in January 1832.

16. Wöhler to Berzelius, 19 December 1830, 24 November 1831, and 1 December 1831, in Wallach, *BWB*, **1**, 325–327, 381, and 387; Liebig to Wöhler, 15 June 1832, in Hofmann, *LWB*, **1**, 53–54.

17. Liebig to Wöhler, 1 February 1836, in Hofmann, *LWB*, **1**, 84–85; Keen, "Life and Work," pp. 98–100. In addition to Gmelin and Berzelius, support from Hausmann, Gauss, and Weber was expressed from within the Göttingen faculty.

18. Wöhler to Berzelius, 27 April 1836, in Wallach, *BWB*, **1**, 652–654.

19. Wöhler to Berzelius, 25 May and 8 September 1836, 30 March and 14 October 1838, ibid., **1**, 656 and 663–664, and **2**, 18 and 66.

20. Keen ("Life and Work," p. 87) was unable to locate any information about a Wöhler practicum in Kassel.

21. GUA, 4 I, Nr. 47 ("Übersichten der Zuhörerzahl"); Wöhler to Berzelius, 27 May and 22 November 1838, 14 February and 10 August 1839, 22 May 1840, 3 November 1841, and 13 January 1842, in Wallach, *BWB*, **2**, 30, 70–72, 96, 125–126, 175, 266, and 275–277; Wöhler to Liebig, 30 June 1838, 23 January and 2 March 1839, 21 May and 25 July 1841, in Hofmann, *LWB*, **1**, 121, 134–135, 140–141, 184, and 253–254. Examples of published work by Wöhler's students includes A. Stürenburg, *Annalen*, **29** (1839), 291–293; F. Weppen, ibid., pp. 317–319; G. Schnedermann, ibid., **45** (1843), 277–286; K. Voelckel, ibid., **33** (1840), 220–222; ibid., **35** (1840), 306–309; and ibid., **38** (1841), 314–320. The biographical information about Wöhler's students is taken from Wilhelm Ebel, ed., *Die Matrikel der Georg-August-Universität zu Göttingen, 1837–1900* (Hildesheim: Lax, 1974), pp. 9, 10, 21, 24, and 28; on Voelckel, see also Wallach, *BWB*, **2**, 349n.

22. An instance of the first category is Wöhler, "Eigenschaften der Tantalsäure," *Annalen*, **31** (1839), 120–124; that an (unnamed) student actually did the analysis was mentioned in Wöhler's letter to Berzelius of 8 June 1839, in Wallach, *BWB*, **2**, 115. An instance of the second sort is Wöhler, "Arsenikgehalt des Zinns," *Annalen*, **29** (1839), 216–217, where Stürenburg's name is mentioned. Some examples of the third category are cited in the previous note.

23. Wöhler to Berzelius, 19 May, 25 July, 12 August, 11 September, and 21 October 1841, in Wallach, *BWB*, **2**, 244, 253–255, 259, and 261–262; Wöhler to Liebig, 21 May 1841, in Hofmann, *LWB*, **1**, 184. Biographical matters are taken from Ebel, *Matrikel*, pp. 6, 30, and 37.

24. Wöhler to Berzelius, 3 November 1841 and 13 January 1842, in Wallach, *BWB*, **2**, 266 and 276–277; GUA, 4 I, Nr. 47 ("Übersichten der Zuhörerzahl"). This document begins only in the winter semester of 1842/43 and includes a separate category of numbers of "tägl. Praktikanten" only for the first five semesters.

25. The data are from Ebel, *Matrikel*, pp. 2, 4, 7–10, 15, 17, 20–22, 24, 28, 30, 36–37, 46, 48, 52, 56, and 66.

26. Wöhler to Berzelius, 24 April, 23 June, 22 September, and 12 November 1842 and 6 June 1843, in Wallach, *BWB*, **2**, 293–294, 300–302, 330, 345–349, and 418–419; Valentin, *Wöhler*, pp. 105–108.

27. The standard biography is Jacob Volhard, *Justus von Liebig*, 2 vols. (Leipzig: Barth, 1909). On Liebig's laboratory research and teaching, see also

J. B. Morrell, "The Chemist Breeders: The Research Schools of Liebig and Thomas Thomson," *Ambix*, **19** (1972), 1–46; Bernard Gustin, "The Emergence of the German Chemical Profession, 1790–1867" (Ph.D. dissertation, Univ. of Chicago, 1975); R. S. Turner, "Justus Liebig versus Prussian Chemistry: Reflections on Early Institute Building in Germany," *Historical Studies in the Physical Sciences*, **13** (1982), 129–162; F. L. Holmes, "The Complementarity of Teaching and Research in Liebig's Laboratory," *Osiris*, [2] **5** (1989), 121–164; and Joseph S. Fruton, "The Liebig Research Group—A Reappraisal," *Proceedings of the American Philosophical Society*, **132** (1988), 1–66.

28. Liebig, "Eigenhändige biographische Aufzeichnungen," in Hertha von Dechend, ed., *Justus von Liebig in Eigenen Zeugnissen*, 2d ed. (Weinheim: Verlag Chemie, 1963), pp. 13–27 (on pp. 13–17). These reminiscences are carefully evaluated in Pat Munday, "Social Climbing through Chemistry: Justus Liebig's Rise from the *Niederer Mittelstand* to the *Bildungsbürgertum*," *Ambix*, **37** (1990), 1–19.

29. Ernst Berl, *Briefe von Justus Liebig, nach neuen Funden* (Giessen: Gesellschaft Liebig-Museum, 1928), pp. 30, 31, and 34–35.

30. Liebig, "Aufzeichnungen," pp. 17–22; Berl, *Briefe*, pp. 43–71.

31. Berl, *Briefe*, pp. 30 and 34; Gustin, "Chemical Profession," pp. 66–102.

32. Volhard, *Liebig*, **1**, 51–63 and 83; Berl, *Briefe*, pp. 75–81.

33. Holmes, "Liebig's Laboratory," pp. 122–132.

34. This contra Holmes, ibid., pp. 126–128. Here it is important to note that the winter semester of 1826/27 was the first time the chemical course was offered in the new pharmaceutical institute. Liebig said he learned from his initial university practica that intensive lab work was necessary to create a proficient chemist.

35. For example, compare Berzelius' judgment of Liebig in his letters to Wöhler of 9 July and 14 October 1830, just before and just after his first meeting with Liebig, in Wallach, *BWB*, **1**, 304 and 315.

36. Volhard, *Liebig*, **1**, 63–85; Holmes, "Liebig's Laboratory," pp. 146–156; Fruton, "Liebig Research Group." The phrase quoted is from Liebig, "Aufzeichnungen," p. 23.

37. Holmes, "Liebig's Laboratory," pp. 155–156 and 159–162.

38. Ibid., p. 162.

39. Volhard, *Liebig*, 1, 83–84.

40. Turner, in "Liebig versus Prussian Chemistry," p. 158, stresses this point. Turner's analysis anticipates and supports the more detailed subsequent studies of Liebig's research lab (by Holmes and by Fruton) and Wöhler's lab (see my discussion above) in their essential aspects.

41. H. E. Roscoe, "Bunsen Memorial Lecture," *JCS*, **77** (1900), 513–554; Theodor Curtius, "Gedächtnissrede," *JpC*, **169** (1900), 381–407; Heinrich Debus, *Erinnerungen an Robert Wilhelm Bunsen* (Kassel: Fischer, 1901); and Georg Lockemann, *Robert Wilhelm Bunsen* (Stuttgart: Wissenschaftliche Verlagsgesellschaft, 1949).

42. *Verzeichniss der Vorlesungen* (Göttingen: Dieterich, 1835–1836); salary information at Kassel is derived from HSA, 153/4, Nr. 21, p. 20.

43. This according to Wöhler's letter to Berzelius of 27 April 1836, in Wallach, *BWB*, **1**, 654.

44. See Berzelius' *JB* for 1834, **15** (1836), 218; *JB* for 1835, **16** (1837), 126–129; *JB* for 1836, **17** (1838), 160; *JB* for 1837, **18** (1839), 487–502; *JB* for 1839, **20** (1841), 526–537; *JB* for 1840, **21** (1842), 495–503 (Bunsen has "immortalized" his name by his research, on p. 496). By this time Bunsen had begun reporting his results directly to Berzelius by letter to facilitate their early inclusion in the *Jahresberichte*.

45. Roscoe, "Bunsen," p. 513.

46. L. W. McCay, "My Student Days in Germany," *Journal of Chemical Education*, **7** (1930), 1081–1099 (on p. 1095), regarding Bunsen's lectures in 1882.

47. Indispensable sources for Bunsen's teaching in Kassel and Marburg are (for both periods) Debus, *Bunsen*, pp. 5–28, and (for Marburg only) Christoph Meinel, *Die Chemie an der Universität Marburg seit Beginn des 19. Jahrhunderts* (Marburg: Elwert, 1978), pp. 20–48. Debus' book provides valuable details regarding Bunsen's strong Berzelian proclivities during the 1840s, as do C. Glück's student notes at Marburg in 1850, in UBM, Mscr. 501 and 502.

48. Meinel, *Chemie*, p. 20.

49. Ibid., pp. 30–31; Debus, *Bunsen*, pp. 18–20, 158; Bunsen report to Universitäts-Deputation, 2 May 1848, HSA, *16* Rep. VI, Kl. 1, Nr. 25, pp. 63–69.

50. Debus, pp. 25–27 and 144–150; Max Bodenstein, "Robert Wilhelm Bunsens Stellung zur organischen Chemie," *Die Naturwissenschaften*, **24** (1936), 193–196.

51. Debus, pp. 19–20, 27, 157–158; Curtius, "Gedächtnissrede," p. 403; and H. Goldschmidt, "Erinnerungen an Robert Wilhelm Bunsen," *Zeitschrift für angewandte Chemie*, **24** (1911), 2137–2140.

52. On laboratories and institutes at German universities in the early nineteenth century, see G. Lockemann, "Der chemische Unterricht an den deutschen Universitäten im ersten Viertel des neunzehnten Jahrhunderts," in Ruska, ed., *Studien* (see n. 9), pp. 148–158; Coleman and Holmes, eds., *Investigative Enterprise*; Gustin, "Chemical Profession"; Morrell, "Chemist Breeders"; McClelland, *State, Society and University*; the several works of R. S. Turner cited in this chapter, and especially his "Liebig versus Prussian Chemistry"; and Kathryn Olesko, *Physics as a Calling: Discipline and Practice in the Königsberg Seminar for Physics* (Ithaca, NY: Cornell Univ. Press, 1991).

53. Hofmann, "Wöhler," pp. 76–77; Hufbauer, *German Chemical Community*, pp. 202 and 244–245; Lockemann, "Chemischer Unterricht," pp. 151–152; and Lockemann and R. E. Oesper, "Friedrich Stromeyer and the History of Chemical Laboratory Instruction," *Journal of Chemical Education*, **30** (1953), 202–204.

54. Maurice Crosland, *Gay-Lussac: Scientist and Bourgeois* (Cambridge: Cambridge Univ. Press, 1978); L. J. Klosterman, "A Research School of Chemistry in the Nineteenth Century: Jean Baptiste Dumas and His Research Students," *Annals of Science*, **42** (1985), 1–80. Academic science in nineteenth-century France is also dealt with in such works as R. Fox and G. Weisz, eds.,

The Organization of Science and Technology in France, 1808–1914 (New York: Cambridge Univ. Press, 1980); Harry W. Paul, *From Knowledge to Power: The Rise of the Science Empire in France, 1860–1939* (New York: Cambridge Univ. Press, 1985); and Mary Jo Nye, *Science in the Provinces* (Berkeley: Univ. of California Press, 1986).

55. Revealing testimony on the state of academic chemistry in Britain ca. 1840 and early evidence of agitation for change is provided by D. B. Reid, *Remarks on the Present State of Practical Chemistry and Pharmacy* (Edinburgh: Neill, 1838); and William Gregory, *Letter to the Right Honorable George, Earl of Aberdeen . . . on the State of the Schools of Chemistry in the United Kingdom* (London: Taylor & Walton, 1842). Gregory lamented (pp. 18–22 and 28–29) that due to the deficiencies and expense of British universities, and the excellence and inexpensiveness of German academies, most ambitious British chemistry students were going to Göttingen or Giessen for their educations. Nineteenth-century British academic science is covered in D. S. L. Cardwell, *The Organisation of Science in England* (London: Heinemann, 1957), and by Robert Bud and G. K. Roberts, *Science Versus Practice: Chemistry in Victorian Britain* (Manchester: Manchester Univ. Press, 1984).

56. Frederick Gregory, "Kant, Schelling, and the Administration of Science in the Romantic Era," *Osiris*, [2] **5** (1989), 17–35, idem, "Kant's Influence on Natural Scientists in the German Romantic Period," in R. Visser et al., eds., *New Trends in the History of Science* (Amsterdam: Rodopi, 1989), pp. 53–66; see also Max Lenz, *Geschichte der königlichen Friedrich-Wilhelms-Universität zu Berlin*, 3 vols. (Halle: Waisenhaus, 1910–1918), **1**, 305ff, 570–571, and **2**, 1, 3ff, 224–230, 509–510.

57. Christa Jungnickel and Russell McCormmach, *The Intellectual Mastery of Nature*, vol. 1 (Chicago: Univ. of Chicago Press, 1986), pp. 23–26; and Kenneth Caneva, "From Galvanism to Electrodynamics: The Transformation of German Physics and Its Social Context," *Historical Studies in the Physical Sciences*, **9** (1978), 63–159.

58. Turner, "Liebig versus Prussian Chemistry," pp. 133–138 and 144–147 and the pages cited in Lenz, *Geschichte* (see n. 56).

59. Jungnickel and McCormmach, *Intellectual Mastery of Nature*, p. 22n. and chap. 4; Olesko, *Physics as a Calling*; and David Cahan, "The Institutional Revolution in German Physics, 1865–1914," *Historical Studies in the Physical Sciences*, **15** (1985), 1–65.

60. Coleman and Holmes, eds., *Investigative Enterprise*, especially the articles by Coleman and by Arleen Tuchman.

61. Meinel, "Chemistry's Place"; see also Hufbauer, *German Chemical Community*; Gustin, "German Chemical Profession"; and Turner, "Liebig."

62. In addition to the sources cited in the previous note, see also Erika Hickel, "Der Apothekerberuf," *Medizinhistorisches Journal*, **13** (1978), 259–276.

63. Turner, "Liebig"; Holmes, "Liebig's Laboratory."

64. See sources cited in n. 10.

65. Liebig to Emil Erlenmeyer, 27 March 1861, in Emil Heuser, ed., *Justus von Liebig und Emil Erlenmeyer in ihren Briefen von 1861–1872* (Mannheim:

Bionomica, 1988), p. 11. I have found the same sentiment expressed in two other letters from Liebig's pen.

66. See citations in n. 8.

67. Coleman, "Prussian Pedagogy," p. 47.

68. Cited in Turner, "Liebig versus Prussian Chemistry," p. 160.

69. A number of polytechnic schools were established in pre-Napoleonic Germany; see Helmuth Albrecht, *Technische Bildung zwischen Wissenschaft und Praxis* (Hildesheim: Ohms, 1987), pp. 25–40.

70. Peter Borscheid, *Naturwissenschaft, Staat und Industrie in Baden (1848–1914)* (Stuttgart: Klett, 1976); Tuchman, "Scientific Medicine at Heidelberg." The practical benefits of chemistry were by no means ignored by state ministries even in the eighteenth century. A number of teaching positions involving what we would call applied chemistry existed (mostly at Gewerbeschulen, mining academies, and polytechnics), as did a significant student clientele interested in industrial careers—for which see Meinel, "Chemistry's Place," and Hufbauer, *German Chemical Community*. But interest in the ministries dramatically increased after mid-century.

71. A. J. Rocke, "Berzelius' Animal Chemistry: From Physiology to Organic Chemistry, 1805–1814," in Frängsmyr and Melhado, *Berzelius*, pp. 107–131.

72. Debus, *Bunsen*, pp. 144–151.

73. Liebig, "Aufzeichnungen," pp. 13–17.

74. Debus, *Bunsen*, p. 145.

75. Liebig to Wöhler, 26 June 1848, in Carrière, *BLB*, p. 265.

76. Berzelius to Wöhler, 20 August 1839, in Wallach, *BWB*, **2**, 134.

77. Wöhler to Liebig, 12 November 1863, in Hofmann, *LWB*, **2**, 149–150.

78. Caneva, "German Physics."

Notes to Chapter 2

1. Götz von Selle, *Die Georg-August-Universität zu Göttingen, 1737–1937* (Göttingen: Vandenhoek & Ruprecht, 1937); Charles E. McClelland, *State, Society, and University in Germany, 1700–1914* (New York: Cambridge Univ. Press, 1980); and R. Steven Turner, "University Reformers and Professorial Scholarship in Germany, 1760–1806," in L. Stone, ed., *The University in Society*, 2 vols. (Princeton, N.J.: Princeton Univ. Press, 1974), **2**, 495–531.

2. See Selle, pp. 261–281.

3. Cited without reference by Hajo Holborn, *A History of Modern Germany, 1840–1945* (Princeton, N.J.: Princeton Univ. Press, 1969), p. 27.

4. See Selle, p. 281.

5. Charles E. McClelland, "Die deutschen Hochschullehrer als Elite, 1815–1850," in Klaus Schwabe, ed., *Deutsche Hochschullehrer als Elite* (Boppard: Boldt, 1988), pp. 27–53 (on p. 43–53).

6. *Die Religion in Geschichte und Gegenwart*, 2d ed., **4** (Tübingen: Mohr, 1930), 1123, s.v. "Pfarrer."

7. G. W. Kolbe's name does not appear in Götz von Selle, ed., *Die Matrikel der Georg-August-Universität zu Göttingen* (Hildesheim: Lax, 1937). He

is briefly described in obituaries of Carl Kolbe: *Vierteljährliche Nachrichten von Kirchen- und Schulsachen*, 1870, pp. 154–155, and P. Meyer, ed., *Die Pastoren der Landeskirchen Hannovers und Schaumburg-Lippes seit der Reformation*, 2 vols. (Göttingen: Vandenhoek & Ruprecht, 1941–1942), **1**, 250, and **2**, 116 and 414. His residence and profession in 1808, 1813, and 1816 are cited in brothers Carl, G. C. A., and F. Kolbe's matriculation entries at the University of Göttingen (Selle, *Matrikel*, **1**, 476, 575, and 766). Information on his situation in 1821 is contained in the dedication to Carl's book (see n. 14). I wish to thank Dr. Günther Beer, Göttinger Museum der Chemie, and Herr Karl-Heinz Bielefeld, director of the Kirchenkreisarchiv Göttingen, for helpful correspondence and for assistance during my visit to their institutions in 1990.

8. See sources cited in the previous note. I thank Herr Leenders of the Landeskirchliches Archiv Hannover for a report on the content of Carl Kolbe's correspondence with his Konsistorium. The Göttingen Gymnasium has no records dating to the early nineteenth century, and the university has only a Matrikel of students. Auguste Hempel's full name is given in the marriage record (written by the pastor, her groom) in the Elliehäuser Kirchenbuch, held by the Kirchenkreisarchiv Göttingen; her precise birth and death dates are on her gravestone in Lutterhausen.

9. On Hempel, see A. C. P. Callisen, ed., *Medicinisches Schriftsteller-Lexicon*, **28** (Copenhagen, 1840), 466–467; August Hirsch, ed., *Biographisches Lexikon der hervorragendsten Aerzte*, **3** (Vienna, 1886), 146; Selle, *Göttingen*, p. 223; *Neuer Nekrolog der Deutschen*, **12** (1834), pt. 1 (Weimar: Voigt, 1836), pp. 194–195. The marriage record of 1816 referred to in the previous note indicates that Louise Hempel was then deceased.

10. Hermann Ost, HK, 118; Ernst von Meyer, HK, 418–420; Georg Lockemann, HK, 124. Ost and Meyer were students of Kolbe; moreover, Ost was his nephew and Meyer his son-in-law; they doubtless had good documentary and oral sources for their biographies. As for Lockemann (1871–1959), he grew up, died, and is buried in the neighboring village to Stöckheim, where Kolbe's family lived from 1826 to 1840. Although when Kolbe died Lockemann was but a Gymnasium student, their lives touched repeatedly even if indirectly: in the early 1890s Lockemann studied at the Hanover Technische Hochschule, where Ost taught chemistry; his father owned (and in the mid-1890s he worked as a chemist at) the same salt-brine works that Kolbe used to visit as a boy; and in 1898 he became assistant to Kolbe's former assistant Ernst Beckmann. Ost was still alive when Lockemann, a prolific historian of chemistry, was writing his biography of Kolbe, and Ost may have shared sources with him. Certainly he conferred with Kolbe's daughter Johanna von Meyer and used documents in her possession: Lockemann, "Aus dem Briefwechsel von Hermann Kolbe," *Zeitschrift für angewandte Chemie*, **41** (1928), 623. In any case, subsequent biographies of Kolbe are all derivative from these three, as this one is in part. See also Grete Ronge, "Hermann Kolbe," *Neue deutsche Biographie*, **12** (Berlin: Duncker and Humblot, 1980), 446–451, and Claus Priesner, "Georg Lockemann," *Neue deutsche Biographie*, **15** (1987), 6–7.

11. Most of the information in this paragraph is derived from conversation

and correspondence with Karl Heinz Thiel, retired pastor of the Elliehausen congregation, to whom I offer my thanks. See also n. 6.

12. Franz Dehme, *Das Kirchspiel Stöckheim im Leinetal* (Northeim: Röhrs, 1928), pp. 7, 34, and 46–47; Stöckheimer Kirchenbuch. I thank Pastor Peter Dortmund for his help, especially in directing me to these sources.

13. The information in this paragraph comes from the Lutterhausen Kirchenbuch and from conversations with Pastor Hermann Charbonnier, for whose kind assistance I am grateful.

14. Carl Kolbe, *Handbuch zum sittlich-religiösen Jugendunterrichte über den Hannoverischen Landes-Katechismus* (Göttingen, 1822). Relevant information on Lutheranism in eighteenth- and nineteenth-century Hanover may be found in Johannes Meyer, *Kirchengeschichte Niedersachsens* (Göttingen, 1939), pp. 183–188 and 191–197, and in *Die Religion*, 3d ed., **3**, 67–72, and **5**, 271–306.

15. Ost, HK, p. 119.

16. Frankland, *Sketches from the Life of Edward Frankland* (London: Spottiswoode, 1902), pp. 44–50.

17. C. A. Russell, *Lancastrian Chemist: The Early Years of Sir Edward Frankland* (Milton Keynes: Open Univ. Press, 1986). Professor Russell has emphasized this point to me in conversations and correspondence.

18. HSA, 16. Rep. VI, Kl. 8, Nr. 25, first leaf written and dated in Kolbe's hand. Kolbe's initial problems with his future father-in-law were related in a letter to his friend and publisher Eduard Vieweg, 25 March 1853, VA 51.

19. Lockemann, "Ernst Beckmann," *Berichte*, **61** (1928), 87–130A (on p. 92). Beckmann, former assistant to Kolbe, was Wislicenus' guide; he told the same story, without the religious commentary, in his biography of Wislicenus: ibid., **37** (1904), 4861–4946 (on pp. 4887–4888).

20. Kolbe, *Das chemische Laboratorium der Universität Leipzig* (Braunschweig: Vieweg, 1872), xxxix–lx; Kolbe, "Zur Erinnerung an Justus von Liebig," *JpC*, **116** (1873), 428–458 (on p. 441).

21. Kolbe to Vieweg, 6 September 1853, VA 57. Zeller and Kolbe later had a falling-out. Zeller's book was published in 1854, but not by Vieweg.

22. For a discussion of the status of the *Bildungsbürgertum* in Germany during the nineteenth century, with citations to the recent secondary literature, see R. Steven Turner, "The Bildungsbürgertum and the Learned Professions in Prussia, 1770–1830: The Origins of a Class," *Histoire Sociale—Social History*, **13** (1980), 105–135; and K. Jarausch, *Students, Society, and Politics in Imperial Germany* (Princeton, N.J.: Princeton Univ. Press, 1982), pp. 81–89, 120–126.

23. Fritz Ringer, "Das gesellschaftliche Profil der deutschen Hochschullehrerschaft 1871–1933," in Schwabe, ed., *Deutsche Hochschullehrer*, pp. 93–104; McClelland, *State, Society, and University*, p. 96.

24. In an undated letter whose scientific content requires a date of 1850, Hofmann consoled Kolbe over an unstated familial death, which must have been Emma's (SSDM 3553); her birth and death dates are given on her headstone, positioned between those of her parents. On Georg Ost, see P. Meyer, *Pastoren*, **1**, 251; also Bertha Ost to Frankland, 1 April 1885,

Frankland Archive 01.04.1603, who informed Frankland that her husband died in 1875.

25. On (Johann) Carl Friedrich Kolbe, see HSA, *305a* A IV, b. 2, Nr. 65, and *305a* II, Nr. 11, 13 November 1851; Kolbe to Bertha and Georg Ost, 4 and 17 October and 10 and 16 December 1870, SSDM 6792, 6793, 6794, and 6795. On Kolbe's renovation of the three graves, see his letter to Hermann Ost, 25 September 1878, SSDM 6801.

26. See letters from Kolbe to Vieweg of 30 December 1851, 20 October 1855, 30 March 1861, and 31 August 1865, VA 35, 110, 167, and 234. In the first of these letters, Kolbe wrote, "My father is now, thank goodness, out of danger, but he is still so weak that he cannot yet read again, and therefore needs company to entertain him. He appears to appreciate it when I stay by him, and so I intend to hold out here as long as my time permits."

27. Kolbe to Vieweg, 8 February 1865, VA 217.

28. Meyer, HK, p. 418; Lockemann, HK, pp. 124–125; Ost, HK, pp. 118–119. The information on Kolbe's residences in Göttingen was kindly provided by Dr. Günther Beer, director of the Museum der Göttinger Chemie. The Knesebecks were a well-known family of the German nobility, with branches in Hanover and Prussia, but I have not been able to learn more about this particular Knesebeck. Selle (see n. 1, p. 262) reported that Knesebeck's father had written a panegyric on the Hanoverian nobility, which had made him highly unpopular in 1831, but the political climate was very different by 1837.

29. Ost, HK, p. 119.

30. Wilhelm Ebel, ed., *Die Matrikel der Georg-August-Universität zu Göttingen (1837–1900)* (Hildesheim: Lax, 1974), pp. 8 and 56; UAG, Abgangszeugnis Nr. 374 (for Hermann Kolbe, dated 22 October 1842).

31. It was Berzelius who broached the subject of the "Göttinger Sieben" in their correspondence: Berzelius to Wöhler, 1 January 1838, in Wallach, *BWB*, **2**, 1. Berzelius agreed with the dissidents' sentiments but disapproved of their action; Wöhler's only concerns were the possible impact on attendance at the university and the difficulty of finding a worthy successor to Weber: Wöhler to Berzelius, 13 January 1838, ibid., 4–5.

32. Wöhler, "Darstellung des Ameisenäthers," *Annalen*, **35** (1840), 238.

33. Kolbe, "Über die Zusammensetzung des Getreidefuselöls," *Annalen*, **41** (1842), 53–56; Wöhler to Berzelius, 25 July 1841, in Wallach, *BWB*, **2**, 254; Berzelius, *JB* for 1842, **23** (1844), 456–457; Meyer, HK, p. 419.

34. Lockemann, HK, pp. 125–126; 200 Hessian thalers were equivalent to about $150 in the United States at that time.

35. John Tyndall, *New Fragments* (New York: Appleton, 1892), p. 238; Tyndall gives a fascinating description of his period of study in Marburg (pp. 232–243).

36. Kolbe related the poisoning incident in a letter to Max von Pettenkofer, 27 January 1884, Bayerische Staatsbibliothek, Pettenkoferiana II 2. He had been assisting Bunsen in an hour-long combustion experiment over a charcoal fire, but fortunately had been more resistant to the effects of the carbon monoxide. Bunsen stressed what a "*true friend*" (Bunsen's emphasis) Kolbe had always been, in a letter to Hermann Ost, 14 January 1885, SSDM 3623.

37. Heinrich Debus, *Erinnerungen an Robert Wilhelm Bunsen* (Kassel: Fisher, 1901), pp. 10–21; C. Glück, student notebook, UBM, Mscr. 501 and 502. The notes reveal a striking Berzelian orientation, although Bunsen also mentioned Gay-Lussac, Thenard, Liebig, Wöhler, and Dumas in the organic introduction. Bunsen also gave special emphasis to Kolbe's work on conjugated acids, carried out in his lab (ibid., esp. Mscr. 502, pp. 1–22 and 122–185).

38. Mulder, trans. Kolbe, *Versuch einer allgemeinen physiologischen Chemie*, vol. 1 (Braunschweig: Vieweg, 1844–1846).

39. Many passages in his letters to Berzelius from 1832 to 1840 (Wallach, *BWB*, **1**, 381, 497, 520, and 604; and **2**, 71n., 113, and 164) suggest that Wöhler was indeed becoming uncomfortable with the theoretically labile field of organic chemistry.

40. Wöhler to Berzelius, 26 July 1842, in Wallach, *BWB*, **2**, 304; Berzelius to Wöhler, 9 August 1842, Wallach, *BWB*, **2**, 317–318. In the second formula, Berzelius actually wrote double prime marks over the carbon to indicate the presence of two sulfur atoms, and used superscripts rather than subscripts.

41. Wöhler to Liebig, 7 August 1842, in Hofmann, *LWB*, **1**, 203; Wöhler to Berzelius, 16 September 1842, in Wallach, *BWB*, **2**, 326.

42. Berzelius to Wöhler, 7 October 1842, in Wallach, *BWB*, **2**, 333; Berzelius, *JB* for 1842, **23** (1844), 77–80. The + symbol here indicates combination of radicals in a compound; Berzelius later substituted a centered dot or period to avoid ambiguity (see examples below).

43. Heinrich [sic] Kolbe, "Über die Einwirkung des Chlors auf Schwefelkohlenstoff," *Annalen*, **45** (1843), 41–46; Berzelius, *JB* for 1842, **23** (1844), 77–80. During this period when Liebig was feuding with Berzelius, he refused to obtain barred letters for Berzelius and Wöhler, and simply doubled the number of atoms for barred letters.

44. In a letter to Wöhler of 25 August 1843, Berzelius mentioned a communication he had received from Kolbe (in Wallach, *BWB*, **2**, 428).

45. Berzelius, *JB* for 1842, **23** (1844), 77–80 and 456–457; *JB* for 1844, **25** (1846), 90–96; *JB* for 1845, **26** (1847), 77–93 and 405–412; Berzelius to Wöhler, 21 January 1845 and 27 February 1846 (in Wallach, *BWB*, **2**, 520–521 and 574–575); Berzelius to Bunsen, 24 January 1845 and 13 February 1846 (transcriptions made in 1851 and preserved HSA, 16. Rep. VI, Kl. 8, Nr. 25, folio 19).

46. The letter is quoted in Ost, HK, p. 120, and Kolbe expressed his veneration of it in *JpC*, **131** (1881), 309.

47. For general background on the theoretical development of organic chemistry in the early nineteenth century, see A. J. Ihde, *Development of Modern Chemistry* (New York: Harper and Row, 1964), chaps. 4–8; J. R. Partington, *A History of Chemistry*, vol. 4 (London: Macmillan, 1964); C. A. Russell, *The History of Valency* (Leicester: Leicester Univ. Press, 1971); and A. J. Rocke, *Chemical Atomism in the Nineteenth Century* (Columbus: Ohio State Univ. Press, 1984). The meanings and historical origins of "two-volume" and "four-volume" formulas, and other notational and conventional matters important for the following discussion, are described in the latter work.

48. Johnston to Charles Daubeny, 1 June 1840, Daubeny Papers, Magdalen

College Oxford, cited in R. F. Bud, "The Discipline of Chemistry: The Origins and Early Years of the Chemical Society of London," Ph.D. dissertation (Philadelphia: Univ. of Pennsylvania, 1980), p. 124.

49. S. C. H. Windler [F. Wöhler], "Über das Substitutionsgesetz und die Theorie der Typen," *Annalen*, **33** (1840), 308–310.

Notes to Chapter 3

1. For background and literature citations for the following material on the copula theory, see J. R. Partington, *A History of Chemistry*, vol. 4 (London: Macmillan, 1964), pp. 372–375 and 506–512.

2. A. Laurent, *Méthode de chimie* (Paris: Mallet & Bachelier, 1854), pp. 249–250; trans. W. Odling, *Chemical Method* (London: Cavendish Society, 1855), pp. 204–205 (quoted here).

3. R. Bunsen, "Untersuchungen über die Kakodylreihe," *Annalen*, **37** (1841), 1–57; **42** (1842), 14–46; **46** (1843), 1–48; Berzelius to Wöhler, 29 January 1841, in Wallach, *BWB*, **2**, 220.

4. Kolbe, "Notiz über einige gepaarte Verbindungen der Chlorkohlenstoffe," *Annalen*, **49** (1844), 339–341. These are (mostly) Berzelian four-volume formulas, with $C = 12$ and $O = 16$. Here Kolbe used barred symbols to indicate double atoms and superscripts rather than subscripts, in accordance with Berzelius' preference. The equation represents a composite of several that are found in the paper.

5. Kolbe, "Beiträge zur Kenntniss der gepaarten Verbindungen," *Annalen*, **54** (1845), 145–188.

6. Ibid., p. 146.

7. Ibid., pp. 186–188.

8. Ibid., pp. 145 and 181–186.

9. J. H. Brooke, "Wöhler's Urea and Its Vital Force?—A Verdict from the Chemists," *Ambix*, **15** (1968), 84–114; C. A. Russell, "The Changing Role of Synthesis in Organic Chemistry," *Ambix*, **34** (1987), 169–180.

10. Berzelius, *JB* for 1844, **25** (1846), 90–96; ibid. for 1845, **26** (1847), 77–93 and 405–412 (quote on p. 410).

11. Berzelius, "Ansichten in Betreff der organischen Zusammensetzung," *Annalen der Physik und Chemie*, [3] **8** (1846), 161–188, esp. pp. 161–163, 176, and 185–188.

12. Hofmann, "Erinnerungen an Peter Griess," *Berichte*, **24** (1891), 1007–1057 (on pp. 1022–1023); Jacob Volhard, *August Wilhelm von Hofmann: Ein Lebensbild* (Berlin: Friedländer, 1902), pp. 30–39.

13. Edward Frankland, reminiscences of Kolbe sent to Hermann Ost shortly after Kolbe's death, SSDM 3576. Doris Street no longer exists; in the nineteenth century, it lay half a mile north of the Oval, south of the Thames. Belvedere Road runs between Westminster and Waterloo Bridges, parallel to and not far from the right bank of the Thames. Neither Hofmann nor Frankland mentioned Kolbe's other residence. Because Hofmann married in August 1846, I assume that Kolbe lived first with Hofmann, then moved near Frankland.

14. Frankland, *Sketches from the Life of Edward Frankland* (London: Spottiswoode, 1901), p. 167. This is the rare (unexpurgated) first edition.

15. Frankland to Hermann Ost, 20 December 1884, SSDM 3575.

16. Georg Lockemann, HK, p. 126.

17. Frankland to Ost, SSDM 3576; Wöhler to Kolbe, 21 November 1849, SSDM 3591.

18. Kolbe to Bunsen, 11 February 1846, SSDM 3632; Bunsen to Kolbe, 18 February 1846, SSDM 3621; H. T. De la Beche, Lyon Playfair, and Warington Smyth, *Reports on the Gases and Explosions in Collieries* (London: Clowes, 1847), especially pp. 5–6; O. Krätz, "Historische Experimente (1846), Hermann Kolbe und Robert Wilhelm Bunsen: Eudiometrische Analysen von Grubengas," *Chem. Exp. Didakt.*, **3** (1977), 31–36. Kolbe found that the flammable portion of firedamp consists almost entirely of methane, consistent with earlier English research and inconsistent with work by other Europeans. His name was not mentioned in Playfair's report.

19. Useful details on Playfair and the Museum of Economic Geology are in C. A. Russell, *Lancastrian Chemist: The Early Years of Sir Edward Frankland* (Milton Keynes: Open Univ. Press, 1986), pp. 141–145.

20. Dated entry in MS. notebook, SSDM 3811, unpaginated; this is the earliest of six of Kolbe's laboratory notebooks donated to the Deutsches Museum by Kolbe's daughter, Johanna von Meyer.

21. Ibid. By 7 December 1846 he had sufficiently promising results publicly to announce the subject of a future article on organic electrolyses: "Observations on the Oxidizing Power of Oxygen when disengaged by means of the Voltaic Electricity," *Memoirs and Proceedings of the Chemical Society*, **3** (1846), 285–287 (on p. 287).

22. Kolbe, "On the Decomposition of Valerianic Acid by the Voltaic Current," *Memoirs and Proceedings of the Chemical Society*, **3** (1847), 378–380 ($H = 1$, $C = 6$, $O = 8$, four volumes). Kolbe was by no means the first to study the electrolysis of organic compounds, but he was certainly the earliest to achieve consistent success. See Alexander Moser, *Die elektrolytische Prozesse der organischen Chemie* (Halle: Knapp, 1910), pp. 1–13. The saturated hydrocarbon described by Kolbe was presumably the dimer of the radical of valeric acid, octane; the olefinic by-product was probably isobutylene.

23. Frankland, *Sketches*, p. 70.

24. Frankland and Kolbe, "Upon the Chemical Constitution of Metacetonic Acid, and Some Other Bodies Related to It," *Memoirs and Proceedings of the Chemical Society*, **3** (1847), 386–391.

25. Frankland to Ost, 20 December 1884, SSDM 3575. Bunsen's letter to Kolbe inviting Frankland to his lab is dated 9 April 1847, SSDM 3622.

26. Frankland, *Sketches*, pp. 72–78 and 261–264.

27. Ibid.; SSDM 3576; Lockemann, HK, p. 126. In letters to Frankland after Kolbe's death, Ost told him of the death of this (unnamed) sister in 1881, and Bertha Ost, née Kolbe, reminisced about Frankland's 1847 visit: H. Ost to Frankland, 25 December 1884, and B. Ost to Frankland, 1 April 1885, Frankland Archive, 01.04.1527 and 01.04.85.

28. Frankland and Kolbe, "Über die chemische Constitution der Säuren

der Reihe $(C_2H_2)_nO_4$ und der unter dem Namen 'Nitrile' bekannten Verbindungen," *Annalen*, **65** (1848), 288–304.

29. Frankland and Kolbe, "Über die Zersetzungsproducte des Cyanäthyls durch Einwirkung von Kalium," *Annalen*, **65** (1848), 269–287.

30. Edward Frankland diary, 27 January and 7 February 1848, Royal Society, MS. 221 XII b. 9.

31. Frankland, *Sketches*, p. 175.

32. Frankland, "On the Isolation of the Organic Radicals," *JCS*, **2** (1849), 263–296 (on p. 263).

33. Ibid.; Frankland, "Researches on the Organic Radicals," *JCS*, **3** (1850), 30–52, 322–347 (on pp. 46–47); Frankland, *Experimental Researches in Pure, Applied, and Physical Chemistry* (London, 1877), pp. 67–118. As Hofmann noted, Liebig had predicted in 1834 that potassium and ethyl iodide might be used to prepare ethyl: see Hofmann, "The Life-Work of Liebig," in *Zur Erinnerung an vorangegangene Freunde* (Braunschweig: Vieweg, 1888), **1**, 274.

34. Liebig to Hofmann, 8 December 1849, in Brock, *LHB*, p. 88.

35. Ernst Dreyer, *Friedr. Vieweg & Sohn in 150 Jahren deutscher Geistesgeschichte* (Braunschweig: Vieweg, 1936), pp. 3–32; Margarete and Wolfgang Schneider, eds., *Justus von Liebig: Briefe an Vieweg* (Braunschweig: Vieweg, 1986). The company moved to Wiesbaden in 1967; the Vieweghaus has been renovated and converted into a state museum.

36. Liebig to Vieweg, 28 April 1847, in M. and W. Schneider, *Justus von Liebig*, pp. 215–216.

37. Kolbe to E. Vieweg, 12 August 1847, VA 18. Wöhler told Kolbe on 16 January 1847 that he had recommended him for the position, but he had not yet heard whether Vieweg was going to take his advice (SSDM 3579).

38. Friedrich Knapp, "Friedrich Varrentrapp," *Berichte*, **10** (1877), 2291–2297.

39. Ost, HK, 121; Lockemann, "Kolbe," p. 126; Frankland's 1884 reminiscences, in SSDM 3576; Frankland diary, 31 May to 3 June 1849, Royal Society, MS. 221 XII b. 9. Ost replied to Frankland's reminiscence by informing him that Franziska von Spilker ended up marrying badly and "ist verkommen und verschollen": Ost to Frankland, 25 December 1884, Frankland Archive 01.04.1527.

40. Helmuth Albrecht, *Technische Bildung zwischen Wissenschaft und Praxis: Die Technische Hochschule Braunschweig 1862–1914* (Hildesheim: Ohms, 1987), pp. 37–40.

41. Ost, HK, p. 121; Dreyer, *Vieweg*, pp. 32–34; Frankland diary, 1 June 1849, Royal Society, MS. 221 XII b. 9.

42. Kolbe to Frankland, 1 August 1848, Frankland Archive 01.02.1297.

43. Ost, HK, p. 121; "Pfaff, Adam," *Allgemeine deutsche Biographie*, **25** (Leipzig, 1887), p. 580.

44. Kolbe to Wöhler, 2 February 1848, with Wöhler's reply on same sheet, SSDM 3577; Liebig to Kolbe, 9 February 1848, SSDM 3525; Wöhler to Kolbe, 25 January, 19 February 1848, and no date, SSDM 3581, 3580, and 3583.

45. A. J. Rocke, *Chemical Atomism in the Nineteenth Century* (Columbus: Ohio State Univ. Press, 1984), pp. 177–182 and passim.

46. Kolbe, "Formeln, chemische," *Handwörterbuch*, **3** (1848), 174–178 (on p. 176). The publication date of the fascicle in which this article appears was determined to be September 1848 by Oberleiter Lücke of Vieweg Verlag, noted on Kolbe's letter to Lücke of 1 January 1878, VA 397.

47. Ibid., p. 177.

48. Kolbe, "Gepaarte Verbindungen," *Handwörterbuch*, **3** (1848), pp. 439–444; publication date was determined by Lücke as November 1848, Kolbe to Lücke, 5 January 1878, VA 398.

49. In 1881 Kolbe drew attention to the importance of this shift in his thinking (*JpC*, **131**, 312–313).

50. Bunsen to Kolbe, 25 December 1847, SSDM 3496.

51. Kolbe, *Ausführliches Lehrbuch der organischen Chemie*, 2 vols. (Braunschweig: Vieweg, 1854–1864). There was a third volume as well as a second edition, but these were produced by Kolbe's students and others.

52. Frankland, SSDM 3576; Ost, HK, p. 121; Kolbe to Frankland, 24 November 1863, Frankland Archive 01.04.73. However, Kolbe's letter to Georg Liebig ("Freitag Nachmittag" [14 June 1850], Bayerische Staatsbibliothek, Ana 377, II B) mentions his intention to work collaboratively with Varrentrapp in the latter's laboratory.

53. Bunsen to Frankland, 1 August 1848, Frankland Archive 01.02.1288, regarding Kolbe's stay in 1848; Kolbe to Frankland, 1 August 1848, from Marburg, Frankland Archive 01.02.1297; Kolbe to E. Vieweg, 7 August and 27 August 1848, both from Marburg, VA 23 and 24; Frankland, SSDM 3376, regarding late summer 1849; Frankland diary, 24 July 1849, reporting on a letter from Kolbe suggesting his arrival in Marburg would be ca. 7 August, Royal Society, MS. 221 XII b. 9; Kolbe to Vieweg, 16 August 1850, from Giessen, VA 26; Kolbe to Georg Liebig, 23 July and 31 August 1850, Bayerische Staatsbibliothek, Ana 377, II B.

54. Frankland, SSDM 3576.

55. Kolbe to Vieweg, 7 August 1848, VA 23.

56. Kolbe, "Untersuchungen über die Elektrolyse organischer Verbindungen," *Annalen*, **69** (1849), 257–294; "Researches on the Electrolysis of Organic Compounds," *JCS*, **2** (1849), 157–184.

57. Liebig to Kolbe, 18 February 1849, SSDM 3527. Since at least 1847, and probably even earlier, Liebig had been following Kolbe's work with great interest and admiration. See, e.g., Liebig to Kolbe, 17 May 1848, SSDM 3526; Liebig to Vieweg, 20 January 1848, in M. and W. Schneider, eds., *Briefe an Vieweg*, p. 225; and Liebig to Hofmann, 25 April 1847, 2 February 1848, and 1 February 1849, in Brock, *LHB*, pp. 72, 77, and 83.

58. Kolbe, "Über die chemische Natur und Constitution der organischen Radicale," *Annalen*, **75** (1850), 211–239, and **76** (1850), 1–73; "On the Chemical Constitution and Nature of Organic Radicals," *JCS*, **3** (1850), 369–405, and **4** (1851), 41–79. Kolbe's letter to Liebig of 11 March 1850 is referred to in Liebig's reply of 12 April 1850, SSDM 3528; Hofmann to Kolbe, no date [ca. March 1850], SSDM 3553.

59. Kolbe, "Chemische Natur," pp. 211–217, 29.

60. Ibid., pp. 29, 57, 65–73.

61. Ibid., pp. 234–235, 36.

62. He was usually perfectly frank and public about such failed predictions, which greatly assists the historian. Four examples of such explicit predictive failures are "Beiträge" (see n. 5), pp. 160 and 183; "Zersetzungsproducte" (see n. 29), pp. 269–270; and "Researches" (see n. 56), pp. 157–158.

63. Kolbe, "Chemische Natur," pp. 65–73.

64. Ibid., p. 69. I have used with slight alteration the English translation (see n. 58), p. 76.

65. Hofmann, "Researches on the Volatile Organic Bases, pt. II," *JCS*, **1** (1848), 269–281 (on p. 280).

66. Ibid.; ibid., pt. I, *JCS*, **1** (1848), 159–173; ibid., pt. III, *JCS*, **1** (1848), 285–317 (on pp. 312–313 and 317).

67. Ibid., pt. V, *JCS*, **2** (1849), 300–335 (on p. 334–335).

68. Hofmann, "Researches Regarding the Molecular Constitution of the Volatile Organic Bases," *PTRS*, **140** (received 26 December 1849, read 17 January 1850), 93–131.

69. Hofmann to Kolbe, no date [ca. March 1850], SSDM 3553. Hofmann also told Liebig how much he enjoyed this research: Hofmann to Liebig, 29 January 1850, in Brock, *LHB*, p. 91.

70. Ibid.

71. Frankland, "On a New Series of Organic Bodies Containing Metals and Phosphorus," *JCS*, **2** (1849), 297–299.

72. Frankland, "Researches" (see n. 33), pp. 48–51; "Researches on the Organic Radicals," *JCS*, **3** (1850), 322–347 (on p. 324); and "On a New Series of Organic Bodies Containing Metals," *PTRS*, **142** (1852), 417–444 (on pp. 440–442).

73. Kolbe, "Chemische Natur," pp. 45 and 49–51.

74. Ibid., pp. 50–51.

Notes to Chapter 4

1. Academic science in nineteenth-century France is discussed in such works as R. Fox and G. Weisz, eds., *The Organization of Science and Technology in France, 1808–1914* (New York: Cambridge Univ. Press, 1980); Harry W. Paul, *The Sorcerer's Apprentice: The French Scientist's Image of German Science, 1840–1919* (Gainesville: Univ. of Florida, 1972); idem, *From Knowledge to Power: The Rise of the Science Empire in France, 1860–1939* (New York: Cambridge Univ. Press, 1985); and Mary Jo Nye, *Science in the Provinces* (Berkeley: Univ. of California Press, 1986). On Dumas, see especially L. J. Klosterman, "A Research School of Chemistry in the Nineteenth Century: Jean Baptiste Dumas and His Research Students," *Annals of Science*, **42** (1985), 1–80; and Marcel Chaigneau, *Jean-Baptiste Dumas: Sa vie, son oeuvre, 1800–1884* (Paris: Guy Le Prat, 1984).

2. Liebig to Berzelius, 8 May 1831, 2 July 1832, and 17 May 1841, in Carrière, *BLB*, pp. 11, 34, and 230.

3. Liebig to Berzelius, 17 April 1841, ibid., p. 223.

4. For a summary of this dispute, see F. L. Holmes, "Justus Liebig," *DSB*,

8, 333–344, and idem, *Claude Bernard and Animal Chemistry* (Cambridge, Mass.: Harvard Univ. Press, 1974), pp. 34–47.

5. Liebig to Berzelius, 30 May 1833, 14 September 1833, and 31 December 1834, in Carrière, *BLB*, pp. 62, 71, and 99. "The most maddening thing is," he wrote in the latter letter, "somewhat upset by the oxamide business, in my paper on the constitution of ether I permitted myself some expressions of a personal nature against Dumas, which I should not have done. . . . The devil take these accursed affairs." Even earlier (28 December 1831, ibid., p. 25) Liebig expressed great contrition over a published critique of some work of O. B. Kühn: "I will write no more critiques as long as I live," he vowed.

6. Liebig to [C. F. Kuhlmann], 23 April 1850, Archives of the Académie des Sciences, Paris (Dossier Liebig, Fonds Dumas). My translation from the German.

7. Ibid. (additional letters from Liebig to Dumas); Carrière, *BLB*, pp. 276–278; Liebig to Wöhler, 1 June 1850, in Hofmann, *LWB*, **1**, 352–353.

8. Liebig to Gerhardt, 15 August 1839 and 1 March 1840, in Edouard Grimaux and Charles Gerhardt, Jr., *Charles Gerhardt: Sa vie, son oeuvre, sa correspondance 1816–1856* (Paris: Masson, 1900), pp. 38 and 42–43. This is hereafter cited simply as "Grimaux."

9. Ibid., pp. 46–62.

10. Gerhardt, "Recherches sur la classification chimique des substances organiques," *C.r.*, **15** (1842), 498–500; ibid., *Revue scientifique*, **10** (1842), 145–218; Grimaux, pp. 63–68, 317–332, and 456–457; Marc Tiffeneau, ed., *Correspondance de Charles Gerhardt*, 2 vols. (Paris: Masson, 1918–1925) (hereafter cited as "Tiffeneau"), **2**, 19–29.

11. Gerhardt, "Considérations sur les équivalents de quelques corps simples et composés," *Ann. chim.*, [3] **7** (1843), 129–143, **8**, 238–245.

12. Grimaux, pp. 65–68.

13. On Gerhardt's change of heart toward Laurent, see Grimaux, pp. 83–85, 147, 342–345, 451–452, and 459; and Tiffeneau, **1**, i–ii, and **2**, 11 and 38. The religion of Laurent is not known. Wurtz stated that Gerhardt and Laurent "were of the same race" ("Éloge" [see n. 37], p. 2). He may have been privy to oral information that Laurent was a Jew or of Jewish descent since the word "race" would not have been appropriate to express common *nationality*. Extant biographical details about Laurent and his parentage are notably sketchy and inconsistent.

14. Laurent to Gerhardt, December 1844, 25 March and 9 April 1845, in Tiffeneau, **1**, 11, 31, and 34.

15. Grimaux, pp. 117–167; Tiffeneau, **1**, 83–88, 115–131, and 137; Liebig, *Annalen*, **57** (1846), 93–118.

16. Tiffeneau, **1**, 355–356.

17. Ibid., pp. 117–123.

18. Laurent, "Sur les combinaisons organiques azotées," *C.r.*, **20** (1845), 850–855; "Recherches sur les combinaisons azotées," *Ann. chim.*, [3] **18** (1846), 266–298 (on pp. 267–268 and 294); Tiffeneau, **1**, 13–14, 25, 31, 61, 76, 81–82, 92, 95, 99, 101, and 105.

19. Tiffeneau, **1**, 19–23, 36–61; Laurent, *Méthode de chimie* (Paris: Mallet

& Bachelier, 1854, p. xiii. See J. H. Brooke, "Laurent, Gerhardt, and the Philosophy of Chemistry," *Historical Studies in the Physical Sciences*, **6** (1975), 405–429; and N. W. Fisher, "Organic Classification Before Kekulé," *Ambix*, **20** (1973), 106–131, 209–233 (on pp. 215–217).

20. Laurent, "Observations sur le Précis de chimie organique de M. Gerhardt" (unpublished MS), in Tiffeneau, **1**, 270–288; Gerhardt to Drion, 15 May 1856, in Grimaux, p. 340.

21. Laurent to Gerhardt, 6 June and 1 July 1846, in Tiffeneau, **1**, 201–202 and 204.

22. Laurent, "Recherches" (see n. 18), p. 296.

23. Laurent to Gerhardt, 2 February 1847, in Tiffeneau, **1**, 222–225.

24. Laurent to Gerhardt, 4 May 1847, ibid., p. 232.

25. Laurent to Gerhardt, 30 May 1847, ibid., p. 239.

26. Seymour Mauskopf, *Crystals and Compounds: Molecular Structure and Composition in Nineteenth Century French Science* (Philadelphia: American Philosophical Society, 1976); Robert Fox, "The Rise and Fall of Laplacian Physics," *Historical Studies in the Physical Sciences*, **4** (1974), 89–136.

27. Grimaux, pp. 168–195.

28. Gerhardt to Jane Gerhardt, 27 March 1848, in Grimaux, p. 174.

29. Ibid., 2 April and 19 April 1848, pp. 175–180.

30. Ibid., 7 April and 13 June 1849, pp. 192–195; Gerhardt to Chancel, 2 February 1851, in Tiffeneau, **2**, 112, describing Laurent's illness.

31. Gerhardt to Liebig, 18 October 1850, in Grimaux, pp. 202–203.

32. Ibid., p. 210; Liebig to Hofmann, 27 October 1850, in Brock, *LHB*, p. 100.

33. Charles Friedel, "Notice sur la vie et les travaux de Charles-Adolphe Wurtz," *Bulletin de la Société Chimique*, [2] **43** (1885), i–lxxx (also issued separately, Paris, 1885); A. W. Hofmann, "Erinnerungen an Adolph Wurtz," *Berichte,* **20** (1887), 815–996, reprinted in *Zur Erinnerung an vorangegangene Freunde* (Braunschweig: Vieweg, 1888), **3**, 171–431; J. H. Brooke, *DSB*, **14**, 529–532; J. R. Partington, *A History of Chemistry*, **4** (London: Macmillan, 1964), 477–488. Friedel misdates Wurtz' stay in Giessen, an error that has been propagated throughout much of the secondary literature.

34. For example, by Hofmann, who should have known better: *Erinnerung*, p. 408 (see previous note).

35. Grimaux, pp. 13–15; Tiffeneau, **2**, 302.

36. Grimaux, pp. 76–77; Tiffeneau, **2**, 303; Hofmann, *Erinnerung*, p. 218. Gerhardt, trans. Wurtz, *Grundriss der organischen Chemie*, 2 vols. (Strasbourg, 1844–1846).

37. Wurtz, "Éloge de Laurent et de Gerhardt," *Moniteur scientifique*, **4** (1862), 482–513; undated separate, p. 3.

38. Wurtz, "Ueber die Constitution der unterphosphorigen Säure," *Annalen*, **43** (1842), 318–334; "Sur l'hydrure de cuivre," *Ann. chim.*, [3] **11** (1844), 250–252.

39. Wurtz, "Recherches sur l'acide sulfophosphorique et le chloroxyde de phosphore," *C.r.*, **24** (1847), 288–290 (290); extended paper, same title, *Ann. chim.*, [3] **20** (1847), 472–481 (on pp. 480–481).

40. Wurtz, "Recherches sur les éthers cyaniques et sur le cyanurate de méthylene," *C.r.*, **27** (1848), 241–243.

41. Wurtz, "Sur une série d'alcalis organiques homologues avec l'ammoniaque," *C.r.*, **28** (1849), 223–226; Wurtz, "Recherches sur les ammoniaques composées," *C.r.*, **29** (1849), 169–172.

42. Hofmann, *Erinnerung*, pp. 217–218.

43. Ibid., pp. 341–342; Wurtz, "Sur une série . . ."; Liebig, "Organische Basen," in *Handwörterbuch*, **1**, 697–699. This fascicle appeared in 1840, not 1837 as Partington states (see n. 33, p. 437).

44. Liebig to Hofmann, 23 April 1849, in Brock, *LHB*, p. 84.

45. Liebig to Hofmann, 8 December 1849 and 17 January 1850, ibid., pp. 88–89. The transcription "den Feind Herrn Wurtz" should read "den Freund Herrn Wurtz."

46. Hofmann, *Erinnerung*, pp. 226–243; Hofmann to Liebig, undated but securely datable by internal references between 23 February and 12 April 1850, in Brock, *LHB*, p. 93. Curiously, in his biography Hofmann gives the year as 1851.

47. J. B. Dumas, "Rapport sur un mémoire de M. Wurtz, relatif à des composés nouveaux analogues à l'ammoniaque," *C.r.*, **29** (1849), 203–206; Grimaux, pp. 197–198 and 379–380; Tiffeneau, **2**, 303–304. Gerhardt Jr. (Grimaux, pp. 197–198) goes so far as to accuse Wurtz of lacking integrity. Gerhardt Jr. was in general an excellent biographer for his father, but he was understandably partisan in his opinions.

48. Gerhardt to Chancel, 1 October 1852, in Tiffeneau, **2**, 130.

49. Wurtz, "Mémoire sur une série d'alcaloides homologues avec l'ammoniaque," *Ann. chim.*, [3] **30** (1850), 443–507, esp. 444–446 and 495–503. He first used the word "structure" in 1859: "Mémoire sur les glycols ou alcools diatomiques," *Ann. chim.* [3] **55**, 400–478 (on p. 478).

50. Grimaux, pp. 201–204, 208–254, 259, 264–265, 285–286, and 432–435.

51. Williamson to Gerhardt, 16 August 1851, in Grimaux, p. 220. On Williamson's syntheses and their significance, see Partington (see n. 33), pp. 444–460; J. Harris and W. H. Brock, "From Giessen to Gower Street: Towards a Biography of Alexander Williamson," *Annals of Science*, **31** (1974), 95–130; and A. J. Rocke, *Chemical Atomism in the Nineteenth Century* (Columbus: Ohio State Univ. Press, 1984), pp. 215–229. Williamson's work is described in greater detail in chap. 6.

52. Gerhardt and G. Chancel, "Sur la constitution des composés organiques," *Comptes rendus des travaux de chimie*, **7** (1851), 65–84; Gerhardt, *Traité de chimie organique*, 4 vols. (Paris, 1853–1856).

53. Gerhardt, "Recherches sur les acides organiques anhydres," *C.r.*, **34** (1852), 755–758 and 902–905; *Ann. chim.*, [3] **37** (1853), 285–342.

54. Grimaux, pp. 229–241 and 403–409; Gerhardt to Chancel, 19 June 1852, in Tiffeneau, **2**, 119–121 and 123.

55. Gerhardt to Chancel, 19 June 1852, in Tiffeneau, **2**, 124.

56. Grimaux, p. 241; Chancel to Gerhardt, 22 May 1852, in Tiffeneau, **2**, 121.

57. Grimaux, p. 247. Gerhardt later told Hofmann in glowing terms about

the warmth of this reception: Hofmann, "The Life-Work of Liebig," *Erinnerung*, **1**, 195–305 (on p. 290).

58. Gerhardt to Chancel, 28 November 1852, in Tiffeneau, **2**, 131. In the event, all mention of Berzelius was omitted from the title page and spine.

59. Tiffeneau, **2**, 141; Liebig to Eduard Vieweg, 12 December 1853, in Margarete and Wolfgang Schneider, eds., *Justus von Liebig: Briefe an Vieweg* (Braunschweig: Vieweg, 1986), pp. 267–268.

60. Gerhardt to Cahours, 24 February 1856, and Gerhardt to Wurtz, 6 April 1856, in Tiffeneau, **2**, 57 and 312.

61. Grimaux, pp. 416–435; Gerhardt to Williamson, 24 June and 8 November 1853, Harris Collection, Archives, Univ. College London.

62. Gerhardt, *Traité*, **1**, i–iii.

63. Gerhardt and Chiozza, "Recherches sur les amides," *C.r.*, **37** (1853), 86–90; Wurtz, "Histoire des doctrines chimiques depuis Lavoisier," in Wurtz, ed., *Dictionnaire de chimie pure et appliquée*, **1** (Paris: Hachette, 1868), pp. i–xciv (on p. liv).

64. Wurtz, "Sur les dédoublements des éthers cyaniques," *C.r.*, **37** (1853), pp. 180–183; ibid., "Sur la théorie des amides," pp. 246–250.

65. Wurtz, "Sur les dédoublements," p. 182n.

66. Auguste Scheurer-Kestner, "Charles Gerhardt, Laurent, et la chimie moderne," *Revue alsacienne*, August 1884; cited in Grimaux, p. 198.

67. Gerhardt, "Note sur la théorie des amides," *C.r.*, **37** (1853), 281–284.

68. Wurtz, "Nouvelles observations sur la théorie des amides," *C.r.*, **37** (1853), 357–361.

69. Wurtz, *Leçons de chimie professées en 1863* (Paris: Hachette, 1864), p. 93.

70. A. W. Williamson, "Sur la théorie de l'étherification," *Ann. chim.*, [3] **40** (1854), 98–114. For Williamson's work on etherification, see Harris and Brock, "From Giessen to Gower Street" (see n. 51).

71. Wurtz to Williamson, 18 April 1854, Harris Collection. This is the earliest correspondence between Wurtz and Williamson of which I am aware.

72. Williamson, "On the Constitution of Salts," *Chemical Gazette*, **9** (1851), 334–339, reprinted in *Papers on Etherification and on the Constitution of Salts* (Edinburgh: Alembic Club Reprint no. 16, 1902), pp. 42 and 45–46.

73. Wurtz, "Éloge" (1862), pp. 25–26; *Leçons* (1864), p. 88; *Cours de philosophie chimique* (Paris, privately publ., 1864), pp. 30–31; *Traité élémentaire de chimie médicale*, 2 vols. (Paris: Masson, 1864–1865), **2**, 54–55. In another letter to Williamson (27 May 1863, Harris Collection), Wurtz spoke of "la grande influence" that Williamson's work had exercised on the science.

74. Wurtz, "Théorie des combinaisons glycériques," *Ann. chim.*, [3] **43** (1855), 492–496, esp. p. 493. In Wurtz' "Histoire générale des glycols," in *Leçons de chimie professées en 1860* (Paris: Hachette, 1861), pp. 103–105, he was even more explicit regarding his debt to Williamson for the ideas in this paper.

75. Wurtz, "Sur une nouvelle classe de radicaux organiques," *Ann. chim.*, [3] **44** (1855), 275–313, esp. 300–313.

76. Wurtz, "Sur un nouveau mode de formation de l'éther carbonique,"

C.r., **32** (1851), 595–596; Wurtz, "Sur l'alcool butylique," *C.r.*, **35** (1852), 310–312.

77. For details, see Rocke, *Chemical Atomism*, passim.

78. R. Anschütz, *August Kekulé*, 2 vols. (Berlin: Verlag Chemie, 1929), **2**, 61–62.

79. Williamson to H. E. Roscoe, 5 December 1853, Roscoe Collection.

80. Wurtz to Liebig, 3 February 1858, Liebigiana IIB.

Notes to Chapter 5

1. Wöhler to Liebig, 10 May 1851, in Hofmann, *LWB*, **1**, 364; Bunsen to Debus, [May or June] 1851, in Heinrich Debus, *Erinnerungen an Robert Wilhelm Bunsen* (Kassel: Fischer, 1901), p. 160.

2. Bunsen to Marburg Faculty, 9 February 1851, HSA, Marburg, 16. Rep. VI, Kl. 8, Nr. 25, f. 17.

3. Liebig to Hofmann, 10 May and 25 October 1851, in Brock, *LHB*, pp. 112–113 and 117; Liebig to Wöhler, 19 May and 8 July 1851, in Hofmann, *LWB*, **1**, 365 and 370.

4. Liebig to Hofmann, 10 May 1851, and Hofmann to Liebig, 18 May 1851, in Brock, *LHB*, pp. 112–114.

5. Liebig to Hofmann, 25 October and 15 November 1851, ibid., pp. 117–119.

6. Liebig to Hofmann, 15 April 1852, ibid., p. 127.

7. Liebig to Hofmann, 17 June 1852, ibid., p. 138.

8. Ibid.; Hofmann to Liebig, no date (ca. July 1852), and Liebig to Hofmann, 18 July 1852, in Brock, *LHB*, pp. 142–143. See also Hofmann to Liebig, 18 May 1851, ibid., p. 113, where Hofmann also commented that a reason for declining Marburg was not to harm his friendship with Kolbe, who he knew very much wanted the call.

9. In letters of 16 January 1847, 25 January 1848, and 21 November 1849 (SSDM 3579, 3581, and 3591) Wöhler tried to respond to Kolbe's job search.

10. Hofmann to Nasse, 19 January 1851, HSA, 16. Rep. VI, Kl. 8, Nr. 25, ff. 15–16.

11. Bunsen to Nasse, 9 February 1851; Wöhler to Bunsen, 26 December 1850 and 1 January 1851; Liebig to Bunsen, 8 January 1851; Berzelius to Bunsen, 24 January 1845 and 13 February 1846; transcriptions made in 1851 and preserved in the HSA, 16. Rep. VI, Kl. 8, Nr. 25, ff. 17–21.

12. Ibid., f. 46; HSA, *305a*. A IV, 4b., Nr. 94; *305a*. A IV, 4c. ϵ 2, Nr. 5; *307d*. 1e., 15 February 1851.

13. Ibid.; HSA, *305a*. A IV, 4c. e.1, Nr. 6; 153/4, Nr. 21, pp. 1 and 25 (salaries for Wöhler and Bunsen); Christoph Meinel, *Die Chemie an der Universität Marburg seit Beginn des 19. Jahrhunderts* (Marburg: Elwert, 1978), pp. 46–50. Six hundred Kurhessian thalers were equivalent to about $425, which was then a typical janitor's salary in Boston, where prices and wages were admittedly much higher than in Germany (Margaret Rossiter, *The Emergence of Agricultural Science* [New Haven: Yale Univ. Press, 1975], p. 78).

14. Kolbe to Vieweg, 12 May 1851, VA 27.

15. Meinel, *Chemie*, pp. 12–31 and 51–64; Bunsen to Universitäts-Deputation, 2 May 1848, HSA, 16. Rep. VI, Kl. 1, Nr. 25, pp. 63–69; Edward Frankland, *Sketches from the Life of Edward Frankland* (London: Spottiswoode, 1901), p. 74; Tyndall to T. A. Hirst, 2 July 1849, in A. S. Eve and C. H. Creasey, *The Life and Work of John Tyndall* (London: Macmillan, 1845), p. 22.

16. Meinel, *Chemie*, pp. 50, 472, and 478–479; Frankland, *Sketches*, pp. 112 and 262.

17. F. Guthrie to H. E. Roscoe, 5 August 1854, Roscoe Collection.

18. I.e., "curse of Hesse," "Hesse's hatred and curse," and "cashbox curse in whore-Hesse."

19. Karl Demandt, *Geschichte des Landes Hessen*, 2d ed. (Kassel: Bärenreiter, 1972), pp. 552–561.

20. Kolbe to Eduard Vieweg, 15 November 1851, VA 33; Kolbe to Heinrich Vieweg, 30 April 1867, VA 249.

21. *Verzeichnisse der Vorlesungen* (Marburg: Bayrhoffer, 1851–1865).

22. Kolbe to Vieweg, 27 June 1851, VA 28.

23. These data and those that follow are based upon my collation of information in the handwritten Marburg Zuhörer-Verzeichnisse, winter 1850/51 to summer 1858 (HSA, *305a.* A IV, b.2, Nr. 65 and 68); the Honorar Einnahme-Manual, winter 1857/58 to summer 1865 (acc. 1902/8); and the Album [Matrikel] der Universität, 1849–1868 (HSA, *305a.* II, Nr. 11). Each of these alone provides only partial and not fully trustworthy data; even when they are used in conjunction, a complete picture of Kolbe's Marburg students is not possible. Meinel (*Chemie*, pp. 470–472) gives overall student statistics, without some of the details on nationality and field of study provided here.

24. Kolbe to Vieweg, 21 November 1853, VA 61.

25. For my purposes here, I define "foreign" as coming from outside the German Confederation. Sources for these data are given in n. 23.

26. For example, Steven Turner, "Justus Liebig versus Prussian Chemistry: Reflections on Early Institute-Building in Germany," *Historical Studies in the Physical Sciences*, **13** (1982), 129–162 (on pp. 142–145), estimates that eighty-five percent of students in Prussian chemical practica in 1855 were studying medicine or pharmacy. Even after excluding the pharmacists from these figures (who in Marburg were required to study under the pharmacist Zwenger), Turner's numbers suggest that two-thirds of the Prussian chemistry clientele in that year came from medicine.

27. Kolbe to Emil Erlenmeyer, 2 July 1864, Dingler Nachlass; HSA, acc. 1902/8.

28. However, wide variations occurred from semester to semester, from a low of 50 thalers in winter 1856/57 to a high of 658 thalers in 1865, his last summer semester in Marburg. These figures were calculated from data in HSA, acc. 1902/8; see also Kolbe to Vieweg, 4 January 1857, VA 122.

29. Kolbe to Emil Erlenmeyer, 6 March 1864, Dingler Nachlass.

30. Even with the chronically low enrollments, close to half of this income

came from student fees, contrary to Turner's rough estimate of twenty to twenty-five percent ("Institute-Building," pp. 155–156). The assumption of the ministries in nineteenth-century Germany was that the great majority of a professor's income would be derived from student fees, thereby justifying the sometimes very low salaries.

31. Kolbe to Vieweg, 14 December 1859, VA 149.

32. Kolbe to Vieweg, 31 December 1860, VA 164.

33. Kolbe to Vieweg, VA 50, 52, 61, 91, 122, 137, and 148 (1853–1859); Kolbe to Vieweg, 14 February 1859, VA 149. Hofmann also lent him money in Marburg; see Kolbe to Hofmann, 3 June 1866, Chemiker-Briefe.

34. For Kopp and Will, see Max Speter, "'Vater Kopp': Bio-, Biblio- und Psychographisches von und über Hermann Kopp (1817–1892)," *Osiris*, **5** (1938), 392–460 (on p. 414); for Prussia in 1834, see Charles McClelland, *State, Society and University in Germany, 1700–1914* (New York: Cambridge Univ. Press, 1980), p. 208; for Thomson, see J. B. Morrell, "The Chemist Breeders: The Research Schools of Liebig and Thomas Thomson," *Ambix*, **19** (1972), 1–46 (on p. 44); for Kekulé, see Richard Anschütz, *August Kekulé* (Berlin: Verlag Chemie, 1929), **1**, 151n.; for Liebig at Giessen, see J. Volhard, *Justus von Liebig* (Leipzig: Barth, 1909), **1**, 80; for Horsford, see Rossiter, *Agricultural Science*, p. 81; for Hofmann and for Liebig at Munich, see Brock, *LHB*, pp. 13 and 136.

35. Meinel, *Chemie*, pp. 51–61 and 438–444; HSA, *305n.*, Nr. 1045; Kolbe to Kurhessisches Ministerium des Innern (KMI), 29 October 1853, HSA, 16. VI, Kl. 13, Nr. 4, Bd. I; Kolbe to KMI, 3 August 1854, 8 September 1855, and 12 September 1860, ibid., Bd. II.

36. In 1842 Liebig was charging his two-days-a-week Praktikanten 13 florins and his full-time advanced Praktikanten 39 florins per semester: William Gregory, *Letter to the Right Honourable George, Earl of Aberdeen . . . on the State of the Schools of Chemistry in the United Kingdom* (London: Taylor & Walton, 1842), p. 22. The latter figure was equivalent to about 23 thalers per semester, thirty-five percent more than Kolbe's fee of 17 thalers during the following decade. The Grand Duchy's government paid for fuel and replacement apparatus; Liebig's fee may, however, have included supplies. His lab budget was 1900 florins after 1843 (equivalent to 1100 thalers). Lab budgets in some Prussian universities are given in Turner, "Institute-Building," pp. 153–155. For costs in England and the United States, see Morrell, "Chemist Breeders," p. 18n., and Rossiter, *Agricultural Science*, n. 13.

37. Meinel, *Chemie*, p. 84; Turner, "Liebig," p. 154.

38. Meinel, *Chemie*, pp. 60–63.

39. Meinel names the philosopher Th. Waitz (ibid., p. 50); Ernst von Meyer identifies a few more (Meyer, HK, pp. 463–464).

40. Grete Ronge, "Hermann Kolbe," *Neue deutsche Biographie*, **12** (Berlin: Duncker & Humblot, 1980), 447.

41. Kolbe to Vieweg, 12 March 1853, VA 50.

42. Kolbe to Vieweg, 25 March 1853, VA 51; Kolbe to Varrentrapp, 11 March 1875, VA 326.

43. Kolbe to Vieweg, 28 April and 22 June 1853, VA 52 and 53.

44. Kolbe to Vieweg, 24 January 1853, 25 March 1853, and 14 February 1859, VA 48, 51, and 149.

45. Ernst von Meyer, *Lebenserinnerungen* (n.p., n.d., ca. 1918), pp. 30 and 97.

46. Kolbe to Vieweg, 18 December 1855 to 9 March 1856, VA 112–115; 13 September 1861 to 21 July 1862, VA 176–184; and 30 March to 23 December 1865, VA 222 to 238.

47. Kolbe to Vieweg, 8 May 1854, VA 78.

48. Guthrie to Roscoe, 5 August 1854, Roscoe Collection.

49. Kolbe to Eduard Vieweg, 14 February 1854 and 20 October 1855; Kolbe to Heinrich Vieweg, 20 October 1855, VA 68, 109, and 110.

50. Kolbe to Vieweg, 20 February 1857, VA 123.

51. Kolbe to Vieweg, 12 July and 4 August 1858, 11 July 1860, 9 April 1863, 20 March 1865, and 5 March 1868, VA 140, 141, 171, 189, 222, and 253; Kolbe to Varrentrapp, 1 January 1874, VA 316; printed announcement of Kolbe's death, with family members' names, 26 November 1884, Frankland Archive 01.04.1527.

52. Kolbe to Vieweg, 5 June 1865, VA 227. He said he and his wife had for years not partaken of the "gesellschaftlichen Strudel" in Marburg.

53. Kolbe to Vieweg, 24 October 1859, VA 153.

54. For example, Kolbe to Vieweg, 5 June and 15 September 1865, VA 227 and 235.

55. Lockemann, HK, p. 125; Frankland, reminiscences of Kolbe sent to Hermann Ost on 20 December 1884, SSDM 3576.

56. Kolbe to Vieweg, 19 December 1853, 20 March 1854, 6 and 19 January and 14 March 1855, VA 62, 70, 95, 96, and 98.

57. Kolbe to Vieweg, 23 December 1855, VA 113.

58. Kolbe to Frankland, 4 April 1871, Frankland Archive 01.03.596; Kolbe to Bertha Ost, 20 April 1876, SSDM 6799.

59. Kolbe to Vieweg, 24 March 1857, VA 125.

60. Kolbe to Vieweg, 12 April to 26 May 1857, VA 126 to 129.

61. Kolbe to Vieweg, 17 June and 31 August 1857, VA 130 and 132.

62. Kolbe to Vieweg, 20 December 1857 and 15 February to 23 June 1858, VA 134 and 136 to 139.

63. Kolbe to Vieweg, 8 July 1861 and 15 July to 5 August 1861, VA 170 and 172 to 174.

64. Kolbe to Vieweg, 24 October 1859, VA 153.

65. Kolbe to Vieweg, 3 and 9 April 1860, VA 156 and 157; Kolbe to Liebig, 16 April 1860, Liebigiana IIB, no. 5.

66. Kolbe to Vieweg, 15 October 1859 and 22 October 1860, VA 152 and 161.

67. Kolbe to Vieweg, 16 October 1860, VA 160.

68. Kolbe to Vieweg, 8 and 15 July 1861, VA 170 and 172.

69. Kolbe to Vieweg, 8 April 1862, VA 180.

70. For example, Kolbe to Varrentrapp, 10 October 1872 and 3 August 1873, VA 292 and 312.

71. Meinel, *Chemie*, pp. 83–119, 470–472, 480–486, and 524–528.

72. Guthrie to Roscoe, 5 August 1854, Roscoe Collection.

73. Kolbe to Vieweg, 16 July 1855, VA 106; W. H. Brock, *H. E. Armstrong and the Teaching of Science* (New York: Cambridge Univ. Press, 1973), pp. 60 and 64.

74. Graebe to his parents, 9 and 20 May 1862, SSDM 1933-78/17. The second of these passages is cited in Elisabeth Vaupel, *Carl Graebe (1841–1927): Leben, Werk und Wirken* (Ph.D. dissertation, Univ. of Munich, 1987), pp. 29 and 32.

75. D. Vorländer, "Jacob Volhard," *Berichte*, **45** (1912), 1855–1902 (on p. 1865).

76. Armstrong later reminisced (more than once) that he arrived in Leipzig in 1868. This appears to have been a trick of memory, for there is no question that he traveled to Leipzig a year earlier.

77. H. E. Armstrong, "The Doctrine of Atomic Valency," *Nature*, **125** (1930), 807–810 (on p. 808–809); idem, "The Riddle of Benzene: August Kekulé," *Journal of the Society of Chemical Industry*, **48** (1929), 914–918 (on pp. 914–915); and idem, "Persönliche Erinnerungen und Gedanken," *Chemiker-Zeitung*, **51** (1927), 114–116.

78. Meyer, "Kolbe," p. 464; idem, *Lebenserinnerungen*, p. 120.

79. Ost, HK, p. 133.

80. At Göttingen Kolbe took courses in physics, mineralogy & geology, mathematics, and metaphysics. At Leipzig, Ernst von Meyer, a man of admittedly much broader interests than his mentor, took physics, mineralogy & crystallography, zoology, logic, political economy, political history, and the history of art and literature over the course of three semesters before concentrating exclusively on chemistry. See chap. 2 and Meyer, *Lebenserinnerungen*, pp. 28–29.

81. The following discussion is based on Kolbe's report (8 October 1863) to the Marburg University Senate on the activities of the Chemical Institute, in HSA, *305a*. A IV, c. ϵ 1, Nr. 12 (a transcript is printed in Meinel, *Chemie*, pp. 435–438, and Kolbe revised it to form part of the introduction to his *Das chemische Laboratorium der Universität Marburg* [Braunschweig: Vieweg, 1865], pp. 17–28); on Kolbe's final report to his ministry, 5 July 1865, HSA, *305n.*, Nr. 1045 (transcript printed in Meinel, *Chemie*, pp. 438–444); on Kolbe, *Das chemische Laboratorium der Universität Leipzig* (Braunschweig: Vieweg, 1872), pp. xxxiv–xlvi; on Meyer, *Lebenserinnerungen*, pp. 27–30; and on Ost, HK, pp. 128–129.

82. Cf. Emil Erlenmeyer's pedagogical goals, namely, to teach students to "chemisch denken lernen" and "die chemische Sprache lesen und schreiben lernen"; see *Die Aufgabe des chemischen Unterrichts* (Munich: Königliche Akademie der Wissenschaften, 1871), esp. pp. 11–17. For a detailed case study from the history of physics, see Kathryn Olesko, *Physics as a Calling: Discipline and Practice in the Königsberg Seminar for Physics* (Ithaca, N.Y.: Cornell Univ. Press, 1991).

83. For which, see especially Michael Polanyi, *Personal Knowledge* (London: Routledge, 1958), and J. R. Ravetz, *Scientific Knowledge and its Social Problems* (Oxford: Clarendon Press, 1971).

84. Harold Hartley, quoting Armstrong without reference, in *Studies in the History of Chemistry* (Oxford: Clarendon Press, 1971), pp. 219–220.

85. A. F. Plate, G. V. Bykov, and M. S. Eventova, *Vladimir Vasil'evich Markovnikov: ocherk zhizni i deiatel'nosti, 1837–1904* (Moscow: Izdatel'stvo Akademii Nauk SSSR, 1962), p. 30, quoting from a letter from Markovnikov to Butlerov with no date cited. The translation is that of H. M. Leicester, "Controversies on Chemical Structure from 1860 to 1870," in O. T. Benfey, ed., *Kekulé Centennial* (Washington: American Chemical Society, 1966), pp. 13–23 (on p. 21).

86. A. Crum Brown to Frankland, 5 June 1866, Frankland Archive 01.04.1266.

87. Kolbe, *Laboratorium der Universität Leipzig*, p. xlviii.

88. Liebig to Vieweg, 28 March 1855, in Margarethe and Wolfgang Schneider, eds., *Justus Liebig: Briefe an Vieweg* (Braunschweig: Vieweg, 1986), p. 288.

89. Kolbe to Vieweg, 31 August 1857, VA 132.

90. Liebig, trans. C. Gerhardt, *Traité de chimie organique*, 3 vols. (Paris, 1841–1844); Karl Löwig, *Chemie der organischen Verbindungen*, 2d ed., 2 vols. (Braunschweig: Vieweg, 1845–1847); J. E. Schlossberger, *Lehrbuch der organischen Chemie*, 2d ed. (Stuttgart: Müller, 1852); V. Regnault, ed. A. Strecker, *Kurzes Lehrbuch der organischen Chemie* (Braunschweig: Vieweg, 1851); 2d ed., 1853. The latter is essentially a rewrite by Strecker of Regnault's French original.

91. Kolbe to Vieweg, 24 January 1853, VA 48.

92. Kolbe to Vieweg, 12 February and 12 March 1853, VA 49 and 50.

93. *Gregory-Gerding's organische Chemie* (Braunschweig: Schwetschke, 1854), from Gregory's *Outlines of Organic Chemistry*, 3d ed. (London, 1852).

94. Kolbe to Vieweg, [16 October 1853], VA 59; undated, but letter no. 60 (17 October 1853) mentions this letter as having been written "yesterday." C. Gerhardt, trans. R. Wagner, *Lehrbuch der organischen Chemie* (Leipzig, 1854–1855).

95. Ibid.

96. Kolbe to Vieweg, 6 January 1854, VA 63. An even more direct statement to this effect is in Kolbe to Vieweg, 21 November 1853, VA 61.

97. Kolbe to Vieweg, 1 March 1853 [sic for 1854], VA 69 (quoted passage); 24 and 27 April 1854, VA 75 and 76. Kolbe received his first reaction to the published installment by the end of June: Kolbe to Vieweg, 1 July 1854, VA 80.

98. Kolbe to Vieweg, 4 April 1854, VA 71; Kolbe, *Lehrbuch*, pp. v–vii.

99. Kolbe to Vieweg, 18 April 1854, VA 74.

100. Kolbe to Vieweg, 24 and 27 April 1854, VA 75 and 76.

101. Kolbe to Vieweg, 1 July 1854, 12 and 23 August 1854, and 5 February 1855, VA 80, 82, 83, and 97.

102. Kolbe to Vieweg, 15 May 1854, VA 79.

103. Kolbe to Vieweg, 5 February 1855, VA 97.

104. Kolbe to Vieweg, 1 June 1856 and 4 August 1858, VA 119 and 141.

105. Kolbe to Heinrich Vieweg, 11 December 1879, VA 447. In this letter

he actually named a second "French" book as well, the Strecker–Regnault text, but as this was a complete rewrite by Strecker of the original, it can hardly be considered French.

106. Kolbe to Vieweg, 1 July 1854 and 6 January 1855, VA 80 and 95; Liebig to Vieweg, 28 March 1855, in Margarete and Wolfgang Schneider, eds., *Briefe an Vieweg*, p. 288.

Notes to Chapter 6

1. See S. H. Mauskopf, "The Atomic Structural Theories of Ampère and Gaudin: Molecular Speculation and Avogadro's Hypothesis," *Isis*, **60** (1969), 61–74.

2. Dumas explained his concept of mechanical types in "Mémoire sur la loi des substitutions et la théorie des types," *C.r.*, **10** (1840), 149–178 (on pp. 163–164). He depicted the molecule as a "building" in which the "framework" could be altered but not removed without dissolution of the compound (pp. 174–175).

3. Laurent to Gerhardt, 4 July 1844, 12 February and 24 February 1845, in M. Tiffeneau, ed., *Correspondance de Charles Gerhardt*, 2 vols. (Paris: Masson, 1918–1825), **1**, 5 and 19–20.

4. P. Thenard, "Suite des recherches sur le phosphore," *C.r.*, **25** (1847), 892–894.

5. J. Harris and W. H. Brock, "From Giessen to Gower Street: Towards a Biography of Alexander W. Williamson," *Annals of Science*, **31** (1974), 95–130.

6. Williamson, "Theory of Aetherification," *Philosophical Magazine*, [3] **37** (1850), 350–356, reprinted in *Papers on Etherification and on the Constitution of Salts* (Edinburgh: Alembic Club Reprint no. 16, 1902), pp. 5–17.

7. Williamson, "Theory of Aetherification," and more explicitly in "On the Constitution of Salts," *JCS*, **4** (1851), 350–355, reprinted in *Papers*, pp. 41–49 (on pp. 45–47). For a discussion see Colin A. Russell, *History of Valency* (Leicester: Leicester Univ. Press, 1971), pp. 50–54.

8. Williamson, "Note on the Decomposition of Sulphuric Acid by Pentachloride of Phosphorus," *Proceedings of the Royal Society*, **7** (1854), 11–15. Williamson mentioned this paper and its results in a letter dated 10 March 1854 to Henry Roscoe, then in Heidelberg (Roscoe Collection).

9. Gerhardt, "Remarques sur un travail de M. Hofmann sur les radicaux," *Comptes rendus des travaux de chimie*, **6** (1850), 233–236.

10. Hofmann, "Note on the Action of Heat upon Valeric Acid," *JCS*, **3** (1850), 121–134.

11. B. C. Brodie, "Observations on the Constitution of the Alcohol-Radicals, and on the Formation of Ethyl," *JCS*, **3** (1850), pp. 405–411, esp. p. 411.

12. A. Wurtz, "Sur une nouvelle classe de radicaux organiques," *Ann. chim.*, [3] **44** (1855), 275–313.

13. In an editorial note, Liebig mentioned Hofmann's priority when he printed a German summary of Wurtz' paper: "Über eine neue Klasse organi-

scher Radicale," *Annalen*, **96** (1855), 364-375 (on p. 365n.). Wurtz himself mentioned no predecessor other than Williamson.

14. Wurtz, "Sur l'hydrure de cuivre," *Ann. chim.*, [3] **11** (1844), 250-252; Dumas, *Traité de chimie appliquée aux arts*, **1** (Paris, 1828), xxxviii.

15. Wurtz, "Radicaux organiques," p. 306.

16. Frankland, "Researches on the Organic Radicals," *JCS*, **3** (1850), 30–52 (on pp. 46–51); Hofmann, "Researches Regarding the Molecular Constitution of the Volatile Organic Bases," *PTRS*, **140** (1850), 93–131 (on pp. 93 and 95–97); Wurtz, "Mémoire sur une série d'alcaloides homologues avec ammoniaque," *Ann. chim.*, [3] **30** (1850), 443-507 (on pp. 444–445 and 501–502); Williamson, "Theory of Aetherification," reprint p. 14.

17. Gerhardt to Liebig, 18 October 1850, in C. Gerhardt, Jr., and E. Grimaux, *Charles Gerhardt: Sa vie, son oeuvre, sa correspondance 1816–1856* (Paris: Masson, 1900), pp. 202–203 and 241; Liebig's letter is excerpted in Tiffeneau, ed., *Correspondance*, **2**, 121.

18. Frankland, "On a New Series of Organic Bodies Containing Metals," *PTRS*, **142** (1852), 417–444 (on p. 441).

19. Wurtz, "Radicaux organiques," pp. 308-313.

20. A. W. Hofmann, "Life-Work of Liebig," in *Zur Erinnerung an vorangegangene Freunde*, 3 vols. (Braunschweig: Vieweg, 1888), **1**, 195–305 (on p. 273); Hofmann to Liebig, 10 May 1852, in Brock, *LHB*, p. 132.

21. Kolbe to Vieweg, 1 March 1853 [sic for 1854], VA 69. For Liebig's indifference to theory, cf. Liebig to Wöhler, 15 April 1857, 27 February 1865, and March 1870, in Hofmann, *LWB*, **2**, 42, 179, and 280; also Kolbe to Vieweg, 16 March 1850, 18 April 1854, and 24 October 1859, VA 26, 74, and 153.

22. See Kolbe to Vieweg, 9 April 1860, VA 157.

23. See Erdmann to Gerhardt, 4 February 1855, in Grimaux, *Gerhardt*, p. 265. Erdmann had taken Gerhardt into the editorial board of the *Journal für praktische Chemie* at the beginning of 1852. When the first installment of Kolbe's *Lehrbuch* appeared, Erdmann protested loudly in a letter to Kolbe, which is not preserved (see Kolbe to Vieweg, 1 July 1854, VA 80). After Kolbe transferred to Leipzig, their relationship improved greatly.

24. A. Strecker, "Über einige Verbindungen der Milchsäure," *Annalen*, **91** (1854), 352–367; Will referred to Strecker as a convert in 1854: *Annalen*, **91**, 265. See also Kolbe to Vieweg, 3 May and 1 June 1856, 20 February 1857, and 15 April 1860, VA 118, 119, 123, and 158. For Strecker's participation in Karlsruhe, see R. Anschütz, *August Kekulé* (Berlin: Verlag Chemie, 1929), **1**, 680 and 686–687. See also R. Wagner, "Adolph Strecker," *Berichte*, **5** (1872), 125–131.

25. H. Kopp, "Über die specifischen Volume flüssiger Verbindungen," *Annalen*, **92** (1854), 1–32 (on pp. 24 and 28); *Jahresbericht über die Fortschritte der Chemie*, **7** (1854), 370–373; ibid., **10** (1857), 269–270; ibid., **13** (1860), 218–222; Kolbe to Vieweg, 1 July and 23 October 1854, VA 80 and 87; Kopp, *Die Entwickelung der Chemie in der neueren Zeit* (Munich: Oldenbourg, 1873), pp. 750–753 and 763. That Kopp as well as Will taught the reformed chemistry while Volhard was a student at Giessen (1852–1855) is stated in Vorländer, "Jacob Volhard," *Berichte*, **45** (1912), 1855–1902 (on pp. 1860–1861).

26. F. Wrightson, "On the Atomic Weight and Constitution of the Alcohols," *Philosophical Magazine*, [4] **6** (1853), 88–99.

27. Liebig's editorial notes to Williamson, "Über Aetherbildung," *Annalen*, **81** (1851), 73–87 (on pp. 73n. and 76n).

28. Williamson, "Note on the Preparation of Propionic and Caproic Acids," *Philosophical Magazine*, [4] **6** (1853), 204–206. Williamson later stated that he had tried in vain to talk Wrightson out of publishing the paper (*JCS*, **7** [1854], 122). Williamson and Wrightson had first met in Liebig's laboratory in Giessen in the late 1840s.

29. Wrightson, "Remarks on Professor Williamson's Othyle Theory," *Philosophical Magazine*, [4] **6** (1853), 418–420.

30. Kolbe to Vieweg, no date, but 16 October 1853 by context, VA 59.

31. Kolbe, "Kritische Bemerkungen zu Williamson's Wasser-, Aether- und Säure-Theorie," *Annalen*, **90** (1854), 46–61; "Critical Observations on Williamson's Theory of Water, Ethers, and Acids," *JCS*, **7** (1854), 111–121.

32. Henry Watts to H. E. Roscoe, no date, but February 1854 by context, Roscoe Collection.

33. Williamson to Roscoe, 9 February 1854, Roscoe Collection.

34. Williamson, "On Dr. Kolbe's Additive Formulae," *JCS*, **7** (1854), 122–139; "Über Kolbe's chemische Formeln," *Annalen*, **91** (1854), 201–228.

35. Williamson, "Kolbe's Formulae," pp. 123 and 132–135. When Kolbe visited London fourteen years later, he was graciously and warmly received by Williamson. In 1881, in the context of a violent diatribe against all of his scientific foes from Dumas to Baeyer, Kolbe mentioned his respect and affection for the Englishman, claiming that their earlier disagreement had been due to a mere misunderstanding (*JpC*, **131** [1881], 311n.).

36. Gerhardt, "Über Wasser-, Aether- und Säure-Theorie," *Annalen*, **91** (1854), 198–200.

37. Ibid. [Liebig, editorial note], p. 199n.

38. Frankland, "Organic Bodies"; "Über eine neue Reihe organischer Körper, welche Metalle enthalten," *Annalen*, **85** (1853), 329–373.

39. Kolbe, *Lehrbuch*, **1**, 12–14 (the first installment consisted of pp. 1–176).

40. Ibid., pp. 14–18.

41. Ibid., pp. 20–22, 24, and 43–44. Cf. Kolbe, "Notiz über das Cyanbenzoyl," *Annalen*, **90** (1854), 52–63, and "Über eine neue Bildungsweise des Benzoylwasserstoffs und die chemische Constitution der Aldehyde," *Annalen*, **98** (1856), 344–349.

42. Kolbe, *Lehrbuch*, **1**, 24–27.

43. Ibid., pp. 29–31.

44. Ibid., pp. 22–23.

45. Ibid., p. 35.

46. Ibid., pp. 40–41.

47. Ibid., pp. 41–43.

48. Ibid., pp. 49–53.

49. Ibid., pp. 51–52; Kolbe to Vieweg, 1 March 1853 [sic for 1854], VA 69.

50. Hofmann, "Heinrich Will: Ein Gedenkblatt," *Berichte*, **23R** (1890), 852–899 (on p. 857).

51. Anschütz, *Kekulé*, **1**, 16–17, from a letter to Anschütz, date not cited but probably shortly after the turn of the century, from Reinhold Hoffmann.

52. Volhard, *Justus von Liebig* (Leipzig: Barth, 1909), **1**, 351; Vorländer, "Volhard," pp. 1860–1861.

53. Will, "Zur Theorie der Constitution organischer Verbindungen," *Annalen*, **91** (1854), 257–292.

54. Hofmann, "Will," pp. 881–882; idem, *Erinnerung*, **3**, 249.

55. Hofmann, *Introduction to Modern Chemistry* (London: Walton & Maberley, 1865), p. v; Schorlemmer to Roscoe, 5 November 1881, Roscoe Collection.

56. Kolbe, "Radicale; Radicaltheorie," *Handwörterbuch*, **6** (1854), 802–807. The German summary of Wurtz' paper was "Über eine neue Klasse organischer Radicale," *Annalen*, **96** (1855), 364–375. See also Kolbe to Vieweg, 23 December 1855, 5 January 1856, and 20 February 1857, VA 113, 114, and 123.

57. Frankland, *Experimental Researches in Pure, Applied, and Physical Chemistry* (London: J. van Voorst, 1877), pp. 147–148; *Sketches from the Life of Edward Frankland* (London: Spottiswoode, 1901), pp. 193–196.

58. Kolbe, "Radicale." To be precise, Kolbe did not explicitly write that carbon has a maximum combining capacity of four, and he did not say that methyls "substitute" directly for the oxygen of carbonic acid. The argument continued to be phrased in terms of the copula theory. He was, however, quite clearly and consciously applying Frankland's law, which he now accepted, to carbon.

59. Kolbe, "Über die chemische Constitution der elementaren Moleküle," *JpC*, **115** (1873), 119–126 (on pp. 124–125n.).

Notes to Chapter 7

1. Williamson mentioned that he had established this convention for himself and his students in a letter of 5 December 1853 to H. E. Roscoe (Roscoe Collection).

2. For background on this French tradition, see S. H. Mauskopf, *Crystals and Compounds: Molecular Structure and Composition in Nineteenth Century French Science* (Philadelphia: American Philosophical Society, 1976). On Dumas' ideas on submolecularity, see A. J. Rocke, *Chemical Atomism in the Nineteenth Century* (Columbus: Ohio State Univ. Press, 1984), pp. 115–118. The passage quoted is from Dumas, *Leçons de philosophie chimique* (Paris: Bechet, 1837), p. 290.

3. Laurent, "Sur les combinaisons organiques azotées," *C.r.*, **20** (1845), 850–855 (854); idem, "Recherches sur les combinaisons azotées," *Ann. chim.*, [3] **18** (1846), 266–298 (on pp. 267–268 and 294–295); idem, *Chemical Method* (London, 1855), pp. 46–48 and 69.

4. Laurent, *Method*, p. 101.

5. Ibid., p. 103.

6. Laurent, "Sur la série naphthalique," *Revue scientifique*, **14** (1843), 74–113 (on pp. 102–103); *Method*, pp. 103–107; Gerhardt, *Traité de chimie orga-*

nique, 4 vols. (Paris: Didot, 1853–1856), **4**, 602–604, 712. Laurent's speculations are discussed by Mauskopf, *Crystals and Compounds*, pp. 50–51. Laurent wrote Berzelius on this subject on 5 January 1844: H. G. Söderbaum, ed., *Berzelius Bref*, 6 vols. (Uppsala: Almqvist & Wiksell, 1912–1961), **3**:2, 199–200. On this letter, see J. H. Brooke, "Chlorine Substitution and the Future of Organic Chemistry," *Studies in the History and Philosophy of Science*, **4** (1973), 47–94 (on pp. 60–64).

7. Laurent, *Method*, pp. 1–16. This is a major thesis of my *Chemical Atomism in the Nineteenth Century*.

8. For Williamson, see the preceding chapter; William Odling, "On the Constitution of Acids and Salts," *JCS*, **7** (1854), 1–22; Edward Frankland, "On a New Series of Organic Bodies Containing Metals," *PTRS*, **142** (1852), 417–444.

9. Williamson, "Note sur la trinitroglycerine," *Ann. chim.*, [3] **43** (1855), 492 (originally published in *Proceedings of the Royal Society*, 7 [June 1854], 130–138); Wurtz, "Théorie des combinaisons glycériques," *Ann. chim.*, [3] **43** (1855), 492–496.

10. Wurtz, "Sur le glycol ou alcool diatomique," *C.r.*, **43** (1856), 199–204. The historical comments, including the date of the experiment, were given in Wurtz, "Histoire générale des glycols," in Société Chimique de Paris, ed., *Leçons de chimie professées en 1860* (Paris: Hachette, 1861), pp. 101–139 (103–109). The assertion that Wurtz had made himself a prediction in this manner was also stated in Wurtz, *Ann. chim.*, [3] **55** (1859), 401.

11. Wurtz, "Sur une nouvelle classe de radicaux organiques," *Ann. chim.*, [3] **44** (1855), 275–313 (on pp. 304–309). The idea was first broached in his "Théorie des combinaisons glycériques," pp. 495–496.

12. Wurtz, "Radicaux organiques," pp. 306–307 and 306–307n.

13. Wurtz, *Répertoire de chimie pure*, **3** (1861), 419; *Leçons de philosophie chimique* (Paris: Hachette, 1864), pp. 114–115; *Cours de philosophie chimique* (Paris: Renou et Maulde, 1864), pp. 74–76; "Histoire des doctrines chimiques depuis Lavoisier," in Wurtz, ed., *Dictionnaire de chimie pure et appliquée*, 3 vols. in 5 (Paris: Hachette, 1868–1878), **1**, lxx; *La théorie atomique* (Paris: Baillière, 1879), pp. 148–149.

14. Rocke, *Chemical Atomism*, pp. 11–12, 299–307.

15. For details, see Rocke, "Subatomic Speculations and the Origin of Structure Theory," *Ambix*, **30** (1983), 1–18.

16. The following biographical details are largely based on Richard Anschütz' monumental, partisan, but indispensable biography *August Kekulé*, 2 vols. (Berlin: Verlag Chemie, 1929), and on two autobiographical speeches that Kekulé gave in his old age, reprinted in ibid., **2**, 937–947 and 947–952. For convenience, I will usually cite Anschütz' collection (cited as *Kekulé*) rather than the original literature.

17.*Kekulé*, **1**, 16–17 (from a letter to Anschütz, no date cited but probably shortly after the turn of the century, from Reinhold Hoffmann).

18. *Kekulé*, **1**, 40.

19. Ibid., p. 41.

20. Ibid., **2**, 950.

21. Ibid., **1**, 664.
22. Kekulé to Planta, 3 March 1854, August-Kekulé-Sammlung.
23. *Kekulé*, **2**, 943–944.
24. Kekulé, "On a New Series of Sulphuretted Acids," *Proceedings of the Royal Society*, **7** (1854), 37–40; "Notiz über eine neue Reihe schwefelhaltiger organischer Säuren," *Annalen*, **90** (1854), 309–316.
25. Odling, "On the Constitution of the Hydro-Carbons," *Proceedings of the Royal Institution*, **2** (1855), 63–66. Anschütz (**1**, 109) comments that Kekulé could not have been aware of Odling's article because this journal was unknown in Germany, but Kekulé was still in London when Odling presented his paper (16 March 1855), and as Kekulé and Odling were friends, it is probable that Kekulé knew the details of this paper.
26. *Kekulé*, **2**, 941–942.
27. This is a thesis argued in Rocke, "Subatomic Speculations."
28. Kekulé, "Über die Constitution des Mesitylens," *ZfC*, **10** (1867), 214–218 (on p. 217); *Kekulé*, **2**, 530.
29. *Kekulé*, **2**, 102.
30. That the July 1855 number of the *Annales de chimie et de physique* did in fact appear in July is documented by its mention in the *Comptes rendus*, **41**, 158 (dated 30 July 1855).
31. *Kekulé,* **1**, 60–63. Kekulé's relations with his brother Charles are described in his letters to Planta of 28 October and 28 December 1854 and especially 9 February 1856, August-Kekulé-Sammlung. The words in quotation marks are August Kekulé's.
32. On Erlenmeyer, see Otto Krätz, "Emil Erlenmeyer, 1825–1909," *Chemie in unserer Zeit*, **6** (1972), 52–58, and idem, ed., *Beilstein-Erlenmeyer: Briefe zur Geschichte der chemischen Dokumentation und des chemischen Zeitschriftenwesens* (Munich: Fritsch, 1972).
33. *Kekulé*, **2**, 940.
34. Kekulé, "Über die Constitution des Knallquecksilbers," *Annalen*, **101** (1857), 200–213, and **105** (1858), 279–286; Adolf Baeyer, "Über die organischen Arsenverbindungen," *Annalen*, **105** (1858), 265–276. Gerhardt recognized the same distinction by disowning the "mechanical type" concept of Dumas, which implied arrangement: *Traité de chimie organique*, **4** (Paris: Didot, 1856), 586.
35. Limpricht and von Uslar, "Über die Sulfobenzoësäure," *Annalen*, **102** (1857), 239–259; Mendius, "Über gepaarte Säuren und insbesondere über Sulfosalicylsäure," *Annalen*, **103** (1857), 39–80.
36. *Kekulé*, **2**, 80–82, 97–102.
37. Ibid., p. 101.
38. Kekulé's criticism of Limpricht's formulas was that, with an atomic weight for oxygen of eight, his water type had to be H_2O_2, which meant that there was no longer a single integral oxygen atom that could form a material link between two hydrogens. As for Gerhardt, both in this paper (*Kekulé*, **2**, 84n.) and two years later in his *Lehrbuch der organischen Chemie* (2 vols. [Erlangen: Enke, 1859–1866], **1**, 94), Kekulé singled out formulas of Gerhardt that could not be expressed in a valence-linked manner, calling them "inconsis-

tent" and "inadmissible" according to the theory of polyatomic radicals (such formulas as appear, for instance, in Gerhardt, *C.r.*, **36** [1853], 1053, and *Traité*, **4**, 629 and 749). Kekulé's disapproval of similar nonlinkable type formulas in Odling's 1854 paper can be inferred. Kekulé's offprint copy of this paper is preserved in the August-Kekulé-Sammlung, and there are penciled marginal annotations in Kekulé's hand ("sic!" and "!!!") every time such a structurally impossible formula appears (Odling, "Constitution," pp. 7–9).

39. Kekulé, "Über die s.g. gepaarten Verbindungen und die Theorie der mehratomigen Radicale," *Annalen*, **104** (1857), 129–150; *Kekulé*, **2**, 80–96.

40. What we now call valence was referred to by Kekulé variously as "basicity" (from Williamson, 1851), "atomicity" (from Wurtz and others, 1856), "affinity," "affinity units," "chemical units," and "units." Other designations include "value" ("Wertigkeit," Erlenmeyer, 1860), "monaffin," etc. (Wislicenus, 1863), "monad," etc. (Odling, 1864, following Laurent), "hydrogen equivalence" (Gerhardt, 1856), "quantivalence," "monovalent," etc. (Hofmann, 1865), and "valence" (Claus, 1866). The latter term gained currency only in the course of the 1870s. See C. A. Russell, *The History of Valency* (Leicester: Leicester Univ. Press, 1971), pp. 83–89, some of which is modified by the previous data. Russell's book is excellent on the development of valence ideas from Dalton through the first third of the twentieth century.

41. *Kekulé*, **2**, 83n.

42. Limpricht, "Einige Bemerkungen zu der von A. Kekulé veröffentlichten Abhandlung 'Uber die s.g. gepaarten Verbindungen und die Theorie der mehratomigen Radicale,'" *Annalen*, **105** (1858), 177–186.

43. Kekulé, "Über die Constitution und die Metamorphosen der chemischenVerbindungen und über die chemische Natur des Kohlenstoffs," *Annalen*, **106** (1858), 129–159; *Kekulé*, **2**, 97–119. An annotated English translation was published by O. T. Benfey, in his edited work *Classics in the Theory of Chemical Combination* (New York: Dover, 1963), pp. 109–131.

44. *Kekulé*, **2**, 102.

45. Ibid., pp. 114–116.

46. Ibid., pp. 109, 116–119, 138–141, 153, and 204; *Lehrbuch der organischen Chemie*, **1** (1859), 131, 156–157, 164, 174, and 224; **2** (1864), 244–245.

47. *Kekulé*, **2**, 110n. and 112.

48. Ibid., pp. 116–117.

49. Ibid., pp. 118–119. Wurtz' words were, "In conclusion, I must say that I do not attach more importance than necessary to the ideas I have tried to develop, and that I am very far from considering them as the absolute expression of truth. In the physical sciences theories must not aim that high. The best are those that embrace the largest number of facts, that account for them in the most satisfactory fashion, and that give rise to predictions of new facts. Chemical theories that tend to prevail today appear to me to fit this pattern: they are good to the extent that they are fruitful." Wurtz, "Radicaux organiques" (see n. 11), p. 313.

50. This information is from Erlenmeyer's letter to Roscoe, no date but ca. 1859, Roscoe Collection.

51. Krätz remarks ("Erlenmeyer," p. 52) that Erlenmeyer, too, was an avid cigar smoker. Kekulé's salary in Ghent, noted Anschütz (**1**, 151n.), was 6000 francs. The Belgian franc was equal in value to the French franc, about a fourth of a thaler, so Kekulé's salary was two and a half times Kolbe's, more than sufficient for a bachelor with a modest lifestyle.

52. Kekulé to Erlenmeyer, 29 January 1859, Dingler Nachlass; a copy is held in the August-Kekulé-Sammlung, and most of the letter is printed in *Kekulé*, **1**, 150–151.

53. Cited from Fittig's diary in F. Fichter, "R. Fittig," *Berichte*, **44** (1911), 1339–1401 (on p. 1361).

54. Kolbe, *JpC*, **131** (1881), 37.

55. Couper, "Sur une nouvelle théorie chimique," *C.r.*, **46** (1858), 1157–1160; "On a New Chemical Theory," *Philosophical Magazine*, [4] **16** (1858), 104–116. On seeing the paper, Kekulé fired off some "Remarques de M. A. Kekulé à l'occasion d'une note de M. Couper sur une nouvelle théorie chimique," *C.r.*, **47** (1858), 378, in which he criticized some of Couper's ideas and also pointed out that *both* of his papers were earlier than Couper's first publication (the second even if only by a month).

56. Rocke, "Subatomic Speculations," pp. 9–10.

57. Odling, "Remarks on the Doctrine of Equivalents," *Philosophical Magazine*, [4] **16** (1858), 37–45 (on pp. 43–44).

58. He said this in a letter to Wurtz of 1 July 1859 (*Kekulé*, **1**, 158), and in an unpublished document written in 1883, found among Kekulé's papers by Anschütz and printed in his biography, **1**, 540–569 (on p. 554).

59. Wurtz, *Répertoire de chimie pure*, **1** (1858), 20–24 (on pp. 24 and 24n.).

60. Kekulé to Wurtz, 15 February 1859, August-Kekulé-Sammlung, printed in *Kekulé*, **1**, 146–148. Anschütz reproduces a facsimile of the letter, which shows he made two small errors in transcription.

61. Wurtz to Kekulé, 7 March 1859, August-Kekulé-Sammlung; *Kekulé*, **1**, 148–149; Wurtz, *Ann. chim.*, [3] **55** (1859), 470n.

62. Kekulé to Erlenmeyer, 16 June 1859, Dingler Nachlass; printed in *Kekulé*, **1**, 152.

63. A. Ladenburg, *Lebenserinnerungen* (Breslau: Trewendt, 1912), p. 26.

64. Kekulé to Wurtz, 1 July 1859, in *Kekulé*, **1**, 157–159.

65. Kekulé, *Lehrbuch*, **1**, 94.

66. Wurtz to Kekulé, 21 July 1859, August-Kekulé-Sammlung, in *Kekulé*, **1**, 159.

67. Wurtz, *Répertoire de chimie pure*, **2** (1860), 354–359; ibid., **3** (1861), 418–421; "Histoire des glycols" (1861), pp. 138–139; *Leçons* (1864), pp. 100, 113–114, and 120–121; *Cours* (1864), pp. 50–51, 56, and 74–76; "Histoire des doctrines" (1868), p. lxx; Wurtz to Williamson, 25 December 1868, Harris Collection, Univ. College London Archives; *Théorie atomique* (1879), pp. 145–149.

68. Wurtz, "Mémoire sur les glycols ou alcools diatomique," *Ann. chim.*, [3] **55** (presented 3 January, published April 1859), 400–478 (on pp. 471 and 474–478). Wurtz did indeed use the French word *structure* here, which he may

have picked up from Butlerov. The word had been used occasionally before Butlerov, for example, by Frankland (*Proceedings of the Royal Institution*, **2** [1854], 454), Gaudin, and Berzelius.

69. The Société des Amis de la Science was essentially a monetary fund formed after the death of Laurent, and increased in size after Gerhardt's death, to support the dependents of the two chemists. The Archives of the Académie des Science has some documents regarding this society in the Fonds Dumas. A letter from Wurtz to Dumas of 22 July 1864 takes the part of Jane Gerhardt, which suggests that Wurtz was Madame Gerhardt's chief link to the society.

70. Wurtz, "Histoire des glycols" (1861); "Éloge de Laurent et de Gerhardt" (13 March 1862); "On Oxide of Ethylene, Considered as a Link Between Organic and Mineral Chemistry," *JCS*, **15** (1862), 387–406; *Leçons* (1864); *Cours* (1864); *Leçons élémentaires de chimie moderne* (Paris: Masson, 1867–1868); *Dictionnaire de chimie* (1869–1878); *Théorie atomique* (1879).

71. The best biography in English is G. V. Bykov, "A. M. Butlerov," *DSB*, **2**, 620–625. Butlerov's February 1858 paper was never published, but it is described in Jacques, "Boutlerov, Couper et la Société Chimique de Paris," *BSC*, 1953, pp. 528–530; Butlerov's commentary on Couper is "Bemerkungen über A. S. Couper's neue chemische Theorie," *Annalen*, **110** (1859), 51–66. For further literature and historiography, see Rocke, "Kekulé, Butlerov, and the Historiography of the Theory of Chemical Structure," *British Journal for the History of Science*, **14** (1981), 27–57. A perceptive reply and commentary on this article is Bykov, "K istoriografii teorii khimicheskogo stroeniia," *Voprosy istorii estestvoznaniia i tekhniki*, **1982**:4, 121–130, which was completed shortly before his death.

72. Butlerov, "Einiges über die chemische Struktur der Körper," *ZfC*, **4** (1861), 549–560.

73. Brodie, "On the Conditions of Certain Elements at the Moment of Chemical Change," *PTRS*, **140** (1850), 759–804; Couper, "New Chemical Theory," pp. 112–113.

74. Kekulé, **2**, 944.

Notes to Chapter 8

1. The best brief source for Edward Frankland is presently W. H. Brock's biography in the *DSB*, **5**, 124–127; C. A. Russell is at work on the second volume of his detailed biography.

2. Frankland, "On a New Series of Organic Bodies Containing Metals and Phosphorus," *JCS*, **2** (1849), 297–299.

3. Frankland, "On a New Series of Organic Bodies Containing Metals," *PTRS*, **142** (1852), 417–444 (on p. 432).

4. Frankland to Kolbe, 9 March 1862, Frankland Archive 01.03.429.

5. Hofmann, "Note on the Action of Heat upon Valeric Acid," *JCS*, **3** (1850), 121–134; Brodie, "Observations on the Constitution of the Alcohol-Radicals, and on the Formation of Ethyl," ibid., pp. 405–411; Wurtz, "Sur une nouvelle classe de radicaux organiques," *Ann. chim.*, [3] **44** (1855), 275–313.

6. Frankland, "Organic Bodies Containing Metals," pp. 438–442; Gerhardt and L. Chiozza, "Recherches sur les amides," *C.r.*, **37** (1853), 86–90.

7. Frankland, "Researches on Organo-Metallic Bodies—Second Memoir. Zincethyl," *PTRS*, **145** (1855), 259–275; "On Some Organic Compounds Containing Metals," *British Association for the Advancement of Science, Reports*, **25**(2) (1855), 62; *Sketches from the Life of Edward Frankland* (London: Spottiswoode, 1901), p. 193; Kolbe, *Das chemische Laboratorium der Universität Marburg* (Braunschweig: Vieweg, 1865), p. 33n.

8. Frankland, "Researches on Organo-Metallic Bodies—Third Memoir. On a New Series of Organic Acids Containing Nitrogen," *PTRS*, **147** (1856), 59–78.

9. Frankland, "On Some Organic Compounds Containing Metals," in n. 7.

10. Frankland, *Experimental Researches in Pure, Applied, and Physical Chemistry* (London: Van Voorst, 1877), pp. 146–148 and 154; *Sketches*, pp. 188–191 and 201–203.

11. Frankland, "Zincethyl," pp. 266, 271, and 274, in n. 7.

12. Frankland, "On the Dependence of the Chemical Properties of Compounds upon the Electrical Character of their Constituents," *Proceedings of the Royal Institution*, **1** (1854), 451–454 (on p. 454).

13. Frankland, *Experimental Researches*, pp. 147–148; *Sketches*, pp. 193–197; Kolbe, "Über die rationelle Zusammensetzung der fetten und aromatischen Säuren, Aldehyde, Acetone u.s.w., und ihre Beziehungen zur Kohlensäure," *Annalen*, **101** (1857), 257–265 (on pp. 259–260); Kolbe, *Das chemische Laboratorium*, p. 33; Kolbe to Frankland, 24 November 1863, Frankland Archive 01.04.73.

14. Kolbe, "Radicale, Radicaltheorie," *Handwörterbuch*, **6**, 802–807. A description of this article in the context of Wurtz' work on radicals is provided at the end of chap. 6.

15. Kolbe, "Rationelle Zusammensetzung."

16. No letters between Frankland and Kolbe appear to have survived from any of the years 1855–1861. Kolbe complained to Frankland of the latter's years of silence in his letters of 2 January and 5 February 1862, Frankland Archive 01.08.567 and 01.03.426.

17. Kolbe to Frankland, 5 February 1862 and 24 November 1863, Frankland Archive 01.03.426 and 01.04.73.

18. Kolbe to Frankland, 24 November 1863, Frankland Archive 01.04.73.

19. R. Piria, "Über die Umwandlung organischer Säuren in die entsprechenden Aldehyde," *Annalen*, **100** (1856), 104–106.

20. This at least is the import of Frankland's letter draft to Kolbe, 9 March 1862, Frankland Archive 01.03.429 and 01.03.424. The actual letter has not survived. Kolbe did not soon respond to the matter raised in this letter. Apparently Frankland referred to it the following year, for in October 1863 (01.04.69), Kolbe answered that he did not remember receiving such a letter and asked him to repeat what he had written. Frankland must have done this, for Kolbe's next communication was the long self-exoneration of 24 November 1863. Thus, Kolbe may never have received the letter of 9 March 1862; it may never even have been mailed.

21. Kolbe to Vieweg, 4 January 1857, VA 122.

22. Frankland to Kolbe, 9 March 1862, Frankland Archive 01.03.429.

23. Kolbe, *Das chemische Laboratorium*, pp. 29–36.

24. Kolbe to Hermann Ost, 20 December 1884, SSDM 3575.

25. Frankland, *Experimental Researches*, p. 148.

26. E. von Meyer, trans. G. M'Gowan, *A History of Chemistry* (London: Macmillan, 1891), p. 298n.

27. Kolbe, *JpC*, **131** (1881), 363 and 367.

28. Kolbe to Vieweg, 24 March 1857, VA 125. Kolbe did not use the word "Revolution" or "Umwälzung"; as is fitting for the son of a Lutheran pastor, he wrote, "Es giebt eine kleine Reformation in der organischen Chemie."

29. James Wanklyn, "On Some New Ethyl-Compounds Containing the Alkali-Metals," *Proceedings of the Royal Society*, **9** (1858), 341–345.

30. Frankland to Wanklyn, 8 March 1858, SSDM 3574.

31. Kolbe to Frankland, 24 November 1863, Frankland Archive 01.04.73; Kolbe, "Über den natürlichen Zusammenhang der organischen mit den unorganischen Verbindungen," *Annalen*, **113** (1860), 293–332 (on pp. 298–299n).

32. Bunsen to Kolbe, 15 March 1858, SSDM 3499.

33. Kolbe to Frankland, 5 February 1862, Frankland Archive 01.03.426.

34. W. H. Brock, "James Wanklyn," *DSB*, **14**, pp. 168–170.

35. Frankland, "On the Artificial Formation of Organic Compounds," *Proceedings of the Royal Institution*, **2** (1858), 538–544.

36. Frankland, "On Organo-Metallic Bodies," *JCS*, **13** (1860), 177–235.

37. Frankland, "Researches on Organo-Metallic Bodies—Fourth Memoir," *PTRS*, **149** (1859), 401–415 (on p. 411).

38. Kolbe to Vieweg, 1 June 1856 and 4 August 1858, VA 119 and 141.

39. Kolbe to Vieweg, 24 October 1859, VA 153.

40. The fourth installment constitutes pp. 481–672 of vol. 1 of Kolbe's *Ausführliches Lehrbuch der organischen Chemie* (Braunschweig: Vieweg, 1854); the section "Theoretische Ansichten über die Zusammensetzung der Alkohole, der Acetone, Aldehyde, und der zugehörigen Säuren" fills pp. 567–575.

41. Kolbe to Vieweg, 31 August 1858, VA 132.

42. Regarding the dating of the writing of this installment, see Kolbe to Vieweg, 13 August, 31 August, 8 November, and 20 December 1857, VA 131–134. A publication date of either December 1857 or January 1858 is consistent with these letters. In an unpublished priority claim, Kekulé cited Hinrich's *Bücherverzeichnis* to argue that it appeared in the first half of 1858 (R. Anschütz, *August Kekulé*, 2 vols. [Berlin: Verlag Chemie, 1929], **1**, 560n.). But Kolbe claimed that it appeared before the end of 1857 (*JpC*, **131** [1881], 369), and Kayser's *Bücherlexicon* supports this dating.

43. Kolbe to Vieweg, 20 December 1857, VA 134.

44. Kolbe to Vieweg, 23 June 1858, VA 139.

45. Kolbe, *Über die chemische Constitution organischer Verbindungen* (Marburg: Elwert, 1858).

46. Double fascicle 8/9 was the fifth installment of vol. 1, pp. 673–848. Sheet 42 consisted of pp. 657–672; one of the copies at my disposal contains both versions of the sheet.

47. Regarding the dating of the writing of the fifth installment, see Kolbe to Vieweg, 10 October, 20 October, 8 November, and 24 December 1858 and 5 January 1859 (VA 142, 143, 144, 146, and 147). Kolbe claimed that this installment was published late in 1858 (*JpC*, **131** [1881], 370), but both Hinrich's *Bücherverzeichnis* and Kayser's *Bücherlexicon* list the work as appearing in the first half of 1859, as Kekulé maintained (Anschütz, **1**, 560n.).

48. Wurtz, "Mémoire sur la constitution et sur la vrai formule de l'acide oxalique," *C.r.*, **44** (1857), 1306–1310; idem, "Recherches sur l'acide lactique," *C.r.*, **46** (1858), 1228–1232; idem, "Sur les éthers du glycol," *C.r.*, **47** (1858), 346–350; H. Debus, "On the Oxidation of Glycol, and on Some Salts of Glyoxylic Acid," *Proceedings of the Royal Society*, **9** (1859), 711–716.

49. F. Guthrie and Kolbe, "Über die Verbindungen des Valerals mit Säuren," *Annalen*, **109** (1859), 296–300.

50. Kolbe, *Lehrbuch*, pp. 672–681 and 733–738.

51. Ibid., pp. 672, 675–676, and 734.

52. Ibid., pp. 738–740.

53. "Theoretische Betrachtungen über die Sättigungscapacität der einfachen und zusammengesetzten Radicale," ibid., pp. 740–749.

54. Ibid., pp. 740–742.

55. Ibid., pp. 744–745. Square brackets are added here for greater clarity; these brackets are consistent with Kolbe's later usage.

56. Ibid., pp. 746–749.

57. These sentiments are expressed in the letters from Kolbe to Vieweg cited in n. 47 (October 1858 through January 1859).

58. Kolbe, *Lehrbuch*, p. 673; "Zusammenhang," p. 294n.

59. Kolbe to Vieweg, 15 and 24 October 1859, VA 152 and 153.

60. Anschütz, *Kekulé*, **1**, 175.

61. Kolbe, "Zusammenhang," pp. 293–295 and 332; reprinted under same title, E. von Meyer, ed., as Ostwalds Klassiker Nr. 92 (Leipzig: Engelmann, 1897).

62. Kolbe, "Zusammenhang," pp. 313–316.

63. Kolbe to Vieweg, 1 March 1860, VA 155. This is the first time Kekulé's name appears in any of Kolbe's surviving correspondence.

64. Wurtz, *Répertoire de chimie pure*, **2** (1860), 354–359; ibid., **3** (1861), 418–421; *Leçons de philosophie chimique* (Paris: Hachette, 1864), pp. 113–114.

65. Kopp, *Jahresbericht über die Fortschritte der Chemie*, **13** (1860), 218–222; ibid., **10** (1857), 269–270.

66. Liebig to Kolbe, 3 April 1860, SSDM 3603.

67. Liebig to Wöhler, 15 April 1857, 27 February 1865, and March 1870, in Hofmann, *LWB*, **2**, 42, 179, and 280; Wöhler to Kolbe, 5 December 1862 and 17 October 1878, SSDM 3538 and 3542; Kolbe to Wöhler, 15 October 1878, Wöhler Nachlass.

68. Liebig to Frankland, 28 January 1866, Liebigiana IIB; Liebig to Kolbe, 15 July 1861, SSDM 3606.

69. E. von Meyer, "Die Karlsruhe Chemiker-Versammlung im Jahre 1860," *Journal für praktische Chemie*, **191** (1911), 182–189; Anschütz, *August*

Kekulé, **1**, 183–209 and 671–688; A. Stock, *Der internationale Chemiker-Kongress* (Berlin: Verlag Chemie, 1933); C. deMilt, "Carl Weltzien and the Congress at Karlsruhe," *Chymia*, **1** (1948), 153–169; idem, "The Congress at Karlsruhe," *Journal of Chemical Education*, **28** (1951), 421–424; A. J. Ihde, "The Karlsruhe Congress: A Centennial Retrospect," *Journal of Chemical Education*, **38** (1961), 83–86; and Rocke, *Chemical Atomism in the Nineteenth Century* (Columbus: Ohio State Univ. Press, 1984), pp. 287–311.

70. Pebal to Roscoe, 25 May 1860, Roscoe Collection.

71. Brodie to Kekulé, 27 May 1860, and Williamson to Kekulé, 24 April 1860, August-Kekulé-Sammlung, Darmstadt.

72. L. Meyer, "Leopold von Pebal," *Berichte*, **20** (1887), 997–1015 (on p. 1000).

73. Meyer to Roscoe, 6 July 1860, Roscoe Collection.

74. Kolbe to Weltzien, 17 April 1860, August-Kekulé-Sammlung, Darmstadt; printed in Anschütz, *August Kekulé*, **1**, 188.

75. Kolbe to Vieweg, 16 October 1860, VA 160.

76. Cannizzaro, ed. and trans. L. Meyer, *Abriss eines Lehrganges der theoretischen Chemie* (Leipzig: Ostwalds Klassiker Nr. 30, Engelmann, 1891), p. 59.

77. Cited in J. R. Partington, *A History of Chemistry*, **4** (London: Macmillan, 1964), 489.

78. Frankland, *Sketches*, pp. 191, 201, and 203.

79. See Otto Krätz, ed., *Beilstein-Erlenmeyer Briefe* (Munich: Fritsch, 1972), passim.

80. Kekulé to Meyer, 23 October 1860, August-Kekulé-Sammlung, Darmstadt; printed in Anschütz, *August Kekulé*, **1**, 205.

81. Kolbe, "Moden der modernen Chemie," *JpC*, **112** (1871), 241–271 (on p. 246).

82. Kekulé, *Lehrbuch der organischen Chemie*, 2 vols. (Erlangen: Encke, 1859–1867), **1**, 736–737; **2**, 247–249.

83. Butlerov, "Über die verschiedenen Erklärungsweisen einiger Fälle von Isomerien," *ZfC* **6** (1863), 500–534 (on p. 513); Kolbe, "Moden," p. 247; Frankland, *Experimental Researches*, pp. 153–154.

84. Anschütz, *Kekulé*, **1**, 556–558.

85. Kolbe, "Über die realen Typen der organischen Chemie," *Das Chemische Laboratorium*, pp. 515–519.

Notes to Chapter 9

1. Data for this table are taken from the numbers and information in Christoph Meinel, *Die Chemie an der Universität Marburg seit Beginn des 19. Jahrhunderts* (Marburg: Elwert, 1978), pp. 471–472 and 522–525.

2. This does not refer to the number of papers *authored* by Praktikanten, but rather is simply the total number of papers per year produced by the institute as a whole, divided by the average number of Praktikanten. This manner of accounting sidesteps the tricky question of authorship and suggests at least a rough *intensive* measure of research productivity.

3. Kolbe to Vieweg, 24 and 29 December 1858 and 5 January 1859, VA 145, 146, and 147; *Annalen*, **109** (1859), 257–304.

4. Kolbe to Vieweg, 15 March 1860 (incorrectly dated 15 March 1856) and 16 October 1860, VA 116 and 160; Meinel, *Chemie*, p. 472.

5. The following information is based on a variety of direct and indirect sources, including Kolbe correspondence, publications, standard biographical reference works, and records in the Hessisches Staatsarchiv (especially *305a.* A IV, b. 2, Nr. 65 and 68; *305a.* II, Nr. 11; and acc. 1902/8).

6. G. N. Vis, "Adolph Claus," *JpC*, **170** (1900), 127–133.

7. Kolbe to Vieweg, 23 June 1858, VA 139; cf. also 9 and 29 May 1858, VA 137 and 138.

8. Kolbe to Vieweg, 15 and 24 October 1859, 3 and 9 April 1860, 16 and 22 October 1860, 8 and 15 July 1861, VA 152, 153, 156, 157, 160, 161, 170, and 172; Kolbe to Liebig, 16 April 1860, Liebigiana IIB. See chap. 5 for details on these events.

9. Kolbe to Vieweg, 29 May 1858, VA 138.

10. Wurtz, "Mémoire sur la constitution et sur la vrai formule de l'acide oxalique," *C.r.*, **44** (1857), 1306–1310. In this paper, Wurtz' formulas were written with conventional equivalents and brackets separating "typical" atoms. His groupings showed slightly more structural detail than is indicated here.

11. Wurtz, "Sur la propylglycol," *C.r.*, **45** (1857), 306–309. Nicholas Fisher, "Wislicenus and Lactic Acid," in O. B. Ramsay, ed., *van't Hoff-Le Bel Centennial* (Washington, D.C.: American Chemical Society, 1975), pp. 33–54, provides details on the development of our understanding of this curious and chemically problematical substance.

12. Wurtz, "Recherches sur l'acide lactique" and "Sur un nouvel acide lactique," *C.r.*, **45** (1857), pp. 1228–1232 and 1232–1234.

13. H. Debus, "On the Oxidation of Glycol, and on Some Salts of Glyoxylic Acid," *Proceedings of the Royal Society*, **9** (1859), 711–716; R. Hoffmann, "Über Monochloressigsäure," *Annalen*, **102** (1857), 1–20; A. Kekulé, "Bildung von Glycolsäure aus Essigsäure," *Annalen*, **105** (1858), 286–292; A. Strecker, "Über die künstliche Bildung der Milchsäure und einen neuen, dem Glycocoll homologen Körper," *Annalen*, **75** (1850), 27–45.

14. Kolbe, *Lehrbuch*, **1**, 672–688 and 733–740.

15. Kolbe, "Über die chemische Constitution der Milchsäure," *Annalen*, **109** (1859), 257–268 (on p. 259).

16. A. Crum Brown, "On the Theory of Isomeric Compounds," *Transactions of the Royal Society of Edinburgh*, **23** (1864), 707–719 (on p. 711). This was probably the first positive assertion of the existence of carbon–carbon double bonds. Erlenmeyer, Kekulé, and Loschmidt had all made hints in that direction, but none had clearly asserted it.

17. C. Ulrich, "Umwandlung der Milchsäure in Propionsäure," *Annalen*, **109** (1859), 268–272.

18. Kolbe, "Constitution der Milchsäure."

19. Wurtz, "Mémoire sur les glycols ou alcools diatomiques," *Ann. chim.*, [3] **55** (1859), 400–478.

20. Ibid., pp. 401, 438, and 474; quote from p. 463; use of term "structure

moléculaire" on p. 478.

21. Wurtz, "Sur l'oxyde d'éthylène," *C.r.*, **48** (1859), 101–105 (on p. 104).

22. Wurtz, "Glycols," p. 478.

23. From Wurtz' review of Couper's paper on structure theory, in *Répertoire de chimie pure*, **1** (1858), 49–52.

24. Wurtz, "Glycols," pp. 474–475.

25. Wurtz, "Recherches sur la constitution de l'acide lactique," *BSC*, **1** (1859), 36–46; "Sur la basicité des acides," *Ann. chim.*, [3] **56** (1859), 342–349; "Nouvelles recherches sur l'acide lactique," *C.r.*, **48** (1859), 1092–1094.

26. Kekulé, "Bildung von Glycolsäure."

27. Wurtz, "Recherches sur la constitution," pp. 45–46.

28. Kekulé to L. Meyer, 23 October 1860, printed in R. Anschütz, *August Kekulé*, 2 vols. (Berlin: Verlag Chemie, 1929), **1**, 205.

29. Kekulé, *Lehrbuch der organischen Chemie*, 2 vols. (Erlangen: Enke, 1859–1867), **1**, 129–131, 174–175 (1859), and 524 (1861).

30. Ibid., pp. 524 and 729–741 (1861); Kekulé, "Note sur les acides itaconique et pyrotartrique," *Bulletin de l'Académie Royale de Belge*, [2] **11** (1861), 662–677; W. H. Perkin, "On the Molecular Constitution of Glycollic and Lactic Acids," *Chemical News*, **3** (1861), 81–83; A. Crum Brown, "On the Theory of Chemical Combination," M.D. dissertation (Univ. of Edinburgh, 1861; printed 1879), p. 23.

31. E. Lautemann, "Über direkte Umwandlung der Milchsäure in Propionsäure," *Annalen*, **113** (1860), 217–220; Kolbe, "Über die Rückbildung des Alanins aus Milchsäure," *Annalen*, **113** (1860), 220–223; Kolbe, "Über die Constitution und Basicität der Milchsäure," *Annalen*, **113** (1860), pp. 223–238.

32. Kolbe, "Über die chemische Constitution der Isäthionsäure und des Taurins," *Annalen*, **112** (1859), 241–243 (on p. 241).

33. Kolbe, "Constitution und Basicität der Milchsäure," pp. 227–236 (on pp. 225–226).

34. Wurtz, "Recherches sur l'acide lactique," *Ann. chim.*, [3] **59** (1860), 161–191 (on pp. 182–187).

35. Wurtz and C. Friedel, "Mémoire sur l'acide lactique," *Ann. chim.*, [3] **63** (1861), 101–124.

36. Kolbe and Lautemann, "Über die Constitution und Basicität der Salicylsäure," *Annalen*, **115** (1860), 157–206 (on p. 161).

37. Five years later he repudiated the *general* implication of the derogatory statement, while reaffirming the specifics: Kolbe, *Das chemische Laboratorium der Universität Marburg* (Braunschweig: Vieweg, 1865), p. 152n.

38. H. Debus, "Bemerkungen zu Kolbe's und Lautemann's Ansichten über die Natur des Glycols und Glyoxals," *Annalen*, **118** (1861), 253–256.

39. E. Drechsel, "Beobachtungen über die Glycolsäure," *Annalen*, **127** (1863), 150–158; Kolbe, "Bemerkungen zu vorstehenden Abhandlung," *Annalen*, **127** (1863), pp. 159–161; Kolbe to Frankland, [early] October 1863, Frankland Archive 01.04.69; Kolbe to Erlenmeyer, 30 October, 23 November, and 6 and 17 December 1863, Dingler Nachlass.

40. Kolbe, "Ist die Oxalsäure eine zweibasische Säure," paper draft dated 23 December 1866, SSDM 3810.

41. F. Guthrie and Kolbe, "Über die Verbindungen des Valerals mit Säuren," *Annalen*, **109** (1859), 296–300.

42. Kolbe, *Das Laboratorium*, p. 101n.

43. Kolbe, *JpC*, **131** (1881), 322–323; idem, "Ist die Oxalsäure eine zweibasische Säure."

44. Kolbe, *JpC*, **130**, (1880), 156 ("Ich galt damals. . . als Sonderling"); ibid., **131** (1881), 310; ibid., **136** (1883), 371–372.

45. Kolbe, "Über den natürlichen Zusammenhang der organischen mit den unorganischen Verbindungen," *Annalen*, **113** (1860), 293–332 (on pp. 313–316).

46. R. Schmitt, "Über die Umwandlung der Weinsäure und Apfelsäure in Bernsteinsäure," *Annalen*, **114** (1860), 106–111.

47. Liebig, "Über die Bildung von Weinsäure aus Milchzucker in Gummi," *Annalen*, **113** (1860), 1–19.

48. Liebig to Kolbe, 3 April 1860, SSDM 3603.

49. Kolbe to Vieweg, 9 March 1856, 29 December 1858, and 5 January 1859, VA 115, 145, and 147; Liebig to Vieweg, 28 March 1855, in M. and W. Schneider, eds., *Justus von Liebig: Briefe an Vieweg* (Braunschweig: Vieweg, 1986), p. 288.

50. Kolbe to Liebig, 16 April 1860, Liebigiana IIB; Kolbe to Vieweg, 9 April 1860, VA 157.

51. Kolbe to Vieweg, 9 April (quoted) and 15 April 1860, VA 157–158.

52. Liebig to Kolbe, 2 December 1860 and 15 July 1861, SSDM 3604 and 3606.

53. Liebig to Vieweg, 17 May 1860, in M. and W. Schneider, *Briefe an Vieweg*, p. 335; Liebig to H. von Fehling, 19 May 1860, Liebigiana IIB (I thank Dr. Elisabeth Vaupel for drawing my attention to this letter); Liebig to Volhard, 2 July 1862, cited in D. Vorländer, "Jakob Volhard," *Berichte* **45** (1912), 1855–1902 (on p. 1866).

54. Liebig, *Annalen*, **121** (1862), 163–164n.; Liebig to Kolbe, 30 October 1861, SSDM 3607; Kopp to Kolbe, 4 December 1861, SSDM 3608; Kolbe to Liebig, 30 December 1861 and 10 December 1862, Liebigiana IIB; Kolbe to Vieweg, 11 July 1860 and 30 December 1861, VA 171 and 177.

55. Liebig to Fehling, 19 May 1860, Liebigiana IIB.

56. Ibid.; Liebig, *Annalen*, **121** (1862), 163–164n.; Liebig to Erlenmeyer, 27 March 1861, in Emil Heuser, ed., *Justus von Liebig und Emil Erlenmeyer in ihren Briefen von 1861–1872* (Mannheim: Bionomica, 1988), p. 11; Liebig to Kolbe, 16 March 1870, SSDM 3612.

57. Kolbe, "Natürlicher Zusammenhang" (1860), p. 307; also in his *Lehrbuch*, **1**, 569, written in 1857 and in print by the beginning of 1858.

58. Friedel, "Transformation des aldéhydes et des acétones en alcools," *C.r.*, **55** (1862), 53–58.

59. Kolbe to Erlenmeyer, 13 November 1862, Dingler Nachlass.

60. Kolbe, "Über die chemische Constitution des aus dem Aceton durch nascirenden Wasserstoff erzeugten Alkohols," *ZfC*, **5** (1862), 687–690; Kolbe to Erlenmeyer, 16 November 1862, Dingler Nachlass.

61. Friedel, "Sur l'alcool dérivé de l'acétone par l'action de l'hydrogène

naissant," *BSC*, **2** (1863), 247–248; Kolbe, "Nachtrag," *ZfC*, **7** (1864), 38–40. Kolbe's rebuttal was six months late, since he continued to ignore the French literature. Erlenmeyer told him of Friedel's second paper on a visit to Marburg in October 1863: Kolbe to Erlenmeyer, 25 and 30 October 1863, Dingler Nachlass.

62. Butlerov, "Studien über die einfachsten Verbindungen der organischen Chemie," *ZfC* **6** (1863), 484–497 (on pp. 490–495); idem, "Über den tertiären Pseudobutylalkohol (den trimethylirten Methylalkohol)," *ZfC*, **7** (1864), 385–402 (on pp. 394–395); idem, "Berichtigung zur Abhandlung, 'Über den tertiären Pseudobutylalkohol,'" *ZfC*, **7** (1864), p. 702; idem, "Über die tertiären Alkohole," *ZfC*, **8** (1865), 614–618. Butlerov identified his new tertiary alcohol in a letter to Wurtz, ca. March 1864, in G. V. Bykov and J. Jacques, "Deux pionniers de la chimie moderne, Adolphe Wurtz et Alexandre M. Boutlerov, d'après une correspondance inédite," *Revue d'histoire des sciences*, **13** (1960), 115–134 (on p. 125).

63. V. H. de Luynes, "Recherches sur l'érythrite et ses dérivés," *Ann. chim.*, [4] **2** (1864), 385–429 (on p. 422); Lieben and Rossi, "Sur l'alcool butylique primaire et normal," *C.r.*, **68** (1869), 1561–1564.

64. Wurtz, "Sur l'alcool butylique," *C.r.*, **35** (1852), 310–312.

65. Kolbe, "Prognose neuer Isomerien," *ZfC*, **7** (1864), 30–40 (on pp. 36–37).

66. Kolbe to Erlenmeyer, 6 December 1863 and 20 October 1864, Dingler Nachlass; Erlenmeyer, "Studien über das Glycerin in seiner Eigenschaft als mehratomiger Alkohol," *ZfC*, **7** (1864), 642–653 (on pp. 651–653); idem, "Über die relative Constitution des Gährungs-, Butyl- und Amylalkohols und des Amylenhydrats," *ZfC*, **10** (1867), 117–118 (also published in slightly revised form in *Annalen*, Supplementband **5** [1867], 337–339); V. V. Markovnikov, "Über die Isobuttersäure," *ZfC*, **8** (1865), 107–110; idem, "Über die Isobuttersäure und den Pseudopropyl-Aethyl-Aether," *Annalen*, **138** (1866), 361–375.

67. Frankland and B. F. Duppa, "Synthetical Researches on Ethers. No. 2. Action of Sodium and Isopropylic Iodide upon Ethylic Acetate," *JCS*, **20**, (1867), 102–116.

68. Wurtz, "Sur un isomère de l'alcool amylique," *C.r.*, **55** (1862), 370–375; idem, "Sur les hydrates des hydrogènes carbonées," *C.r.*, **56** (1863), 715–718; idem, "Note sur l'hydrate d'amylène," *C.r.*, **56** (1863), pp. 793–796; idem, "Sur quelques dérivés de l'hydrate d'amylène," *C.r.*, **57** (1863), 479–482; idem, "Sur les produits d'oxydation de l'hydrate d'amylène et sur l'isomérie dans les alcools," *C.r.*, **58** (1864), 971–974; idem, "Mémoire sur l'isomérie dans les alcools et dans les glycols," *Ann. chim.*, [4] **3** (1864), 129–186.

69. Erlenmeyer and Wanklyn, "Über Hexylverbindungen," *ZfC*, **6** (1863), 564–575; Erlenmeyer, "Relative Constitution," pp. 117–118.

70. Erlenmeyer and Wanklyn, "Über Hexylverbindungen," *ZfC*, **6** (1863), p. 574; Kolbe, "Über die sekundären Alkohole," *Annalen*, **132** (1864), 102–117.

71. Kolbe to Erlenmeyer, 6 and 11 December 1863, 12 February and 31

July 1864, Dingler Nachlass; Kolbe to Liebig, 7 February 1864, Liebigiana IIB; Kolbe to Frankland, 18 January and 24 February 1864, Frankland Archive 01.02.1493 and 01.02.1511; Kolbe, "Sekundäre Alkohole."

72. Wurtz, "Sur les produits d'oxydation," p. 972; "Mémoire sur l'isomèrie," p. 144. The chemistry of these reactions is complex.

73. Kolbe to Frankland, [early] October 1863 and 27 May 1866, Frankland Archive 01.04.69 and 01.02.1558; Frankland, "On the Synthesis of Leucic Acid," *Proceedings of the Royal Society*, **12** (1863), 396–398; Frankland and Duppa, "Notes of Researches on the Acids of the Lactic Series—No. 1.," *Proceedings of the Royal Society*, **13** (1864), 140–142; idem, "Researches on Acids of the Lactic Series—No. 1. Synthesis of Acids of the Lactic Series," *PTRS*, **156** (1866), 309–359 (on p. 312).

74. Hans Hübner, "Über Cyanacetyl," *Annalen*, **124** (1862), 315–324; M. Simpson, "On Cyanide of Ethylene and Succinic Acid," *Proceedings of the Royal Society*, **10** (1860), 574–576; idem, "Synthesis of Tribasic Acids," *Proceedings of the Royal Society*, **12** (1863), 236–239; idem, "On the Synthesis of Succinic and Pyrotartaric Acids," *PTRS*, **151** (1861), 61–67.

75. Kolbe to Frankland, 9 July and [early] October 1863, 18 January and 5 and 12 February 1864, Frankland Archive 01.03.431, 01.04.69, 01.02.1493, 01.02.1497, and 01.02.1501; Kolbe to Erlenmeyer, 12 February 1864, Dingler Nachlass; Kolbe to Liebig, 7 February 1864, Liebigiana IIB; and Kolbe to Volhard, 7 February 1864, SSDM 3653.

76. Kolbe, "Umwandlung der Monocarbonsäuren in die zugehörenden kohlenstoffreicheren Dicarbonsäuren," *Annalen*, **131** (1864), 348–349; "Conversion of Monocarbon-Acids into the Corresponding More Highly Carbonated Dicarbon-Acids," *JCS*, **17** (1864), 109.

77. Beilstein to Kekulé, 21 February 1864, August-Kekulé-Sammlung; a similar but complementary account is related in Beilstein to Baeyer, 4 June 1864, Baeyer Collection.

78. Müller to Kekulé, 28 February 1864, Kekulé-Sammlung.

79. Kolbe to Frankland, 24 February 1864, Frankland Archive 01.02.1511.

80. Frankland to Kolbe, 22 February 1864, SSDM 3563; Kolbe to Erlenmeyer, 6 March 1864, Dingler Nachlass (reporting on a later letter from Frankland to Kolbe received that day); Müller to Kekulé, 22 April and 9 July 1864, Kekulé-Sammlung; Müller, "On a New Formation of Malonic and Succinic Acids," *JCS*, **17** (1864), 109–111; idem, "Über eine neue Bildungsweise der Malonsäure und der Bernsteinsäure," *ZfC*, **7** (1864), 146–148.

81. Beilstein to Baeyer, 4 June 1864, Baeyer Collection.

82. Müller, "Vorläufige Notiz über einen rothen Körper, der sich bei der Einwirkung von Cyankalium auf Chloressigäther als Nebenproduct bildet," *ZfC*, **7** (1864), 382–383; Kolbe to Erlenmeyer, 2 July 1864, Dingler Nachlass.

83. Müller to Kekulé, 28 February 1864, Kekulé-Sammlung.

84. Kolbe to Erlenmeyer, 2 July 1864, Dingler Nachlass.

85. Beilstein to Butlerov, 30 August 1865, in Bykov and Bekassova, "Beiträge zur Geschichte der Chemie der 60-er Jahre des XIX. Jahrhunderts: II. F. Beilsteins Briefe an A. M. Butlerov," *Physis*, **8** (1966), 267–285 (on p. 275).

86. Cited by Krätz, in *Beilstein-Erlenmeyer Briefe* (Munich: Fritsch, 1972), p. 79.

87. Liebig to Hofmann, 24 January 1868, in Emil Heuser and Regine Zott, eds., *Justus von Liebig und August Wilhelm Hofmann in ihren Briefen* (Mannheim: Bionomica, 1988), p. 45.

88. These numbers are calculated from data given in Meinel, *Chemie*, p. 472.

89. Kolbe to Vieweg, 31 December 1860, VA 164. He reported the presence of only one foreigner that semester, a Dane. This was Emil Ruge, who (as Kolbe reported in *Das Laboratorium*, p. 363n.) died in October 1864 in Copenhagen. The previous six semesters he had not had a single non-German foreigner.

90. Meinel, *Chemie*, p. 472; Elisabeth Vaupel, "Carl Graebe (1841–1927)—Leben, Werk und Wirken," Doctoral dissertation, University of Munich, 1987.

91. Vorländer, "Volhard" (see n. 53); Volhard, "Über Sarkosin," *Annalen*, **123** (1862), 261–265.

92. J.W., "Alexander Crum Brown," *JCS*, **123** (1923), 3422–3431.

93. E. Fischer, "Edmund Drechsel," *Berichte*, **30** (1897), 2168–2173.

94. S. N. Vinogradov, "Chemistry at Kazan University in the Nineteenth Century: A Case History of Intellectual Lineage," *Isis*, **56** (1965), 168–173; B. Menshutkin, "N. A. Menshutkin," *Berichte*, **40** (1907), 5087–5098. I gathered the Russian names from the university's Matrikel (HSA, *305a*. II, Nr. 11), and then transformed them from German into standard Anglo-American transliterations.

95. Henry James, *Charles W. Eliot: President of Harvard University*, 2 vols. (Boston: Houghton Mifflin, 1930), **1**, 135–137 and 145–147. I thank Dr. Jun Fudano for this reference. As far as I could determine, Eliot's name appears nowhere in the university's records.

96. Kolbe to Vieweg, 9 November 1863, 22 June 1864 and 6 November 1864, VA 197, 206, 211; Kolbe to Frankland, no date (but ca. 27 October 1864) and 12 November 1864, Frankland Archive 01.04.84 and 01.04.89; HSA, *305a*. II, Nr. 11.

97. Untitled anonymous review of Kolbe's *Das Laboratorium*, in *Westminster Review*, **29** (1866), 548–549.

98. Kolbe to Vieweg, 30 June and 9 November 1863, VA 192 and 197; Kolbe, *Das Laboratorium*, pp. 8–9.

99. Information in this and the next three paragraphs is derived from Meinel, *Chemie*, pp. 16–18, 30–31, 51–63, 98–112, and 435–444, and from Kolbe, *Das Laboratorium*, pp. 1–17.

100. Kolbe lived, as he later commented (*Das neue chemische Laboratorium der Universität Leipzig* [Leipzig: Brockhaus, 1868], p. 27n.) "across the street" from the institute. This may have been in Bunsen's old lodgings, on the first floor of the corner house on Elisabethstrasse, but I have not been able to document Kolbe's residence in Marburg.

101. Kolbe to Vieweg, 19 October and 9 November 1863 and 21 February 1864, VA 196, 197, and 199; Meinel, *Chemie*, p. 111.

Notes to Chapter 10

1. A useful review of the literature on this subject by John Brooke can be found in C. A. Russell, ed., *Recent Developments in the History of Chemistry* (London: Royal Society of Chemistry, 1985), pp. 107–109; the classic analysis is Brooke, "Wöhler's Urea and its Vital Force? A Verdict from the Chemists," *Ambix*, **15** (1968), 84–114. See also C. A. Russell, "The Changing Role of Synthesis in Organic Chemistry," *Ambix*, **34** (1987), 169–180.

2. Wöhler to Berzelius, 22 February 1828, in Wallach, *BWB*, **1**, 205–208; Berzelius to Wöhler, 7 March 1828, ibid., pp. 208–209; Wöhler to Liebig, 13 October 1863, in Hofmann, *LWB*, **2**, 145–146. Isomerism was indeed a hot topic in the years after 1828, but the phenomenon was by no means completely novel; at least seven instances of organic isomerism (and several more in the inorganic realm) had been studied during the years 1811–1826 (A. J. Rocke, *Chemical Atomism in the Nineteenth Century* [Columbus: Ohio State Univ. Press, 1984], pp. 167–174). Berzelius coined the word *isomer* in 1830 in response to Wöhler's and several other recent discoveries.

3. Hermann Kopp, *Geschichte der Chemie*, **1** (Braunschweig: Vieweg, 1843), 442; the assertion is repeated in the same work, **4** (1847), 244.

4. Brooke, "Verdict," pp. 109–113.

5. For example, in Kolbe, *Lehrbuch* (1854), pp. 3–4. Kolbe, who joined Wöhler just ten years after the synthesis, believed in a strong version of the claim (Kolbe to Vieweg, 26 January 1864, VA 198).

6. For the former attitude, see Thomas Thomson, *The History of Chemistry*, 2 vols. (London: Colburn and Bentley, 1830–1831), **2**, 317; Wöhler and Liebig, "Untersuchungen über die Natur der Harnsäure," *Annalen*, **26** (1838), 241–340 (on p. 340) (in a letter of 24 July 1837, Liebig urged that Wöhler should add to one of their joint papers "some clever comments" asserting that it will prove possible to create sugar from "charcoal and rainwater" (Hofmann, *LWB*, **1**, 112); and J. E. Schlossberger, *Lehrbuch der organischen Chemie*, 3d ed. (Stuttgart: Müller, 1854), pp. 27–28. For examples of doubts, see ibid.; and V. Regnault, ed. A. Strecker, *Kurzes Lehrbuch der Chemie* (Braunschweig: Vieweg, 1851), p. 576.

7. A. Strecker, "Über die künstliche Bildung der Milchsäure und einen neuen, dem Glycocoll homologen Körper," *Annalen*, **75** (1850), 27–45 (on p. 28); Kolbe, *Über die chemische Constitution organischer Verbindungen* (Marburg: Elwert, 1858), p. 6.

8. Kolbe, "Über die chemische Constitution der Isäthionsäure und des Taurins," *Annalen*, **112** (1859), 241–243; idem, "Über die chemische Constitution und künstliche Bildung des Taurins," *Annalen*, **122** (1862), 33–47.

9. Kolbe, "Über die Rückbildung des Alanins aus Milchsäure," *Annalen*, **113** (1860), 220–223.

10. J. Volhard, "Über Sarkosin," *Annalen*, **123** (1862), 261–265.

11. Kolbe and Schmitt, "Directe Umwandlung der Kohlensäure in Ameisensäure," *Annalen*, **119** (1861), 251–253.

12. Kolbe and Schmitt, "Rother Farbstoff aus dem Kreosot," *Annalen*, **119** (1861), 169–172. They had first encountered this reaction in February 1859

(Kolbe's laboratory notebook, unpaginated, SSDM 3812).

13. The anecdote is told by Wilhelm Ostwald, regarding his only conversation with Kolbe, in January 1883: *Lebenslinien: Eine Selbstbiographie*, 3 vols. (Berlin, 1926–1927), **1**, 190–191. Kolbe "was at that time in his own estimation the decisive personality in chemical affairs not only for Germany, but for the entire world. Accordingly he acted dignified and reserved."

14. Kolbe indicated to Varrentrapp (28 February 1870, VA 266) that a synthesis of indigo had long been his great desire.

15. N. W. Fisher, "Organic Classification Before Kekulé," *Ambix*, **20** (1973), 106–131, 209–233; M. J. Nye, "Berthelot's Anti-Atomism: A 'Matter of Taste'?" *Annals of Science*, **31** (1981), 585–590; idem, "Explanation and Convention in Nineteenth-Century Chemistry," in R. Visser, et al., eds., *New Trends in the History of Science* (Amsterdam: Rodopi, 1989), pp. 171–186. Nye's book, *From Chemical Philosophy to Theoretical Chemistry: Dynamics of Matter and Dynamics of Discipline, 1800–1950* (Berkeley: Univ. of California Press, 1993), treats many of the same issues as this section.

16. Larry Laudan has written a history of the "method of hypothesis" during the eighteenth and nineteenth centuries: *Science and Hypothesis: Historical Essays on Scientific Methodology* (Boston: Reidel, 1981); a discussion of this subject specific to chemistry is A. J. Rocke, "Methodology and Its Rhetoric in Nineteenth-Century Chemistry: Induction versus Hypothesis," in Elizabeth Garber, ed., *Beyond History of Science: Essays in Honor of Robert E. Schofield* (Bethlehem, P.A.: Lehigh Univ. Press, 1990), pp. 137–155.

17. J. B. Dumas, *Leçons sur la philosophie chimique*, 2d ed. (Paris, 1878), pp. 66–67 (lecture delivered on 23 April 1836). An explicit reference to the predictive function of theories is also contained in his *Traité de chimie appliquée aux arts*, 8 vols. (Paris, 1828–1846), **5** (1835), 72.

18. L. Pearce Williams, "André-Marie Ampère," *Scientific American*, January 1989, pp. 90–97.

19. John F. W. Herschel, *A Preliminary Discourse on the Study of Natural Philosophy* (London: Longman, Green, 1830), pp. 29–34, 150, 186–188, and 197–212.

20. Rocke, "Methodology," pp. 149–151.

21. Kenneth Caneva, "From Galvanism to Electrodynamics: The Transformation of German Physics and Its Social Context," *Historical Studies in the Physical Sciences*, **9** (1978), 63–159, esp. 95–122.

22. Kolbe, "Beiträge zur Kentniss der gepaarten Verbindungen," *Annalen*, **54** (1845), 145–188 (on pp. 160, 183, and 188).

23. This is a principal thesis of my *Chemical Atomism in the Nineteenth Century: From Dalton to Cannizzaro* (Columbus: Ohio State Univ. Press, 1984).

24. For a detailed study exemplifying these trends, see Kathryn M. Olesko, *Physics as a Calling: Discipline and Practice in the Königsberg Seminar for Physics* (Ithaca, N.Y.: Cornell Univ. Press, 1991).

25. Today elemental analyses are done by specialized commercial laboratories, not by individual chemists. I also ignore here instrumental methods of analysis developed during the twentieth century, which are indeed based

on high technology, mathematical theory, physical properties and precision measurements. That is a story that does not enter into our narrative.

26. These examples and quotes are taken from John Servos, *Physical Chemistry from Ostwald to Pauling: The Making of a Science in America* (Princeton, N.J.: Princeton Univ. Press, 1990), pp. 48–51 and 321–324; quotes are from the years 1926–1927.

27. Details of this story can be found in Rocke, *Chemical Atomism*.

28. Wurtz, in *Répertoire de chimie pure*, **3** (1861), 419.

29. Hofmann, *Introduction to Modern Chemistry* (London: Walton and Maberley, 1865), p. v.

30. On Erlenmeyer, see especially Otto Krätz, "Das Portrait: Emil Erlenmeyer, 1825–1909," *Chemie in unserer Zeit*, **6** (1972), 52–58; idem, *Beilstein-Erlenmeyer: Briefe zur Geschichte der chemischen Dokumentation und des chemischen Zeitschriftenwesens* (Munich: Fritsch, 1972); and Rita Meyer, "Emil Erlenmeyer (1825–1909) als Chemietheoretiker und sein Beitrag zur Entwicklung der Strukturchemie," Ph.D. dissertation, University of Munich, 1984.

31. Kekulé to Erlenmeyer, 29 January 1859, in Richard Anschütz, *August Kekulé*, 2 vols. (Berlin: Verlag Chemie, 1929), **1**, 150–151; Kekulé to Erlenmeyer, 6 August 1859, August-Kekulé-Sammlung.

32. Kekulé to Liebig, undated, in Anschütz, *Kekulé*, **1**, 130.

33. Erlenmeyer, *ZfC*, **3** (1860), 1–3. The next year Erlenmeyer noted that the study of chemistry had become "positively fashionable" (ibid., **4** [1861], 217).

34. The originals of these letters do not seem to have survived, but the content is clear from the replies: Kolbe to Erlenmeyer, 19 February 1860, and Wurtz to Erlenmeyer, 22 February 1860, Dingler Nachlass.

35. Kolbe to Erlenmeyer, 19 November and 6 and 11 December 1863, Dingler Nachlass.

36. Kolbe to Erlenmeyer, 17 December 1863, Dingler Nachlass.

37. Erlenmeyer, *Die Aufgabe des chemischen Unterrichts* (Munich: Akademie-Verlag, 1871), p. 27.

38. Erlenmeyer, "Vorläufige Notiz über eine etwas abgeänderte Betrachtungsweise der Alkohole und ihrer Derivate," *ZfC*, **4** (1861), 202–204; ibid., p. 197. In another passage from about this time, Erlenmeyer adds another warning against the effort to judge "the true constitution of a compound" (ibid., pp. 167–168).

39. Erlenmeyer, "Die Theorie," *ZfC*, **5** (1862), 18–32 (on p. 27–31).

40. A. J. Rocke, "Subatomic Speculations and the Origin of Structure Theory," *Ambix*, **30** (1983), 1–18.

41. Erlenmeyer, "Bemerkungen zu der vorstehenden Abhandlung [by Heintz]," *ZfC*, **5** (1862), 218–223.

42. Erlenmeyer, "Vorläufige Notiz über das Verhältniss der Kolbe'schen Betrachtungsweise zu der sog. Typentheorie," *ZfC*, **6** (1863), 728–735.

43. Kolbe to Erlenmeyer, 12 February and 22 November 1864, Dingler Nachlass.

44. Krätz, "Erlenmeyer," p. 55; idem, *Beilstein-Erlenmeyer*, pp. 16–18.

45. Erlenmeyer to Kolbe, 28 June 1871, SSDM 3551.

46. Erlenmeyer, *Aufgabe*, pp. 26–33.

47. Ibid., p. 32n.

48. Ibid., p. 31.

49. Ibid.

50. Kolbe to Varrentrapp, 1 October 1871, VA 268; the article was "Moden der modernen Chemie," *JpC*, **112** (1871), 241–271.

51. Kolbe to Vieweg, 4 October 1871, VA 269.

52. Kolbe to Volhard, 22 June 1873 and 20 November 1874, SSDM 3662 and 3673.

53. Alexander Vucinich, *Science in Russian Culture, 1861–1917* (Stanford, Calif.: Stanford Univ. Press, 1970); Nathan M. Brooks, "The Formation of a Community of Chemists in Russia, 1700–1870," Ph.D. dissertation, Columbia Univ., 1990.

54. G. V. Bykov, "A. M. Butlerov," *DSB*; A. J. Rocke, "Kekulé, Butlerov, and the Historiography of the Theory of Chemical Structure," *British Journal for the History of Science*, **14** (1981), 27–57; Brooks, "Formation," pp. 257–281.

55. Jean Jacques, "Boutlerov, Couper et la Société Chimique de Paris (Notes pour servir à l'histoire des théories de la structure chimique)," *BSC*, **1953**, 528–530.

56. Butlerov, "Einiges über die chemische Structur der Körper," *ZfC*, **4** (1861), 549–560.

57. Erlenmeyer to Butlerov, 4 May 1862, in G. V. Bykov and L. M. Bekassova, "Beiträge zur Geschichte der Chemie der 60-er Jahre des XIX. Jahrhunderts: I: Briefwechsel zwischen E. Erlenmeyer und A. M. Butlerov," *Physis*, **8** (1966), 185–198 (on pp. 187–188). Bykov and Bekassova suggest (p. 186) that the two probably first met at Speyer, but this is not consistent with the passage just cited. It is more probable that they first became acquainted when Butlerov visited Heidelberg (twice) in 1857–1858.

58. G. V. Bykov and Z. I. Sheptunova, "Nemetskii 'Zhurnal khimii' (1858–1871) i russkie khimiki (k istorii khimicheskoi periodiki)," *Trudy instituta istorii estestvoznaniia i tekhniki*, **30** (1960), 97–110; O. Krätz, "Iwan Turgenjew und die russischen Chemiker in Heidelberg," *Chemie in unserer Zeit*, **21** (1987), 89–99; idem, "Erlenmeyer," pp. 54–55.

59. For example, Butlerov, "Ueber die Verwandtschaft der mehraffinen Atome," *ZfC*, **5** (1862), 297–304; idem, "Ueber die verschiedenen Erklärungsweisen einiger Fälle von Isomerie," *ZfC*, **6** (1863), 500–534; idem, "Ueber die systematische Anwendung des Princips der Atomigkeit zur Prognose von Isomerie und Metameriefällen," *ZfC*, **7** (1864), 513–532.

60. Butlerov, "Erklärungsweisen," pp. 501–506 and 509–514; "Systematische Anwendung," p. 513.

61. Ibid., p. 504–505 and 520.

62. Kolbe to Frankland, 9 July 1867, Frankland Archive 01.04.1374.

63. Kekulé, *Annalen*, **130** (1864), 12; idem, *C.r.*, **58** (1864), 510.

64. Rocke, "Historiography of Chemical Structure;" C. A. Russell, *History of Valency* (Leicester: Leicester Univ. Press, 1971).

65. Kekulé to Erlenmeyer, 6 August 1859, and Kekulé to Meyer, 23 Octo-

ber 1860, both in the August-Kekulé-Sammlung, the latter also printed in Anschütz, *Kekulé*, **1**, 204; ibid., p. 290 (citing Kekulé's 1866 benzene theory paper).

66. Kolbe to Vieweg, 1 March and 16 October 1860, 9 November 1863, and 15 May 1866, VA 155, 160, 197, and 244.

67. Kekulé, "Beiträge zur Kenntniss der Salicylsäure und der Benzoësäure," *Annalen*, **117** (1861), 145–164 (on p. 164).

68. Kekulé, "Zwei Berichtigungen zu Kolbe's Abhandlung: 'Ueber die chemische Constitution der Mellithsäure, des Paramids u.s.w.,'" *Annalen*, **125** (1863), 375–376; Kolbe, "Constatirung eines Irrthums," *Annalen*, **126**, 125–126, with editorial note by Kopp.

69. Kolbe, "Vermischte Notizen," *Annalen*, **113** (1860), 238–244 (on p. 244). The details in this paragraph are derived from Kolbe's explanatory note in his *Das chemische Laboratorium der Universität Marburg* (Braunschweig: Vieweg, 1865), pp. 109–111.

70. Kekulé, "Elektrolyse zweibasischer Säuren," *ZfC*, **7** (1864), 293.

71. Kolbe to Erlenmeyer, 26 May and 2 July 1864, Dingler Nachlass.

72. Kolbe, *Laboratorium*, pp. 110–111.

73. Albert Ladenburg, *Lebenserinnerungen* (Breslau: Trewendt & Granier, 1912), pp. 38–39.

74. This thesis is defended in Rocke, "Historiography of Chemical Structure."

75. Kekulé, *Lehrbuch der organischen Chemie*, 2 vols. (Erlangen: Enke, 1859–1866), **2**, 244–249 (fascicle written and published in 1864).

76. Wurtz to Butlerov, 19 February 1864, in G. V. Bykov and J. Jacques, "Deux pionniers de la chimie moderne, Adolphe Wurtz et Alexandre M. Boutlerov, d'après une correspondance inédite," *Revue d'histoire des sciences*, **13** (1960), 115–134 (on pp. 121–122).

77. Butlerov to Wurtz, no date, but ca. March 1864, in ibid., pp. 123–124.

78. Butlerov, *Vvedeniie k polnomu izucheniiu organicheskoi khimii* (Kazan, 1864–1866); *Lehrbuch der organischen Chemie zur Einführung in das specielle Studium derselben* (Leipzig, 1867–1868); Beilstein to Baeyer, 31 January 1864, Baeyer Collection. Butlerov had completed about 100 pages of manuscript by this date and had enlisted Beilstein's help (who was then in Göttingen) to find a German publisher.

79. Butlerov, *Lehrbuch*, pp. 75–78.

80. Beilstein to Butlerov, 14 December 1862, in Bykov and Bekassova, "Beiträge zur Geschichte der Chemie der 60-er Jahre des XIX. Jahrhunderts: II. F. Beilsteins Briefe an A. M. Butlerov," *Physis*, **8** (1966), 267–285 (on p. 268).

81. Erlenmeyer to Butlerov, 9 July 1864, in Bykov and Bekassova, "Briefwechsel," p. 192.

Notes to Chapter 11

1. Hubert Kiesewetter, *Industrialisierung und Landwirtschaft: Sachsens Stellung im regionalen Industrialisierungsprozess Deutschlands im 19. Jahrhun-*

dert (Vienna: Böhlau, 1988), pp. 745–748; idem, *Industrielle Revolution in Deutschland 1815–1914* (Frankfurt: Suhrkamp, 1989), pp. 16–19 and 305–314. Kiesewetter rightly stresses that a proper understanding of industrialization in Germany can only be achieved by examining regional differences and especially interregional competition—a point that is paralleled by many aspects of the present work as well.

2. The first, that is, of any economic significance. The Nürnberg–Fürth line in Bavaria was essentially a demonstration project.

3. Helmut Kretzschmar, "Johann Paul von Falkenstein," *Neue deutsche Biographie*, **5** (Berlin: Duncker and Humblot, 1961), 15–16.

4. The discussion and numerical data in this and the next two paragraphs derive principally from Franz Eulenburg, *Die Entwicklung der Universität Leipzig in den letzten hundert Jahren* (Leipzig: Hirzel, 1909), pp. 11–19, 35–39, 46–48, 190, and 197.

5. Franz Eulenburg, *Die Frequenz der deutschen Universitäten von ihrer Gründung bis zur Gegenwart* (Leipzig: Hirzel, 1904), pp. 250ff.

6. Rektor und Senat der Universität Leipzig, ed., *Festschrift zur Feier des 500 jährigen Bestehens der Universität Leipzig* (Leipzig: Hirzel, 1909), **4**:2, pp. 11–14, 19, and 72; Eulenburg, *Leipzig*, pp. 146–147.

7. Universität Leipzig, *Festschrift*, pp. 70–72; Erdmann to Falkenstein, 7 January 1865, in UAL, Med. Fac., BIII, Nr. 2b, Bd. 2, ff. 33–44, on ff. 35r–35v. See also C. Jungnickel, "Teaching and Research in the Physical Sciences and Mathematics in Saxony, 1820–1850," *Historical Studies in the Physical Sciences*, **10** (1979), 3–47.

8. All these details derive from Kolbe's long letter to Varrentrapp of 26 February 1871, VA 267.

9. Universität Leipzig, *Festschrift*, **4**:2, 85–89, and various other volumes, passim.

10. Joseph Ben-David, *The Scientist's Role in Society*, 2d ed. (Chicago: Univ. of Chicago Press, 1984), chap. 7; Avraham Zloczower, *Career Opportunities and the Growth of Scientific Discovery in Nineteenth-Century Germany* (New York: Arno, 1981).

11. Eulenburg, *Universität Leipzig*, pp. 141–150; Kretzschmar, "Falkenstein."

12. Peter Borscheid, *Naturwissenschaft, Staat und Industrie in Baden (1848–1914)* (Stuttgart: Ernst Klett, 1976); Arleen M. Tuchman, "From the Lecture to the Laboratory: The Institutionalization of Scientific Medicine at the University of Heidelberg," in William Coleman and Frederic L. Holmes, eds., *The Investigative Enterprise: Experimental Physiology in Nineteenth-Century Medicine* (Berkeley: Univ. of California Press, 1988), pp. 65–99, esp. p. 97, n. 58; Tuchman, "Science, Medicine, and the State: The Institutionalization of Scientific Medicine at the University of Heidelberg," Ph.D. dissertation, Univ. of Wisconsin-Madison, 1985.

13. Theodor Curtius, *Geschichte des chemischen Universitäts-Laboratoriums Heidelberg* (Heidelberg, 1908), pp. 4–5.

14. In a suggestive and ambitious article, Steven Turner has made an admirable start on the complex goal of analyzing the Prussian case: "Justus

Liebig versus Prussian Chemistry: Reflections on Early Institute-Building in Germany," *Historical Studies in the Physical Sciences*, **13** (1982), 129–162. On the cost of the Heidelberg lab, see Curtius, p. 14; the cost has been converted from guldens to thalers at the rate of 1.75 guldens per thaler.

15. Erdmann to Falkenstein, 7 January 1865, UAL, Med. Fac., BIII, Nr. 2b, Bd. 2, ff. 37r–38r; Jeffrey Johnson, "Academic Chemistry in Imperial Germany," *Isis*, **76** (1985), 500–524.

16. For which see Turner, "Institute-Building," pp. 146–147 (above, n. 14).

17. The following discussion of the call of Hofmann to Bonn and Berlin is based on Jonathan Bentley, "Hofmann's Return to Germany from the Royal College of Chemistry," *Ambix*, **19** (1972), 197–203; Gert Schubring, "The Rise and Decline of the Bonn Natural Sciences Seminar," *Osiris*, [2] **5** (1989), 57–93; W. H. Brock, ed., *Justus von Liebig und August Wilhelm Hofmann in ihren Briefen (1841–1873)* (Weinheim: Verlag Chemie, 1984); A. W. Hofmann, *The Chemical Laboratories in Course of Erection in the Universities of Bonn and Berlin* (London: Clowes, 1866); and Jacob Volhard, *August Wilhelm von Hofmann: Ein Lebensbild* (Berlin: Friedländer, 1902). However, there are errors and omissions in several of these discussions, and I have tried to determine the most accurate course of events by critical collation and by comparison with a variety of unpublished letters that are not cited by these authors.

18. Bentley, "Return to Germany," pp. 198 and 201–202; Volhard, *Hofmann*, p. 79; Wöhler to Liebig, 27 March and 25 July 1863, in Hofmann, *LWB*, **2**, 132 and 141.

19. Hofmann to Buff, 4 December 1863, in Volhard, *Hofmann*, p. 81; Kekulé to Beilstein, 2 January 1864, August-Kekulé-Sammlung (reporting on a letter dated 21 December 1863 from Hugo Müller, who had spoken with Hofmann two days earlier).

20. Frankland wrote Kolbe on 22 February 1864, "Hofmann has not yet quite made up his mind, but I think there is little doubt that he will go to Berlin, in which case you must of course go to Bonn" (SSDM 3563); Kolbe reported to Vieweg on 22 June that Hofmann had officially accepted the Berlin offer (VA 206).

21. This reported by Kolbe in letters to Vieweg of 22 June 1864 and undated (but datable to 16 July 1865), VA 206 and 232. The latter letter cites a conversation Kolbe had with Kultusminister Olshausen as the source, so it must be regarded as more than mere rumor. Nonetheless, it appears Hofmann kept this detail confidential, for the Bonn succession was the subject of much impatient gossip in chemists' correspondence during 1864, 1865, and 1866.

22. Hofmann, *Laboratories*, p. 52; Volhard, *Hofmann*, p. 195; Kolbe, *Das chemische Laboratorium der Universität Leipzig* (Braunschweig: Vieweg, 1872), p. xix n.

23. Kolbe to Vieweg, 19 April 1865, VA 223.

24. Kolbe to Vieweg, 31 August and 1 October 1865, and 9 July 1866, VA 234, 237, and 246; Kolbe to Erlenmeyer, 26 March 1868, Dingler Nachlass. Regarding the Bonn lab, "Es wird ein chemischer Palast, und hat alle meine Erwartungen übertroffen. Er [Hofmann] weiss nicht, ob er Berlin oder Bonn wählen soll." (VA 234) The lab is "etwas zu luxuriös und weitläufig eingerich-

tet." (Kolbe to Vieweg, 15 September 1865, VA 235.) To Hofmann, he declared that he would have nothing to change were he to come to Bonn: letter of 14 January 1867, Chemiker-Briefe.

25. Johnson, "Academic Chemistry in Imperial Germany."

26. Kolbe to Vieweg, 25 April 1863, 21 and 26 February, 2 March, and 22 June 1864, VA 191, 199, 200, 201, and 206; Kolbe to Frankland, 24 February and no date (but datable to ca. 27 October 1864), Frankland Archive 01.02.1511 and 01.04.84. The translated phrases in quotation marks are from VA 201 and 206.

27. Kolbe to Vieweg, 8 July 1864, VA 207; Kekulé to Baeyer, 16 March 1865, August-Kekulé-Sammlung (latter also printed in Richard Anschütz, *August Kekulé*, 2 vols. [Berlin: Verlag Chemie, 1929], **1**, 366–367); Kolbe to Hofmann, 14 January 1867, Chemiker-Briefe.

28. Kolbe to Vieweg, 9 November 1863, VA 197; Kolbe to Frankland, 5 February 1864, Frankland Archive 01.02.1497.

29. This story was related to Baeyer by Kekulé, who had it from Müller, who had it from Hofmann, who described the conversation (in Berlin in early 1865) but refused to identify his interlocutor: Kekulé to Baeyer, 16 March 1865, August-Kekulé-Sammlung; also in Anschütz, *Kekulé*, **1**, 366–367.

30. Kolbe to Frankland, 9 July 1867, Frankland Archive 01.04.1374; Kolbe to Hofmann, 27 October 1866, Chemiker-Briefe. In the latter letter, Kolbe indicated to Hofmann, for whatever use he wanted to make of the information, that he could be lured from Leipzig by an offer of a new large laboratory.

31. Kolbe to Hofmann, 30 June 1867, Chemiker-Briefe.

32. Anschütz, *Kekulé*, **1**, 369–373.

33. Medical Faculty to Falkenstein, 11 June 1864, UAL, Med. Fac., BIII, Nr. 2b, Bd. 2, ff. 22–24.

34. Wagner to Falkenstein, 10 June 1864, ibid., f. 21. This argument was only implied, not developed explicitly.

35. On Hirzel and Knop, see especially Poggendorff and the *Neue deutsche Biographie*. On the Möckern experiment station, see Mark Finlay, "The German Agricultural Experiment Stations," *Agricultural History*, **62** (1988), 41–50; idem, "Wissenschaft und Praxis in der deutschen Landwirtschaft: Justus von Liebig, Hermann von Liebig, und die landwirtschaftlichen Versuchsstationen," *Zeitschrift für Agrargeschichte und Agrarsoziologie*, in press; and Ursula Schling-Brodersen, "Entwicklung und Institutionalisierung der Agrikulturchemie im 19. Jahrhundert: Liebig und die landwirtschaftlichen Versuchsstationen," Ph.D. dissertation, Technische Universität Braunschweig, 1988.

36. Falkenstein to Medical Faculty, 9 December 1864, UAL, Med. Fac., BIII, Nr. 2b, Bd. 2, ff. 26–28.

37. Medical Faculty to Falkenstein, 24 January 1865, ibid., ff. 45–61; Wagner to Falkenstein, no date, ibid., f. 31; Erdmann to Falkenstein, 7 January 1865, ibid., ff. 33–44; Medical Faculty to Falkenstein, no date, ibid., ff. 29–30.

38. Erdmann to Falkenstein, ibid., f. 37r–37v: "'Bekommen Sie denn nicht Alles was Sie beantragen?'"

39. Ibid., ff. 41v–42r.

40. Falkenstein to Medical Faculty, 28 April 1865, ibid., ff. 62–68.

41. ". . . und mag nicht unterlassen, hierbei unter Anderen auf den Professor Kolbe in Marburg, der von mehreren sachkundigen Männern Ihm [dem Ministerium] empfohlen worden ist, aufmerksam zu machen." Ibid., ff. 67v–68r; these are the last words of Falkenstein's directive. Timothy Lenoir has given another reading to Falkenstein's memo in his "Science for the Clinic: Science Policy and the Formation of Carl Ludwig's Institute in Leipzig," in Coleman and Holmes, eds., *Investigative Enterprise*, pp. 139–178 (on p. 166–169). Contrary to Lenoir, however, Falkenstein barely mentions agricultural chemistry, makes no references to limitations of resources (quite the contrary), and only once in the scores of pages of this series of memos refers vaguely to "dieses für das gesammte wissenschaftliche und praktische Leben, zumal in einem Staate wie Sachsen hochwichtigen Unterrichtszweiges" (f. 65r). Nor could Falkenstein have intended to have suggested Kolbe's name as one who would directly contribute to the Saxon chemical industry, for (among other reasons) Kolbe was known at this time as an intensely *theoretically* inclined chemist, and there was no indication that his orientation might change in the future. In short, the model of state socioeconomic interest as guiding the Kolbe call, which was Lenoir's theme, does not survive close inspection. I doubt if any of Falkenstein's calls will be found to fit this model, except in the very indirect and long-term sense which I argue in the first section of this chapter. State interest was operating here and strongly so, but it was state interest in achieving academic excellence, not immediate economic payoff.

42. Medical Faculty to Falkenstein, 31 May 1865, UAL, Med. Fac., BIII, Nr. 2b, Bd. 2, f. 69; Philosophical Faculty to Falkenstein, 13 June 1865, UAL, PA 645, and UAL, Phil. Fac., 82, pp. 330–332; Kolbe to Vieweg, 1 June 1865, VA 226, regarding Erdmann's letter, just received.

43. Ludwig also told Kolbe that confidential inquiries regarding Kolbe had already been made in Marburg, and favorably answered, and that Ludwig was doing his best to advance Kolbe's candidacy. Since neither Kolbe's nor Kekulé's names ever came up in the internal deliberations of the Medical Faculty, Falkenstein must have been making independent inquiries. The Dekan of the Philosophical Faculty must also have been involved since Ludwig added that his favorite candidate was Kekulé. (I have found no documentation in the Leipzig Universitätsarchiv on these questions.) All this was related in Kolbe's letter to Vieweg of 8 February 1865, VA 217, regarding Ludwig's letter received the day before.

44. Kolbe to Vieweg, 8 February 1865, VA 217; Kolbe to Vieweg, 16 February and 1 June 1865, VA 219 and 226.

45. Kolbe to Vieweg, 5 June 1865 and no date (but datable to 16 July 1865), VA 227 and 232.

46. HSA, *305a*. AIV 4b., Nr. 94.

47. Kolbe to Vieweg, 25 June, 8 and 16 July, [1] and 31 August, 15 September, 1 October, and 7 November 1865, VA 229, 231, 232, 233, 234, 235, 237, and 236; Kolbe to Liebig, 26 July 1865, Liebigiana IIB; Liebig to Kolbe, 31 July 1865, SSDM 3609; Kolbe to Volhard, 11 August 1865, SSDM 3654; Kolbe to Frankland, 21 September 1865, Frankland Archive 01.02.1480; Kolbe to Erlenmeyer, 11 August 1865, Dingler Nachlass. The sticking point in

the negotiations was the new laboratory. Kolbe averred to Vieweg (VA 229) that even if he failed to get the new lab, he would accept the call as a means of escaping Marburg, but would then feel free to accept the subsequent expected call to Bonn. On the other hand, if a new lab came with the Leipzig call, he would feel obliged to refuse the Bonn offer. As it turned out, he got the lab in Leipzig and yet very nearly also accepted the call to Bonn, only refusing because the Bonn lab was to be shared with Landolt.

48. Kolbe to Vieweg, 1 October, 7 November and 23 December 1865, VA 237, 236, and 238; Kolbe to Erlenmeyer, 11 August 1865 and 25 February 1866, Dingler Nachlass; Kolbe to Hofmann, 3 June 1866, Chemiker-Briefe.

49. Kolbe to Vieweg, no date (1 August 1865), 23 and 30 December 1865, 5 May, 9 July, and 28 October 1866, VA 233, 238, 239, 243, 246, and 248.

50. This is equal to 70 by 72 meters.

51. Kolbe to Frankland, 9 July 1867, Frankland Archive 01.04.1374; Kolbe to Liebig, 3 March 1867, Liebigiana IIB; Wilhelm Strube, "Hermann Kolbe (1818–1884)," in Gerhard Harig, ed., *Bedeutende Gelehrte in Leipzig*, 2 vols. (Leipzig: Karl-Marx-Universität, 1965), **2**, 25–35; Kolbe, *Das chemische Laboratorium der Universität Leipzig* (Braunschweig: Vieweg, 1872), pp. xv–xx. Other sources for the history of chemistry at the University of Leipzig are Wilhelm Treibs, "Zur Geschichte der Entwicklung der Chemie an der Universität Leipzig," in Ernst Engelberg et al., eds., *Karl-Marx-Universität Leipzig 1409–1959: Beiträge zur Universitätsgeschichte*, 2 vols. (Leipzig: Verlag Enzyklopädie, 1959), **1**, 464–480; and Burckhardt Helferich, *Das Studium der Chemie an der Universität Leipzig* (Leipzig: Lorentz, 1932).

52. The following description of Kolbe's institute building is based on his lengthy depictions in *Das neue chemische Laboratorium der Universität Leipzig* (Leipzig: Brockhaus, 1868); on Kolbe, *Das chemische Laboratorium der Universität Leipzig* (Braunschweig: Vieweg, 1872), pp. xv–xxxiii and plates; on Kolbe to Volhard, 16 May 1873, SSDM 3663; and on *Festschrift der Universität Leipzig*, **4**:2, 70–84.

53. Kolbe, *Das chemische Laboratorium*, p. xviii. There are no surviving statistics or class lists in the archives of the University of Leipzig, so the sort of analysis carried out in chap. 9 for the case of Marburg is not possible here. On Kolbe's enrollments in summer semester 1866 and summer semester 1867, see Kolbe to Vieweg, 5 May 1866, VA 243, and Kolbe to Frankland, 9 July 1867, Frankland Archive 01.04.1374.

54. Kolbe to Vieweg, 19 July and 8 November 1869, 1 May and 10 and 23 October 1872, VA 258, 263, 280, 292, and 296; Kolbe to Frankland, 1 May and 8 November 1872, Frankland Archive 01.03.593 and 01.03.704; Kolbe to Liebig, 15 December 1869, 1 January and 29 November 1872, Liebigiana IIB. The story sometimes appears in the Kolbe literature that Liebig had warned Kolbe that he was planning far too large an institute. Liebig simply asserted in this letter (25 December 1872), which I have not found in original but is excerpted in Ost, HK, p. 128, that one cannot effectively teach so many Praktikanten at once.

55. Kolbe to Liebig, 29 November 1872, Liebigiana IIB.

56. Kolbe to Liebig, 23 February 1873, ibid; Kolbe to Frankland, 1 January 1873, 20 March and 19 December 1877, Frankland Archive 01.03.604, 01.02.1411, and 01.04.1506; Kolbe to Volhard, 28 September 1873, SSDM 3664.

57. Kolbe to Wöhler, 4 November 1881, Wöhler Nachlass; Ost, HK, p. 133.

58. Eulenburg, *Universität Leipzig*, pp. 112–113.

59. On the lab budget, see *Festschrift*, **4**:2, 84, and *Das chemische Laboratorium Leipzig*, p. xlii; on Kolbe's income, see his letter to Volhard of 16 May 1873, SSDM 3663. In addition to his salary of 2000 thalers (which did in effect represent a raise since after 1868 he no longer paid rent for his residence), he was making "somewhat more than 5000 thalers per year" in honoraria and an additional 500 or 600 from examination and promotion fees. Kolbe's frankness to Volhard was because Volhard was gathering information regarding a possible call for Kolbe to Munich as Liebig's successor. Kolbe's honorarium total is consistent with the supposition that he kept his charges the same as at Marburg, namely, 11 and 22 thalers for the part- and full-day Praktika, respectively, and 6 and 8 thalers for four- and six-hour lecture courses, respectively.

60. Wislicenus to Kolbe, 7 June 1874, SSDM 3549; Volhard to Kolbe, 30 December 1877, SSDM 3515.

61. Kolbe, *Das chemische Laboratorium Leipzig*, pp. xxxviii–xxxix.

62. Ibid., pp. xxxix–xl.

63. Ibid., pp. xlii–xlvi; *Das neue chemische Laboratorium*, pp. 1 and 21.

64. Ibid., *Das chemische Laboratorium Leipzig*, p. xliii.

65. Frankland to Kolbe, 1 January 1873, SSDM 3567.

66. Classic and recent writings on these issues include L. F. Haber, *The Chemical Industry During the Nineteenth Century* (Oxford: Clarendon Press, 1958); John Beer, *The Emergence of the German Dye Industry* (Urbana: Univ. of Illinois Press, 1959); D. H. Wilcox, "Kekulé and the Dye Industry," in O. T. Benfey, ed., *Kekulé Centennial* (Washington, D.C.: American Chemical Society, 1966), pp. 24–71; Borscheid, *Naturwissenschaft, Staat und Industrie in Baden*; G. Meyer-Thurow, "The Industrialization of Invention: A Case Study from the German Chemical Industry," *Isis*, **73** (1982), 363–381; and A. S. Travis, *The Rainbow Makers: The Origins of the Synthetic Dyestuffs Industry in Western Europe* (Bethlehem, P.A.: Lehigh Univ. Press, 1993). I thank Dr. Travis for sharing with me the manuscript of chap. 3 of this work.

67. UAL, Phil. Fac., B 128.

68. *Verzeichniss der . . . auf der Universität Leipzig zu haltenden Vorlesungen* (Leipzig: Edelmann, various years).

69. Carl Graebe, then Privatdozent, wrote Carl Liebermann that 36 of Kolbe's 130 Praktikanten were full-day workers. He added that his own lecture course on organic chemistry was drawing 40 students: Graebe to Liebermann, 7 November 1869, SSDM 1933/1, cited by Elisabeth Vaupel, "Carl Graebe (1841–1927)—Leben, Werk und Wirken," Ph.D. dissertation, Univ. of Munich, 1987, p. 189. Graebe had worked in Kolbe's Marburg lab in 1862. In another letter, he remarked on his friendly relations with Kolbe, but "I will not

come with him into a closer relationship, as this is too little in his nature" (Graebe to his parents, 24 November 1869, SSDM 1933-78/14, cited in Vaupel, "Carl Graebe," p. 189).

70. Kolbe, *Das chemische Laboratorium Leipzig*, pp. xl–xli. Exceptions to this generalization were during the three peak semesters of 1872–1874 when 40 students were entrusted completely to assistants, and during some isolated semesters in the late 1870s and early 1880s when Kolbe was too ill to work in the lab regularly.

71. Ibid., pp. vii and xl.

72. Ernst von Meyer, *Lebenserinnerungen* (n.p., n.d., privately printed ca. 1918), passim. As assistant, Meyer lived in the same building as the Kolbe family; he and Johanna had been playing chamber music together since shortly after their first meeting in 1866. Meyer's relationship with "Vater Kolbe" was extraordinarily cordial and remained undisturbed by Kolbe's increasingly intemperate conduct toward his professional peers. Meyer was inducted into modern structural chemistry by Carl Graebe, who served for one semester (winter 1869/70) as Privatdozent in Leipzig.

73. See discussion of these events, with citations to the literature, in chap. 10.

74. Beilstein to Butlerov, 24 November 1866, in G. V. Bykov and L. M. Bekassova, "Beiträge zur Geschichte der Chemie der 60er Jahre des XIX. Jahrhunderts. II. F. Beilsteins Briefe an A. M. Butlerov," *Physis*, **8** (1966), 267–285 (on p. 280).

75. Anschütz, *Kekulé*, **1**, 256–261.

76. Beilstein to Butlerov, 24 November 1866, in Bykov and Bekassova, "Beiträge," p. 281 (see n. 74).

77. On this tradition, see Otto Krätz, *Beilstein–Erlenmeyer: Briefe zur Geschichte der chemischen Dokumentation und des chemischen Zeitschriftenwesens* (Munich: Fritsch, 1972), p. 18n. and passim.

78. Kolbe to Erlenmeyer, 15 March 1864, Dingler Nachlass. The content of Erlenmeyer's letter to Kolbe, which does not appear to have survived, may be inferred from Kolbe's words.

79. Krätz, *Beilstein–Erlenmeyer*," p. 17n. (above, n. 77).

80. Kolbe to Erlenmeyer, 20 October and 22 November 1864, Dingler Nachlass; Kolbe to Vieweg, 14 November 1864 and 8 February 1865, VA 212 and 217. Erlenmeyer wanted an annual guaranteed honorarium of 1000 florins, equal to about 570 thalers (letter of 20 October).

81. Kolbe to Erlenmeyer, 15 March and 11 December 1864, Dingler Nachlass.

82. For these events, see Anschütz, *Kekulé*, **1**, 404–409, and Krätz, *Beilstein-Erlenmeyer*, passim.

83. Kolbe to Varrentrapp, 10, 21, and 27 October 1879, VA 260, 262, and 261; quote is from the last of these. Considering that in 1864 Eduard Vieweg had turned Erlenmeyer down chiefly because of his request for an honorarium of 570 thalers, it is not surprising that five years later Heinrich Vieweg balked at the prospect of paying 600 thalers per year.

84. Kolbe to Varrentrapp, 21 and 27 October 1869, VA 262 and 261.

85. Kolbe to Varrentrapp, 8 November 1869 and 17 January 1870, VA 263 and 265.

Notes to Chapter 12

1. A. W. Hofmann, "Chemische Untersuchung der organischen Basen im Steinkohlen-Theeröl," *Annalen*, **47** (1843), 37–87.

2. Hofmann, "On Insolinic Acid," *Proceedings of the Royal Society*, **8** (1855), 1–3.

3. See Richard Anschütz' discussion, in his *August Kekulé*, 2 vols. (Berlin: Verlag Chemie, 1929), **1**, 296–305. It is known that Kekulé had read, and disapproved of, the work of both men.

4. Kolbe to Frankland, 3 and 17 March 1852, Frankland Archive 01.08.604 and 01.08.610; Kolbe, "Über Synthese der Salicylsäure," *Annalen*, **113** (1859), 125–127.

5. B. W. Gerland, "Über Anthranilsäure, Benzaminsäure, und Carbanilidsäure," *Annalen*, **86** (1853), 143–156, the work having been done "auf den Rath und unter der Leitung von" Kolbe; note by Kolbe, ibid., pp. 148–149.

6. Gerland, "Über Benzaminsäure," *Annalen*, **91** (1854), 185–198 (on p. 194). We may presume this was Gerland's idea because Schmitt, no doubt reflecting his teacher's view, still expressed faith in the earlier hypothesis in his paper "Vorläufige Notiz über die Einwirkung der salpetrigen Säure auf Sulfanilidsäure," *Annalen*, **112** (1859), 118–121.

7. Schmitt, "Vorläufige Notiz"; Kolbe and Lautemann, "Über die Constitution und Basicität der Salicylsäure," *Annalen*, **115** (1860), 157–206. Lautemann was at work on the project at least as early as 1859, for Kolbe mentions this in his "Synthese der Salicylsäure," p. 127.

8. Kolbe, "Synthese der Salicylsäure."

9. Ibid.

10. This is clear from the language in ibid., and also from a statement in Schmitt's paper "Über die Umwandlung der Weinsäure und Apfelsäure in Bernsteinsäure," *Annalen*, **114** (1860), 106–111 (on p. 111), paper dated 10 March 1860.

11. Kolbe and Lautemann, "Über die Constitution and Basicität der Salicylsäure," *Annalen* **115** (1860), 157–206 (on pp. 158 and 161–163). The matter in parentheses is discussed in chap. 9.

12. Ibid., pp. 168–171; Kolbe to Liebig, 16 April 1860, Liebigiana IIB. Modern chemists consider benzyl alcohol to be $C_6H_5CH_2OH$ and cresol to be $HOC_6H_4CH_3$; "benzyl" in modern vocabulary is $C_6H_5CH_2$–, not C_6H_5 as for Kolbe.

13. Warren De La Rue and Hugo Müller, "On Some Products of the Action of Dilute Nitric Acid on Some Hydrocarbons of the Benzol Series," *JCS*, **14** (1861), 54–57; Müller to Kekulé, 15 February 1861, August-Kekulé-Sammlung.

14. August Kekulé, "Beiträge zur Kenntniss der Salicylsäure und der Benzoësäure," *Annalen*, **117** (1861), 145–164 (on pp. 161–164).

15. Anschütz, *Kekulé*, **1**, 77, 82, 124, and 177–182; **2**, 141, 145–146, and 162–186. For a historical discussion and analysis of these events, see A. J. Rocke, "Hypothesis and Experiment in the Early Development of Kekulé's Benzene Theory," *Annals of Science*, **42** (1985), 355–381, esp. pp. 364–368.

16. Kolbe notebook, December 1860, unpaginated, SSDM 3812; Kolbe to Liebig, 21 February and 7 May 1861, Liebigiana IIB; Kolbe to Vieweg, 30 March 1861, VA 167; Kolbe, "Über die Einführung von Wasserstoff in organische Verbindungen und die Umwandlung der Salicylsäure in Galussäure," *Annalen*, **118** (1861), 122–124.

17. Liebig to Kolbe, 1 March 1861, SSDM 3605.

18. Kolbe and Lautemann, "Salicylsäure," pp. 187–190.

19. Kolbe, *Das chemische Laboratorium der Universität Marburg* (Braunschweig: Vieweg, 1865), pp. 174–175n.

20. S. Cannizzaro, "Über die Zersetzung der Salylsäure durch Aetzbaryt," *Annalen*, Suppl.-Bd. **1** (1861), 274; R. Fittig, "Über die Oxydationsproducte des Toluols durch verdünnte Salpetersäure," *Annalen*, **120** (1861), 214–226 (on pp. 222–223); J. Wilbrand and F. Beilstein, "Über eine neue Reihe isomerer Verbindungen der Benzoëgruppe—Nitrodracylsäure und deren Derivate," *Annalen*, **128** (1863), 257–273 (on p. 273); E. Reichenbach and Beilstein, "Untersuchungen über isomerie in der Benzoëreihe," *Annalen*, **132** (1864), 137–155 (on p. 152); idem, "Dritte Abhandlung. Über die Natur der sogenannten Salylsäure," *Annalen*, **132** (1864), 309–321 (on p. 311).

21. Kolbe, "Bemerkungen zu vorstehenden Abhandlungen," *Annalen*, **127** (1863), 159–161; Max Herrmann, "Über die Einwirkung des nascirenden Wasserstoffs auf Benzoësäure," *Annalen*, **132** (1864), 75–82; idem, "Über die Veränderungen, welche die Hippursäure in saurer Lösung durch nascirenden Wasserstoff erleidet," *Annalen*, **133** (1865), 335–338.

22. Kolbe, *Laboratorium*, pp. 169n. and 174–175n.

23. C. Saytseff, "Über Paraoxybenzoësäure" *Annalen*, **127** (1863), 129–137. Zaitsev had been a student of A. M. Butlerov, who had urged him to study with Kolbe. Although I know of no direct personal relationship between the two men, Butlerov clearly respected Kolbe's work and regarded it as structural in orientation. He may have preferred to send students to Kolbe because he was constantly feuding with Kekulé in these years.

24. G. Fischer, "Über Paranitrobenzoësäure" *Annalen*, **127** (1863), 137–149. Fischer, who was from Frankfurt/Main, was in Kolbe's lab in winter semester 1862/63, when he was already Ph.D., and later worked with Rudolf Schmitt in Dresden: *JpC*, **127** (1879), 318n. I know nothing else about him.

25. Wilbrand and Beilstein, "Nitrodracylsäure," p. 269; H. Hlasiwetz and L. Barth, "Über einen neuen, dem Orcin homologen Körper," *Sitzungsberichte der kaiserlichen Akademie der Wissenschaften*, Wien, **49**:2 (1864), 203–207; N. Sokolov, "Über die Salze der β-Nitrobenzoësäure," *JpC*, **93** (1864), 425–492; Beilstein and F. Schlun, "Vierte Abhandlung: Über die isomeren Chlorbenzoësäuren," *Annalen*, **133** (1865), 239–252.

26. Fischer, "Paranitrobenzoësäure," p. 149; Kolbe, *Laboratorium*, p. 175n.

27. For citations and a detailed discussion, see Rocke, "Kekulé's Benzene Theory" (see n. 15).

28. Kolbe to Erlenmeyer, 25 February 1865, Dingler Nachlass. That Kolbe did not mean this phrase to be complimentary is indicated by his remark to Frankland the following year, playing again on an equestrian figure: Kekulé's "imagination bolted with his understanding long ago" (Kolbe to Frankland, 23 July 1866, Frankland Archive 01.02.1505).

29. Rocke, "Kekulé's Benzene Theory," pp. 373–377.

30. Ladenburg, "Über das Mesitylen," *Berichte*, **7** (1874), 1133–1137; idem, *Theorie der aromatischen Verbindungen* (Braunschweig: Vieweg, 1876), pp. 1 and 28; Kolbe, "Über eine neue Darstellungsmethode und einige bemerkenswerthe Eigenschaften der Salicylsäure," *JpC*, **118** (1874), 89–112 (on p. 90). See Rocke, "Kekulé's Benzene Theory," and Rocke, "Kekulé's Benzene Theory and the Appraisal of Scientific Theories," in A. Donovan, L. Laudan, and R. Laudan, eds., *Scrutinizing Science: Empirical Studies of Scientific Change* (Boston: Kluwer, 1988), pp. 145–161.

31. Kolbe, "Über Ätherschwefelsäuren und ätherschweflige Säuren," *Annalen*, **143** (1867), 64–72. In this article, Kolbe coined and introduced the word "Alkyl," a contraction of "Alkoholradikal."

32. Kolbe, *Über die chemische Constitution der organischen Kohlenwasserstoffe* (Braunschweig: Vieweg, 1869), p. 5.

33. Ibid., pp. 9–12; Kolbe, *Kurzes Lehrbuch der organischen Chemie* (Braunschweig: Vieweg, 1883), pp. 365–368. Kolbe had coined the word "methine" to apply to (CH)‴ earlier that year; it is still used today by organic chemists with that denotation.

34. Rocke, "Appraisal of Scientific Theories" (see n. 30). In this article I characterized Kolbe's theory, I now believe erroneously, as a non-cyclohexamethine theory.

35. Adolf Claus, *Theoretische Betrachtungen und deren Anwendung zur Systematik der organischen Chemie* (Freiburg: Pappen, 1866), p. 208.

36. Claus, "Über die chemische Constitution der Diglycolsäure und der Glycolamidsäuren," *JpC*, **111** (1871), 123–127.

37. It could be that Claus recognized the identity of his and Kolbe's benzene formula, but regarded a priority claim as unfruitful for a question so murky. Considering that Claus later argued that his conception of benzene was not in fact a "diagonal" interpretation but rather the same as the much more empirically justified "centric" formula, it may also be that he did not view his and Kolbe's formulas as identical.

38. Kolbe, *JpC*, **111** (1871), pp. 124n. and 134.

39. Kolbe, *Kurzes Lehrbuch*, pp. 373–376.

40. Kolbe, *Constitution*, pp. 11–22; *Kurzes Lehrbuch*, pp. 367–376.

41. Kolbe, *Constitution*, p. 15.

42. Kolbe to Hofmann, 14 December 1868, Chemiker-Briefe; Kolbe to Liebig, 2 February 1869, Liebigiana IIB.

43. C. Glaser, "Über die Constitution einiger Zimmtsäurederivate," *ZfC*, **12** (1869), 111–113; Kolbe, "Bemerkungen zu Glaser's Abhandlung, *ZfC*, **12** (1869), 160.

44. H. E. Armstrong, "The Riddle of Benzene: August Kekulé," *Journal of the Society of Chemical Industry*, **48** (1929), 914–918 (on p. 914); H. E. Armstrong to Richard Armstrong, 6 February 1870, quoted in J. Vargas Eyre, *Henry Edward Armstrong* (London: Butterworths, 1958), pp. 51–52 ("my views are diametrically opposite to Kolbe's").

45. Lothar Meyer, *Die modernen Theorien der Chemie*, 4th ed. (Breslau: Maruschke & Berendt, 1880), p. 263; R. Meyer, *Erlenmeyers Lehrbuch der organischen Chemie*, Abt. II: *Die Aromatische Verbindungen* (Leipzig: Winter, 1883–1894), pp. 87ff.; Kekulé, "Zur Geschichte der Benzoltheorie," unpubl. MS (1883), first printed in Richard Anschütz, *August Kekulé*, 2 vols. (Berlin: Verlag Chemie, 1929), **1**, 548. Kekulé also pointed out, as I have above, that Kolbe's benzene formula is structurally equivalent to Claus' diagonal formula, "which is now recognized to be untenable" (ibid., p. 544).

46. Kolbe, "Vorläufige Mittheilung," *JpC*, **116** (1873), 41–42; idem, "Synthese der Paraoxybenzoësäure," *JpC*, **116** (1873), p. 336; idem, "Über eine neue Darstellungsmethode und einige bemerkenswerthe Eigenschaften der Salicylsäure," *JpC*, **118** (1874), 89–112.

47. Kolbe, "Über die chemische Natur der Salylsäure," *JpC*, **120** (1875), 151–157. That the crucial test was devised and performed in June 1873 is indicated in Kolbe's letter to Volhard, 22 June 1873, SSDM 3662 (where the quoted phrase is found).

48. Kolbe, "Chemischer Rückblick auf das Jahr 1874," *JpC*, **118** (1874), 449–456 (on p. 451–453); Ost, "Über das Verhalten der Chlorsalylsäure, Salicylsäure und Paraoxybenzoësäure gegen schmelzende Alkalien," *JpC*, **119** (1875), 385–401.

49. Kolbe to Bertha Ost, no date, DM 6803; Kolbe to B. Ost, 4 July [1880], DM 6804.

50. Kolbe to Volhard, 28 September 1873, SSDM 3664. How this view of salylic acid can be reconciled with his crucial test of Beilstein's hypothesis performed three months earlier, and which he regarded then as "unambiguous proof" of the compound's nonexistence, is unclear to me.

51. Kolbe to Volhard, 5 October 1873, SSDM 3665; Kolbe to Varrentrapp, 8 October 1873, VA 315.

52. Kolbe, "Vorläufige Mittheilung" (30 November 1873).

53. In a letter to Volhard of 25 December 1881 (SSDM 3685), Kolbe was still very optimistic that he could identify the pure compound. A letter from Kolbe's assistant G. Schmidt dated 13 March 1884 and sent to his vacation resort on the Italian Riviera (SSDM 3807) reports a number of unsuccessful experiments on this subject. Schmidt concluded that either salylic converts readily to benzoic acid *at room temperature* or Beilstein was right all along and salylic acid is simply impure benzoic acid.

54. Kolbe to Varrentrapp, 1 January 1874, 11 March and 17 July 1875, VA 316, 326, and 332; Kolbe to Frankland, 1 January 1874, Frankland Archive 01.03.604; Ernst von Meyer, *Lebenserinnerungen* (n.p., n.d., ca. 1918), pp.

102–103, 118, and 133–134; and O. Schlenk, *Chemische Fabrik von Heyden, 1874–1934, Erinnerungsblätter aus 6 Jahrzehnten* (Radebeul, 1934), pp. 12–25. In his letter of 11 March, Kolbe asked Varrentrapp if he wanted to be a licensee, commenting that the industrial process was straightforward and very profitable.

55. Wilhelm Vershofen, *Wirtschaftsgeschichte der chemisch-pharmazeutischen Industrie: Eine wirtschaftshistorische Studie*, 3 vols. (Aulendorf: Cantor, 1949–1958), **3**, 52.

56. Kolbe's expected production cost is mentioned in Kolbe to Frankland, 1 January 1874 (Frankland Archive 01.03.604); retail price is mentioned in *JpC*, **119** (1875), 219. By 1878 the price had dropped to around 2½ thalers per pound (Schlenk, *Fabrik*, p. 25). For the sake of comparison, commercial grade salicylic acid today (1990) sells at retail for about $7 per pound.

57. A London entrepreneur attempted to evade fees by patenting the novel process in England and licensing production to Merck in Darmstadt, the product then being imported and sold in England. Kolbe and Heyden sought an injunction, which was granted; Armstrong's testimony for the plaintiffs then won the appeal. Merck had been selling salicylic acid in England at 7s. 6d. (equivalent to around two thalers) per pound. See *Law Times Reports*, **42** (8 May 1880), 300–303.

58. All of this is related in Kolbe's letter to Frankland of 15 November 1878 and 3 May 1879 (Frankland Archive 01.04.381 and 01.07.357); see also Heyden's appreciative letter to Frankland, 26 November 1878 (Frankland Archive 01.04.0386). The Prussian litigation was with E. Schering's firm in Berlin; both sides publicized their dispute in pamphlets published in 1876 and 1877. See Vershofen, *Wirtschaftsgeschichte*, **3**, 25–26; Schlenk, *Fabrik*, p. 25; and Kolbe to Varrentrapp, no date (but ca. 11 April 1875), VA 327.

59. R. Schmitt, "Beitrag zur Kenntniss der Kolbe'schen Salicylsäure-Synthese," *JpC*, **139** (1885), 397–411.

60. Kolbe, "Eigenschaften der Salicylsäure"; "Weitere Mittheilungen über Wirkungen der Salicylsäure," *JpC*, **119** (1875), 9–23; idem, "Ist anhaltender Genuss kleiner Mengen Salicylsäure der Gesundheit nachtheilig?" *JpC*, **125** (1878), 347–350; Kolbe to Vieweg, 12 and 15 October 1878, VA 421 and 422 (quoted passage from VA 421). A gram of salicylic acid is equivalent to about three modern aspirin tablets. Kolbe's habit may well have lengthened his life, if one makes the likely assumption that salicylic acid shares aspirin's anticoagulant properties. Seven years after beginning this daily habit, Kolbe died of a heart attack caused by severe atherosclerosis.

61. Kolbe, "Bemerkenswerthe Eigenschaften," pp. 108–111; "Weitere Mittheilungen," p. 18; Meyer and Kolbe, "Versuche über die gährungshemmende Wirkungen der Salicylsäure und andere aromatischen Säuren," *JpC*, **120** (1875), 133–151; idem, "Über die antiseptischen Wirkungen der Salicylsäure und Benzoësäure in Bierwürze und Harn," *JpC*, **120** (1875), pp. 178–203; Kolbe, "Chemische Winke für praktische Verwendungen der Salicylsäure," *JpC*, **121** (1876), 106–120; idem, "Zerstörende Wirkung der Holzsubstanz auf Salicylsäure," *JpC*, **129** (1880), 443–447 and **130**, 112; idem, "Antiseptische Eigenschaften der Kohlensäure," *JpC*, **134** (1882), 249–255. Otto

Krätz reproduces extensive excerpts on this subject from Kolbe's laboratory notebooks from the years 1874–1880 in his *Historische chemische und physikalische Versuche* (Cologne: Aulis Verlag Deubner, 1979), pp. 198–201.

62. Ost, HK, p. 130.

63. Carl Thiersch, "Klinische Ergebnisse der Lister'schen Wundbehandlung und über den Ersatz der Carbolsäure durch Salicylsäure," *Volkmann's Sammlung*, nos. 84 and 85, 1875.

64. Kolbe, "Bemerkenswerthe Eigenschaften," pp. 111–112; "Weitere Mittheilungen," p. 19; "Weitere Mittheilungen über Wirkungen der Salicylsäure," *JpC*, **119** (1875), pp. 213–215; Kolbe to Pettenkofer, 26 March, 20 November, and 26 December 1874, Pettenkoferiana II 2, Bayerische Staatsbibliothek; Kolbe to Varrentrapp, 3 August 1874, VA 322.

65. Kolbe to Varrentrapp, 10 August and 9 October 1875, VA 333 and 334.

66. W. Walter Cheyne, *Antiseptic Surgery: Its Principles, Practice, History and Results* (London: Smith, Elder, 1882), pp. 136–139 and 404; Just Lucas-Charbonnière, *Antiseptic Surgery*, 2d ed. (Portland: Loring Short & Harmon, 1881), pp. 227–228.

67. Jan R. McTavish, "Aspirin in Germany: The Pharmaceutical Industry and the Pharmaceutical Profession," *Pharmacy in History*, **29** (1987), 103–115.

Notes to Chapter 13

1. A more literal translation might be "bright–born merry children of Jove": UAL, PA 515, f. 3r and 3v, no date, but ca. 29 August 1868; if this is a literary allusion I have not traced it. The document is Erdmann's favorable certification of Graebe's Habilitationsschrift for the University of Leipzig. Erdmann disapproved of Graebe's structural formulas, but thought that such a good chemist as he had proven himself to be would with time be healed of this unfortunate speculative tendency. Kolbe's certificate of approval (ibid., f. 2v) expressed similar thoughts. I thank Dr. E. Vaupel for drawing my attention to these documents.

2. He appeared to concede the issue in E. Frankland and B. F. Duppa, "Synthetische Untersuchungen über Aether," *ZfC*, **11** (1868), 60–64 (on p. 63); however, as late as 1877 he was still hedging his bets: *Experimental Researches in Pure, Applied, and Physical Chemistry* (London: van Voorst, 1877), p. 65. The same ambivalence is found in his letter to Kolbe of 3 December 1871, SSDM 3566, where he writes, "I am quite certain you will never make an Isomalonic acid, unless the 4 bonds of carbon have different values in which case you may get a dozen."

3. Kolbe to Frankland, no date (but ca. 27 October 1864), Frankland Archive 01.04.84.

4. This assumption underlay a great deal of Kolbe's work in predicting new isomers and interpreting existing ones. It was made explicit in his *Kurzes Lehrbuch der organischen Chemie* (Braunschweig: Vieweg, 1883), pp. 6–7.

5. Kolbe, "Chemische Constitution des Benzols und Phenols und einiger Derivate derselben," *JpC*, **122** (1876), 347–355 (on p. 347).

6. Frankland, "On the Synthesis of Diethoxalic Acid," *Proceedings of the*

Royal Society, **12** (1863), 396; Frankland and Duppa, "Synthesis of Butyric and Caproic Ethers from Acetic Ether," *Proceedings of the Royal Society*, **14** (1865), 198–204; "Synthetical Researches on Ethers. No. 2. Action of Sodium and Isopropylic Iodide upon Ethylic Acetate," *JCS*, **20** (1867), 102–116.

7. Frankland and Duppa, "Synthetical Researches on Ethers. No. 1. Synthesis of Ethers from Acetic Ether," *PTRS*, **156** (1866), 37–72; see also idem, "Researches on Acids of the Lactic Series," *PTRS*, **156** (1866), 309–359.

8. Kolbe to Frankland, 27 May 1866, Frankland Archive 01.02.1558.

9. Frankland and Duppa, "Synthetische Untersuchungen," pp. 60–61.

10. Kolbe admitted this in *JpC*, **131** (1881), 377.

11. Kolbe to Frankland, 27 May 1866, Frankland Archive 01.02.1558; Kolbe to Frankland, 23 July 1866, Frankland Archive 01.02.1505 (quote is from this letter). Kolbe constantly used equestrian metaphors connected with Kekulé: other examples are benzene as Kekulé's "show-horse," he "rides a fiery steed" in proposing the theory, he always failed to "rein his imagination," or that he "hat sich einmal mit seinem Benzolring vergaloppirt" (Kolbe to Volhard, 16 December 1874, SSDM 3676; cf. also Kolbe's Pegasus metaphor connected with van't Hoff). This may well have had psychological significance, but I am unwilling to speculate on this topic. Another subject for retrospective psychoanalysis is his repeated implication that graphical formulas are somehow materialistic and irreligious.

12. Kolbe, "Interpretation der Ergebnisse von Frankland's und Duppa's synthetische Untersuchungen über Aether," *ZfC*, **10** (1867), 636–640.

13. Frankland to Crum Brown, 4 June 1866, quoted in J. W., "Alexander Crum Brown," *JCS*, **123** (1923), 3422–3431 (on p. 3425); Crum Brown to Frankland, 5 June 1866, Frankland Archive 01.04.1266.

14. Frankland and Duppa, "Synthetische Untersuchungen."

15. Carl Graebe, "Notizen aus meinem Leben," SSDM 1976-29n, 4, cited in Elisabeth Vaupel, "Carl Graebe (1841–1927): Leben, Werk und Wirken," Ph.D. dissertation, University of Munich, 1987, p. 29.

16. J. Volhard, "Über Sarkosin," *Annalen*, **123** (1862), 261–265.

17. For example, T. Wilm and G. Wischin, "Versuche mit Phosgen und Phosgenäther," *Annalen*, **147** (1868), 150–157, with note by Kolbe on p. 157; H. Byk, "Die isomeren Bernsteinsäuren," *JpC*, **109** (1870), 19–30, with Kolbe's "Bemerkungen zu vorstehender Abhandlung" on pp. 30–33; and R. Bahrmann, "Zur Kenntniss des Amarins und Furfurins," *JpC*, **135** (1883), 295–320, with note by Kolbe on p. 320. The first instance can be explained by the probable supposition that Wilm was a former student of Butlerov, but Byk and Bahrmann were pure Kolbe products. Byk commented that he did the work "auf Veranlassung des Prof. Kolbes." Kolbe's objection to Bahrmann was that he made liberal use of structural formulas; he commented that he had followed the research with interest, but disapproved of the formulas. Finally, Kolbe's trusted assistant for many years, Ernst Carstanjen, published a number of papers using Kekulé's benzene theory and benzene rings. Some additional instances of Kolbe's students publicly disagreeing with him (while still students) are cited in n. 36.

18. A. F. Plate, G. V. Bykov, and M. S. Eventova, *Vladimir Vasil'evich*

Markovnikov: ocherk zhizni i deiatel'nosti, 1837–1904 (Moscow: Izdatel'stvo Akademii Nauk SSSR, 1962), p. 30, quoting from a letter to Butlerov with no date cited. See also H. M. Leicester, "Controversies on Chemical Structure from 1860 to 1870," in O. T. Benfey, ed., *Kekulé Centennial* (Washington, D.C.: American Chemical Society, 1966), pp. 13–23 (on p. 21), and Leicester, "Vladimir Vasil'evich Markovnikov," *Journal of Chemical Education*, **18** (1841), 53–57.

19. Markovnikov, "Vorläufige Notiz über die Identität der Acetonsäure mit Oxyisobuttersäure," *ZfC*, **10** (1867), 434; idem, "Über die Acetonsäure," *Annalen*, **146** (1868), 339–352.

20. Plate et al., *Markovnikov*, p. 31; Markovnikov, in Plate and Bykov, eds., *Vladimir Vasil'evich Markovnikov: Izbrannye trudy* (Moscow: Izdatel'stvo Akademii Nauk SSSR, 1955), pp. 830–831.

21. Kolbe's own account of his conversion to O = 16 is in "Moden der modernen Chemie," *JpC*, **112** (1871), 241–271 (on pp. 246–254).

22. For Graebe's habilitation and teaching at Leipzig, see Vaupel, "Carl Graebe" (see n. 15), pp. 181–190.

23. For example, Meyer to Ostwald, 18 November 1883, Ostwald Nachlass, Akademie der Wissenschaften der DDR, Berlin; Meyer, *Geschichte der Chemie von den ältesten Zeiten bis zur Gegenwart* (Leipzig: Veit, 1889); idem, *Lebenserinnerungen* (n.p., n.d., ca. 1918), pp. 39–40.

24. The chronological division in this table is not always by uniform time increments but rather is determined by events in Kolbe's life; because of the nonuniformity, an intensive measure (papers per year) is also given. The year 1865 is listed twice, as Kolbe transferred to Leipzig in the middle of that year; papers dating from that year were divided according to where the work was performed. A "paper" is any publication, including notes, brief comments, and polemical articles, and the date of the paper is the publication date of the journal issue in which it appeared. "Student papers" represent papers describing research performed in Kolbe's laboratory, whether the author was a student in the strict sense or not, or whether he still resided in Marburg or Leipzig or not. Thus, this category includes independent papers by assistants, Privatdozenten, and postdoctoral workers, and even student papers that had been directed by assistants or Privatdozenten. Because there is no systematic method to ensure completeness in this category, there are probably some papers that I failed to locate.

25. Wilhelm Prandtl, "Das chemische Laboratorium der Bayerischen Akademie der Wissenschaften in München," *Chymia*, **2** (1949), 81–97 (on p. 93); Jeffrey Johnson, "Hierarchy and Creativity in Chemistry, 1871–1914," *Osiris*, [2] **5** (1989), 214–240 (on p. 225).

26. Richard Anschütz, *August Kekulé*, 2 vols. (Berlin: Verlag Chemie, 1929), **2**, 953–960.

27. Namely, the Russians A. Bazarov (1868), H. Byk (1868), S. Byk (1879), C. Fahlberg (1873), and A. M. Zaitsev (1866), and the Britons H. Armstrong (1870), C. Bingley (1854), E. Cook (1865), F. Guthrie (1855), W. James (1882), H. Smith (1877), and F. Wrightson (1853).

28. A convenient source for Fahlberg, Moore, and Norton is Wyndham D.

Miles, ed., *American Chemists and Chemical Engineers* (Washington, D.C.: American Chemical Society, 1976), s.v. It is interesting—to this writer, at least—to note that Norton was a native Clevelander who attended Western Reserve College for one year (1852–1853) and taught high school in Cleveland before spending a semester in Leipzig in 1870.

29. One of Schmidt's doctoral students was Richard Fischer; Fischer taught Henry Schuette, who was Aaron Ihde's doctoral advisor. As I am an Ihde student, I can claim Kolbe as my Doktor-Ur-Ur-Urgrossvater—were I so inclined.

30. Joseph S. Fruton, *Contrasts in Scientific Style: Research Groups in the Chemical and Biochemical Sciences* (Philadelphia: American Philosophical Society, 1990), pp. 32 and 141. The smallest number of German Ordinarien among scholarly progeny any of the six chemists that Fruton studied (Liebig, Baeyer, Emil Fischer, Felix Hoppe-Seyler, Willy Kühne, and Franz Hofmeister) was sixteen for Kühne.

31. This statement cannot be quantified, as records that would allow one to sort Kolbe's students by field of study have not survived.

32. The numbers were derived in the following fashion. Peter Borscheid has estimated [*Naturwissenschaft, Staat und Industrie in Baden* (*1848–1914*) (Stuttgart: Klett, 1976), pp. 84–87 and 234] that there were around 380 university-educated chemists working in German chemical industries in 1851, 900 in 1865, and 2100 in 1884. Taking 1200 as a round-number average for the period 1865–1884, one can presume that the requisite increment for such a work force might be something like 130 new chemists per year (65 representing the average 5.4% growth rate indicated by Borscheid, plus an equal number for replacement due to death, retirement, and so on). Since Leipzig had about sixteen percent of total German university enrollment, Kolbe may have trained an average of about 21 industrial chemists per year. It is true that Kolbe was not the only chemist at Leipzig, but considering Saxony's strength in chemical industry, such a number is not unreasonable. The total over nineteen years is thus something like 400 chemists, or about twenty-five percent of his Praktikanten. The figure for Marburg was derived in a similar fashion. See also Lothar Burchardt, "Die Ausbildung des Chemikers im Kaiserreich," *Zeitschrift für Unternehmungsgeschichte*, **23** (1978), 31–53.

33. In his autobiography (*Lebenserinnerungen*, p. 115), Meyer describes his growing independence and increasing share of direction of Doktoranden. "I could name many chemists here," Meyer concludes, "but I will confine myself to mentioning my especially famous students E. Beckmann, Th. Curtius and Hermann Ost." Other dissertations directed by Meyer include those of Paul Degener (1879), J. William James (1882), M. Wallach (1882), O. Henzold (1883), and G. McGowan (1884); those directed by Ost include A. Klinkhardt (1881) and E. Mennel (1882). One of Joseph Fruton's conclusions after studying six research groups (*Contrasts in Scientific Styles*) was that junior colleagues made far greater contributions to the life of their institutes than has hitherto been appreciated. This pattern holds as well in Leipzig.

34. J. B. Morrell, "The Chemist Breeders: The Research Schools of Liebig and Thomas Thomson," *Ambix*, **19** (1972), 1–46.

35. Gerald Geison, "Scientific Change, Emerging Specialties, and Research Schools," *History of Science*, **19** (1981), 20–40.

36. Here are some examples: Guthrie and Kolbe, "Über die Verbindungen des Valerals mit Säuren," *Annalen*, **109** (1859), 296–300, demonstrating, contrary to Kolbe's prediction, that Wurtz' glycol does not dehydrate to acetaldehyde; Kolbe, "Muthmaassliche Existenz zweier Kohlenoxysulfide," *JpC*, **112** (1871), 381–382, a conjecture that was disproven by his student F. Salomon, "Über Kohlenoxysulfid," *JpC*, **113** (1872), 476–480; Constantin Fahlberg, "Über Oxyessigsäure (Glycolsäure)," *JpC*, **115** (1873), 329–346, demonstrating, contrary to Kolbe, the identity of the two named acids; and Kolbe, "Über die chemische Natur der Salylsäure," *JpC*, **120** (1875), 151–157, disproving the existence of salylic acid (he eventually repudiated this last result in private, but he never found the compound and never repudiated it publicly). Armstrong worked three years on a project under Kolbe's direction, even though his views were "diametrically opposite to Kolbe's": H. E. Armstrong to Richard Armstrong, 6 February 1870, quoted in J. Vargas Eyre, *Henry Edward Armstrong* (London: Butterworths, 1958), pp. 51–52. Three additional instances are cited in n. 17 above.

37. Kolbe, "Über einige Abkömmlinge des Cyanamids," *JpC*, **109** (1870), 288–306 (on pp. 292–294); "Über die Structurformeln und die Lehre von der Bindung der Atome," *JpC*, **111** (1871), 127–136 (on p. 128). In an intriguing new biography of Lord Kelvin (*Energy and Empire: A Biographical Study of Lord Kelvin* [Cambridge Univ. Press, 1989]), Crosbie Smith and M. Norton Wise have drawn substantive parallels between Kelvin's science and the Victorian British imperial age. One might be tempted by Kolbe's illiberal political–military metaphors to make the same sort of connection to the militaristic German empire just then being formed at the expense of France. However, the metaphor that actually succeeded in the Bismarckian and Wilhelmian chemical community at large was a democratic one, namely, structuralist notions, which rather contrasts with the dominant political culture.

38. Kolbe to Frankland, 17 November 1867, Frankland Archive 01.02.1515 ("Ich habe [Kekulés] Zuvorkommenheit auch angenommen und bin es wohl zufrieden, den alten Streit ruhen zu lassen"); Baeyer to V. Meyer, 3 October 1874, SSDM 7020.

39. Kolbe to Varrentrapp, 1 October 1871, VA 268.

40. Kolbe, "Moden der modernen Chemie," *JpC*, **112** (1871), 241–271 (on pp. 257–258).

41. Frankland to Kolbe, 3 December 1871, SSDM 3566.

42. Kolbe to Volhard, 26 June 1871 and 9 June 1876, SSDM 3656 and 3681; Kolbe to Varrentrapp, 29 August 1874, VA 321.

43. He also expressed this view privately to Volhard (16 December 1874, SSDM 3676). "Kekulé belongs to these people the least of all. He blundered once with his benzene ring, but I believe I can say with certainty that he looks with sovereign contempt down upon those chemists who consider his (in the abstract certainly ingenious, but untenable) idea as infallible dogma. Kekulé, of all these people the most sensible, will be the first in due course to disavow his child."

44. Kolbe, "Chemischer Rückblick auf das Jahr 1873," *JpC*, **116** (1874), 417–425 (on pp. 419–420); idem, "Chemischer Rückblick auf das Jahr 1874," *JpC*, **118** (1875), 449–456 (on pp. 449–450 and 456).

45. Kolbe, "Die chemische Synthese, ein chemischer Traum," *JpC*, **126** (1878), 432–455 (on pp. 440n. and 444–445); idem, "Kritisch-chemische Gänge [gegen die transscendentalen Chemiker]: I," *JpC*, **135** (1883), 408–417; idem, "Kritisch-chemische Gänge . . . IV," *JpC*, **136** (1883), 356–382 (on p. 372n.); idem, "Die realen Typen der organischen Verbindungen," *JpC*, **136** (1883), 440–447 (on p. 441n.).

46. J. H. van't Hoff, tr. F. Herrmann, *Die Lagerung der Atome im Raume* (Braunschweig: Vieweg, 1877).

47. Kolbe to Heinrich Vieweg, 31 January and 5 February 1881, VA 467 and 468. Until this time, Vieweg had not realized that his business manager had been regularly sending proofs to Kolbe, having long ago been directed to do so by his deceased father Eduard. According to an annotation on the second of these letters, Heinrich put an immediate end to the practice.

48. Kolbe to Lücke, 9 March 1877, VA 364.

49. Kolbe, "Zeichen der Zeit: II," *JpC*, **123** (1877), 473–477.

50. Kolbe to Vieweg, 9 June 1877, VA 375.

51. Wislicenus to Kolbe, 24 November 1877, SSDM 3550.

52. Van't Hoff, "Über den Zusammenhang zwischen optischer Aktivität und Constitution," *Berichte*, **10** (1877), 1620–1623 (on p. 1620).

53. Those who so speculated include Lothar Meyer (1872), Schorlemmer (1881), and Volhard (1882).

54. Kolbe to Wöhler, 15 and 18 October 1878, Wöhler Nachlass; Wöhler to Kolbe, 17 October 1878, SSDM 3542; Bunsen to Kolbe, 3 November 1873, SSDM 3505.

55. E. von Meyer, *Lebenserinnerung*, p. 120. "These attempts at moderating him usually failed when he claimed the right to call things by their correct name . . . he was a fanatic for the truth." Meyer readily conceded that Kolbe often went too far and became too personal, hurting his own cause. Moreover, the journal itself was badly damaged by these attacks.

56. Dr. R. [Kolbe], "Vertrauliches Schreiben an Professor Kolbe," *JpC*, **124** (1877), 467–472.

57. Kekulé to Kolbe, 5 February 1878, reprinted in *JpC*, **125** (1878), 157–158. Kolbe replied privately to Kekulé (11 February 1878, August-Kekulé-Sammlung), assuring Kekulé that his article would be published, and commenting only that he thought it not gentlemanly (anständig) to appeal to an opponent's sense of fair play (Rechtlichkeit). He repeated the comment in the printed version.

58. Kolbe, "Kritik der Rectoratsrede von Aug. Kekulé: 'über die wissenschaftlichen Ziele und Leistungen der Chemie,'" *JpC*, **125** (1878), 139–156; idem, "Nachtrag zu dem vertraulichen Schreiben des Dr. R.," *JpC*, **125** (1878), 157–163. Kekulé's address was published separately and was subsequently reprinted in Anschütz, **2**, 903–917.

59. A bogus charge, of course. Kolbe had written Volhard on 1 January 1878 (SSDM 3682) to try to find out whether or not Kekulé had had a classical

Gymnasium education, saying it was very important that he find out *soon*, for the sake of a wager.

60. Kolbe, "Kritik," pp. 140, 145, 149–152, and 156.

61. Graebe to Schmitt, 12 April 1878, Staatsbibliothek Preussischer Kulturbesitz, Berlin, Darmstaedter-Sammlung, G 2 1868 (6), cited in Elisabeth Vaupel, "Carl Graebe" (see n. 15), p. 36.

62. Kolbe, "Die chemische Synthese."

63. Kolbe to Wöhler, 15 October 1878, Wöhler Nachlass; Kolbe to Frankland, 27 November 1878, 5 January 1879, and 30 July 1883, Frankland Archive 01.02.1423, 01.02.1432, and 01.02.1533; Kolbe to Volhard, 27 October 1878, SSDM 3683. Frankland was indeed amused, "but, altogether, I dont [*sic*] think the Rede a bad one for a mixed audience. It is of about the calibre of a Friday evening lecture at the Royal Institution" (Frankland to Kolbe, 3 January 1879, SSDM 3569).

64. Volhard to Kolbe, 9 November 1878, SSDM 3516. Volhard told Kolbe of his high regard in his letter of 7 May 1870, SSDM 3620.

65. Kolbe to Volhard, 20 November 1878, SSDM 3684.

66. For example, he thought that the Berzelian "schwefelsaures Bleioxyd" was much superior to the Frenchified "Bleisulfat" (*JpC*, **112**, 242); on another occasion, he excoriated Volhard for using the neologism "ester" (Kolbe to Volhard, 2 July 1874, SSDM 3669). In this regard, Thomas Lounsbury has trenchantly observed, ". . . no one who has once taken the language under his care can ever again be really happy. That way misery lies": *The Standard of Usage in English* (New York: Harper, 1908), p. 11.

67. Kolbe to Frankland, 5 January 1879 and 11 February 1881, Frankland Archive 01.02.1432 and 01.02.1447.

68. Kolbe to Roscoe, 9 February and 29 October 1881, and Schorlemmer to Roscoe, 5 November 1881, Roscoe Collection.

69. Kolbe, "Meine Betheiligung an der Entwickelung der theoretischen Chemie," *JpC*, **131** (1881), 305–323, 353–379, and 497–517; and **132** (1881), 374–425; issued as a separate under the title *Zur Entwicklungsgeschichte der theoretischen Chemie* (Leipzig: Brockhaus, 1881).

70. This comment actually appeared in his "Kritisch-chemische Gänge . . . IV," p. 372n. Kekulé's student and biographer Richard Anschütz also noted Kekulé's physical decline, starting around 1870: *August Kekulé*, **1**, 369 and 415–416.

71. Kolbe, "Betheiligung," p. 405.

72. Kolbe, *Kurzes Lehrbuch der organischen Chemie* (Braunschweig: Vieweg, 1883), pp. vi–vii.

73. Kolbe to Frankland, 23 March [1881] and 1 April 1881, Frankland Archive 01.02.1442 and 01.02.1436.

74. Frankland to Kolbe, 19 May 1881, SSDM 3572; Frankland's self-defense is in his *Experimental Researches*, pp. 146–154.

75. Frankland to Kolbe, 23 September 1883, SSDM 3573.

76. Kolbe to Frankland, 30 July and 28 October 1883, Frankland Archive 01.02.1533 and 01.02.1526.

77. Kekulé's manuscript response was first published by Anschütz (*Kekulé*, **1**, 540–569); a facsimile edition of the manuscript was published in 1965: *Cassirte Kapitel aus der Abhandlung: Ueber die Carboxytartronsäure und die Constitution des Benzols* (Weinheim: Verlag Chemie), from which the quotation is taken (letter of 25 August 1883, unpaginated).

78. Volhard to Frau Baeyer, 2 December 1882, Baeyer Collection.

79. Kolbe to Volhard, 21 July 1884, SSDM 3687.

80. Kolbe to Vieweg, 30 January 1878 and 24 January 1881, VA 408 and 466.

81. Kolbe to Vieweg, 16 October 1883, VA 511.

82. Kolbe to Ost, 8 April 1884, SSDM 6808.

83. Volhard to Kolbe, 4 December 1874, SSDM 3514.

84. For example, Kolbe, "Über die Structurformeln und die Lehre von der Bindung der Atome," *JpC*, **111** (1871), 133; "Moden der modernen Chemie," p. 255; and "Rückblick auf 1874," pp. 453–455.

85. Kolbe, *Über die chemische Constitution der organischen Kohlenwasserstoffe* (Braunschweig: Vieweg, 1869), pp. 8–10.

Notes to Chapter 14

1. The literature on nationalism and internationalism in nineteenth-century science is sparse. Useful discussions include Brigitte Schroeder-Gudehus, "Science, Technology and Foreign Policy," in Ina Spiegel-Rösing and Derek Price, eds., *Science, Technology and Society: A Cross-Disciplinary Perspective* (London: Sage, 1977), pp. 473–506; idem, "Nationalism and Internationalism," in R. C. Olby, G. N. Cantor, J. R. R. Christie, and M. J. S. Hodge, eds., *Companion to the History of Modern Science* (London: Routledge, 1990), pp. 909–919; and Elisabeth Crawford, *Nationalism and Internationalism in Science, 1880–1939* (Cambridge Univ. Press, 1992), esp. chap. 2. Christoph Meinel has a discussion that focuses on chemistry in particular: "Nationalismus und Internationalismus in der Chemie des 19. Jahrhunderts," in Peter Dilg, ed., *Perspektiven der Pharmaziegeschichte* (Graz: Akademische Druck- und Verlagsanstalt, 1983), pp. 225–243.

2. H. E. Armstrong, "The Doctrine of Atomic Valency," *Nature*, **125** (1930), 807–810 (on pp. 808–809).

3. Edward Frankland began to give Kolbe English in return for German lessons soon after their arrival in London, and Kolbe could "soon speak with facility" (Frankland to Hermann Ost, 20 December 1884, SSDM 3576). However, by 1864 Kolbe reported to Frankland that he had forgotten so much that he needed to be allowed to speak German in presenting a lecture to the Chemical Society (Kolbe to Frankland, 4 December 1864, Frankland Archive 01.04.59). The lecture was never given.

4. Boussingault to Dumas, 1 April 1842, Archives of the Académie des Sciences, Paris; cited and translated in Holmes, *Claude Bernard and Animal Chemistry* (Cambridge, Mass.: Harvard Univ. Press, 1974), p. 42.

5. Kolbe to Vieweg, 21 July and 31 December 1862, 19 October 1863, 31

December 1864, and 18 March, 5 and 16 May 1866, VA 184, 187, 196, 213, 242, 243, and 245. The passage quoted is from Kolbe to Frankland, 27 May 1866, Frankland Archive 01.02.1558.

6. Kolbe to Frankland, 23 July 1866, Frankland Archive 01.02.1505; Kolbe expressed similar sentiments in his letters to Vieweg, 9 and 22 July 1866, VA 246 and 247.

7. Wurtz, "Histoire des doctrines chimiques depuis Lavoisier," in Wurtz, ed., *Dictionnaire de chimie pure et appliquée*, 3 vols. in 5 (Paris: Hachette, 1868–1878), **1**, i–xciv (on p. i) (republished monographically in 1869).

8. Kolbe, "Über den Zustand der Chemie in Frankreich," *JpC*, **110** (1870), 173–183.

9. Kolbe to Liebig, 12 November 1870 and 2 December 1870, Liebigiana IIB.

10. Kolbe to Varrentrapp, 26 February 1871, VA 267.

11. Kolbe to Frankland, 26 December 1870, Frankland Archive 01.04.645. Frankland had regretted to see that "there is far too much Gottes Gnaden in [Wilhelm's] nature," and predicted that "in form & constitution the German Despotism will be worse than the French." Still, he thought that in the results of the war, "the rest of the world will be greatly benefitted . . . unless indeed (which is not likely) the German people, excited by victory, turn to be a warlike, instead of a peaceful people" (Frankland to Kolbe, 23 December 1870, SSDM 3564).

12. Kolbe, "Haltung der Pariser Akademie der Wissenschaften," *JpC*, **113** (1872), 225–226; idem, "Chemischer Rückblick auf das Jahr 1872," *JpC*, **114** (1873), 461–470. See also Kolbe's letters to Frankland of 18 March 1872 (Frankland Archive 01.02.948) and to Liebig of 4 April 1872 (see next note).

13. Kolbe to Liebig, 4 April 1872, Liebigiana IIB. For discussions of the contretemps over Pasteur, see Gerald Geison, "Louis Pasteur," *DSB*, **10**, 350–416 (on p. 354), and John Wotiz and Susanna Rudofsky, "Louis Pasteur, August Kekulé, and the Franco-Prussian War," *Journal of Chemical Education*, **66** (1988), 34–36.

14. Liebig to Emma Muspratt, 27 September 1870, Roscoe Collection; Liebig to Wöhler, 25 September 1870, in Hofmann, *LWB*, **2**, 299; Liebig to Kolbe, 2 October 1870, SSDM 3614.

15. Liebig to Wöhler, August 1870, in Hofmann, *LWB*, **2**, 295.

16. "Hundevolk, diese Franzosen," Kekulé to Hans Hübner, 15 July 1870, August-Kekulé-Sammlung, cited in John Wotiz and Susanna Rudofsky, "The Unknown Kekulé," in James G. Traynham, ed., *Essays on the History of Organic Chemistry* (Baton Rouge: Louisiana State, Univ. Press, 1987), pp. 21–34 (on p. 31). Wotiz and Rudofsky use the phrase "sons of bitches" to translate the German "Hundevolk." Here they have committed the common error of preferring a more literal to a connotatively more accurate translation. In fact, "Hundevolk" is undocumented in the German language. "Hunde-" is simply a negatively intensifying prefix and has none of the connotations of profanity that the English expression "son of a bitch" has.

17. Kolbe, "Chemischer Rückblick," pp. 465–466.

18. Kolbe to Varrentrapp, 22 January 1873, VA 304.

19. Kolbe to Volhard, 9 June 1876, SSDM 3681.

20. Kolbe to Vieweg, 6 November 1882, VA 482.

21. Wurtz, "Éloge de Laurent et de Gerhardt," *Moniteur scientifique*, **4** (1862), 482–513 (also an offprint separate); "Histoire générale des glycols," in Société Chimique de Paris, *Leçons de chimie professées en 1860* (Paris: Hachette, 1861), pp. 101–139; "On Oxide of Ethylene, Considered as a Link between Organic and Mineral Chemistry," *JCS*, **15** (1862), 387–406; *Leçons de chimie professées en 1863* (Paris: Hachette, 1864; identical to *Leçons de philosophie chimique*, same publisher and date); *Cours de philosophie chimique* (Paris: privately printed, 1864); *Leçons élémentaire de chimie moderne* (Paris: Masson, 1867–1868); *Dictionnaire de chimie pure et appliquée*, 3 vols. in 5 (Paris: Hachette, 1868–1878).

22. See L. Graham, W. Lepenies, and P. Weingart, eds., *The Functions and Uses of Disciplinary Histories* (Dordrecht: Reidel, 1983). For the discipline of chemistry, see also Jost Weyer, *Chemiegeschichtsschreibung von Wiegleb (1790) bis Partington (1970)* (Hildesheim: Gerstenberg, 1974); and C. A. Russell, "'Rude and Disgraceful Beginnings': A View of History of Chemistry from the Nineteenth Century," *British Journal for the History of Science*, **21** (1988), 273–294 (on pp. 288–294), who has some additional apposite examples.

23. Kopp, *Entwickelung der Chemie in der neueren Zeit* (Munich: Oldenbourg, 1873). The work was published in three parts, beginning in 1871. A discussion of this work in a Kolbean context is my "'Between Two Stools': Kopp, Kolbe, and the History of Chemistry," *Bulletin for the History of Chemistry*, **7** (1990), 19–24.

24. For example, Jacob Volhard, *Justus von Liebig*, 2 vols. (Leipzig: Barth, 1909), **2**, 418–422.

25. Liebig to Wöhler, 30 September 1870, in Hofmann, *LWB*, **2**, 300.

26. Liebig to Wöhler, 7 December 1870, in Hofmann, *LWB*, **2**, 304.

27. Liebig to Wöhler, 24 May 1845, in Hofmann, *LWB*, **1**, 257.

28. Liebig, "Eröffnungsworte . . . nach dem Friedensschluss," 28 March 1871, in *Reden und Abhandlungen* (Leipzig and Heidelberg: Winter, 1874), pp. 331–333; excerpted by Volhard, *Liebig*, **2**, 420–422.

29. W. H. Brock, "Liebig, Wöhler, Hofmann—An English Perspective," in W. Lewicki, ed., *Wöhler und Liebig: Briefe von 1829–1873* (Göttingen: Cromm, 1982), pp. xvi–xviii; this is a photographic one-volume republication, with new front matter, of Hofmann's edition of the Liebig–Wöhler correspondence.

30. S. Kapoor, "Jean-Baptiste Dumas," *DSB*, **4**, 242–248 (on p. 243).

31. Richard Willstätter, *From My Life* (New York: Benjamin, 1965).

32. Albert Ladenburg, *Lebenserinnerungen* (Breslau: Trewendt & Granier, 1912), pp. 51–52. Neither Willstätter nor Ladenburg were *practicing* Jews, and both were fully assimilated Germans. Ladenburg was in fact an atheist; for reasons that he does not explain, he finally underwent baptism in 1891.

33. A sampling of the large literature that pertains to these issues is as follows: Jacob Katz, *From Prejudice to Destruction: Anti-Semitism, 1700–1933* (Cambridge, Mass.: Harvard Univ. Press, 1980), pp. 145–220; George Mosse,

Germans and Jews (New York: Fertig, 1970); Peter Gay, *Freud, Jews, and Other Germans* (New York: Oxford Univ. Press, 1978), pp. 93–168; W. E. Mosse, *Jews in the German Economy* (Oxford: Clarendon Press, 1987); W. E. Mosse, ed., *Juden im Wilhelminischen Deutschland 1890–1914* (Tübingen: Mohr, 1976); R. Rürup, "Emancipation and Crisis: The 'Jewish Question' in Germany, 1850–1890," *Leo Baeck Institute Yearbook*, **20** (London: Secker and Warburg, 1975), 13–25; Fritz Stern, *Gold and Iron: Bismarck, Bleichröder, and the Building of the German Empire* (New York: Knopf, 1977); and David L. Preston, "Science, Society, and the German Jews: 1870–1933," Ph.D. dissertation, Univ. of Illinois, 1971.

34. This paragraph summarizes material in Preston's valuable dissertation; see esp. pp. 21–24, 99–118, and 216–222.

35. It is often difficult to identify Jews, as most obituarists of scientists do not mention religion. The *Neue deutsche Biographie* does, but it is as yet only half complete. The best source for late nineteenth-century German chemists' biographies and obituaries, particularly for less well known figures, is the *Berichte*. Also helpful are such compilations as Salomon Wininger, ed., *Grosse jüdische Nationalbiographie*, 5 vols. (Leipzig: Braun, n.d. [ca. 1925]), and Charlotte Politzer, "Chemie," in Sigmund Kaznelson, ed., *Juden im deutschen Kulturbereich* (Berlin: Jüdischer Verlag, 1959), pp. 429–450, which was completed shortly after 1933, but because of the rise of Nazism, was not published until much later. Unfortunately, such philosemitic compilations do not distinguish between practicing, nonpracticing, and baptized Jews—or even, perhaps, children of baptized Jewish parents.

36. A. W. Hofmann, "Gustav Magnus," in *Zur Erinnerung an vorangegangene Freunde*, 3 vols. (Braunschweig: Vieweg, 1888), **1**, 43–194. All three chemists can be found in the German philosemitic literature.

37. David E. Rowe, "'Jewish Mathematics' at Göttingen in the Era of Felix Klein," *Isis*, **77** (1986), 422–449 (on p. 429).

38. Hans Rosenberg, *Grosse Depression und Bismarckzeit* (Berlin: de Gruyter, 1967), pp. 88–117.

39. Alexander Busch, *Die Geschichte des Privatdozenten* (Stuttgart: Enke, 1959), pp. 148–162; Fritz K. Ringer, *The Decline of the German Mandarins: The German Academic Community, 1890–1933* (Cambridge, Mass.: Harvard Univ. Press, 1969), pp. 135–139.

40. Preston, "German Jews," pp. 110–112. No statistics were ever compiled on the distinctions between professing and baptized Jews in nineteenth-century German academia.

41. Ringer, pp. 5–13; a more recent study is Konrad H. Jarausch, *Students, Society, and Politics in Imperial Germany: The Rise of Academic Illiberalism* (Princeton, N.J.: Princeton Univ. Press, 1982).

42. As related by Frankland to Roscoe, 9 January 1879, Roscoe Collection. Kolbe was aware of the pun, but got it somewhat wrong, as he proudly related to Wöhler (31 March [1880], Wöhler Nachlass) that someone had applied the name "Journal für chemische Polizei."

43. Volhard, "Die Begründung der Chemie durch Lavoisier," *JpC*, **110** (1870), 1–47; Kolbe, "Über den Zustand der Chemie in Frankreich," *JpC*, **110**

(1870), pp. 173–183; N. Zinin, A. Butlerov, D. Mendeleev, and A. Engelhardt, *St. Petersburger Zeitung*, Nr. 271 (4 October 1870), cited in Volhard, "Berichtigung," *JpC*, **110** (1870), 381–384 (on p. 381).

44. Volhard, "Berichtigung"; Liebig to Kolbe, 9 November 1870, SSDM 3615; *Berichte*, **3** (24 October 1870), 873.

45. Kolbe to Baeyer, 29 May 1871, Baeyer Collection.

46. Baeyer to Kolbe, 8 June 1871, SSDM 3628.

47. Kolbe to Baeyer, 5 June 1871, Baeyer Collection; Baeyer to Kolbe, 8 June 1871, SSDM 3628.

48. Kolbe wrote Liebig on 4 January 1870, giving him the news that he had accepted the editorship of the *JpC* (Liebigiana IIB); it is quite clear from the correspondence that he had not asked for prior approval. In his letter to Baeyer of 5 June 1871, Kolbe also responded to Baeyer's implication of hypocrisy by asserting that Liebig had sent him his article on "Fermentation" with the request that Kolbe publish it as one of the first of his tenure as editor. However, the letter of 4 January 1870 has Kolbe asking Liebig for permission to *re*publish the article in his journal, an offprint of which Liebig had simply sent him as a courtesy. Finally, in this letter Kolbe told Liebig that he intended to remain faithful to the *Annalen* by continuing to publish most of his works there. On the contrary, Kolbe never published another paper in the *Annalen*. Baeyer responded to Kolbe's arguments (Baeyer to Kolbe, 25 June 1871, SSDM 3629) by suggesting that Liebig must certainly have been distressed by Kolbe's action in taking over a competitor journal, but that he had been too polite to say so directly. Baeyer probably did not know the details mentioned here, but he was likely correct in his presumption. Kopp, the managing editor of the *Annalen*, was certainly distressed. On 7 and 20 January 1870, Kopp wrote Liebig, worrying about the dangerous competition represented by Kolbe's new journal (cited in Max Speter, "'Vater Kopp,'" *Osiris*, **5** [1938], 392–460, on p. 447).

49. Kolbe to Baeyer, 5 June 1871 (quoted) and 9 June 1871, Baeyer Collection.

50. Hofmann to Kolbe, 12 June 1871, SSDM 3557.

51. Kolbe to Hofmann, 29 May, 5 June, 26 June, and 8 December 1871, Chemiker-Briefe.

52. Kolbe to Baeyer, 27 June 1871; Baeyer to Kolbe, July 1871; Kolbe to Baeyer, 1 August 1871; all printed in *Berichte*, **4** (1871), 993–995. Kolbe to Liebig, 26 January 1872, Liebigiana IIB.

53. Kolbe to Volhard, 26 June 1871, SSDM 3656.

54. Kolbe, "Chemischer Rückblick auf das Jahr 1871," *JpC*, **112** (1871), 464–468; ". . . 1872," *JpC*, **114** (1872), 461–470; ". . . 1873," *JpC*, **116** (1873), 417–425; ". . . 1874," *JpC*, **118** (1874), 449–456.

55. *Berichte*, **4** (1871), 993–995; ibid., **5** (1872), 1114–1116.

56. Kolbe to Varrentrapp, 4 October 1871, 22 January 1873, 24 February 1873, 16 June 1873, and 18 June 1873 (VA 269, 304, 305, 309, and 310); Kolbe to Liebig, 1 January 1872, 26 January 1872, and 23 February 1873 (Liebigiana IIB); Kolbe to Volhard, 23 December 1873, 8 January 1874 and 10 April 1874 (SSDM 3666, 3667, 3668); Kolbe to Kopp, 12 March 1882 (SSDM 3633).

57. Kolbe to Liebig, 26 January 1872 (in n. 56); Liebig to Kolbe, 3 January 1872 and 20 July 1872 (SSDM 3618 and 3617); Bunsen to Kolbe, 10 July 1872 and 3 November 1873 (SSDM 3504 and 3505); L. Meyer to Kolbe, 14 April 1873 (SSDM 3532); Kolbe to Varrentrapp, 16 June 1873 (VA 309); Kolbe to Volhard, 23 December 1873 (SSDM 3666); Kekulé to Erlenmeyer, 26 December 1874, Dingler Nachlass.

58. Beilstein to Erlenmeyer, 5 October 1873, in Otto Krätz, ed., *Beilstein-Erlenmeyer: Briefe zur Geschichte der chemischen Dokumentation und des chemischen Zeitschriftenwesens* (Munich: Fritsch, 1972), pp. 41–45.

59. Meyer to Baeyer, 11 February 1872, Baeyer Collection.

60. Kolbe to Varrentrapp, 4 October 1871, 22 January 1873, and 18 June 1873 (VA 269, 304 and 310); Kolbe to Kopp, 12 March 1882 (SSDM 3633).

61. Kolbe to Varrentrapp, 18 June 1873 and 22 January 1873, VA 310 and 304.

62. This according to letters from Kolbe to Varrentrapp, 8 June and 18 June 1873 (VA 308 and 310); the letter from Kolbe to Hofmann does not seem to have survived.

63. During his stay in Berlin, Kolbe called on Hofmann twice. The first time he was informed that Hofmann was not in, the second time that Hofmann had left town (ibid., VA 308). There exist three letters from Hofmann to Kolbe and one from Kolbe to Hofmann written after 1873, but none are substantive (SSDM 3558, 3560, and 3561, and Kolbe to Hofmann, 2 March 1877, Chemiker-Briefe).

64. These events are described more fully, for example, in Katz, *Anti-Semitism*, pp. 245–252.

65. See Jarausch, *Students*, pp. 208–212 and 264–271; Hermann von Petersdorff, *Die Vereine Deutscher Studenten: Zwölf Jahre akademischer Kämpfe*, 2d ed. (Leipzig: Breitkopf und Härtel, 1895), pp. 7–22.

66. Jarausch, *Students*, pp. 208–212 and 264–271; Petersdorff, *Vereine*, pp. 23–63.

67. Petersdorff, *Vereine*, p. 41.

68. Volhard, *Hofmann*, p. 137. On Hofmann's fight against the antisemites, see also Monika Müller, "Aus dem Leben und Wirken des Chemikers und Hochschullehrers August Wilhelm von Hofmann (1818–1892)," Ph.D. dissertation, Humboldt-Universität Berlin, 1981, pp. 46–52.

69. Volhard, *Hofmann*, pp. 192–195.

70. Postscript by Lothar Meyer to letter of Beilstein to Erlenmeyer, 28 August 1880, in Krätz, *Beilstein-Erlenmeyer*, p. 75. The reader is reminded that Kolbe's laboratory and dwelling were on the Waisenhausstrasse in Leipzig.

71. Kolbe to Volhard, 26 June 1871 and 2 July 1874, SSDM 3656 and 3669. To be precise, Volhard was not a student of Kolbe's but rather worked in Kolbe's Marburg lab for a year after he received his doctorate and before he became Liebig's assistant.

72. D. Vorländer, "Jacob Volhard," *Berichte*, **45** (1912), 1855–1902 (on pp. 1856–1858). Volhard's internationalism is indicated by his agitation in 1873, along with Kekulé and Erlenmeyer, to reserve honorary memberships in the DCG for foreigners only: Richard Anschütz, *August Kekulé*, 2 vols. (Berlin:

Verlag Chemie, 1929), **1**, 419. Volhard also made much of Liebig's internationalism in his biography (*Justus Liebig*, **2**, 418–422). Jacob's father, Karl Ferdinand Volhard, had been a schoolmate of Liebig and was also an intimate friend of the liberal politician Heinrich von Gagern, prime minister of Hesse-Darmstadt and president of the Frankfurt Parliament (this according to Liebig's letter to Wöhler of 29 August 1848, in Hofmann, *LWB*, **1**, 320). There may even have been Jewish blood in Volhard's family, for Liebig mentions in a letter to Hofmann (5 December 1850, in Brock, *LHB*, p. 103) that K. F. Volhard was removed from state service without pension "wegen seiner Abstammung." He then became a private attorney.

73. Kolbe to Volhard, 20 November 1878, SSDM 3684.

74. Kolbe, "Begründung meiner Urtheile über Ad. Baeyer's wissenschaftliche Qualification," *JpC*, **134** (1882), 308–323.

75. Volhard to Frau Baeyer, 2 December 1882, Baeyer Collection.

76. Volhard, *Hofmann*; H. Caro's history of the German dye industry, in which Hofmann features prominently, appeared in *Berichte*, **25** (1892), 955–1105; F. A. Abel, H. E. Armstrong, W. H. Perkin, and L. Playfair, "Hofmann Memorial Lectures," *JCS*, **69** (1896), 575–732; and B. Lepsius, *Festschrift zur Feier des 50jährigen Bestehens der Deutschen Chemischen Gesellschaft und des 100. Geburtstages ihres Begründers August Wilhelm von Hofmann* (Berlin: Friedländer, 1918).

77. Volhard, *Hofmann*, p. 137.

78. Volhard's talk is mentioned in Vorländer, "Volhard," p. 1865; Hofmann's obituary of Kolbe is in *Berichte*, **17** (1884), 2809–2812.

79. Hofmann, *Erinnerung*. There are a total of fifteen biographies here; several, including those for Liebig, Wurtz, Dumas, and Magnus, are book-length. Hofmann also wrote a total of fifty-one obituaries in the *Berichte*.

80. Wurtz, *Geschichte der chemischen Theorien seit Lavoisier bis auf unsere Zeit*, trans. A. Oppenheim (Berlin: R. Oppenheim, 1870). In editorial notes (pp. iii–viii), Oppenheim defended his former teacher and censured his fellow Germans for their excessive chauvinism.

81. Gay, *Freud, Jews, and Other Germans*, pp. 14–16.

82. Elisabeth Vaupel has shown that even so liberal a man as Carl Graebe, whose relationships with Liebermann and other Jews were very close, occasionally betrayed negative prejudices: "Carl Graebe (1841–1927)—Leben, Werk und Wirken im Spiegel seines brieflichen Nachlasses," Ph.D. dissertation, Univ. of Munich, 1987, pp. 281–284.

83. Here are some exceptions to this statement. Vaupel ("Carl Graebe," p. 281) mentions a letter from Otto Witt to Graebe (13 July 1904, SSDM 1976-29 0), wherein Witt inquired about a certain candidate's suspicious racial background, stating that he avoided hiring Jews as assistants. Vaupel has also published a passage from a letter by Liebig (to C. Crämer, 13 May 1860, Liebig-Museum, Giessen) that has distinct antisemitic references: "Justus von Liebig und die Glasversilberung," *Praxis der Naturwissenschaften—Chemie*, **40** (1991), 22–29 (on p. 29). David Cahan cites Friedrich Kohlrausch (1840–1910) directing an inquiry about the Jewishness of a candidate to a friend in 1885: "Kohlrausch and Electrolytic Conductivity," *Osiris*, [2] **5**

[1989], 167–185 (on p. 178). Witt and Kohlrausch, however, strongly denied being antisemites even in the act of revealing their prejudice. What is striking is the *rarity* of such comments from private letters, considering the pervasiveness of antisemitism in German society at large, particularly after 1873. Later, there were a number of open antisemites in the scientific community, Philip Lenard and Johannes Stark being only the most scientifically eminent and socially unsavory examples.

84. This was particularly emphasized by Hofmann, in *Erinnerung*, **1**, 342–344. Not far beneath the surface is the rhetorical point that if Oppenheim could inspire such love in the ultranationalist antisemite Treitschke, he deserves no treasonous opprobrium for having translated Wurtz' *Histoire*.

85. Victor Meyer described to his brother an extremely flattering and generous letter he had received from Kolbe in autumn 1874 ("astonishing, from such a raging tyrant," he commented): Richard Meyer, *Victor Meyer: Leben und Wirken eines deutschen Chemikers und Naturforschers* (Leipzig: Akademische Verlagsgesellschaft, 1917), p. 88. To be sure, Kolbe may not have known Meyer was Jewish. Kolbe also recommended Ladenburg to Vieweg Verlag for editing duties: Ladenburg is "zwar Jude, aber wegen Tüchtigkeit und Gewandtheit in der Darstellung zu empfehlen" (Kolbe to Lücke, 8 June 1876, VA 342); he also praised Ladenburg's history of chemistry in *JpC*, **110** (1870), 175.

86. Hofmann, *The Question of a Division of the Philosophical Faculty*, 2d ed. (Boston: Ginn, 1883), pp. 74–75.

87. William Coleman, "Prussian Pedagogy: Purkyně at Breslau, 1823–1839," in Coleman and F. L. Holmes, eds., *The Investigative Enterprise: Experimental Physiology in Nineteenth-Century Medicine* (Berkeley: Univ. of California Press, 1988), pp. 15–64 (on pp. 28–37 and 47–48).

88. Rowe, "'Jewish Mathematics.'" Kathryn Olesko, "On Institutes, Investigations, and Scientific Training," in Coleman and Holmes, eds., *Investigative Enterprise*, pp. 295–332 (on p. 298), also argues this point.

89. Christoph Meinel, *Karl Friedrich Zöllner und die Wissenschaftskultur der Gründerzeit* (Berlin: Sigma, 1991), p. 5 and passim. Zöllner's famous diatribe is his *Ueber die Natur der Cometen*, 3d ed. (Leipzig: Staackmann, 1883; first published 1872). The classic treatment of antimodernist culture, though not directed toward science, is Fritz Stern's *Politics of Cultural Despair* (Berkeley: Univ. of California Press, 1961).

90. Kolbe to Hofmann, 30 June 1872, Chemiker-Briefe; Kolbe to Heinrich Vieweg, 22 January 1873, VA 304; Kolbe, "Erklärung," *JpC*, **113** (1872), 480–481.

91. Beyerchen, *Scientists Under Hitler* (New Haven, Conn.: Yale Univ. Press, 1977).

92. Volhard, *Hofmann*, pp. 192–195.

93. Quoted in Koppel S. Pinson, *Modern Germany: Its History and Civilization*, 2d ed. (New York: Macmillan, 1966), p. 278.

94. Kolbe to Vieweg, 7 November and 30 December 1865, VA 236 and 239; Wöhler to Kolbe, 5 November 1867, SSDM 3539; Kolbe to Liebig, 10 December 1872, Liebigiana IIB; Kolbe to Varrentrapp, 1 and 16 December 1872, VA 301 and 302.

95. Kolbe to Volhard, 16 and 19 May, 19 June, and 3 August 1873, 8 January, 2 July, and 20 November 1874, and 25 April 1875, SSDM 3658, 3659, 3661, 3663, 3667, 3669, 3673, and 3677; Kolbe to Varrentrapp, 16 and 18 June 1873, VA 309 and 310. Although I have not located Volhard's side of this correspondence, it appears that he was under the impression that Kolbe would have been called had his conditions been more reasonable. Further details on the Munich succession are not known to me. Prandtl's statement ("Laboratorium," pp. 88–89) that Baeyer was third on the faculty's list would suggest that Kolbe was the first choice of the faculty, but that his call was derailed at the ministerial level.

96. According to his letter to Erlenmeyer of 3 February 1875, Kekulé was given a salary increase from 2200 to 3000 thalers per year (printed in Anschütz, *Kekulé,* **1**, 465).

97. Lieben to Erlenmeyer, 11 October 1873, SSDM 1968–190/7. As we have noted, Kolbe actually became Geheimrat a few months before Liebig's death. The reference to the Bible plays on the quotation from the Wisdom of Solomon that Kolbe had placed above the wall-sized chart of the chemical elements in his auditorium.

98. Kolbe to Volhard, 16 May 1873, SSDM 3663.

99. Kolbe to Bertha Ost, 22 January 1872, SSDM 6797. It is interesting that he stated his total teaching income here as 4000 thalers, which would mean only 2000 from student honoraria and fees. To be sure, he had substantially more students three semesters later at the time of his letter to Volhard, but one wonders whether he may have overstated his income to Volhard, or understated it to Bertha, or both.

100. Kolbe to H. Vieweg, 9 July 1877, VA 379.

101. Kolbe, *Das neue chemische Laboratorium der Universität Leipzig* (Leipzig: Brockhaus, 1868); idem, *Das chemische Laboratorium der Universität Leipzig* (Braunschweig: Vieweg, 1872). Many German chemists published such works during the nineteenth century. Perhaps the earliest model, as in so many other respects, was Liebig's *Das chemische Laboratorium der Ludwigs-Universität Giessen* (Heidelberg: Winter, 1842). Probably Kolbe's immediate model, however, was that of the son of Liebig's architect, namely Hofmann, who published two editions of papers from his London lab (1849–1853) and followed them in 1866 with a detailed description of the Berlin and Bonn labs in construction, both of which were being built for his occupancy.

102. Kolbe, E. von Meyer, ed., *Ausführliches Lehr- und Handbuch der organischen Chemie*, 2d ed., 2 vols. (Braunschweig: Vieweg, 1880–1884), esp. **2**, 387–391 and 639.

103. Kolbe, *Kurzes Lehrbuch der anorganischen [organischen] Chemie*, 2 vols. (Braunschweig: Vieweg, 1877–1883).

104. Kolbe to Liebig, 2 February, 17 March, and 15 December 1869, Liebigiana IIB; Kolbe to Varrentrapp, 28 February 1870, VA 266. Kolbe reported that the Uruguayan factory was producing 10,000 kilograms of beef extract per month.

105. Kolbe to Varrentrapp, 2 February 1872 and 24 July 1873, VA 274 and 305; Kolbe to Bertha Ost, 22 January 1872, SSDM 6797; and Kolbe to Zarncke, 3 November [1870] and 18 June 1872, Universitätsbibliothek Leipzig,

Handschriftenabteilung. Several parties or balls hosted by the Kolbes are described in these letters, many in the range of 40 to 60 guests; such activity is consistent with spending 200 thalers a year on entertaining, as he told his sister. Armstrong also described a "bachelors dinner party" at the Kolbes' where "great dissipation" was indulged in, while the "old man" became "immensely lively and entertaining": Henry Armstrong to Richard Armstrong, 14 July 1869, cited in J. Vargas Eyre, *Henry Edward Armstrong, 1848–1937* (London: Butterworths, 1958), p. 49.

106. Carl Kolbe (1855–1909) finished his education with Rudolf Fittig in Strasbourg, earning his Ph.D. in 1882. After two years working at Kalle & Co. in Biebrich, he succeeded the retiring Heyden at the Radebeul salicylic acid works—at a salary of 9000 thalers! (Kolbe to Bertha Ost, 24 November 1884, SSDM 6813; this was Kolbe's last letter, for he died the following day.) He married Emilie Pistor in 1883 or 1884. An obituary is in *Zeitschrift für angewandte Chemie*, **22** (1909), 2272, and further (unhappy) details are in Meyer's *Lebenserinnerungen* (n.p., n.d., ca. 1918), pp. 133–134.

107. These data are taken from Meyer, *Lebenserinnerungen*, and Poggendorff. Maria's eldest son, Ferdinand Hermann Krauss (1889–1938), became ausserordentlicher Professor of chemistry at the Braunschweig Technische Hochschule.

108. In 1882 Kolbe weighed 196 Pfund (i.e., 98 kilograms or 216 pounds): Kolbe to H. Vieweg, 20 March 1882, VA 477. He knew he suffered from "Herzverfettung": Kolbe to Lücke, 11 July 1879, VA 445. He ate red meat at nearly every meal and few vegetables, with the consent and approval of his personal physician, Thiersch: Kolbe to Varrentrapp, 29 July 1874, VA 321; Kolbe to H. Vieweg, 13 June 1880, VA 456. Thiersch was Liebig's son-in-law, and it is relevant to note that Liebig's nutritional ideas placed great emphasis on protein.

109. Kolbe to H. Vieweg, 11 April 1877, VA 369. The course of Charlotte Kolbe's final illness is described in the Kolbe–Vieweg correspondence, as well as in Kolbe to Hofmann, 2 March 1877, Chemiker-Briefe.

110. Meyer to Ostwald, 25 November 1884, Ostwald Nachlass, Zentrales Archiv der Akademie der Wissenschaften der DDR. An autopsy revealed advanced arteriosclerosis: "Hermann Kolbe," *Chemiker-Zeitung*, **8** (30 November 1884), 1725–1726.

Notes to "Issues and Reflections"

1. H. E. Armstrong, "The Doctrine of Atomic Valency," *Nature*, **125** (1930), 807–810 (on p. 808).

2. Alexander Vucinich, *Science in Russian Culture, 1861–1917* (Stanford, Calif.: Stanford Univ. Press, 1970), and Nathan M. Brooks, "The Formation of a Community of Chemists in Russia, 1700–1870," Ph.D. dissertation, Columbia University, 1990.

3. W. H. Brock, *H. E. Armstrong and the Teaching of Science, 1880–1930* (Cambridge Univ. Press, 1973); J. Vargas Eyre, *Henry Edward Armstrong, 1848–1937* (London: Butterworths, 1958), esp. pp. 62–64 and 263–296. The broader context is given in D. S. L. Cardwell, *The Organization of Science in*

England, rev. ed. (London: Heinemann, 1972), passim, esp. p. 167, and in R. Bud and G. K. Roberts, *Science Versus Practice: Chemistry in Victorian Britain* (Manchester: Manchester Univ. Press, 1984).

4. ". . . because I know of no one who is more capable of doing this than you, and no one whose views agree more closely with mine than yours." Wöhler to Kolbe, 5 December 1862, SSDM 3538.

5. Lothar Meyer to Adolf Baeyer, 11 February 1872, Baeyer Collection.

6. Kolbe to Volhard, 19 June 1873, SSDM 3659; J. Volhard, *Justus von Liebig*, 2 vols. (Leipzig: Barth, 1909), **2**, 427.

7. Kolbe to Wöhler, 22 August 1875, Wöhler Nachlass.

8. That Kolbe was the first-born of a large number of siblings instantiates Frank Sulloway's suggestion that birth order correlates strongly to resistance to scientific novelty: Daniel Goleman, "The Link Between Birth Order and Innovation," *New York Times*, 8 May 1990, B5 and B9.

9. One exception to this generalization was Wilhelm Heintz, Wislicenus' doctoral advisor, who was born in 1817; remarkably, Heintz was one of the very few structural chemists with whom Kolbe remained friends.

10. Kolbe, footnote to Heintz, "Noch ein Wort über die Constitution der Diglycolsäure," *JpC*, **111** (1871), 122–123n.; "Ueber die realen Typen der organischen Chemie," *Das chemische Laboratorium der Universität Marburg* (Braunschweig: Vieweg, 1865), pp. 515–519.

11. Kolbe, *JpC*, **132** (1881), 405.

12. Kolbe, "Kritisch-chemische Gänge IV," *JpC*, **136** (1883), 356–582 (on pp. 362–363).

13. For example, in Kolbe, "Reale Typen," pp. 518–519.

14. Kolbe, *Kurzes Lehrbuch der organischen Chemie* (Braunschweig: Vieweg, 1883), pp. vi–vii.

15. Lothar Meyer to Kolbe, 13 October 1871, SSDM 3531. On Neumann's conventionalism, see Kenneth Caneva, "From Galvanism to Electrodynamics: The Transformation of German Physics and Its Social Context," *Historical Studies in the Physical Sciences*, **9** (1978), 63–159, who also has much to say about the growth of hypothetico-deductive theorization in nineteenth-century Germany; also Kathryn Olesko, *Physics as a Calling: Discipline and Practice in the Königsberg Seminar for Physics* (Ithaca, N.Y.: Cornell Univ. Press, 1991).

16. Frankland to Kolbe, 19 April 1871, SSDM 3567. This passage constitutes a concise statement of the conventionalist's creed.

17. See Rocke, "Kekulé's Benzene Theory and the Appraisal of Scientific Theories," in A. Donovan, L. Laudan, and R. Laudan, eds., *Scrutinizing Science: Empirical Studies of Scientific Change* (Boston: Kluwer, 1988), pp. 145–161.

18. Of course, "structure theory" was not monolithic, and there was continual and often bitter controversy within the structuralist camp over details—double bonds, free affinities, variability of valence, the nature of aromaticity, and so on. For these events, see C. A. Russell's fine treatment in *The History of Valency* (Leicester: Leicester Univ. Press, 1971). Here I am most interested in the conflict between Kolbe and the structure theorists as a group, who comprised most of the active collegial community after around 1865.

19. The strong program is defined and characterized in David Bloor, *Knowledge and Social Imagery* (London: Routledge, 1976), pp. 1–19. The quotes by Latour are found in K. D. Knorr-Cetina and Michael Mulkay, eds., *Science Observed: Perspectives on the Social Studies of Science* (London: Sage, 1983), p. 141; and Steve Woolgar, ed., *Knowledge and Reflexivity* (London: Sage, 1988), p. 166. See also Latour and Woolgar, *Laboratory Life: The Social Construction of Scientific Facts* (London: Sage, 1979); and Latour, *The Pasteurization of France* (Cambridge, Mass.: Harvard Univ. Press, 1988). Collins' sentence is in his "Stages in the Empirical Program of Relativism," *Social Studies of Science*, **11** (1981), 3–10 (on p. 3). Barnes and Shapin's statement is in their introduction to their edited volume *Natural Order: Historical Studies of Scientific Culture* (London: Sage, 1979), p. 9.

20. The first of these phrases is used in Paul A. Roth's elaborate critique, *Meaning and Method in the Social Sciences* (Ithaca, N.Y.: Cornell Univ. Press, 1987), pp. 152–225, and the second is in Larry Laudan, "The Pseudo-Science of Science?" *Philosophy of the Social Sciences*, **11** (1981), 173–198; see also Laudan, *Science and Relativism* (Chicago: Univ. of Chicago Press, 1990).

21. The first of these quotes is from Bloor, *Knowledge and Social Imagery*, p. 45; the second is in John A. Schuster, "Constructing Contextual Webs," *Isis*, **80** (1989), 493–496 (on p. 494).

22. For example, Bloor, *Knowledge and Social Imagery*, pp. 13–14; Woolgar, ed., *Reflexivity*; review by Bloor of the latter book, in *Isis*, **81** (1990), 155–156; Steve Woolgar, *Science: The Very Idea* (London: Tavistock, 1988), pp. 43–44; and Laudan, "Pseudo-Science of Science?".

23. But see, for example, Terry Shinn, "Orthodoxy and Innovation in Science: The Atomist Controversy in French Chemistry," *Minerva*, **18** (1980), 539–555; Mary Jo Nye, "Berthelot's Anti-Atomism: A 'Matter of Taste'?," *Annals of Science*, **38** (1981), 585–590; idem, *Science in the Provinces: Scientific Communities and Provincial Leadership in France, 1860–1930* (Berkeley: Univ. of California Press, 1986); Harry W. Paul, *The Sorceror's Apprentice: The French Scientist's Image of German Science, 1840–1919* (Gainesville: Univ. of Florida Press, 1972); and Robert Fox, "Scientific Enterprise and the Patronage of Research in France, 1800–70," in G. L'E. Turner, ed., *The Patronage of Science in the Nineteenth Century* (Leiden: Noordhof, 1976), pp. 9–51.

24. Joseph Ben-David, *The Scientist's Role in Society*, 2d ed. (Chicago: Univ. of Chicago Press, 1984); Avraham Zloczower, *Career Opportunities and the Growth of Scientific Discovery in Nineteenth-Century Germany* (New York: Arno, 1981); Steven Turner, Edward Kerwin, and David Woolwine, "Careers and Creativity in Nineteenth-Century Physiology: Zloczower Redux," *Isis*, **75** (1984), 523–529; Peter Borscheid, *Naturwissenschaft, Staat und Industrie in Baden (1848–1914)* (Stuttgart: Klett, 1976); Jeffrey Johnson, "Academic Chemistry in Imperial Germany," *Isis*, **76** (1985), 500–524; and idem, *The Kaiser's Chemists: Science and Modernization in Imperial Germany* (Chapel Hill: Univ. of North Carolina Press, 1990).

25. A French model of center–periphery competition has recently been explored in Nye, *Science in the Provinces*.

26. On this question, see Fox, "Scientific Enterprise."

27. Note, for example, the historical and historiographical similarities between Martin Rudwick's *The Great Devonian Controversy* (Chicago: Univ. of Chicago Press, 1985), James A. Secord's *Controversy in Victorian Geology: The Cambrian–Silurian Dispute* (Princeton, N.J.: Princeton Univ. Press, 1986), and David Oldroyd's *The Highlands Controversy: Constructing Geological Knowledge through Fieldwork in Nineteenth-Century Britain* (Chicago: Univ. of Chicago Press, 1990). A concise and perceptive discussion is Charles Rosenberg, "Woods or Trees? Ideas and Actors in the History of Science," *Isis*, **79** (1988), 565–570.

Glossary

I. German Vocabulary

Abitur:	Gymnasium (secondary-school) leaving exam or certification
Aufklärung:	Enlightenment
ausserordentlicher Professor = Extraordinarius:	loosely, associate professor
Bildung:	education (as opposed to training or apprenticeship), usually implying inculcation of elite values
Bildungsbürgertum:	university-educated classes
Doktorand:	student engaged in doctoral research
Doktorarbeit:	doctoral dissertation
Doktorvater:	doctoral advisor
Duzfreund:	intimate friend whom one addresses as "Du"
florin = gulden:	at mid-century equaled about 0.57 thaler = U.S. $0.41 = 1 shilling 8 pence
Gewerbeschule:	trade school
Gymnasium:	neohumanist secondary school, preparatory to university
Habilitation:	qualification process for teaching at a university

Habilitationsschrift: dissertation connected with habilitation

Kurhessen = Hesse-Kassel: Electoral Hesse

mark = Reichsmark: one-third of a thaler

Naturforscherversammlung: annual meeting of the Gesellschaft Deutscher Naturforscher und Ärzte (Society of German Scientists and Physicians)

Naturphilosophie: speculative organicist philosophy derived from the writings of Friedrich Schelling and popular at the beginning of the nineteenth century

Naturwissenschaften: natural sciences, science

Neohumanismus: educational philosophy of the Gymnasium and university system in the nineteenth century, emphasizing classical philology and pure scholarship

ordentlicher Professor = Ordinarius: full professor, holder of a "chair"

Praktikant: student laboratory worker

Praktikum: laboratory exercises, practicum

Privatdozent: private lecturer, habilitated and generally a Ph.D., but paid directly by students and not by the university that certified the candidate

Publikum: public university lecture course for which no fees may be charged

Realschule: secondary school with more "modern" subjects and less classical emphasis than the neohumanist Gymnasium

technische Hochschule: technical college, polytechnic institute

thaler = Taler: monetary unit, cognate with "dollar"; at mid-century equaled about U.S. $0.71 or roughly 3 shillings

venia legendi: certification necessary for habilitation

Vormärz: period of German history between the Congress of Vienna and March 1848

Wanderjahr: journeyman period, literally or figuratively

Wissenschaft: scholarship, often restricted to nonapplied fields

Wissenschaftsideologie: scholarly ideology associated with neohumanist reforms

II. Organic Chemistry

A. Precis of Modern Structure Theory

Valence refers to the number of bonds an atom can form with its neighbors. Organic chemists normally regard hydrogen and *halogen* atoms (fluorine, chlorine, bromine, and iodine) as monovalent, oxygen and sulfur as divalent, nitrogen and phosphorus as trivalent, and carbon atoms as tetravalent. These elements provide the building blocks for the large majority of all organic compounds. Following the valence rules, one can schematically construct organic compounds by linking the atoms together. The simplest organic compound is methane, CH_4; the carbon atom is tetravalent and each of the hydrogen atoms is monovalent, so the rules are satisfied and the compound is stable. The first *homolog* (that is, next in the same chemical series) of methane is ethane, constructed by removing one of the four hydrogen atoms of methane (which produces a methyl radical, $-CH_3$, sometimes symbolized Me) and filling the vacant valence with another methyl group: $H_3C–CH_3$. Note that conditions of carbon tetravalence and hydrogen monovalence are still met.

Taking one hydrogen from each carbon atom of ethane will schematically construct ethylene, the simplest of the class of hydrocarbons known as *olefins*: $H_2C=CH_2$. Note that the two unoccupied valences on the two adjacent carbons may be considered to have linked with each other. Olefins by definition have at least one double bond between adjacent carbon atoms, so that each of these atoms has only two remaining valences for attachment to hydrogen atoms. Ethylene chloride results from the addition of a chlorine molecule, Cl_2, to ethylene, with the carbon–carbon double bond thus disappearing: $CH_2CH_2 + Cl_2 \rightarrow CH_2ClCH_2Cl$. (Here the carbon atoms are understood to be bonded directly together, despite the fact that they are not written adjacently; there is no ambiguity if one remembers the valences.) An isomer of ethylene chloride is ethylidene chloride, CH_3CHCl_2. *Isomers* are compounds with the same overall formula but with different structures, which is to say, with different properties.

Substances containing only carbon and hydrogen (*hydrocarbons*) are mostly fairly unreactive and chemically uninteresting compounds. In general, chemical interest is created by the introduction of *heteroatoms*, i.e., atoms other than carbon or hydrogen. The most common and important of these is oxygen. Heteroatoms, or any multiple bonds, tend to create chemically reactive sites within an organic molecule and are called *functional groups*. Most reactions center on such groups of atoms.

Some examples will help. If a hydrogen atom is removed from either carbon atom of ethane (creating ethyl, CH_3CH_2-, often symbolized Et) and is replaced by a hydroxyl radical, $-OH$, one obtains ethyl alcohol, CH_3CH_2OH. if one then extracts the remaining two hydrogens from the second carbon atom of alcohol and replaces them with one double-bonded oxygen atom, we have the formula for acetic acid, CH_3COOH. The $-COOH$ or CO_2H group ($O=C–OH$, with one valence still remaining on the carbon atom) is known as *carboxyl*. An intermediate stage of oxidation between alcohol and acid is aldehyde, $CH_3HC=O$; it is carboxyl without the second oxygen atom. Two carboxyl groups connected together (or, equivalently, acetic acid in which the three hydrogens attached to the first carbon atom are replaced by =O and −OH), is the simplest dibasic (i.e., double) organic acid, oxalic acid. It can be obtained by vigorous oxidation of acetic acid or alcohol. If the oxidation is done more carefully or indirectly, other functional groups can be introduced. Glycol has hydroxyl groups on both carbon atoms; it is a double alcohol. Glyoxal has two aldehyde groups on the two carbons, a double aldehyde. Glycolic acid, a hydroxy-acid, is acetic acid in which one of the hydrogens of the methyl group is replaced by hydroxyl, $HOCH_2CO_2H$. Analogous homologs can be created from the three-carbon hydrocarbon, propane, and from the three-carbon acid, propionic acid, $CH_3CH_2CO_2H$. Propionic acid with a hydroxyl group on the middle carbon is lactic acid; the trialcohol with a hydroxyl on each carbon atom is glycerin.

This gives only a sample of the richness of organic chemical formulas, but it accurately reflects the sort of schematic manipulations that characterize structural organic chemistry.

B. Modern Chemical Definitions[1]

acetic acid = essence of vinegar: CH_3CO_2H

acetoacetic ester: $CH_3COCH_2CO_2CH_2CH_3$, produced by self-condensation of ethyl acetate, $CH_3CO_2CH_2CH_3$

acetone: CH_3COCH_3, the simplest ketone

acetyl: CH_3CO- (the oxygen is double bonded to carbon, which has an unfilled valence)

acetylene: $HC\equiv CH$

acid anhydride: acid from which the elements of water (H_2O) have been extracted

acid chloride: compound of the form RCOCl

alanine: $CH_3CH(NH_2)CO_2H$, aminopropionic acid

alcohol: loosely, a compound with a hydroxyl (−OH) group

aldehyde: any compound whose carbon chain ends in −CH=O

aldehyde (loosely) = acetaldehyde: CH_3CHO

aliphatic compound: non-benzenoid organic compound

1. Organic chemists will note that some degree of generality and rigor has been sacrificed here for the sake of clarity. Terminology sometimes differed in the nineteenth century; such distinctions are explained in the text.

alkyl: saturated hydrocarbon radical, often symbolized generically as "R"

amide: compound of the form $RCONH_2$ (this particular example is a *primary* amide)

amine: compound of the form NR_3, where the R's are hydrogen or hydrocarbon groups

amino acid: compound with both amine and carboxylic acid groups

amyl: a five-carbon saturated hydrocarbon radical

amylene: a five-carbon olefin

aniline: $C_6H_5NH_2$, aminobenzene

anthranilic acid: *ortho*-aminobenzoic acid

aromatic compound: organic compound derived from benzene

-ate: suffix indicating ester or salt of an acid

benzaldehyde: C_6H_5CHO, benzoyl hydride

benzene: C_6H_6

benzoic acid: $C_6H_5CO_2H$, benzenecarboxylic acid

benzonitrile: C_6H_5CN

benzoyl: C_6H_5CO-, phenyl carbonyl

benzyl alcohol: $C_6H_5CH_2OH$

butane: a four-carbon saturated hydrocarbon

butyl: a four-carbon saturated hydrocarbon radical

butylene: a four-carbon olefin

butyric acid: a four-carbon carboxylic acid

caproic acid: a six-carbon carboxylic acid

carbohydrate: a compound exhibiting a multiple of the formula CH_2O

carbonyl: functional group consisting of a carbon atom double bonded to oxygen

carboxyl: $-COOH$, confers acid character on a molecule

carboxylic acid: loosely, an organic acid, or a compound with a carboxyl group

chloroform: $CHCl_3$

condensation: amalgamation of two molecules into one, often with loss of H_2O

cresol: methyl phenol, MeC_6H_4OH

cyanogen: NC−CN

cyclohexamethine: any proposed structure for benzene exhibiting a cyclical array of six trivalent CH groups

cyclohexatriene:	Kekulé's hexagonal benzene formula, possessing three carbon–carbon double bonds and three carbon–carbon single bonds; an example of a cyclohexamethine theory
dibasic acid:	acid having two replaceable hydrogen atoms; a double acid
dimer:	compound formed of two identical parts, a "doubled" molecule
dithionic acid = hyposulfuric acid:	$H_2S_2O_6$
ester:	compound of the form RCO_2R, resulting from condensation of an organic acid and an alcohol
ethane:	CH_3CH_3
ether:	compound of the form R–O–R, where the R's are hydrocarbon radicals
ether (loosely) = diethyl ether:	$CH_3CH_2OCH_2CH_3$ or EtOEt
ethyl:	CH_3CH_2–, often symbolized Et
ethyl alcohol = (loosely) alcohol:	CH_3CH_2OH or EtOH
ethylamine:	$EtNH_2$
ethylene:	$CH_2{=}CH_2$, the simplest olefin or alkene
ethylidene:	$CH_3CH{=}$ (a radical, or incomplete molecule)
fatty acid:	large organic acids found as components of fats
formic acid:	HCO_2H
functional group:	any site on an organic molecule that has a multiple bond or a heteroatom
glyceride:	ester of fatty acid(s) plus glycerin; triglycerides are fats
glycerin:	$CH_2OHCHOHCH_2OH$, a trialcohol
glycine = glycocoll:	aminoacetic acid
glycol:	compound with two hydroxyl groups, a dialcohol
glycol (loosely) = ethylene glycol:	CH_2OHCH_2OH
glycolic acid:	CH_2OHCO_2H
glyoxal:	OHCCHO
halide:	chloride, bromide, iodide, or fluoride
halogen:	chlorine, bromine, iodine, or fluorine

heteroatom: in an organic molecule, any atom other than hydrogen or carbon

homolog: a member of a homologous series

homologous series: series of organic compounds that differ only by successive CH_2 (methylene) groups, each methylene representing a link in the carbon chain

hydracid: compound whose acidity is unconnected with its oxygen content

hydride: compound of the form X−H, where X may be almost any atom or radical

hydrocarbon: a compound consisting only of hydrogen and carbon

hydrolysis: reaction with water, often splitting the reactant molecule

hydroxyl: −OH, confers alcohol character

iso-: prefix indicating either (1) a new isomer of a familiar compound hitherto thought to have no isomers; or (2) a branched-chain hydrocarbon

isoamyl: $Me_2CHCH_2CH_2-$

isobutyl: Me_2CHCH_2-

isobutyric acid: Me_2CHCO_2H

isomers: compounds with the identical formula but different structures, that is, compounds that have the same sorts and numbers of atoms but arranged differently

isopropyl alcohol: $CH_3CH(OH)CH_3$

ketone: compound of the form RCOR, that is, carbonyl flanked by two alkyl radicals

lactic acid: $CH_3CH(OH)CO_2H$

leucic acid: a six-carbon hydroxy acid

leucine: a six-carbon amino acid

malic acid: a four-carbon hydroxy diacid; hydroxysuccinic acid

malonic acid: $HO_2CCH_2CO_2H$, a three-carbon diacid

methane: CH_4, called "marsh gas" in the nineteenth century

methine: CH, a trivalent radical

methyl: CH_3-, often symbolized Me

methylamine: $MeNH_2$

methylene:	CH_2, a divalent radical
methylsulfonyl chloride:	CH_3SO_2Cl
methylsulfonic acid:	CH_3SO_3H
moiety:	organic-chemical term indicating a portion of a molecule
nitrile = cyanide:	compound with the group $-CN$
normal:	possessing a straight (unbranched) carbon chain, symbolized *n*-
octane:	an eight-carbon saturated hydrocarbon
olefin:	a hydrocarbon with a carbon–carbon double bond; alkene
ortho/meta/para:	referring to aromatic isomers distinguished by virtue of two substituents occupying different positions on the benzene hexagon–on adjacent carbons (1,2), separated by one carbon (1,3), or on opposing carbons (1,4), respectively
oxalic acid:	HO_2CCO_2H
oxyacid:	compound whose acidity is related to its oxygen content
paraffin:	a hydrocarbon with no multiple bonds; alkane
pentane:	a five-carbon saturated hydrocarbon
phenol = carbolic acid:	phenyl hydroxide or hydroxybenzene, C_6H_5OH
phenyl:	C_6H_5-
phosgene = carbonyl chloride:	$COCl_2$
polymer:	compound formed of many identical parts
primary alcohol or amine:	compound of the form RCH_2OH or RNH_2
propane:	$CH_3CH_2CH_3$
propionic acid:	$CH_3CH_2CO_2H$
propyl = *n*-propyl = normal propyl:	$CH_3CH_2CH_2-$
propylene:	$CH_3CH=CH_2$
salicylic acid:	*ortho*-hydroxybenzoic acid
saturated:	organic compound possessing no carbon–carbon multiple bonds
secondary alcohol or amine:	compound of the form R_2CHOH or R_2NH
succinic acid:	a four-carbon saturated dicarboxylic acid

sulfonic acid: a compound of the form RSO_3H

tartaric acid: a four-carbon dihydroxy diacid; dihydroxysuccinic acid

tertiary alcohol or amine: a compound of the form R_3COH or R_3N

tertiary butyl alcohol = *t*-butyl alcohol: Me_3COH

toluene: methylbenzene

unsaturated: possessing carbon–carbon multiple bonds (can absorb hydrogen gas)

valence: number of bonds a given kind of atom can form to other atoms

valeric acid: a five-carbon carboxylic acid

Bibliography

I. Manuscript Sources

Aschaffenburg
Emil Erlenmeyer Nachlass, Hugo-Dingler-Stiftung, Hofbibliothek

Berlin
Zentrales Archiv, Akademie der Wissenschaften der DDR (newly renamed Akademie der Wissenschaften zu Berlin Brandenburg)

Darmstadt
August-Kekulé-Sammlung, Institut für Organische Chemie, Technische Hochschule

Giessen
Justus-Liebig-Museum

Göttingen
Cod. Ms. F. Wöhler, 74, Niedersächsische Staats- und Universitätsbibliothek
Kirchenkreisarchiv Göttingen
Universitätsarchiv Göttingen

Hannover
Landeskirchliches Archiv Hannover

Hardegsen
Lutterhäuser Kirchenbuch

Leipzig
Universitätsarchiv Leipzig
Handschriftenabteilung, Universitätsbibliothek Leipzig

London
Royal Institution
Royal Society

H. E. Roscoe Collection, Royal Society of Chemistry
Archives, University College London
Marburg
Hessisches Staatsarchiv
Erbengemeinschaft Justi, Universitätsmuseum für Kunst- und Kulturgeschichte
Universitätsbibliothek Marburg
Milton Keynes
Edward Frankland microfilmed papers, Open University
Munich
Adolf Baeyer Collection, privately held
Handschriftenabteilung, Bayerische Staatsbibliothek
Sondersammlungen, Deutsches Museum
Paris
Archives de l'Académie des Sciences
Stöckheim
Stöckheimer Chronik
Wiesbaden
Vieweg Verlag Archiv

II. Texts and Monographs by Hermann Kolbe

Ausführliches Lehrbuch der organischen Chemie, 3 vols. Braunschweig: Vieweg, 1854–1878; with second title page indicating as vols. 3–5 of *Graham-Otto's ausführliches Lehrbuch der Chemie*, 3d ed. Second edition published as Kolbe, ed. E. von Meyer, *Ausführliches Lehr- und Handbuch der organischen Chemie*, 2 vols. Braunschweig: Vieweg, 1880–1884.

Beleuchtung von Virchow's Schrift: Nach dem Kriege. Leipzig: Barth, 1872.

Das chemische Laboratorium der Universität Leipzig und die seit 1866 darin ausgeführten chemischen Untersuchungen. Braunschweig: Vieweg, 1872.

Das chemische Laboratorium der Universität Marburg und die seit 1859 darin ausgeführten chemischen Untersuchungen. Braunschweig: Vieweg, 1866.

Das neue chemische Laboratorium der Universität Leipzig. Leipzig: Brockhaus, 1868.

The Electrolysis of Organic Compounds: Papers by Hermann Kolbe (1845–1868). Edinburgh: Alembic Club, 1900.

Kurzes Lehrbuch der anorganischen [organischen] Chemie, 2 vols. Braunschweig: Vieweg, 1877–1883; 2d ed., 1883–1884.

A Short Textbook of Inorganic Chemistry, trans. T. S. Humpidge. London: Longmans, Green & Co., 1884; 2d ed., 1888; 3d ed., rev. H. L. Snape, 1892; 4th ed., 1893; New York: Wiley, 1884.

Ueber den natürlichen Zusammenhang der organischen mit den unorganischen Verbindungen, ed. E. von Meyer. Leipzig: Engelmann, 1897.

Ueber die chemische Constitution der organischen Kohlenwasserstoffe. Braunschweig: Vieweg, 1869.

Ueber die chemische Constitution organischer Verbindungen. Marburg: Elwert, 1858.

Ueber die Producte der Einwirkung des Chlors auf Schwefelkohlenstoff. Marburg: Bayrhoffer, 1845.

Ueber eine neue Bildungsweise des Benzoylwasserstoffs. Marburg: Elwert, 1857.

Uchebnik' neorganicheskoi khimii, trans. Kartsev and Sabaneiev. Moscow, 1882.

Uitvoerig leerboek der organische scheikunde, trans. J. Rendler. Haarlem, 1854–1855.

Zur Entwickelungsgeschichte der theoretischen Chemie. Leipzig: Brockhaus, 1881.

III. Obituaries and Biographical Literature on Hermann Kolbe

[Anon.] "Hermann Kolbe." *American Journal of Science*, [3] **29** (1885), 84.

[Anon.] "Hermann Kolbe." *Chemiker-Zeitung*, **8** (1884), 1725–1726.

[Anon.] "The Late Professor Kolbe." *Chemical News*, **50** (1884), 282–283.

Armstrong, H. E. "Persönliche Erinnerungen und Gedanken." *Chemiker-Zeitung*, **51** (1927), 114–116.

Glemser, Oskar. "Hermann Kolbe, Chemiker." *Göttinger Jahrbuch*, 1985, pp. 207–208.

Hartley, Harold. "Henry Armstrong (1848–1937) and Some of the Great Figures of Nineteenth Century Organic Chemistry." In idem, *Studies in the History of Chemistry* (Oxford: Clarendon Press, 1971), pp. 195–222.

Heinig, K. "Hermann Kolbe and the 'Journal für praktische Chemie.'" *Proceedings of the XI International Congress of the History of Science*, **4** (1965, publ. 1968), 96–100.

Hofmann, A. W. "Hermann Kolbe." *Berichte der Deutschen Chemischen Gesellschaft*, **17** (1884), 2809–2812.

Kasiwagi, Hazime. "Some Marked Characteristics of Hermann Kolbe's Conception on the Chemical Constitution." *Proceedings of the XIV International Congress of the History of Science*, **2** (1974, publ. 1975), 401–404.

Kerkovius, Wilhelm. "Kolbes Angriffe gegen 'die modernen chemischen Anschauungen und Lehren.'" *Chemiker-Zeitung*, **35** (1911), 1117–1119 and 1142–1144.

Kolbe, Hermann. "Meine Betheiligung an der Entwickelung der theoretischen Chemie." *Journal für praktische Chemie*, **131** (1881), 305–323, 353–379, and 497–517, and **132** (1881), 374–425.

Krätz, Otto. "Zur Geschichte der organisch-chemischen Formelschreibweise: Ein Brief von C. W. Blomstrand an H. Kolbe." *Physis*, **15** (1973), 157–177.

———. "Historische Experimente (1846): Hermann Kolbe und Robert Wilhelm Bunsen, Eudiometrische Analysen von Grubengas." *Chemie, Experiment, und Didaktik*, **3** (1977), 31–36.

Leicester, Henry. "Hermann Kolbe." *Dictionary of Scientific Biography*, **7** (New York: Scribners, 1973), 450–453.

Lockemann, Georg. "Aus dem Briefwechsel von Hermann Kolbe." *Zeitschrift für angewandte Chemie*, **41** (1928), 623.

———. "Hermann Kolbe." In G. Bugge, ed., *Das Buch der grossen Chemiker*, 2 vols. (Berlin: Verlag Chemie, 1930), **2**, 124–135.

Meyer, Ernst von. "Zur Erinnerung an Hermann Kolbe." *Journal für praktische Chemie*, **138** (1885), 417–466.

Ost, Hermann. "Hermann Kolbe: Ein Lebensbild." *Westermann illustrierte deutsche Monatshefte*, **30** (1885), 118–133.

[Perkin, W. H.] "Hermann Kolbe." *Journal of the Chemical Society*, **47** (1885), 323–327.

Phillips, J. P. "Liebig and Kolbe, Chemical Editors." *Chymia*, **11** (1966), 89–97.

Remane, Horst. "Hermann Kolbe—einer der hervorragendsten Vertreter der organischen Chemie des 19. Jahrhunderts." *Mitteilungsblatt der Chemischen Gesellschaft der DDR*, **1984**, pp. 236–242.

Remane, Horst, Hantschmann, Achim, and Weissenfels, Manfred. "Hermann Kolbe und sein Beitrag zur Chemie des 19. Jahrhunderts." *Zeitschrift für Chemie*, **24** (1984), 393–403.

Rocke, A. J. "Kolbe versus the 'Transcendental Chemists': The Emergence of Classical Organic Chemistry." *Ambix*, **34** (1987), 156–168.

———. "'Between Two Stools': Kopp, Kolbe, and the History of Chemistry." *Bulletin for the History of Chemistry*, **7** (1990), 19–24.

———. "Research Groups and Group Research in German Chemistry: Kolbe's Marburg and Leipzig Institutes." *Osiris*, [2] **8** (1993), 51–79.

Ronge, Grete. "Hermann Kolbe." *Neue Deutsche Biographie*, **12** (Berlin: Duncker & Humblot, 1980), 446–451.

Strigel, A. "Hermann Kolbe." *Allgemeine Deutsche Biographie*, **51** (Leipzig: Duncker & Humblot, 1906), 321–329.

Strube, Wilhelm. "Hermann Kolbe (1818–1884)." In G. Harig, ed., *Bedeutende Gelehrte in Leipzig*, 2 vols. (Leipzig: Karl-Marx-Universität, 1965), **2**, 25–35.

Voit, C. von. "Nekrologe auf . . . H. Kolbe . . . und andere." Munich: Akademie-Verlag, 1884.

Walker, O. J. "The Kekulé–Kolbe Valency Controversy." *Chemistry in Britain*, **2** (1966), 241.

IV. Selected Secondary Literature

Abel, F. A., et al. "Hofmann Memorial Lectures." *Journal of the Chemical Society*, **69** (1896), 575–732.

Albrecht, Helmuth. *Technische Bildung zwischen Wissenschaft und Praxis: Die Technische Hochschule Braunschweig 1862–1914*. Hildesheim: Ohms, 1987.

Anschütz, Richard. *August Kekulé*, 2 vols. Berlin: Verlag Chemie, 1929.

Armstrong, H. E. "The Riddle of Benzene: August Kekulé." *Journal of the Society of Chemical Industry*, **48** (1929), 914–918.

———. "The Doctrine of Atomic Valency." *Nature*, **125** (1930), 807–810.

Ben-David, Joseph. *The Scientist's Role in Society*, 2d ed. Chicago: Univ. of Chicago Press, 1984.

Benfey, O. T., ed. *Kekulé Centennial*. Washington, D.C.: American Chemical Society, 1966.

Bentley, Jonathan. "Hofmann's Return to Germany from the Royal College of Chemistry." *Ambix*, **19** (1972), 197–203.

Borscheid, Peter. *Naturwissenschaft, Staat und Industrie in Baden (1848–1914)*. Stuttgart: Klett, 1976.

Brock, William H. *H. E. Armstrong and the Teaching of Science*. Cambridge: Cambridge Univ. Press, 1973.

Brooke, John H. "Wöhler's Urea and Its Vital Force?—A Verdict from the Chemists." *Ambix*, **15** (1968), 84–114.

———. "Organic Synthesis and the Unification of Chemistry—A Reappraisal." *British Journal for the History of Science*, **5** (1971), 363–392.

———. "Chlorine Substitution and the Future of Organic Chemistry." *Studies in the History and Philosophy of Science*, **4** (1973), 47–94.

———. "Laurent, Gerhardt, and the Philosophy of Chemistry." *Historical Studies in the Physical Sciences*, **6** (1975), 405–429.

———. "Avogadro's Hypothesis and its Fate: A Case Study of the Failure of Case-Studies." *History of Science*, **19** (1981), 235–273.

Brooks, Nathan M. "The Formation of a Community of Chemists in Russia, 1700–1870." Ph.D. dissertation, Columbia Univ., 1990.

Bud, Robert, and Roberts, Gerrylynn K. *Science Versus Practice: Chemistry in Victorian Britain*. Manchester: Manchester Univ. Press, 1984.

Burchardt, Lothar. "Die Ausbildung des Chemikers im Kaiserreich." *Zeitschrift für Unternehmungsgeschichte*, **23** (1978), 31–53.

———. "Professionalisierung oder Berufskonstruktion? Das Beispiel des Chemikers im wilhelminischen Deutschland." *Geschichte und Gesellschaft*, **6** (1980), 326–348.

Bykov, G. V. "The Origin of the Theory of Chemical Structure." *Journal of Chemical Education*, **39** (1962), 220–224.

Bykov, G. V., and Bekassova, L. M. "Beiträge zur Geschichte der Chemie der 60-er Jahre des XIX. Jahrhunderts: I. Briefwechsel zwischen E. Erlenmeyer und A. M. Butlerov." *Physis*, **8** (1966), 185–198.

———. "Beiträge zur Geschichte der Chemie der 60-er Jahre des XIX. Jahrhunderts: II. F. Beilsteins Briefe an A. M. Butlerov." *Physis*, **8** (1966), 267–285.

Bykov, G. V., and Jacques, J. "Deux pionniers de la chimie moderne, Adolphe Wurtz et Alexandre M. Boutlerov, d'après une correspondance inédite." *Revue d'histoire des sciences*, **13** (1960), 115–134.

Caneva, Kenneth. "From Galvanism to Electrodynamics: The Transformation of German Physics and Its Social Context." *Historical Studies in the Physical Sciences*, **9** (1978), 63–159.

Cardwell, D. S. L. *The Organisation of Science in England*. London: Heinemann, 1957.

Chemnitius, Fritz. *Die Chemie in Jena von Rolfinck bis Knorr (1629–1921)*. Jena: Böhlaus, 1929.

Coleman, William. "Prussian Pedagogy: Purkyně at Breslau, 1823–1839." In Coleman and Holmes, *Investigative Enterprise*, s.v., pp. 15–64.

Coleman, William, and Holmes, F. L., eds. *The Investigative Enterprise: Experimental Physiology in Nineteenth-Century Medicine*. Berkeley: Univ. of California Press, 1988.

Debus, Heinrich. *Erinnerungen an Robert Wilhelm Bunsen*. Kassel: Fischer, 1901.

Dreyer, Ernst. *Friedr. Vieweg & Sohn in 150 Jahren deutscher Geistesgeschichte*. Braunschweig: Vieweg, 1936.

Eulenburg, Franz. *Die Frequenz der deutschen Universitäten von ihrer Gründung bis zur Gegenwart*. Leipzig: Hirzel, 1904.

———. *Die Entwicklung der Universität Leipzig in den letzten hundert Jahren*. Leipzig: Hirzel, 1909.

Eyre, J. Vargas. *Henry Edward Armstrong*. London: Butterworths, 1958.

Festschrift zur Feier des 500 jährigen Bestehens der Universität Leipzig, ed. Rektor und Senat der Universität Leipzig. Leipzig: Hirzel, 1909.

Fisher, Nicholas W. "Organic Classification Before Kekulé." *Ambix*, **20** (1973), 106–131 and 209–233.

———. "Kekulé and Organic Classification." *Ambix*, **21** (1974), 29–52.

———. "Wislicenus and Lactic Acid." In Ramsay, *van't Hoff-Le Bel Centennial*, s.v., pp. 33–54.

Fox, Robert. "Scientific Enterprise and the Patronage of Research in France, 1800–70." In G. L'E. Turner, ed., *The Patronage of Science in the Nineteenth Century* (Leiden: Noordhof, 1976), pp. 9–51.

Fox, Robert, and Weisz, George, eds. *The Organization of Science and Technology in France, 1808–1914*. Cambridge: Cambridge Univ. Press, 1980.

Frankland, Edward. *Sketches from the Life of Edward Frankland*. London: Spottiswoode, 1901; expurgated ed. 1902.

Friedel, Charles. "Notice sur la vie et les travaux de Charles-Adolphe Wurtz." *Bulletin de la Société Chimique*, [2] **43** (1885), i–lxxx.

Fruton, Joseph S. "The Liebig Research Group—A Reappraisal." *Proceedings of the American Philosophical Society*, **132** (1988), 1–66.

———. *Contrasts in Scientific Style: Research Groups in the Chemical and Biochemical Sciences*. Philadelphia: American Philosophical Society, 1990.

Geison, Gerald. "Scientific Change, Emerging Specialties, and Research Schools." *History of Science*, **19** (1981), 20–40.

Graebe, Carl. *Geschichte der organischen Chemie*. Berlin: Springer, 1920.

Gregory, Frederick. "Kant, Schelling, and the Administration of Science in the Romantic Era." *Osiris*, [2] **5** (1989), 17–35.

———. "Kant's Influence on Natural Scientists in the German Romantic Period." In R. Visser et al., eds., *New Trends in the History of Science* (Amsterdam: Rodopi, 1989), pp. 53–66.

Grimaux, Edouard, and Gerhardt, Charles, Jr. *Charles Gerhardt: Sa vie, son oeuvre, sa correspondance 1816–1856*. Paris: Masson, 1900.

Gustin, Bernard. "The Emergence of the German Chemical Profession, 1790–1867." Ph.D. dissertation, Univ. of Chicago, 1975.

Harris, J., and Brock, W. H. "From Giessen to Gower Street: Towards a Biography of Alexander Williamson." *Annals of Science*, **31** (1974), 95–130.

Helferich, Burckhardt. *Das Studium der Chemie an der Universität Leipzig*. Leipzig: Lorentz, 1932.

Hjelt, Edvard. *Geschichte der organischen Chemie von ältester Zeit bis zur Gegenwart*. Braunschweig: Vieweg, 1916.

Hofmann, August Wilhelm. *The Chemical Laboratories in Course of Erection in the Universities of Bonn and Berlin*. London: Clowes, 1866.

———. *Zur Erinnerung an vorangegangene Freunde*, 3 vols. Braunschweig: Vieweg, 1888.

Holmes, F. L. "The Complementarity of Teaching and Research in Liebig's Laboratory." *Osiris*, [2] **5** (1989), 121–164.

Ihde, Aaron J. *Development of Modern Chemistry*. New York: Harper and Row, 1964.

Jacques, Jean. "Boutlerov, Couper et la Société Chimique de Paris." *Bulletin de la Société Chimique*, 1953, pp. 528–530.

Jarausch, Konrad. *Students, Society, and Politics in Imperial Germany: The Rise of Academic Illiberalism*. Princeton, N.J.: Princeton Univ. Press, 1982.

Johnson, Jeffrey. "Academic Chemistry in Imperial Germany." *Isis*, **76** (1985), 500–524.

———. "Hierarchy and Creativity in Chemistry, 1871–1914." *Osiris*, [2] **5** (1989), 214–240.

———. *The Kaiser's Chemists: Science and Modernization in Imperial Germany*. Chapel Hill: Univ. of North Carolina Press, 1990.

Jungnickel, Christa. "Teaching and Research in the Physical Sciences and Mathematics in Saxony, 1820–1850." *Historical Studies in the Physical Sciences*, **10** (1979), 3–47.

Jungnickel, Christa, and McCormmach, Russell. *The Intellectual Mastery of Nature*, vol. 1. Chicago: Univ. of Chicago Press, 1986.

Keas, Michael. "The Structure and Philosophy of Group Research: A. W. Hofmann's Research Program in London." Ph.D. dissertation, Univ. of Oklahoma, 1992.

Keen, Robin. "The Life and Work of Friedrich Wöhler." Ph.D. dissertation, Univ. College London, 1976.

Kiesewetter, Hubert. *Industrialisierung und Landwirtschaft: Sachsens Stellung im regionalen Industrialisierungsprozess Deutschlands im 19. Jahrhundert*. Vienna: Böhlau, 1988.

———. *Industrielle Revolution in Deutschland 1815–1914*. Frankfurt: Suhrkamp, 1989.

Klosterman, L. J. "A Research School of Chemistry in the Nineteenth Century: Jean Baptiste Dumas and His Research Students." *Annals of Science*, **42** (1985), 1–80.

Kopp, Hermann. *Geschichte der Chemie*, 4 vols. Braunschweig: Vieweg, 1843–1847.

———. *Die Entwickelung der Chemie in der neueren Zeit*. Munich: Oldenbourg, 1873.

Krätz, Otto, "Emil Erlenmeyer, 1825–1909." *Chemie in unserer Zeit*, **6** (1972), 52–58.

———. *Beilstein–Erlenmeyer: Briefe zur Geschichte der chemischen Dokumentation und des chemischen Zeitschriftenwesens*. Munich: Fritsch, 1972.

———. "Der Chemiker in den Gründerjahren." In Schmauderer, *Der Chemiker*, s.v., pp. 259–284.

Ladenburg, Albert. *Vorträge über die Entwicklungsgeschichte der Chemie in den letzten hundert Jahren*. Braunschweig: Vieweg, 1869.

———. *Lebenserinnerungen*. Breslau: Trewendt & Granier, 1912.

Leicester, H. M. "Controversies on Chemical Structure from 1860 to 1870." In Benfey, *Kekulé Centennial*, s.v., pp. 13–23.

Lenoir, Timothy, "Science for the Clinic: Science Policy and the Formation of Carl Ludwig's Institute in Leipzig." In Coleman and Holmes, *Investigative Enterprise*, s.v., pp. 139–178.

Lepsius, B. *Festschrift zur Feier des 50jährigen Bestehens der Deutschen Chemischen Gesellschaft und des 100. Geburtstages ihres Begründers August Wilhelm von Hofmann*. Berlin: Friedländer, 1918.

Levere, Trevor. *Affinity and Matter*. Oxford: Clarendon Press, 1971.

Lockemann, Georg. "Der chemische Unterricht an den deutschen Universitäten im ersten Viertel des neunzehnten Jahrhunderts." In J. Ruska, ed., *Studien zur Geschichte der Chemie* (Berlin: Springer, 1927), pp. 148–158.

———. *Robert Wilhelm Bunsen*. Stuttgart: Wissenschaftliche Verlagsgesellschaft, 1949.

Mauskopf, Seymour. "The Atomic Structural Theories of Ampère and Gaudin: Molecular Speculation and Avogadro's Hypothesis." *Isis*, **60** (1969), 61–74.

———. *Crystals and Compounds: Molecular Structure and Composition in Nineteenth Century French Science*. Philadelphia: American Philosophical Society, 1976.

McClelland, Charles. *State, Society, and University in Germany, 1700–1914*. New York: Cambridge Univ. Press, 1980.

———. "Die deutschen Hochschullehrer als Elite, 1815–1850." In Schwabe, *Deutsche Hochschullehrer*, s.v., pp. 27–53.

Meinel, Christoph. *Die Chemie an der Universität Marburg seit Beginn des 19. Jahrhunderts*. Marburg: Elwert, 1978.

———. "Nationalismus und Internationalismus in der Chemie des 19. Jahrhunderts." In Peter Dilg, ed., *Perspektiven der Pharmaziegeschichte* (Graz: Akademische Druck- und Verlagsanstalt, 1983), pp. 225–243.

———. "*Artibus Academicis Inserenda*: Chemistry's Place in Eighteenth and Early Nineteenth Century Universities." *History of Universities*, **7** (1988), 89–115.

———. *Karl Friedrich Zöllner und die Wissenschaftskultur der Gründerzeit*. Berlin: Sigma, 1991.

Meyer, Ernst von. *Geschichte der Chemie von ältesten Zeiten bis zur Gegenwart*. Leipzig: Veit, 1889.

———. *Lebenserinnerungen*. N.p., n.d., ca. 1918.

Meyer, Rita. "Emil Erlenmeyer (1825–1909) als Chemietheoretiker und sein Beitrag zur Entwicklung der Strukturchemie." Ph.D. dissertation, Univ. of Munich, 1984.

Morrell, J. B. "The Chemist Breeders: The Research Schools of Liebig and Thomas Thomson." *Ambix*, **19** (1972), 1–46.

Müller, Monika. "Aus dem Leben und Wirken des Chemikers und Hochschullehrers August Wilhelm von Hofmann (1818–1892)." Ph.D. dissertation, Humboldt-Universität Berlin, 1981.

Munday, Pat. "Social Climbing through Chemistry: Justus Liebig's Rise from the *Niederer Mittelstand* to the *Bildungsbürgertum*." *Ambix*, **37** (1990), 1–19.

Nye, Mary Jo. "Berthelot's Anti-Atomism: A 'Matter of Taste'?" *Annals of Science*, **31** (1981), 585–590.

———. *Science in the Provinces: Scientific Communities and Provincial Leadership in France, 1860–1930*. Berkeley: Univ. of California Press, 1986.

———. "Explanation and Convention in Nineteenth-Century Chemistry." In R. Visser, et al., eds., *New Trends in the History of Science* (Amsterdam: Rodopi, 1989), pp. 171–186.

———. *From Chemical Philosophy to Theoretical Chemistry: Dynamics of Matter and Dynamics of Discipline, 1800–1950*. Berkeley: Univ. of California Press, 1993.

Olesko, Kathryn. "On Institutes, Investigations, and Scientific Training." In Coleman and Holmes, *Investigative Enterprise*, s.v., pp. 295–332.

———. *Physics as a Calling: Discipline and Practice in the Königsberg Seminar for Physics*. Ithaca, N.Y.: Cornell Univ. Press, 1991.

Partington, J. R. *A History of Chemistry*, vol. 4. London: Macmillan, 1964.

Paul, Harry W. *The Sorcerer's Apprentice: The French Scientist's Image of German Science, 1840–1919*. Gainesville: Univ. of Florida Press, 1972.

Ramsay, O.B., ed. *van't Hoff-Le Bel Centennial*. Washington, D.C.: American Chemical Society, 1975.

———. *Stereochemistry*. London: Heyden, 1981.

Ringer, Fritz. *The Decline of the German Mandarins: The German Academic Community, 1890–1933*. Cambridge, Mass.: Harvard Univ. Press, 1969.

———. "Das gesellschaftliche Profil der deutschen Hochschullehrerschaft 1871–1933." In Schwabe, *Deutsche Hochschullehrer*, s.v., pp. 93–104.

Rocke, A. J. "Kekulé, Butlerov, and the Historiography of the Theory of Chemical Structure." *British Journal for the History of Science*, **14** (1981), 27–57.

———. "Subatomic Speculations and the Origin of Structure Theory." *Ambix*, **30** (1983), 1–18.

———. *Chemical Atomism in the Nineteenth Century: From Dalton to Cannizzaro*. Columbus: Ohio State Univ. Press, 1984.

———. "Hypothesis and Experiment in the Early Development of Kekulé's Benzene Theory." *Annals of Science*, **42** (1985), 355–381.

———. "Kekulé's Benzene Theory and the Appraisal of Scientific Theories." In A. Donovan, L. Laudan, and R. Laudan, eds., *Scrutinizing Science: Empirical Studies of Scientific Change* (Boston: Kluwer, 1988), pp. 145–161.

———. "Methodology and Its Rhetoric in Nineteenth-Century Chemistry: Induction versus Hypothesis." In Elizabeth Garber, ed., *Beyond History of Science: Essays in Honor of Robert E. Schofield* (Bethlehem, Pa.: Lehigh Univ. Press, 1990), pp. 137–155.

———. "Berzelius' Animal Chemistry: From Physiology to Organic Chemistry, 1805–1814." In T. Frängsmyr and E. Melhado, eds., *Enlightenment Science in a Romantic Age: Berzelius and His Science in International Context* (Cambridge Univ. Press, 1992), pp. 107–131.

Roscoe, H. E. "Bunsen Memorial Lecture." *Journal of the Chemical Society*, **77** (1900), 513–554.

Russell, Colin A. *The History of Valency*. Leicester: Leicester Univ. Press, 1971.

———. *Lancastrian Chemist: The Early Years of Sir Edward Frankland.* Milton Keynes: Open Univ. Press, 1986.

———. "The Changing Role of Synthesis in Organic Chemistry." *Ambix*, **34** (1987), 169–180.

Schmauderer, E., ed. *Der Chemiker im Wandel der Zeiten*. Weinheim: Verlag Chemie, 1973.

Schorlemmer, Carl. *The Rise and Development of Organic Chemistry*. London: Macmillan, 1894.

Schütt, Hans-Werner. *Eilhard Mitscherlich*. Munich: Oldenbourg, 1992.

Schwabe, Klaus, ed. *Deutsche Hochschullehrer als Elite*. Boppard: Boldt, 1988.

Selle, Götz von. *Die Georg-August-Universität zu Göttingen, 1737–1937*. Göttingen: Vandenhoek & Ruprecht, 1937.

Servos, John. *Physical Chemistry from Ostwald to Pauling: The Making of a Science in America*. Princeton, N.J.: Princeton Univ. Press, 1990.

Speter, Max. "'Vater Kopp': Bio-, Biblio- und Psychographisches von und über Hermann Kopp (1817–1892)." *Osiris*, **5** (1938), 392–460.

Stern, Fritz. *Politics of Cultural Despair*. Berkeley: Univ. of California Press, 1961.

Travis, Anthony S. *The Rainbow Makers: The Origins of the Synthetic Dyestuffs Industry in Western Europe*. Bethlehem, Pa.: Lehigh Univ. Press, 1993.

Treibs, Wilhelm. "Zur Geschichte der Entwicklung der Chemie an der Universität Leipzig." In Ernst Engelberg et al., eds., *Karl-Marx-Universität Leipzig 1409–1959: Beiträge zur Universitätsgeschichte*, 2 vols. (Leipzig: Verlag Enzyklopädie, 1959), **1**, 464–480.

Tuchman, Arleen. "From the Lecture to the Laboratory: The Institutionalization of Scientific Medicine at the University of Heidelberg." In Coleman and Holmes, *Investigative Enterprise*, s.v., pp. 65–99.

Turner, R. Steven. "The Growth of Professorial Research in Prussia, 1818–1848—Causes and Context." *Historical Studies in the Physical Sciences*, **3** (1971), 137–182.

———. "University Reformers and Professorial Scholarship in Germany, 1760–1806." In Lawrence Stone, ed., *The University in Society*, 2 vols. (Princeton, N.J.: Princeton Univ. Press, 1974), **2**, 495–531.

———. "The Bildungsbürgertum and the Learned Professions in Prussia, 1770–1830: The Origins of a Class." *Histoire Sociale—Social History*, **13** (1980), 105–135.

———. "The Prussian Professoriate and the Research Imperative." In H. N. Jahnke and M. Otte, eds., *Epistemological and Social Problems of the Sciences in the Early Nineteenth Century* (Dordrecht: Reidel, 1981), pp. 109–121.

———. "Justus Liebig versus Prussian Chemistry: Reflections on Early Institute Building in Germany." *Historical Studies in the Physical Sciences*, **13** (1982), 129–162.

Turner, R. Steven, Kerwin, E., and Woolwine, D. "Careers and Creativity in Nineteenth-Century Physiology: Zloczower Redux." *Isis*, **75** (1984), 523–529.

Tyndall, John. *New Fragments*. New York: Appleton, 1892.

Valentin, Johannes. *Friedrich Wöhler*. Stuttgart: Wissenschaftliche Verlagsgesellschaft, 1949.

Vaupel, Elisabeth. "Carl Graebe (1841–1927): Leben, Werk und Wirken." Ph.D. dissertation, Univ. of Munich, 1987.

Volhard, Jacob. *August Wilhelm von Hofmann: Ein Lebensbild*. Berlin: Friedländer, 1902.

———. *Justus von Liebig*, 2 vols. Leipzig: Barth, 1909.

Vucinich, Alexander. *Science in Russian Culture, 1861–1917*. Stanford, Calif.: Stanford Univ. Press, 1970.

Wurtz, Adolphe. "Éloge de Laurent et de Gerhardt." *Moniteur scientifique*, **4** (1862), 482–513.

———. "Histoire des doctrines chimiques depuis Lavoisier." In Wurtz, ed., *Dictionnaire de chimie pure et appliquée*, 3 vols. in 5 (Paris: Hachette, 1868–1878), **1**, i–xciv. Republished monographically in 1869. *Geschichte der chemischen Theorien seit Lavoisier bis auf unsere Zeit*, trans. A. Oppenheim (Berlin: R. Oppenheim, 1870).

Yoshida, Akira. "Charles Adolphe Wurtz et la théorie atomique." *Japanese Studies in the History of Science*, **16** (1977), 129–135.

Zloczower, Avraham. *Career Opportunities and the Growth of Scientific Discovery in Nineteenth-Century Germany*. New York: Arno, 1981.

Index

Designer:	U.C. Press Staff
Compositor:	Asco Trade Typesetting Ltd.
Text:	10/12 Times Roman
Display:	Times Roman
Printer:	BookCrafters
Binder:	BookCrafters